CLIMATE CHANGE EFFECTS ON GROUNDWATER RESOURCES

INTERNATIONAL CONTRIBUTIONS TO HYDROGEOLOGY

27

Series Editor: Dr. Nick S. Robins
Editor-in-Chief, IAH Book Series
British Geological Survey
Wallingford, UK

 INTERNATIONAL ASSOCIATION OF HYDROGEOLOGISTS

Climate Change Effects on Groundwater Resources

A Global Synthesis of Findings and Recommendations

Editors

Holger Treidel & Jose Luis Martin-Bordes
UNESCO, International Hydrological Programme, Paris, France

Jason J. Gurdak
San Francisco State University, California, USA

CRC Press
Taylor & Francis Group
Boca Raton London New York

CRC Press is an imprint of the
Taylor & Francis Group, an **informa** business

A BALKEMA BOOK

Published by: CRC Press/Balkema
                P.O. Box 447, 2300 AK Leiden, The Netherlands
                e-mail: Pub.NL@taylorandfrancis.com
                www.crcpress.com – www.taylorandfrancis.com – www.balkema.nl

First issued in paperback 2020

© 2012 Taylor & Francis Group, London, UK
CRC Press/Balkema is an imprint of the Taylor & Francis Group, an informa business

No claim to original U.S.Government works

ISBN-13: 978-0-415-68936-6 (pbk)
ISBN-13: 978-0-367-57682-0 (hbk)

Visit the Taylor & Francis Web site at
http://www.taylorandfrancis.com

and the CRC Press Web site at
http://www.crcpress.com

Library of Congress Cataloging-in-Publication Data

Climate change effects on groundwater resources : a global synthesis of findings and recommendations / editors: Holger Treidel, Jose Luis Martin-Bordes, Jason Gurdak. — 1st ed.
    p. cm.
   Includes bibliographical references and index.
    ISBN 978-0-415-68936-6 (hardback : alk. paper)   1. Groundwater.   2. Climatic changes—Environmental aspects—Case studies.   I. Treidel, Holger.   II. Martin-Bordes, José Luis.   III. Gurdak, Jason J.

   TD403.C576 2012
   553.7′9—dc23

                                                                        2011036931

Typeset by MPS Limited, a Macmillan Company, Chennai, India

TABLE OF CONTENTS

**4   Groundwater discharge as affected by land use change
in small catchments: A hydrologic and economic case study
in Central Brazil**

*Henrique M.L. Chaves, Ana Paula S. Camelo &
Rejane M. Mendes*

**5   Effects of storm surges on groundwater resources,
North Andros Island, Bahamas**

John Bowleg & Diana M. Allen

Dry (Arid and Semiarid) Climates

Temperate Climates

Continental Climates

17   Possible effects of climate change on hydrogeological systems: results from research on Esker aquifers in northern Finland   305

Bjørn Kløve, Pertti Ala-aho, Jarkko Okkonen & Pekka Rossi

Polar Climates

18   Impacts of climate change on groundwater in permafrost areas: case study from Svalbard, Norway   323

Sylvi Haldorsen, Michael Heim & Martine van der Ploeg

Various Climates

ABOUT THE EDITORS

Holger Treidel is an environmental scientist and works as project coordinator with UNESCO's International Hydrological Programme in Paris. His work is related to the sustainable management of groundwater resources under the effects of climate change & variability, with particular focus on the complex challenges related to the management of transboundary aquifer systems. He is coordinating the UNESCO project Groundwater Resources Assessment under the Pressures of Humanity and Climate Change (GRAPHIC) and global and regional transboundary groundwater management projects in cooperation with the Global Environmental Facility (GEF).

Jose Luis Martin-Bordes is a civil engineer specialized in groundwater resources management and works as project coordinator in the International Hydrological Programme (IHP) within the Division of Water Sciences of UNESCO, Paris, France. He provides support to the coordination of the IHP Groundwater activities including the Groundwater Resources Assessment under the Pressures of Humanity and Climate Change (GRAPHIC), the International Shared Aquifer Resources Management Initiative (ISARM), Groundwater Dependent Ecosystems and Groundwater for Emergency Situations (GWES).

Jason J. Gurdak is Assistant Professor of hydrogeology in the Department of Geosciences at San Francisco State University, California, USA. He and his research group address basic and applied questions about sustainable groundwater management, vadose zone and soil-water processes that affect recharge and contaminant transport, groundwater vulnerability to contamination and climate extremes, and the effects of climate change and interannual to multidecadal climate variability on water resources. Since 2004 he has served on the UNESCO project Groundwater Resources Assessment under the Pressures of Humanity and Climate Change (GRAPHIC) that promotes science, education, and awareness of the coupled effects of climate change and human stresses on global groundwater resources.

ACKNOWLEDGEMENTS

Compiling this book was a collaborative effort. We are sincerely grateful to all authors for their contributions. Their enthusiastic involvement and insightful feedback have allowed us to put together an interesting and valuable publication.

The preparation of this publication would have not been possible without the support of UNESCO's International Hydrological Programme (IHP) and its *Groundwater Resources Assessment under the Pressures of Humanity and Climate Change* (GRAPHIC) project, which has helped create an active and global group of scientists dedicated to unraveling groundwater and climate interactions and raising attention for a topic that has received only little attention previously. We would like to thank in particular Alice Aureli for her guidance and overall coordination and to Timothy Green for his continued support of the GRAPHIC expert group. Our thanks also go to the many cooperating universities, institutions, and organizations – too many to mention – that support GRAPHIC.

The Editors are grateful to the following people and many anonymous reviewers for their assistance with the external peer review of papers submitted for publication in this volume:

Giovanni Barrocu	University of Cagliari, Department of Land Engineering, Italy
John Bloomfield	British Geological Survey, UK
Elisabetta Carrara	Water Resource Assessment, Climate & Water Division/ Bureau of Meteorology, Melbourne, Australia
Dioni Cendon Sevilla	ANSTO Institute for Environmental Research, Australia
Jianyao Chen	Department of Water Resource and Environment, School of Geographical Science and Planning, Sun Yatsen University, China
Ian Ferguson	U.S. Bureau of Reclamation, Lakewood, Colorado, USA
Timothy Green	U.S. Department of Agriculture, Agricultural Systems Research Unit, USA
Ian Holman	Cranfield Water Science Institute (CWSI), Cranfield University, UK
Neno Kukuric	International Groundwater Resources Assessment Centre (IGRAC), The Netherlands
James Terry	Department of Geography, National University of Singapore, Kent Ridge, Singapore
Tristan Wellman	U.S. Geological Survey, Lakewood, Colorado, USA
Kamel Zouari	Laboratory of Radio-Analysis and Environment of the National School of Engineers, Sfax, Tunisia

CHAPTER 1

Introduction

1.1 RATIONALE

Groundwater is an essential part of the hydrological cycle and is a valuable natural resource providing the primary source of water for agriculture, domestic, and industrial uses in many countries. Groundwater is now a significant source of water for human consumption, supplying nearly half of all drinking water in the world (WWAP 2009) and around 43 percent of all water effectively consumed in irrigation (Siebert et al. 2010). Groundwater also is important for sustaining streams, lakes, wetlands, and ecosystems in many countries.

The use of groundwater has particular relevance to the availability of many potable-water supplies because groundwater has a capacity to balance large swings in precipitation and associated increased demands during drought and when surface water resources reach the limits of sustainability. During extended droughts the utilization of groundwater for irrigation is expected to increase, including the intensified use of non-renewable groundwater resources, which may impact the sustainability of the resource. However, global groundwater resources may be threatened by human activities and the uncertain consequences of climate change.

Global change encompasses changes in the characteristics of inter-related climate variables in space and time, and derived changes in terrestrial processes, including human activities that affect the environment. Changes in global climate are expected to affect the hydrological cycle, altering surface-water levels and groundwater recharge to aquifers with various other associated impacts on natural ecosystems and human activities. Also groundwater discharge, storage, saltwater intrusion, biogeochemical reactions, and chemical fate and transport may be modified by climate change. Although the most noticeable impacts of climate change could be changes in surface water levels and quality, there are potential effects on the quantity and quality of groundwater. While recognizing that groundwater is a major source of water across much of the world, particularly in rural areas in arid and semi-arid regions, the Intergovernmental Panel on Climate Change (IPCC) 3rd and 4th Assessment Reports state that there has been very little research on the potential effects of climate change (IPCC 2001, 2007; Bates 2008). In recent decades, a wide array of scientific research has been carried out to better understand how water resources might respond to global change (Green et al. 2011). Recent research has been focused predominantly on surface-water systems, due to their visibility, accessibility and more obvious recognition of surface waters being affected by global change. However, little is known about how subsurface waters in the vadose zone and groundwater might respond to climate change and affect the current availability and future sustainability of groundwater resources (UNESCO 2008). It is important to mention that in the past ten years the number of peer-reviewed journal paper publications addressing groundwater and climate change has increased considerably as shown in Fig. 1.1. Also only recently, water resources managers and politicians are progressively recognising the important role of groundwater resources in meeting the demands for drinking water, agricultural and industrial activities, and sustaining ecosystems, as well as in the adaptation to and mitigation of the impacts of climate change and coupled human activities (Green et al. 2011).

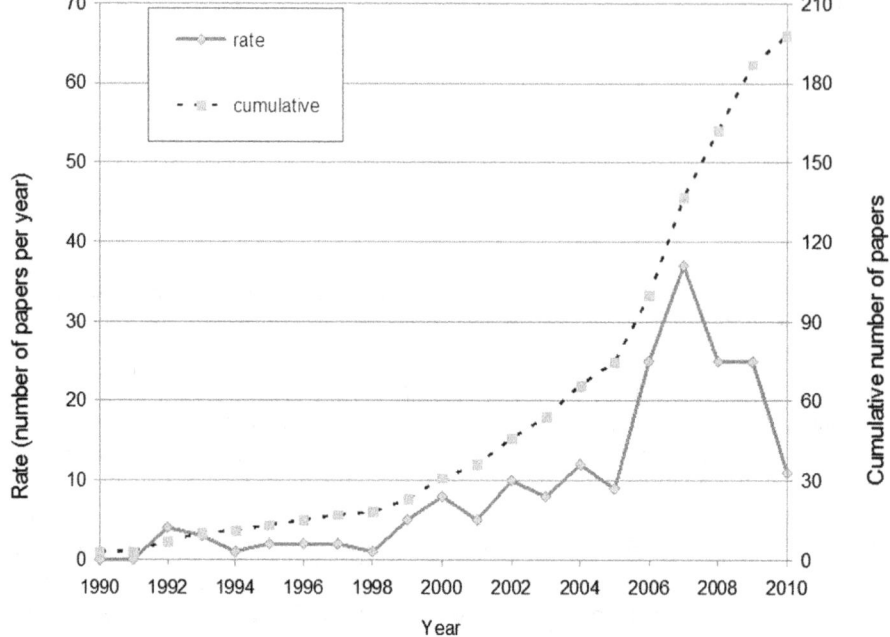

Figure 1.1.   Rate of peer-reviewed journal paper publications addressing groundwater and climate change from 1990 to 2010. A total of 198 papers addressing subsurface water and climate change are included. Final references were compiled in February 2011, so some papers published late in 2010 may be missing (modified from Green et al. 2011).

Besides the direct impacts of climate change on the natural processes of the global hydrological cycle, it is crucial to also consider the indirect impacts. These are human responses to the direct impacts, such as increased utilization of groundwater in times of drought and non-availability of surface water and may lead to increased and unsustainable abstraction and utilization of groundwater resources, including non-renewable groundwater reserves. Thus, there are urgent and ongoing needs to address the expected coupled effects of human activities and climate change on global groundwater resources.

To address these concerns, the United Nations Educational, Scientific, and Cultural Organisation (UNESCO) International Hydrological Programme (IHP) initiated the GRAPHIC project (*Groundwater Resources Assessment under the Pressures of Humanity and Climate Change*) in 2004. GRAPHIC seeks to improve our understanding of how groundwater interacts within the global water cycle, supports ecosystems and humankind and, in turn, responds to complex and coupled pressures of human activities and climate change. To successfully achieve these objectives within a global context, GRAPHIC was developed to incorporate a collaborative effort and umbrella for international research and education. GRAPHIC outlines areas of desired international investigations covering major geographical regions, groundwater resource topics, and methods to help advance the combined knowledge needed to address scientific and social aspects (UNESCO 2008).

The GRAPHIC project was designed with the understanding that groundwater resources can have nonlinear responses to atmospheric conditions associated with climate change and/or terrestrial-surface conditions associated with human activities. Therefore, groundwater assessments under the coupled pressures of human activities and climate change and variability involve the exploration of complex-system interactions. GRAPHIC incorporates a multidisciplinary scientific approach as the most rigorous platform to address such complexity. Furthermore, the GRAPHIC project extends investigations beyond physical, chemical, and biological interactions to include human systems of resource management and governmental policies. The structure of the GRAPHIC project has been divided into subjects, methods, and regions. The subjects encompass (i) groundwater quantity (recharge, discharge, and storage), (ii) quality, and (iii) management aspects. A variety of scientific methods and tools are being applied in the framework of GRAPHIC, including analysis of field data, geophysics, geochemistry, paleohydrology, remote sensing (in particular GRACE satellite gravimetry), information systems, modelling, and simulation. GRAPHIC consists of regional components (Africa, Asia and Oceania, Europe, Latin America, and the Caribbean and North America) where case studies have been identified and carried out.

The management of groundwater resources under the coupled pressures of climate change and human activities is a challenge. Sound understanding of the functioning of groundwater systems and their interactions with numerous and interlinked external factors is an indispensable basis for informed management. GRAPHIC strives to facilitate cooperation between scientists of different disciplines and from different countries. The basin/aquifer scale case studies presented in this book have been selected in each region by local scientists and experts of the respective subject.

1.2 OVERVIEW OF THE BOOK

Climate Change Effects on Groundwater – A Global Synthesis of Findings and Recommendations is a compilation of 20 case studies from more 30 different countries that have been carried out under the framework of the UNESCO-IHP GRAPHIC project. The approximate location of each case study is displayed on the "Groundwater Resources of the World" map (WHYMAP 2008) (Fig 1.2).

The case studies presented in this volume represent aquifers from all the major climate regions of the world. The studies address groundwater resources in a range of hydrogeological settings from mountainous to coastal aquifer systems, including unconfined, semi-confined, and confined aquifers in unconsolidated to fractured-rock material. More details on each case study location, climate, hydrogeological setting, land use, groundwater use, as well as subjects addressed and methods applied are presented in the overview table (Table 1.1).

This volume is organized by case study according to the major climate groups of the Köppen-Geiger climate classification scheme (Köppen 1936): tropical, dry (arid and semi-arid), temperate, continental, and polar climates. Three chapters that cover several study areas and different climate groups are presented under "various climates" and are displayed in Figure 1.2 as one large circle or multiple circles indicating the regional scope of the respective chapter. The case study chapters (Chapters 2 to 21) each follow a similar organization and structure. The introduction of each chapter describes the purpose and scope, study area, methodology, and relevance to the GRAPHIC project. The results and discussion are followed by recommendations for water managers and planners, as well as policy and decision makers. Finally, the continuation of research activities and future work are outlined.

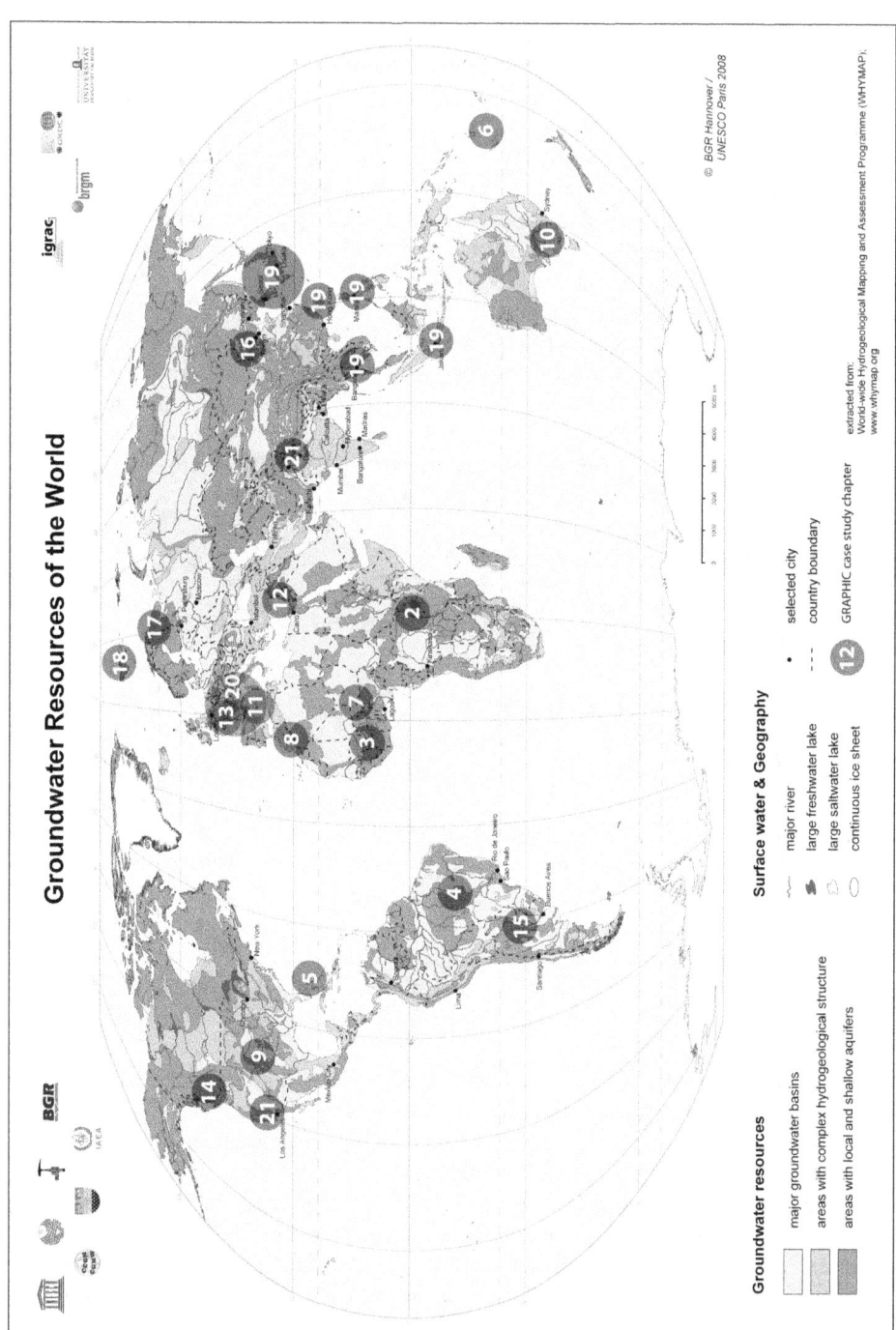

Figure 1.2. Approximate location of case study displayed on the "Groundwater Resources of the World" map (WHYMAP 2008). Numbers refer to the chapters in this volume. Case studies that cover several study areas and different climate groups are displayed as one large circle or multiple circles indicating the regional scope of the respective chapter.

Table 1.1   Overview of case studies.

Location	Climate	Hydrogeological setting	Land use	Groundwater use	Quantity or Quality	Methods
Chapter 2: The Impacts of Climate Change and Rapid Development on Weathered Crystalline Rock Aquifer Systems in the Humid Tropics of sub-Saharan Africa: Evidence from South-Western Uganda						
East Africa, South-western Uganda, River Mitano Basin	Tropical (humid)	Deeply weathered, crystalline rock aquifers	Agriculture, grassland, small areas of wetland, forest and plantations	Irrigation, livestock, drinking	<u>Quantity</u>: recharge, discharge, storage	Modelling
Chapter 3: Groundwater Recharge and Storage Variability in Southern Mali, Africa						
Western Sub-Saharan Africa, southern Mali, Niger river basin	Tropical (wet and dry), and partly dry (semiarid)	Clayey laterites overlying unconfined/semi-confined fractured sandstone aquifers	Savanna, shrubland, agriculture	Drinking, agriculture, livestock	<u>Quantity</u>: recharge, storage	GRACE, Modelling, Monitoring
Chapter 4: Groundwater Discharge as Affected by Land Use Change in Small Catchments: A Hydrologic and Economic Case Study in Central Brazil						
South America, central Brazil, Pipiripau river basin	Tropical (humid)	Deep, well drained soils (red oxisols and ultisols), underlain by quartzites, phyllites, and rhythmites	Agriculture, pastureland, natural savannah, woodland, grassland	Support aquatic ecosystems and hydrological services	<u>Quantity</u>: base flow discharge	Data correlation, empirical method
Chapter 5: Effects of Storm Surges on Groundwater Resources, North Andros Island, Bahamas						
The Caribbean, The Bahamas, North Andros Island	Tropical (humid)	Shallow, fresh groundwater lens in limestone and limesand aquifers	Forest, shrubland, rural communities	Local drinking and domestic needs; primary water supply for New Providence Island	<u>Quantity</u>: recharge, storage <u>Quality</u>: saltwater intrusion, salinity, septic systems	Monitoring

(Continued)

Table 1.1 Continued

Chapter 6: Reducing Groundwater Vulnerability in Carbonate Island Countries in the Pacific

Central and southern Pacific Ocean, small island nations	Tropical/Sub-Tropical	Shallow, fresh groundwater lens in permeable coral sand and karst limestone aquifers	Forest, shrubland, urban	Drinking, agriculture	Quantity: recharge, abstraction, storage; Quality: saltwater intrusion	Modelling, Monitoring

Chapter 7: Groundwater Resources Increase in the Iullemmeden Basin, West Africa

West Africa, Nigeria and Niger, Iullemmeden Basin	Dry (semiarid)	Sedimentary basin, largely unconfined. Several confined aquifers exists at depth. (Continental Terminal aquifer – unconfined)	Mainly rainfed agriculture, livestock breeding (in the North)	Drinking, livestock breeding. Use for irrigation very limited spatially	Quantity: groundwater dynamics and recharge	Remote sensing, subsurface geophysics, environmental geochemistry hydrodynamics, monitoring, numerical modeling at various scales

Chapter 8: Climate Change and its Impacts on Groundwater Resources in Morocco: the Case of the Souss-Massa Basin

North Africa, Morocco, Souss-Massa basin	Dry (arid to semiarid)	Shallow aquifer of the Souss-Massa plain, coastal aquifer	Irrigated agriculture	Irrigation, drinking, industry	Quantity: storage, recharge Quality: salinization, nitrate	Trend analyses (precipitation and temperature), monitoring (gw level), hydrochemical and isotopic tracers

Chapter 9: Vulnerability of Groundwater Quality to Human Activity and Climate Change and Variability, High Plains Aquifer, USA

North America, central United States, Great Plains province	Dry (semiarid)	High Plains aquifer: primarily unconsolidated, unconfined aquifers	Irrigated and dryland agriculture, rangeland	Irrigation, livestock, drinking	<u>Quality</u>: nitrate, other chemical constituents <u>Quantity</u>: recharge, abstraction	Age dating, GIS, Modelling, Monitoring

Chapter 10: Groundwater Change in the Murray Basin from Long-Term In-Situ Monitoring and GRACE Estimates (Australia)

Southeastern Australia, Murray Basin	Dry (semiarid)	Unconsolidated sediments and sedimentary rocks. Confined and unconfined. Specific aquifers: Murray Group, Pliocene Sands aquifer, Shepparton Formation	Farming land, native and plantation forests, livestock production (cattle and sheep)	Irrigation, livestock, drinking	<u>Quantity</u>: recharge, discharge; <u>Quality</u>: salinization	GRACE, Monitoring

Chapter 11: Impact Assessment of Combined Climate and Management Scenarios on Groundwater Resources. The Inca-Sa Pobla Hydrogeological Unit (Majorca, Spain)

Europe, Mediterranean Balearic island, Majorca, Spain	Mediterranean climate, temperate/semi-arid	Four different hydrostratigraphic units and three aquitard units, grouped into an upper and lower aquifer system	Agriculture	Irrigation, tourism, ecosystems	<u>Quantity</u>: recharge, discharge, exploitation	Modelling, simulations, management

(Continued)

Table 1.1 Continued

Chapter 12: The Effect of Climate and Sea Level Changes on Israeli Coastal Aquifers

Mediterranean, coastal aquifers and Dead Sea, Israel	Mediterranean climate, dry (arid and semiarid)	Israeli coastal aquifer: inter-layered sandstone, calcareous sandstone, siltstone, and red loam Dead Sea coastal aquifer: Upper Cretaceous Judea Group Aquifer and the Quaternary alluvial coastal aquifer	Agriculture	Irrigation, domestic	Quantity: recharge Quality: saltwater intrusion, salinization	Modelling, simulations, monitoring

Chapter 13: Land Subsidence and Sea-Level Rise Threaten Freshwater Resources in the Coastal Groundwater System of the Rijnland Water Board, The Netherlands

Europe, Coastal groundwater system, Rijnland, The Netherlands	Temperate, Continental	Quaternary deposits, intersected by loamy aquitards and overlain by a Holocene aquitard of clay and peat	Agriculture	Irrigation, domestic and industrial	Quality: saltwater intrusion, salinization	Modelling, simulations

Chapter 14: Climate Change Impacts on Valley-Bottom Aquifers in Mountain Regions: Case Studies from British Columbia, Canada

North America, western Canada, mountain regions British Columbia	Dry (semi-arid to arid)	Okanagan Basin, Grand Forks: valley-bottom unconsolidated aquifers	Forest, shrubland, urban	Drinking, agriculture, industry	Quantity: recharge	GCM downscaling, Modelling, GIS

Chapter 15: Possible Effects of Climate Change on Groundwater Resources in the Central Region of Santa Fe Province, Argentina

South America, Argentina, Santa Fe Province	Temperate (humid)	Upper unconfined aquifer: aeolian sedimentary deposits Semi-unconfined aquifer: sands of fluvial origin	Agriculture, livestock, rearing	Drinking, food production (agriculture, livestock rearing), industry	Quantity: recharge, discharge Quality: chemical compound input, salinization	Modelling

Chapter 16: Impacts of Drought on Groundwater Depletion in the Beijing Plain, China

East Asia, China, Beijing Plain	Continental (dry)	Sedimentary (alluvial), shallow aquifer mainly unconfined, deep aquifers confined	Agriculture, industry, drinking	Irrigation from shallow aquifer; drinking, industry mainly from deep aquifer	Quantity: recharge, storage	Monitoring, modelling

Chapter 17: Possible Effects of Climate Change on Hydrogeological Systems: Results from Research on Esker Aquifers in Northern Finland

Europe, northern Finland	Continental (polar)	Esker aquifers: unconsolidated, unconfined or confined	Forest, peatland	Ecosystems, drinking, recreation	Quantity: recharge, discharge Quality: temperature, dissolved oxygen, salts	Monitoring, modelling

Chapter 18: Climate Change Effects on Groundwater in Permafrost Areas – Case Study from the Arctic Peninsula of Svalbard, Norway

Europe, Norway, Svalbard peninsula	Polar (arctic)	Subpermafrost groundwater	none (60% covered by glaciers, large part is declared National Park)	Drinking (very limited)	Quantity: recharge, discharge	Monitoring, rock cores, simulation and modelling

Chapter 19: Groundwater Management in Asian Cities under the Pressures of Human Impacts and Climate Change

Asian coastal cities: Tokyo, Osaka, Seoul, Taipei, Bangkok, Jakarta and Manila	Temperate, Continental Tropical	Coastal alluvial plain, urban subsurface soil	Urban	Domestic use, industry	Quantity: recharge, storage Quality: contamination	GRACE, modelling, GIS

(Continued)

Table 1.1 Continued

Chapter 20: Evaluation of Future Climate Change Impacts on European Groundwater Resources

Northern and southern Europe, centred on the Å (Denmark), Medway (UK), Seine (France), Guadalquivir (Spain) and Po (Italy) river basins	Temperate, Continental Mediterranean	River Å: glacial sands and gravels River Medway: Cretaceous Chalk and Lower Cretaceous Sands River Seine: Cretaceous Chalk and Lower Cretaceous Sands River Guadalquivir: dolomitic limestone and alluvial deposits River Po: alluvial sediments	River Å: agriculture, industry River Medway: agriculture, pasture, urban River Seine: agriculture, urban, semi-urban River Guadalquivir irrigated agriculture River Po: irrigated agriculture, urban, industry	Drinking water, irrigation	Quantity: recharge, water-stress	Modelling, simulations

Chapter 21: Sustainable Groundwater Management for Large Aquifer Systems: Tracking Depletion Rates from Space

North America, western US, California, Central Valley aquifer; and northern India	Central Valley: Temperate (Mediterranean climate); northern India: Dry-Continental	Central Valley and northern India: confining units and unconfined, semiconfined, and confined aquifers	Agriculture	Irrigated agriculture, drinking, and industry	Quantity: discharge, storage	GRACE, monitoring, and modelling

Tropical climate case studies (Chapters 2 to 6) include those from Africa (Uganda and Mali), Latin America (Brazil), the Caribbean (The Bahamas), and Pacific Island countries. Based on findings from south-western Uganda, Chapter 2 addresses whether intensive groundwater abstraction from weathered crystalline rock aquifers is a viable option to meet rapidly rising demand for domestic and agricultural water in Sub-Saharan Africa. The chapter also analyses projections of climate change impacts on groundwater resources and discusses opportunities and risks of their application to inform decision making. Chapter 3 describes the combined application of several methodologies, including measured field data, remote sensing, and modelling for estimating groundwater recharge and storage variability in southern Mali. The integration of these methods may be a promising tool for assessing groundwater resources in data scarce regions. The chapter also provides a preliminary assessment of the impacts of future climate change on groundwater recharge. The case study from Brazil (Chapter 4) uses an empirical method to assess the hydrological and economical effects of land-use change on groundwater discharge in a small tropical catchment.

Groundwater is the main source of freshwater on many islands. The resource is particularly vulnerable to extreme climate events, sea-level rise, and human-induced perturbations. Chapter 5 describes a storm surge from Hurricane Frances in 2004 that contaminated the groundwater supply on North Andros Island, The Bahamas. Chapter 6 presents key climatic, hydrogeological, physiographic, and management factors that influence groundwater quantity and saline intrusion into freshwater lenses beneath small Pacific Island countries.

Dry (arid and semiarid) climate case studies (Chapters 7 to 10) focus on the effects of climate change and human activities on groundwater resources in Africa (Morocco, Niger, and Nigeria), the United States (US), and Australia. Chapter 7 describes large-scale land clearing in the southern part of the Iullemmeden Basin that experiences increased groundwater recharge and rising water levels over the past several decades. Management responses to outcropping water tables and salinization of soils are discussed. The Morocco case study (Chapter 8) analyses trends in temperature and precipitation and the effects of projected changes on groundwater recharge and water quality in the arid Souss-Massa Basin.

The quality of groundwater is often as critically important as its quantity in terms of groundwater sustainability. Chapter 9 presents the coupled effects of human and climate stresses on groundwater quality in the High Plains aquifer, which is the most heavily used aquifer in the US and supplies about 30% of the groundwater used for irrigation in the US. Focusing, in turn, mainly on groundwater quantity aspects, Chapter 10 shows the complex and coupled effects of human activity (land clearing) on groundwater (increase of recharge and groundwater levels), and subsequent multi-year drought (decrease of groundwater levels) in the Murray Basin in south-eastern Australia. A comparison of borehole data with space gravimetry (GRACE) and soil moisture estimates from hydrological models is used to test the capability of the GRACE mission and provide regional estimates of change in groundwater storage so that it can be applied for the monitoring of insufficiently instrumented regions.

Temperate climate case studies (Chapters 11 to 15) include those from coastal aquifers in Spain, Israel, and The Netherlands, mountain regions of British Columbia, and the Santa Fe Province of Argentina. The Mediterranean region faces an increasing water demand for agriculture and tourism, while climate change projections forecast an increase of temperature, decrease of precipitation, and increased occurrence of extreme events. Chapter 11 analyses combinations of climate scenarios and management strategies on the island of Majorca (Spain) in view of preserving groundwater resources under predicted climate change.

Seawater intrusion into coastal aquifers is a concern in the Mediterranean. Chapter 12 describes the coupled effect of climate and anthropogenic sea level changes on Israeli coastal aquifers of the Mediterranean Sea and the Dead Sea. Chapter 13 presents the impacts of land subsidence and sea-level rise on freshwater resources in coastal groundwater systems of The Netherlands. In these systems, saline groundwater comes from the sea and from deep saline aquifers, and subsequently intrudes near-surface coastal groundwater systems. The salinization of the subsoil is caused by human-driven processes of land subsidence that have been going on for nearly a millennium.

Mountain watersheds or basins are unique high-relief environments that are important sources of water for local and downstream ecosystems and human population. Chapter 14 provides an overview of hydrogeological processes in temperate mountain regions as a basis for understanding how climate change may influence the groundwater systems. Case study examples of two valley-bottom aquifer systems in southern British Columbia, Canada highlight the complex interactions that need to be considered for climate change impact and adaptation assessment. Applying a modelling approach, the chapter explores recharge mechanisms and evaluates how the magnitude and timing of recharge may change under future climate conditions.

In the temperate central region of the Santa Fe Province in Argentina (Chapter 15) groundwater is the only source of water supply for all regional demands. The case study analyses available hydrogeological data to describe the aquifer system and quantify present groundwater availability. Future recharge to the aquifer system is estimated, and incorporated into a numerical groundwater flow model to assess future groundwater availability for drinking and food production under different climate scenarios.

Continental climate case studies (Chapters 16 and 17) include those from China and Finland. Chapter 16 analyses the impacts of prolonged drought on groundwater resources in the Beijing Plain where the combined effects of decreasing natural recharge and increasing abstraction have caused rapid depletion of groundwater storage. The chapter elaborates on direct and indirect impacts of climate change and proposes management responses based on simulations of groundwater depletion under various scenarios. Chapter 17 describes possible effects of climate change on esker aquifers in northern Finland. Eskers are an important source of potable groundwater in Finland and support many ecosystem services. However, groundwater in eskers is threatened by peatland drainage, agriculture, roads, and other land uses. This chapter describes the possible impacts of climate change and land use on esker groundwater systems with focus on the impact of peatland drainage in the esker discharge zone.

The polar climate case study (Chapter 18) is from Svalbard, Norway. Polar regions are sparsely populated, but have gained a lot of interest in the discussions about climate change because high-latitude areas are predicted to experience the most dramatic global climate change in this century. Moreover, large parts of these areas are regarded as pristine, with unique and highly specialized habitats for animals and plants. Groundwater forms part of this system that is – and will be – highly impacted by climate change. Chapter 18 presents a case study that examines climate change impacts on arctic sub-permafrost groundwater from the Arctic Peninsula of Svalbard, Norway.

Chapters 19 to 21 present case studies that encompass different climatic zones. Chapter 19 assesses the effects of climate change and human activities on urban subsurface environments and groundwater, which is an important but largely unexamined field of human-environment interactions. In this chapter, the subsurface environments of seven Asian coastal cities are studied with respect to water shortage, land subsidence, groundwater storage and contamination, thermal anomalies, and the urban heat island effect.

Similar to other regions of the world, groundwater in Europe is a substantial economic resource that is threatened by over-abstraction and contamination from surface-derived pollutants, which could be exacerbated by climate change. Chapter 20 evaluates future climate change effects on European groundwater resources in five study areas in northern and southern Europe, centred on the Å (Denmark), Medway (UK), Seine (France), Guadalquivir (Spain), and Po (Italy) river basins.

Chapter 21 describes the application of satellite gravimetry (GRACE) for characterizing groundwater storage changes in large aquifer systems – a method that provides new opportunities for water-resources monitoring, particularly in data sparse regions. Two case studies of groundwater depletion are presented, one in the relatively data-rich Central Valley aquifer of California (US) and in the other in more data-poor northern India.

The last chapter, Chapter 22, summarizes the main findings of the book in terms of new scientific insight and policy recommendations. This chapter, in particular, is expected to be of great interest to water resource managers, planners, and decision makers entrusted with the management of a valuable resource. In the light of global change, and climate change in particular, groundwater will continue to be an important resource that supports human health and livelihoods and many natural ecosystems. A sound understanding of the resource and current and future pressures from climate and human activities are necessary to guide adaptive management towards long-term groundwater sustainability.

REFERENCES

Bates, B., Kundzewicz, Z.W., Wu, S. & Palutikof, J.P. (2008) Climate change and water. *Technical Paper VI of the Intergovernmental Panel on Climate Change*. Geneva, Intergovernmental Panel on Climate Change Secretariat. 210 pp.

Green, T.R., Taniguchi, M., Kooi, H., Gurdak, J.J., Allen, D.M., Hiscock, K.M., Treidel, H. & Aureli, A. (2011) Beneath the surface of global change: Impacts of climate change on groundwater. *Journal of Hydrology*. doi:10.1016/j.jhydrol.2011.05.002 [Online] Available from: http://www.sciencedirect.com/science/article/pii/S0022169411002988. Accessed 1 October 2011

IPCC. (2001): Working Group II: *Climate Change 2001, Impacts, Adaptation and Vulnerability*. [Online] Available from: http://www.grida.no/publications/other/ipcc_tar/?src=/climate/ipcc_tar/wg2/377.htm. Accessed 30 September 2011

IPCC. (2007) Climate change 2007: The physical science basis. *Contribution of Working Group I to the Fourth Assessment Report of the Intergovernmental Panel on Climate Change*. Cambridge, United Kingdom and New York, NY, Cambridge University Press.

Köppen, W. (1936) Das geographische System der Klimate. In: Köppen, W. and Geiger, G. (eds.) *Handbuch der Klimatologie*. Vol. 1, part C. Berlin, Gebr, Borntraeger. pp. 1–44.

Siebert, S., Burke, J., Faures, J.M., Frenken, K., Hoogeveen, J., Döll, P. & Portmann, F.T. (2010) Groundwater use for irrigation – a global inventory. *Hydrology and Earth System Sciences*.

[Online] 14, 1863–1880. Available from: www.hydrol-earth-syst-sci.net/14/1863/2010/doi:10.5194/hess-14-1863-2010 (direct link to the paper: http://www.hydrol-earth-syst-sci.net/14/1863/2010/hess-14-1863-2010.pdf) . Accessed 1 October 2011

UNESCO. (2008) *International Hydrological Programme of the United Nations Educational, Scientific and Cultural Organization (UNESCO)*. GRAPHIC framework document. [Online] Available from: http://unesdoc.unesco.org/images/0016/001631/163172e.pdf [Cited 20th July 2011].

WHYMAP. (2008) *Worldwide Hydrogeological Mapping and Assessment Programme, Groundwater Resources of the World 1:25,000,000*. BGR/UNESCO. [Online]. Available from: http://www.whymap.org/whymap/EN/Downloads/Global_maps/globalmaps_node_en.html. Accessed 1 October 2011

WWAP. (2009) *The United Nations World Water Development Report 3: Water in a Changing World, World Water Assessment Programme*. Paris, UNESCO Publishing, UNESCO 2009. 349 p.

Tropical Climates

CHAPTER 2

The impacts of climate change and rapid development on weathered crystalline rock aquifer systems in the humid tropics of sub-Saharan Africa: evidence from south-western Uganda

Richard Taylor & Callist Tindimugaya

ABSTRACT

Deeply weathered crystalline rock aquifer systems underlie much of the continental land masses in the tropics including sub-Saharan Africa, India and South America. Detailed catchment-scale studies in south-western Uganda reveal three key characteristics of the impacts of climate change and rapid development on weathered crystalline rock aquifer systems in the humid tropics. First, rapid development of groundwater resources is expected over the next two decades to have a more pronounced impact on groundwater resources than climate change. Second, projected changes in the seasonality and intensity of rainfall, though rarely considered, substantially influence the timing and magnitude of groundwater recharge. Third, quantified uncertainty in climate change impacts on groundwater discharges is substantial and arises primarily from the structure of general circulation models to simulate future rainfall and algorithms to estimate climate change impacts on potential evapotranspiration. In light of substantial uncertainty in current hydrological projections, groundwater management in the humid tropics will necessarily need to be adaptive and informed by groundwater and meteorological monitoring systems that reveal the response of groundwater systems to increased abstraction and climate change. A concerted, collaborative research effort between climate and water scientists is also required to reduce current uncertainty in hydrological projections; research in Uganda shows furthermore that employed models of groundwater recharge need to explicitly consider projected changes in rainfall intensity.

2.1 INTRODUCTION

2.1.1 Purpose and scope

The tropics[1] is home to ~40% of the world's population, the highest levels of population growth, and the majority of the world's poor. It is also where most of the sun's energy that drives global climate is absorbed and, as such, where changes in the water-holding capacity of the atmosphere as a result of anthropogenic warming are highest (Allen and Ingram 2002; Trenberth et al. 2003). Consequently, it is in the tropics where the impacts of rapid development and climate change on water resources are expected to be the most severe and where the need for sustainable adaptive management strategies in the water sector is greatest. Perversely, it is the tropics where the human and institutional capacity

---

[1]The region between latitudes 23.45°N and 23.45°S.

and hydrological knowledge base to devise adaptive water strategies are the most limited. Groundwater has for decades enabled communities across the tropics to adapt to seasonal or perennial shortages in surface water by providing water for drinking, watering livestock, and more recently irrigation. It is unclear, however, whether more intensive groundwater abstraction to meet rapidly rising demand for domestic and agricultural water is viable. Furthermore, a quantitative understanding of the impact of climate change on groundwater resources in the tropics remains elusive.

Large areas of the tropics including 40% of Sub-Saharan Africa (SSA) are underlain by deeply weathered crystalline rock aquifers (Taylor and Howard 2000; Maréchal et al. 2004; MacDonald et al. 2005). Aquifers comprising fractured bedrock (saprock) and a primarily *in situ* weathered, unconsolidated regolith (saprolite) possess low transmissivities typically ranging from 1 to $10\,m^2{\cdot}day^{-1}$ (e.g. Houston and Lewis 1988; Howard et al. 1992; Briz-Kishore 1993; Owoade 1995; Chilton and Foster 1995) Higher transmissivities exceeding $10\,m^2{\cdot}day^{-1}$ have recently been observed in tropical regoliths where alluvial and fluvio-laustrine deposits are present (Bradley 2011). Saprock and saprolite aquifers can operate as an integrated aquifer system wherein the overlying, more porous saprolite provides storage to underlying more transmissive fracture bedrock (Rushton and Weller 1985; Sekhar et al. 1994; Chilton and Foster 1995; Taylor and Howard 2000; Taylor et al. 2010). Further research is, however, required to resolve more completely the groundwater flowpaths and storage characteristics of this aquifer system.

In SSA, over half of the population depends in part or in whole on groundwater resources from saprolite-saprock aquifer systems for domestic water supplies. Of major concern is whether this aquifer system can sustain greater and more widespread groundwater withdrawals for rising domestic water demand and irrigation. The latter is recommended by both the Agricultural Water for Africa Initiative (AgWA) and Comprehensive Africa Agriculture Development Programme (CAADP) to improve food production and the resilience of the still predominantly rural population of small-scale farmers in SSA to climate variability and change.

At present, the duration and coverage of monitoring data for saprolite-saprock aquifer systems in the tropics are limited. With few exceptions (e.g. Groundwater Resilience in Africa[2]), research into these aquifer systems has focused on specific locations or basins (e.g. Nkotagu 1996; Taylor and Howard 2000; Maréchal et al. 2004; Giertz et al. 2006a) that are intended to be representative of more widespread hydrogeological conditions in saprolite and saprock aquifers. Here, we review evidence from a series of recent studies focused on the River Mitano Basin (Tindimugaya 2008; Mileham et al. 2008, 2009; Kingston et al. 2009; Kingston and Taylor 2010) of south-western Uganda which features the longest continuous record of daily river discharge in Uganda (1965 to present) and a time series of groundwater level and abstraction observations. Following a description of the basin, we describe three key outcomes from this research regarding the impacts of rapid development and climate change on groundwater resources in saprolite-saprock aquifer systems of the humid tropics.

2.1.2   Description of the study area: the River Mitano Basin

The River Mitano Basin is located within the humid, inner tropics just south of the equator in south-western Uganda (Fig. 2.1a). The basin is underlain by deeply weathered

---

[2]www.bgs.ac.uk/gwresilience

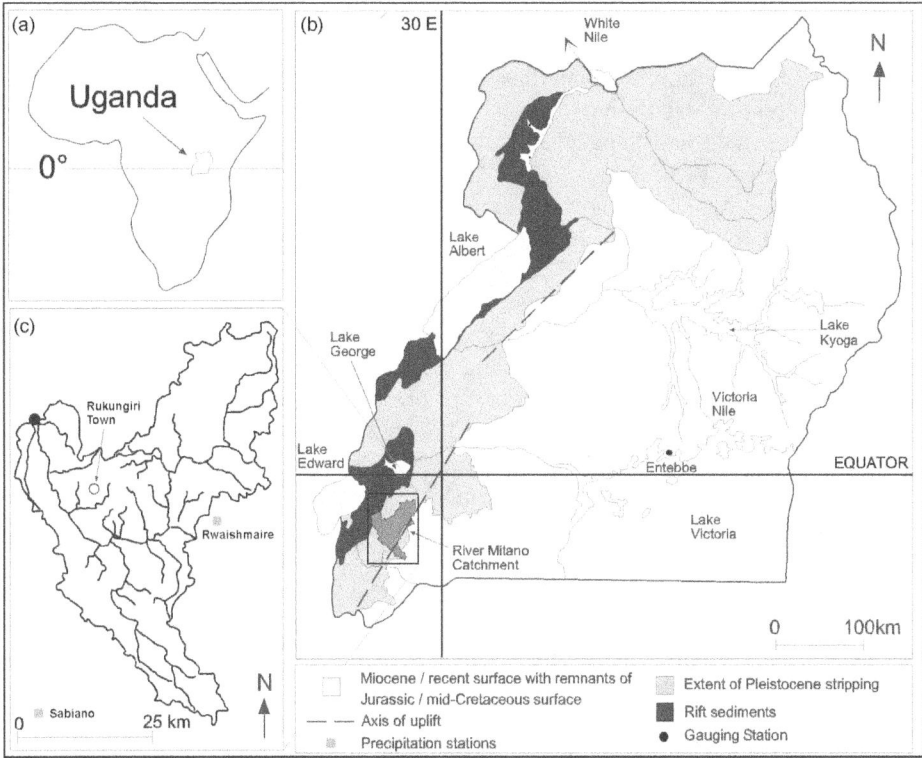

Figure 2.1. (a) Location of Uganda; (b) location of the basin within the weathered land surfaces of Uganda (Taylor & Howard, 1999a); and (c) map of the catchment drainage system showing gauging stations.

Precambrian crystalline rocks including gneisses, schists, phyllites and granites. The River Mitano flows in a north-westerly direction from upland areas, 2500 m above mean sea level (mamsl) in the south of the catchment, to the depression (graben) containing Lake Edward (975 mamsl) in the western arm of the East African Rift (Fig. 2.1b–c). The river's gauging station, located 20 km upstream of Lake Edward, represents an area of 2098 km^2. Detailed descriptions of the basin's geology, geomorphology and hydro-climatology are given in Tindimugaya (2008), Mileham et al. (2009), and Kingston and Taylor (2010). The weathered overburden (or regolith) comprises commonly truncated profiles composed primarily of coarse, less weathered material and there are frequent exposures of the Precambrian bedrock (Taylor and Howard 1999a). Fractures within the bedrock have developed from regional tectonic activity associated with rifting and more commonly from isostatic uplift resulting from the long-term weathering processes (Taylor and Howard 2000; Tindimugaya 2008). Groundwater from the weathered over-burden (where it exists) and underlying fractured bedrock commonly form an integrated aquifer system that discharges into the River Mitano drainage network. The basin's high relief and incised drainage reflect, however, a runoff-dominated regime (Taylor and Howard 1999b). Land use is primarily agrarian (79%). Principal crops are millet, cassava, sugarcane, simsim, maize, groundnuts, soybeans, bananas, rice, maize, cotton, coffee and tobacco. Grassland dominates the remainder of the catchment (17%) with small areas of wetland (3%), forest and plantations.

Similar to other humid equatorial environments, the River Mitano basin experiences a bi-modal rainfall regime in which the dominant modes (wet seasons) occur in March–May (MAM) and September–November (SON) as a result of the movement of the Inner-Tropical Convergence Zone (Fig. 2.2). Mean annual catchment precipitation observed between 1965 and 1979 ranges from 963 mm in the east at Rwaishmaire to 1699 mm in the southwest at Sabiano (Fig. 2.1c). Mean annual pan evaporation observed over the period 1967–1977 at Mbarara, approximately 50 km to the east of the catchment, is 1535 mm. Monthly pan evaporation is relatively constant throughout the year, varying by less than 15% and exceeds precipitation in all months except peak wet season precipitation. Diurnal mean temperatures range from 13 to 26°C and this range is significantly greater than the variation (2°C) in mean monthly maximum and minimum temperature. Discharge records (1965–1979) for the River Mitano reflect bi-modal precipitation (Fig. 2.2) but lag peak precipitation by approximately 2 to 6 weeks. Mean river discharge is highest during SON ($35\,\mathrm{m^3 \cdot s^{-1}}$, equivalent to a specific discharge of $520\,\mathrm{mm \cdot a^{-1}}$) and lowest during the dry season after MAM ($6\,\mathrm{m^3 \cdot s^{-1}}$, equivalent to $90\,\mathrm{mm \cdot a^{-1}}$).

Both the regolith and fractured bedrock form relatively weak aquifers. Of 51 tested wells in the basin, well yields range from 0.1 to $23\,\mathrm{m^3 \cdot hour^{-1}}$. Half of these wells possess yields of $<2\,\mathrm{m^3 \cdot hour^{-1}}$ whereas 15 wells had yields exceeding $5\,\mathrm{m^3 \cdot hour^{-1}}$. Low yielding boreholes are predominantly located in upland areas where the regolith is thin whereas high yielding boreholes are most commonly located in low-lying areas underlain by a comparatively thicker regolith. Across the basin, low-intensity (handpump) abstraction of groundwater ($<0.7\,\mathrm{m^3 \cdot hour^{-1}}$ per borehole) has taken place for decades from boreholes installed primarily into the fractured bedrock aquifer to supply potable water to the overwhelmingly rural population. Consequently, the predominance of higher yields observed in boreholes in low catena positions likely reflects one or both of enhanced bedrock transmissivities associated with greater fracturing and enhanced groundwater storage, sustaining higher yields, where the regolith is thicker. More recently, intensive

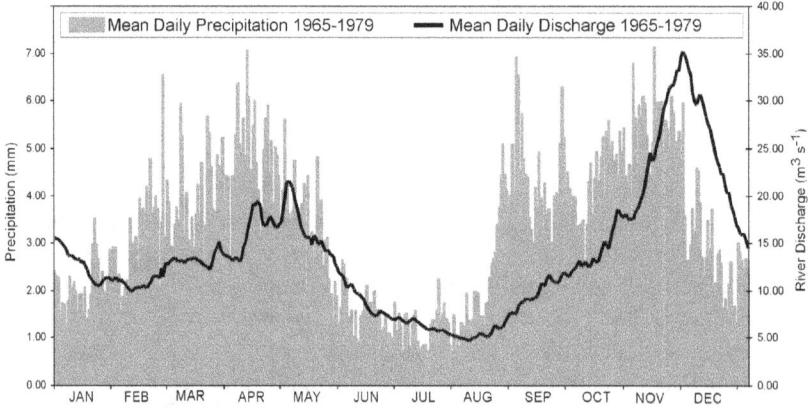

Figure 2.2.   Mean daily rainfall (20 stations) and river discharge in the River Mitano basin from 1965 to 1980 (Mileham 2008).

groundwater abstraction from the saprock-saprolite aquifer system by motorised pumps ($>3.6\,m^3\cdot hour^{-1}$ per borehole) has been conducted to supply the rapidly growing town of Rukungiri located in the north-central part of the basin.

2.2 RESULTS AND DISCUSSION

2.2.1 Impacts of intensive groundwater abstraction

Intensive groundwater abstraction in Sub-Saharan Africa (SSA) is primarily conducted for town and city water supplies as either a primary (e.g. Dodoma, Tanzania; Lusaka, Zambia) or supplementary (e.g. Nairobi, Kenya; Dar es Salaam, Tanzania) source of potable water. Smaller towns throughout SSA can be entirely dependent on groundwater as a source of potable water because such systems do not require the expense and expertise of sophisticated water treatment. For many towns, groundwater is their only source of water for some or all of the year. Intensive groundwater abstraction was promoted in many parts of SSA during the 1990s under a number of schemes including the Small Towns Water and Sanitation Project funded among others by the World Bank, DANIDA, the European Union, and African Development Bank. Intensive groundwater abstraction for irrigation in SSA is currently minimal and restricted to few locations (e.g. Kabwe, Zambia). Regionally, less than 5% of the arable land in SSA is under irrigation from either surface water or groundwater withdrawals (Giordano 2006). Substantial increases in groundwater withdrawals for irrigation are, however, proposed (World Bank 2007) to improve food production and the resilience of agricultural systems in SSA to climate variability and change.

The town of Rukungiri in the River Mitano Basin with an estimated population of 18,600 in 2006 relies exclusively upon groundwater for its piped water supply that is based on two high yielding boreholes. Motorised abstraction from one borehole commenced in 1998 but substantially increased after February 2003 to between 150 and $200\,m^3\cdot day^{-1}$ using both boreholes following a sharp rise in demand (MWLE 2006). This piped water supply is unable to meet demand and currently supplemented by 14 hand-pumped boreholes and 47 springs in the town in addition to another motorised borehole privately operated by a local hospital. The percentage of the population with access to safe water in the town is 71%. Of this, 56% are supplied by the piped water supply system whereas 15% are supplied by the point water sources (MWE 2006).

Both production boreholes are located in a wellfield near the base of a valley and installed into a thick (60 to 70 m) sequence of not only *in situ* weathered rocks but also colluvial and possibly alluvial fine, medium and coarse-grained sands. Surface geophysical investigations employing using electrical resistivity indicate that the wellfield is bounded geologically by dense, relatively unfractured bedrock (Tindimugaya 2008). The discontinuous nature of aquifers in the unconsolidated regolith is expected on stripped surfaces where the rejuvenation of drainage promotes colluvial and alluvial erosion (Taylor and Howard 2000). In Rukungiri Town and Mitano Basin, erosion to a new base level for drainage was triggered by downfaulting of the western arm of the East African Rift and creation of Lake Edward during the Pleistocene (Fig. 2.1b). On low-relief surfaces where the regolith is regionally extensive, it remains unclear whether aquifers within the regolith are regional or effectively discontinuous due to substantial variability in the regolith's lithology and hydraulic conductivity.

Combined rainfall and groundwater-level monitoring around the two production wells in Rukungiri Town was initiated after the commissioning of the town's piped water supply system (Fig. 2.3). Monitoring records for the production borehole (Ruk 5) that was unpumped show initially the response of the aquifer from September 2001 to January 2003 to estimated rainfall-fed recharge (Mileham et al. 2008) and subsequently intensive abstraction from February 2003 to August 2005. Small positive deflections in the groundwater-level record reveal episodic recharge events that are restricted to heavy rainfall events $(>10\,\text{mm}\cdot\text{d}^{-1})$ during monsoons. This pattern has been observed elsewhere in Uganda (Taylor and Howard 1996; Owor et al., 2009) and is discussed further in the next section.

A sharp drop and sustained decline in the observed groundwater level from 10 m below ground level (mbgl) in February 2003 to 25 mbgl by August 2004 (Fig. 2.3) occur in response to groundwater abstraction in the order of $120\,\text{m}^3\cdot\text{day}^{-1}$. This substantial ($\sim$15 m) and sustained decline in observed groundwater level suggests that abstraction has drawn from long-term storage. Strong positive deflections in the hydrograph after February 2003 not only correspond well with estimated recharge events but suggest further that the pumping borehole captures more recharge than under previous, non-pumping conditions. The extent to which positive deflections in groundwater

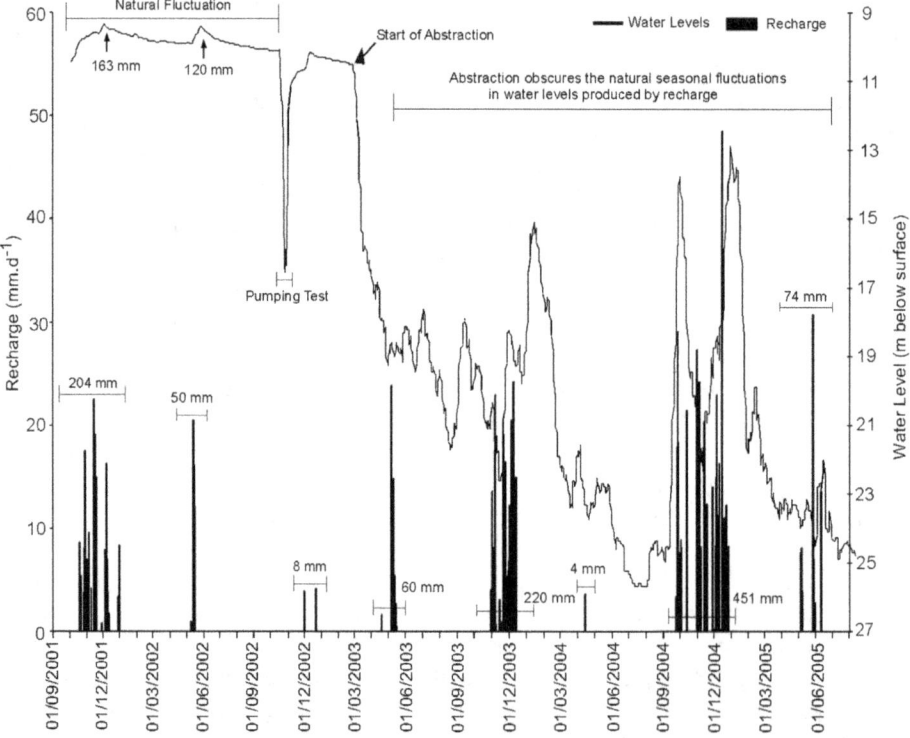

Figure 2.3.   Groundwater-level hydrograph for production borehole, Ruk 5 (Tindimugaya 2008) and estimated recharge (Mileham 2008) in Rukungiri Town from September 2001 to August 2005.

levels arise from variability in daily abstraction is, however, unclear. Continued pumping has led to shorter but more frequent periods of pumping as maximum pumping depths (determined by the depth of the submersible pump and borehole itself) are realised more quickly. A third production borehole, approximately 2 km away, has recently been commissioned to supplement the water from the two existing boreholes but the town's current water demand still remains unmet. Monitoring of groundwater levels in Rukungiri Town highlights how the localised nature of aquifers in deeply weathered crystalline rock environments constrains the long-term sustainability of intensive abstraction using submersible pumps (i.e. rates exceeding >1 $L \cdot s^{-1}$ per borehole).

2.2.2 Impact of climate change on groundwater recharge

An important, generic impact of human-induced warming on the global hydrological system is an increased frequency of very heavy rainfall events, specifically those in the uppermost quantiles of the rainfall distribution (Allen and Ingram 2002; Trenberth et al. 2003; Pall et al. 2007). The shift to more intensive rainfall is explained by the fact that water-holding capacity of the atmosphere increases according to the Clausius–Clapeyron relation ($\sim 6.5\%$ K^{-1} rise in air temperature) and the observation that heaviest rainfall events tend to deplete air of all of its available moisture. Recent analyses (Allan and Soden 2008; Min et al. 2011) verify this conceptual model and suggest further that the amplification of very heavy rainfall events under a warmer atmosphere is greater than that currently projected by General Circulation Models (GCMs). As warmer air temperatures in the humid tropics lead to larger absolute rises in the moisture content of the atmosphere, it is here where increased rainfall intensities are expected to be especially pronounced.

In the humid tropics, effective precipitation (i.e. that which contributes to groundwater recharge and runoff) depends upon heavy rainfall events when the rate of incoming rainfall temporarily exceeds high rates (>4 mm $\cdot$ day^{-1}) of potential evapotranspiration (PET) and soil moisture deficits that have accrued from this flux. In tropical Africa, there is a growing body of evidence from stable isotope tracers (Vogel and Urk 1975; Taylor and Howard 1999a), soil-moisture balance models (Taylor and Howard 1996; Eilers et al. 2007; Mileham et al. 2008) and borehole hydrographs (Owor et al. 2009) that demonstrates the importance of heavy rainfall events (>10 mm day^{-1}) in determining the magnitude and timing of direct, rainfall-fed recharge. Using a dataset of coincidental, daily observations of rainfall and groundwater levels remote from abstraction at 4 stations in the Upper Nile Basin of Uganda over the period 1999 to 2008, Owor et al. (2009) show that the magnitude of observed recharge events is better related to the sum of heavy rainfalls that exceed a threshold of 10 mm $\cdot$ day^{-1} compared to that of all daily rainfall events. Consequently, a shift toward more frequent heavy rainfall events under an anthropogenically warmed climate should promote rather than restrict groundwater recharge in similar environments of the tropics.

Substantial (70%) declines in groundwater recharge have been projected in northeast Brazil and southwest Africa in association with higher air temperatures and lower rainfall (Döll and Florke 2005, cited in IPCC Fourth Assessment Report) yet these projections failed to consider changes in the distribution of projected daily precipitation (Kundzewicz et al. 2008). Indeed, this oversight is commonly perpetuated through the use of "delta factors" to simulate climate change impacts on basin hydrology (e.g. Tate

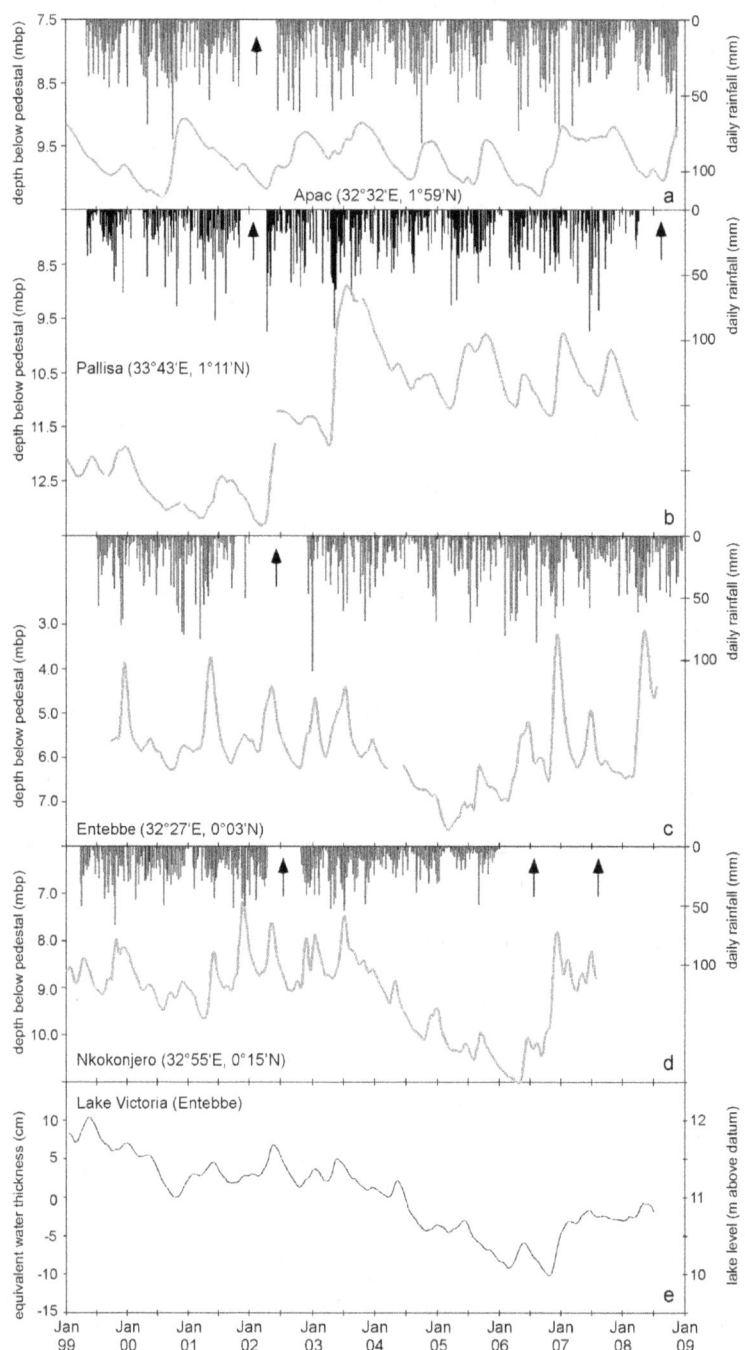

Figure 2.4.   Daily groundwater-level and rainfall observations over the period 1999 to 2008 from (a) Apac, (b) Pallisa, (c) Entebbe and (d) Nkokonjero in the Upper Nile Basin of Uganda together with (e) changes in the level of Lake Victoria at Entebbe. Arrows in plots (a) to (d) indicate gaps in the rainfall record.

et al. 2004). Delta factors represent proportional or absolute changes in rainfall from baseline (e.g. 1960–1990) that are applied to historical data. Using this approach, scaled and baseline scenarios differ only in terms of their respective means, maxima and minima; all other properties of the data such as the range and variability are assumed to remain constant. The procedure thus fails to account for changes in either the number of rain days or the spatial distribution of precipitation.

Research in the River Mitano Basin shows that failure to consider projected changes in rainfall intensities influences both the direction and magnitude of the climate change signal for groundwater recharge (Mileham et al. 2009). Projections of daily rainfall by the HadCM3 GCM under the A2 (SRES) greenhouse-gas emissions scenario were dynamically downscaled using the regional climate model (RCM), PRECIS (Jones et al. 2004), and applied to a semi-distributed soil-moisture balance model (Taylor and Howard 1999a). Under this scenario, the projected rainfall distribution demonstrates a substantial shift toward more intensive rainfall events relative to the observed 1965–1974 baseline (Fig. 2.5). Critically, projections of groundwater recharge that employ an historical (baseline) distribution of daily rainfall and higher PET associated with warmer air temperatures are 55% lower than the baseline period (1961–1990); transformation of the rainfall distribution to account for projected changes in rainfall intensity results in a 53% increase in recharge relative to the same baseline period (Fig. 2.6). These projected changes in rainfall intensity also increase the projected rise in runoff, relative to the baseline (1961–1990), from 86% to 137% (Fig. 2.6).

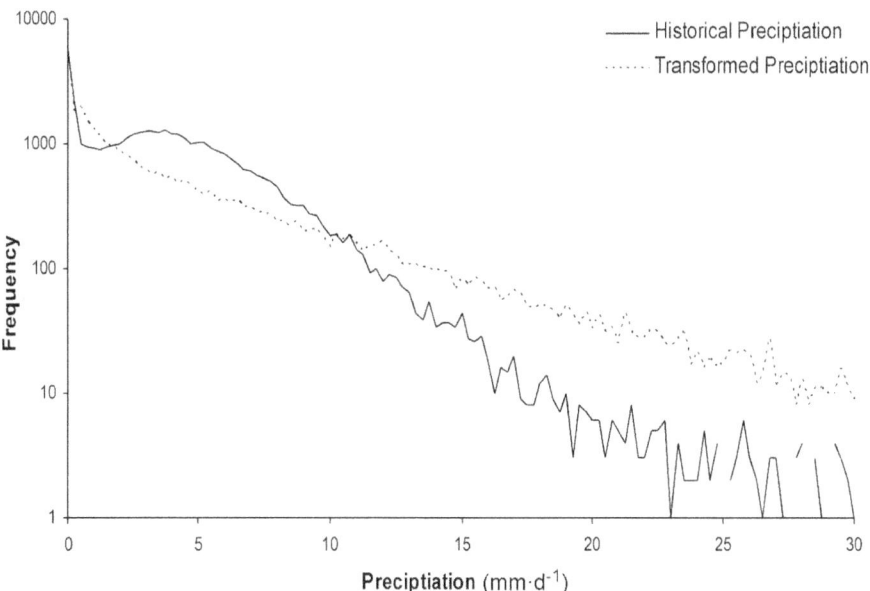

Figure 2.5. Frequency distribution of precipitation events in the River Mitano catchment for: (a) historical gridded station precipitation (1965–1974); and (b) historical precipitation transformed to represent the projected distribution in daily precipitation (2070–2100) under A2 emissions scenario using PRECIS RCM of the HadCM3 GCM (adapted from Mileham et al. 2009).

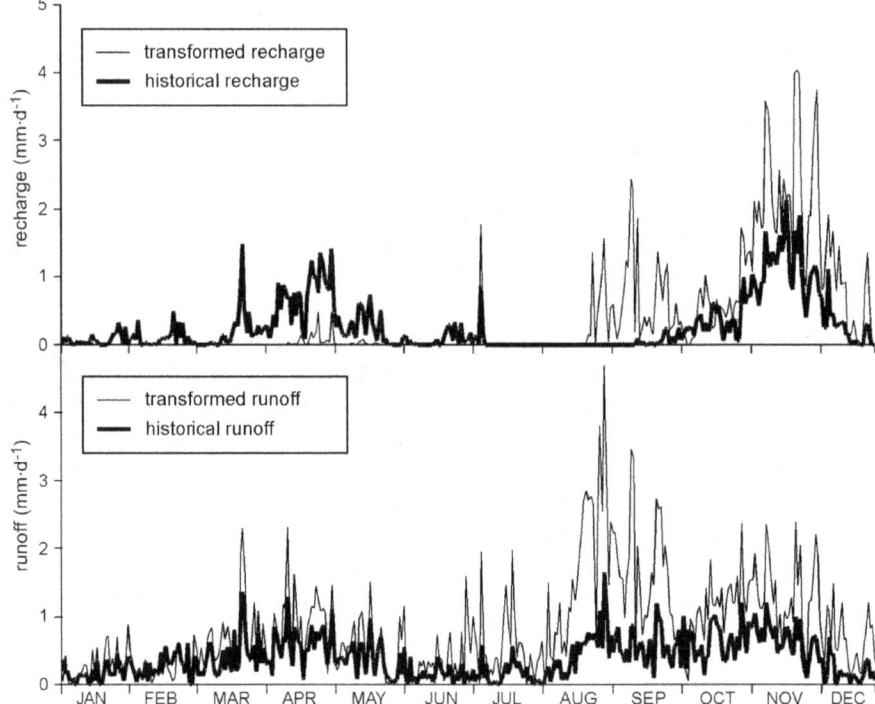

Figure 2.6.   Mean annual modelled recharge and runoff $(mm \cdot d^{-1})$ using transformed gridded River Mitano station precipitation and delta factor forced Mbarara PET for the River Mitano catchment for the period 2070–2100 relative to historical gridded River Mitano station precipitation derived recharge for the period 1965–1974 (adapted from Mileham et al. 2009).

2.2.3   Uncertainty in climate change impacts on groundwater resources

Quantitative evaluations of the impact of climate change on groundwater resources in saprolite-saprock aquifer systems in the tropics are few. Published studies have, to date, relied primarily upon climate projections from individual GCMs including HadCM3 (e.g. Mileham et al. 2009; Nyenje and Batelaan 2009) and ECHAM (Giertz et al. 2006b) despite substantial uncertainty in projected precipitation that is associated with GCM structure (Kundzewicz et al. 2007). Kingston and Taylor (2010) recently quantified uncertainty in hydrological projections the River Mitano Basin that is associated with GCM structure as well as the magnitude of anthropogenic warming and parameterisation of the hydrological model, Soil and Water Assessment Tool (SWAT, Arnold et al. 1998). In this study, future climate scenarios were generated using the ClimGen pattern-scaling technique described in Todd et al. (2011). Scenarios were generated for prescribed warming in global mean air temperature of 1, 2, 3, 4, 5, and 6°C using the UKMO HadCM3 GCM, and for 2°C warming, the proposed threshold of 'dangerous' climate change, with six additional GCMs: CCCMA CGCM31, CSIRO Mk30, IPSL CM4, MPI ECHAM5, NCAR CCSM30, and UKMO HadGEM1. These 7 GCMs derive from the World Climate Research Programme (WCRP) Coupled Model Intercomparison Project phase 3 (CMIP-3) and comprise a priority subset that spans the range of projected changes in precipitation.

As highlighted by Kingston and Taylor (2010), modelling of climate change impacts on the River Mitano reveals substantial seasonal changes in the relative contributions of runoff (quickflow) and groundwater (baseflow) to river discharge as a result of anthropogenic warming (Figure 2.7). Under baseline conditions, the majority (55%) of river discharge during the first wet season (MAM) derives from runoff whereas baseflow is responsible for most (80%) of the river discharge during the subsequent dry season during June and July. Under prescribed increases in global mean air temperature (up to 6°C), the proportion of river discharge derived from groundwater during the first half of the year (February–July) progressively declines as global mean temperature rises (Figure 2.7). This comparative decrease in baseflow contributions substantially reduces projected MAM river discharge under anthropogenic warming of 4 to 6°C so that formerly bimodal (MAM, SON) river discharge becomes unimodal (SON).

To assess the relative importance of projected changes in catchment precipitation and air temperature (and hence evapotranspiration) on the discharge of the River Mitano, Kingston and Taylor (2010) conducted modelling experiments employing projected air temperatures and baseline precipitation and vice versa (projected precipitation and baseline air temperatures). Projected reductions in baseflow and thus river discharge during the first half of the year were found to result primarily from increasing evapotranspiration associated with rising air temperatures. A key limitation in this analysis, discussed in section 4, is the fact that projections of daily rainfall employed a weathered generator informed by the historical distribution in daily rainfall (Kingston and Taylor 2010, Todd et al. 2011) and thus did not consider projected changes in the intensity of daily rainfall. There remains, therefore, some doubt about the projected shift from a bimodal to unimodal distribution in river discharge. Such a shift, should it occur, has important implications for water management. First, the longer period of time between recharge pulses (i.e. annual rather twice a year) would necessarily result in greater groundwater-level declines assuming abstraction remains the same or increases. Second, enhanced basin storage, be it natural or constructed, would be required to sustain allocations of river water, should they be developed in future for irrigation, livestock watering or hydroelectric power generation, during a longer period of low river discharge.

Under a 2°C rise in global mean air temperature (i.e. dangerous climate change), substantial disparities are projected in mean annual groundwater discharges between GCMs that dwarf estimated uncertainty (±5%) associated with SWAT model parameterisation (Kingston and Taylor 2010). There is little consistency among the GCMs in either magnitude or direction of change (Figure 2.8). For example, simulations based on the CSIRO GCM projections result in a 17% reduction in groundwater discharge under a 2°C rise in global mean air temperature whereas simulations employing NCAR GCM projections result in a 83% increase in groundwater discharge. Simulations based on CCMA and HadGEM1 GCMs project increased groundwater discharges of 25% and 6%, respectively; projections derived from HadCM3 (+0.3%), MPI (−5%) and IPSL (+2%) are within estimated model uncertainty. Resolution of the climate change signal for each of the GCMs derived from experiments changing temperature and precipitation independently (as above) reveals consistency in the 2°C temperature signal between GCMs. Consequently, uncertainty in projections of precipitation among the 7 GCMs is primarily responsible for variations in projected groundwater discharges under the 2°C warming scenario. These results highlight the severe limitations of climate change impact assessments based on projections from a single GCM.

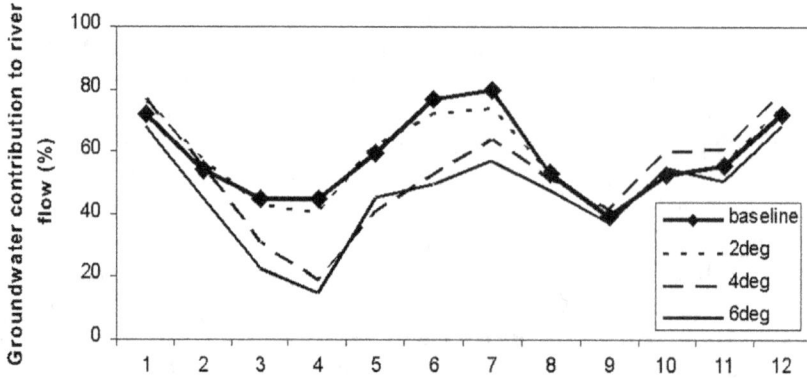

Figure 2.7.   Mean monthly baseflow contributions as proportion of mean monthly river discharge in the River Mitano Basin under HadCM3 prescribed warming scenarios (adapted from Kingston and Taylor 2010).

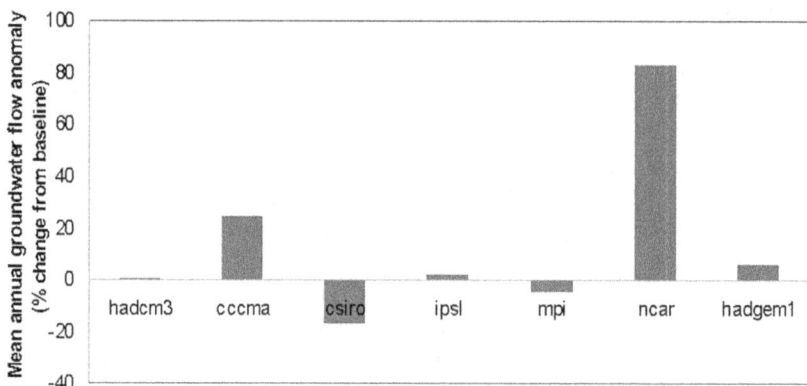

Figure 2.8.   Projected changes in mean annual groundwater flow in the River Mitano Basin, relative to the 1961–1990 baseline period, under a 2°C rise in global mean air temperature for the priority subset of 7 CMIP-3 GCMs.

A commonly overlooked aspect of uncertainty in hydrological projections pertains to the response of empirically estimated PET to climate change. Kingston et al. (2009) investigated the response of six different methods (Penman-Monteith, Hamon, Hargreaves, Priestley-Taylor, Blaney-Criddle and Jensen-Haise) of estimating PET to a 2°C rise in global mean temperature simulated by the priority subset of 7 GCMs on water surplus (P-PET) in East Africa. Differences in PET climate change signal of over 100% are found between methods; this uncertainty is several times that (20 to 40%) observed between methods over the baseline period (1961–1990). In East Africa, the choice of PET method can determine the direction of change in water surplus. A quantitative analysis of the impact of the choice of PET method on simulated catchment water balances and groundwater recharge has yet to be undertaken in the River Mitano Basin and the humid tropics more generally. It should be noted that land-cover changes in the River

Mitano Basin were not considered in any of the simulations reported above. Land cover is primarily agrarian (79%); forest cover is <1%. Although the area of urbanisation is likely to increase, quantification of this and other future changes in land cover in the basin remains problematic.

2.3   CONCLUSIONS AND RECOMMENDATIONS

The shift toward more intensive rainfall as a result of anthropogenic warming will result in more variable river discharge and soil moisture. The former will exacerbate intra-annual freshwater shortages and the risk of flooding whereas the latter threatens food security through reduced crop yields (Challinor et al. 2007). Groundwater resources, which are better distributed than surface waters and account for over 90% of the world's accessible freshwater (Shiklomanov and Rodda 2003), are expected to feature prominently in low-cost strategies to adapt to changing freshwater availability and demand throughout the tropics.

The River Mitano Basin in south-western Uganda is one of series of case studies within the global GRAPHIC programme of UNESCO-IHP[3] that seeks to provide an indication of the regional impacts of human activities and climate change on groundwater resources and opportunities for groundwater-based adaptation strategies. To date, catchment-scale studies in the River Mitano Basin of south-western Uganda reveal three key characteristics regarding the impacts of climate change and rapid development on deeply weathered crystalline rock aquifer systems comprising saprolite and saprock in the humid tropics and sub-Saharan Africa (SSA) in particular (GRAPHIC-Africa). First, intensive groundwater abstraction from the saprolite-saprock aquifer system for town water supplies has recently led to a localised reduction in groundwater storage evident from a substantial ($\sim$15 m), sustained decline in groundwater levels. As a result, continued development of groundwater resources for forecasted increases in domestic and agricultural water demand are expected to have a more pronounced impact on groundwater resources than the projected impacts of climate change over the next two decades (e.g. rises of $\leq 2°$ in global mean air temperature). Second, projected changes in the intensity of rainfall greatly influence the timing and magnitude of estimated groundwater recharge. A substantial decline (55%) in groundwater recharge as a result of climate change (dynamically downscaled SRES A2 emissions scenario in HadCM3) is projected using delta factors that employ historical (daily) rainfall distributions whereas a substantial (53%) increase in recharge is projected if the rainfall distribution is transformed to account for projected changes in rainfall intensity. Third, quantified uncertainty in GCM structure associated with the projected impact of a $2°$ in global mean air temperature on groundwater discharges is considerable and ranges from a 17% decline to an 83% rise relative to the 1961–1990 baseline period. This uncertainty overshadows that estimated in the parameterisation ($\pm 5\%$) of the applied hydrological model (SWAT) but is similar in magnitude to that derived from the choice of algorithms to estimate climate change impacts on potential evapotranspiration.

From a water policy perspective, rapid development and climate change in the humid tropics will continue to have considerable impacts on groundwater resources in

---

[3]http://www.unesco.org/new/en/natural-sciences/environment/water/ihp/ihp-programmes/graphic/

saprolite-saprock aquifer systems as domestic and agricultural water demands rise and variability in surface water resources increases as a result of a shift toward more intensive rainfall. Evidence from both the River Mitano Basin and adjacent Upper Nile Basin suggest that a shift towards more intensive rainfall – a robust, generic climate change impact – serves to increase groundwater recharge (and surface runoff) but reduce soil moisture in saprolite-saprock aquifer systems on surfaces of low relief within cratonic areas of SSA. Groundwater-fed irrigation is consequently a logical strategy to adapt to reduced soil moisture (and crop yields) associated with more intensive rainfall. Research in the River Mitano Basin shows furthermore that estimated uncertainty in current projections of climate change impacts on groundwater resources is so large (e.g. 17% decline to 83% rise) to be of limited use to water management. A concerted effort involving collaborations between groundwater and climate scientists is required to investigate ways of reducing this uncertainty. This GRAPHIC case study shows, however, that employed models of groundwater recharge need to explicitly consider projected changes in rainfall intensity. Due, in part, to substantial uncertainty in current hydrological projections, groundwater management will necessarily become both adaptive ('learn by doing') and informed. The latter will require the expansion and, in many cases, establishment of groundwater and meteorological monitoring systems at the national or basin scale to assess how groundwater systems respond to abstraction and climate variability. Such monitoring systems will ideally test and scale up results from localised, basin-scale studies described herein.

ACKNOWLEDGEMENTS

The authors gratefully acknowledge the substantial contribution made by colleagues (Daniel Kingston, Michael Owor, Christine Mukwaya, Pule Johnson, Lucinda Mileham, Martin Todd) to the understanding of the impacts of rapid development and climate change on groundwater resources presented herein. The authors also gratefully acknowledge the substantial support from the Ministry of Water and Environment (Uganda), Natural Environment Research Council (UK), International Atomic Energy Agency (Vienna), and Department for International Development (UK) that has made the research discussed herein possible.

REFERENCES

Allan, R.P. & Soden, B.J. (2008) Atmospheric warming and the amplification of precipitation extremes. *Science*, 321, 1481–1484.
Allen, M.R. & Ingram, W.J. (2002) Constraints on future changes in climate and the hydrologic cycle. *Nature*, 419, 224–232.
Arnold, J.G., Srinivasan, R., Muttiah, R. & Williams, J.R. (1998) Large area hydrologic modelling and assessment, Part 1 model development. *Journal of the American Water Resources Association*, 34, 73–89.
Bradley, B. (2011) A synthesis of geological history and assessment of the hydrogeology of the Katonga Valley in southwest Uganda with evidence from apatite fission track analysis and electrical resistivity tomography. *PhD Thesis*. University College London, London (unpublished).
Briz-Kishore, B.H. (1993) Assessment of yield characteristics of granitic aquifers in south India. *Ground Water*, 31, 921–928.

Challinor, A., et al. (2007) Assessing the vulnerability of food crop systems in Africa to climate change. *Climate Change*, 83, 381–99.

Chilton, P.J., Foster, S.S.D. (1995) Hydrogeological characterisation and water supply potential of basement aquifers in tropical Africa. *Hydrogeology Journal*, 3, 36–49.

Döll, P. & Florke, M. (2005) Global-scale estimation of diffuse groundwater recharge. *Frankfurt Hydrology Paper 03*. Frankfurt am Main, Germany, Institute of Physical Geography, University of Frankfurt.

Eilers, V.H.M., Carter, R.C. & Rushton, K.R. (2007) A single layer soil water balance model or estimating deep drainage (potential recharge): An application to cropped land in semi-arid North-east Nigeria. *Geoderma*, 140, 119–131.

Giertz, S., Diekkrüger, B. & Steup, G. (2006a) Physically modelling of hydrological processes in a tropical headwater catchment in Benin (West Africa) – process representation and multi-criteria validation. *Hydrology and Earth System Sciences*, 10, 829–947.

Giertz, S., Diekkrüger, B., Jaeger, A. & Schopp, M. (2006b) An interdisciplinary scenario analysis to assess water availability and water consumption in the Upper Ouémé catchment in Benin. *Advances in Geosciences*, 9, 3–13.

Giordano, M. (2006) Agricultural groundwater use and rural livelihoods in sub-Saharan Africa: A first-cut assessment. *Hydrogeology Journal*, 14, 310–318.

Houston, J.F.T. & Lewis, R.T. (1988) The Victoria Province drought relief project II. Borehole yield relationships. *Ground Water*, 26, 418–426.

Howard, K.W.F., Hughes, M., Charlesworth, D.L. & Ngobi, G. (1992) Hydrogeologic evaluation of fracture permeability in crystalline basement aquifers of Uganda. *Hydrogeology Journal*, 1, 55–65.

Jones, R.G., Noguer, M., Hassell, D.C., Hudson, D., Wilson, S.S., Jenkins, G.J. & Mitchell, J.F.B. (2004) *Generating High Resolution Climate Change Scenarios Using PRECIS*. Exeter, UK, Met Office Halley Centre.

Kingston, D., & Taylor, R.G. (2010) Projected impacts of climate change on groundwater and stormflow in a humid, tropical catchment in the Ugandan Upper Nile Basin. *Hydrology and Earth System Sciences*, 14, 1297–1308.

Kingston, D., Todd, M., Taylor, R.G., Thompson, J.R. & Arnell, N. (2009) Uncertainty in PET estimation under climate change. *Geophysical Research Letters*, 36, L20403.

Kundzewicz, Z. W., Mata, L. J., Arnell, N. W., Döll, P., Kabat, P., Jiménez, B., Miller, K. A., et al. (2007) Freshwater resources and their management. In: M.L. Parry, O.F. Canziani, J.P. Palutikof, P.J. van der Linden, and C.E. Hanson, (Eds.), *Climate Change 2007: Impacts, Adaptation, and Vulnerability. Contribution of Working Group II to the Fourth Assessment Report of the Intergovernmental Panel on Climate Change* (173–210). Cambridge, Cambridge University Press.

Kundzewicz, Z.W., Mata, L.J., Arnell, N., Döll, P., Jiménez, B., Miller, K., Oki, T., Şen, Z. & Shiklomanov, I. (2008) The implications of projected climate change for freshwater resources and their management. *Hydrological Sciences Journal*, 53(1), 3–10.

MacDonald, A.M., Davies, J., Calow, R.C. & Chilton, P.J. (2005) *Developing Groundwater – a Guide to Rural Water Supply*. Rugby, UK, Practical Action Publishing. 358 pp.

Maréchal, J.C., Dewandel, B. & Subrahmanyam, K. (2004) Use of hydraulic tests at different scales to characterise fracture network properties in weathered fractured layer of a hard rock aquifer. *Water Resources Research*, 40, 1–17.

Mileham, L. (2008) Impact of climate change on the terrestrial hydrology of a humid, equatorial catchment in Uganda. *PhD Thesis*. University College London, London (unpublished).

Mileham, L., Taylor, R.G., Thompson, J., Todd, M. & Tindimugaya, C. (2008) Impact of rainfall distribution on the parameterisation of a soil-moisture balance model of groundwater recharge in equatorial Africa. *Journal of Hydrology*, 359, 46–58.

Mileham, L., Taylor, R.G., Todd, M., Tindimugaya, C. & Thompson, J. (2009) Climate change impacts on the terrestrial hydrology of a humid, equatorial catchment: Sensitivity of projections to rainfall intensity. *Hydrological Sciences Journal*, 54(4), 727–738.

Min, S.-K., Zhang, X., Zwiers, F.W. & Hegerl, G.C. (2011) Human contribution to more-intense precipitation extremes. *Nature*, 470, 378–381.

MWE. (2006) *Water and Sanitation Sector performance report 2006*. Ministry of Water and Environment (unpublished).

MWLE. (2006) *Water Authorities Division quarter four and annual inspection report*. Ministry of Water, Lands and Environment, Directorate of Water Development (unpublished).

Nkotagu, H. (1996) Application of environmental isotopes to groundwater recharge studies in a semi-arid fractured crystalline basement area of Dodoma, Tanzania. *Journal of African Earth Sciences*, 22, 443–457.

Nyenje, P.M. & Batelaan, O. (2009) Estimating effects of climate change on groundwater recharge and base flow in the upper Ssezibwa catchment, Uganda. *Hydrological Sciences Journal*, 54(4), 713–726.

Owoade, A. (1995) The potential for minimizing drawdowns in groundwater wells in tropical aquifers. *Journal of African Earth Sciences*, 20, 289–293.

Owor, M., Taylor, R.G., Tindimugaya, C. & Mwesigwa, D. (2009) Rainfall intensity and groundwater recharge: Evidence from the Upper Nile Basin. *Environmental Research Letters*, 4, 035009.

Pall, P., Allen, M.R. & Stone, D.A. (2007) Testing the Clausius–Clapeyron constraint on changes in extreme precipitation under CO_2 warming. *Climate Dynamics*, 28, 351–363.

Rushton, K.R. & Weller, J. (1985) Response to pumping of a weathered–fractured granite aquifer. *Journal of Hydrology*, 80, 299–309.

Sekhar, M., Mohan Kumar, M.S. & Sridharan, K. (1994) A leaky aquifer model for hard rock aquifers. *Hydrogeology Journal*, 3, 32–39.

Shiklomanov, I.A. & Rodda, J.C. (2003) *World Water Resources at the Beginning of the Twenty-First Century*. Cambridge, Cambridge University Press.

Tate, E., Sutcliffe, J.V., Conway, D. & Farquharson, F. (2004) Water balance of Lake Victoria: Update to 2000 and climate change modelling to 2100. *Hydrological Sciences Journal*, 49, 563–574.

Taylor, R.G. (2009) Rethinking water scarcity: Role of storage. *EOS, Transactions, American Geophysical Union*, 90(28), 237–238.

Taylor, R.G., Howard, K.W.F. (1996) Groundwater recharge in the Victoria Nile basin of East Africa: Support for the soil–moisture balance method using stable isotope and flow modelling studies. *Journal of Hydrology*, 180, 31–53.

Taylor, R.G. & Howard, K.W.F. (1999a) Lithological evidence for the evolution of weathered mantles in Uganda by tectonically controlled cycles of deep weathering and stripping. *Catena*, 35, 65–94.

Taylor, R.G. & Howard, K.W.F. (1999b) The influence of tectonic setting on the hydrological characteristics of deeply weathered terrains: Evidence from Uganda. *Journal of Hydrology*, 218, 44–71.

Taylor, R.G. & Howard, K.W.F. (2000) A tectono-geomorphic model of the hydrogeology of deeply weathered crystalline rock: Evidence from Uganda. *Hydrogeology Journal*, 8, 279–294.

Taylor, R.G., Tindimugaya, C., Barker, J.A., Macdonald, D. & Kulabako, R. (2010) Convergent radial tracing of viral and solute transport in gneiss saprolite. *Ground Water*, 48, 284–294.

Tindimugaya, C. (2008) Groundwater flow and storage in weathered crystalline rock aquifer systems of Uganda: Evidence from environmental tracers and aquifer responses to hydraulic stress. *PhD Thesis*. University College London, London (unpublished).

Todd, M.C., Taylor, R.G., Osborn, T., Kingston, D., Arnell, N.W. & Gosling, S. (2011) Quantifying the impact of climate change on water resources at the basin scale on five continents – a unified approach. *Hydrology and Earth System Sciences*, 15, 1035–1046.

Trenberth, K.E., et al. (2003) The changing character of precipitation. *Bulletin of the American Meteorological Society*, 84, 1205–1217.

Vogel, J.C. & Van Urk, H. (1975) Isotopic composition of groundwater in semi-arid regions of southern Africa. *Journal of Hydrology*, 25, 23–26.

World Bank. (2007) Investment in agricultural water for poverty reduction and economic growth in Sub-Saharan Africa Collaborative programme of AFDB, FAO, IFAD, IWMI, and World Bank. Washington, DC, World Bank. p. 234.

CHAPTER 3

Groundwater recharge and storage variability in southern Mali

Chris M. Henry, Harm Demon, Diana M. Allen & Dirk Kirste

ABSTRACT

Assessment and sustainable management of groundwater resources require estimates of groundwater recharge. This case study in southern Mali, Africa compares approaches for estimating groundwater recharge and understanding recharge processes using a variety of methods encompassing groundwater level–climate data analysis, **G**ravity **R**ecovery **A**nd **C**limate **E**xperiment (GRACE) satellite data analysis, and recharge modelling. Analysis of water level measurements from 15 wells throughout the region (1982–2001) revealed that, on average, water levels are at their lowest in May or June, rise rapidly between July and September, and peak in September or October. The average annual recharge calculated using the water table fluctuation method for individual wells, ranged from 78.6 mm to 236.6 mm. The GRACE terrestrial water storage (TWS) anomalies show lows and peaks in May and September, respectively. Soil moisture-corrected GRACE data (using the Global Land Data Assimilation System – GLDAS) predict a recharge of 149.7 mm, with an offset in timing of peak recharge, likely due to inaccuracy of the GLDAS in this area. Recharge simulation results show good agreement between the timing and magnitude of the mean monthly simulated recharge and the regional mean monthly storage anomaly hydrograph generated from all monitoring wells. Annual recharge is projected to decrease by 8% for areas with luvisols and by 11% for areas with nitosols. Given this potential reduction in groundwater recharge, there may be added stress placed on an already stressed resource.

3.1 INTRODUCTION

3.1.1 Purpose and scope

Groundwater constitutes the main potable water supply source for human consumption and irrigation of agriculture worldwide (Bear et al. 1999). This is especially true of arid or semi-arid regions that are prone to drought and may have limited surface water resources (de Vries and Simmers 2002), and in developing nations where lack of sanitation has led to contamination of surface waters and where rapid population growth has increased the demand for freshwater (McMurray 2007; Rosen and Conly 1998).

Assessment and sustainable management of groundwater resources require estimates of groundwater recharge. Unfortunately, the rate of aquifer replenishment is one of the most difficult aspects of groundwater resource evaluation to derive with confidence, and estimates are often plagued by uncertainties (Simmers 1997). In addition, climate change is predicted to exacerbate demands placed on groundwater by increasing drought frequency and altering precipitation and temperature patterns (Calow et al. 1997; Christensen et al. 2007). The effects of climate change on groundwater recharge are particularly uncertain.

This paper describes several approaches, spanning groundwater level–climate data analysis, GRACE satellite data analysis, and recharge modelling, for estimating groundwater recharge and groundwater storage variability. The integration of these various methods is particularly suited to regions where groundwater information may be incomplete, such as in the case study area in southern Mali. Groundwater is a vital resource in the region, but groundwater data in Mali are limited, and little has been done with existing data in regards to characterizing recharge processes. As well, given the uncertainty inherent to recharge estimation, the different approaches used give greater confidence in the results.

The paper also provides a preliminary assessment of the impacts of future climate change on groundwater recharge. The approach considers the median climates predicted in the future from a sub-set of global climate models (median projection for the ensemble) for one emission scenario. As such, uncertainty is not addressed; however, this preliminary work highlights the sensitivity of recharge to changes in climate in southern Mali.

3.1.2   Study area description: southern Mali

The study area is southern Mali, situated in Sub-Saharan Africa, which lies within the Niger River Basin. The principal study region, with an area of 54,971 km^2 (Fig. 3.1), has a population of 612,321 in 862 villages according to a 1998 census (ARP Développement 2003). Apart from the 23,535 people living in the City of Bougouni, the remaining population is almost entirely rural. The broader study region includes the cities of Bamako and Sikasso and has an estimated population of 3.3 million people, representing a significant proportion of the country's 12.3 million people.

The physiography of southern Mali is that of a plain, with low hills, gentle slopes, and broad valleys. There is little topographic relief in the region, which ranges between 350 and 450 metres above sea level (ARP Développement 2003). Natural vegetation consists of wooded savannas, shrublands, or grasslands of the Sudanian eco-zone (United States Geological Survey 2003; Tappan and McGahuey 2007); however, the region is increasingly cultivated. The climate of southern Mali can be characterized as a tropical wet and dry regime and is generally divided into three seasons: a hot and dry period from February to June, a warm and rainy season from June to October, and a relatively cooler dry season from October to February. Although the rainy season extends over a five month period, precipitation events are sporadic (yet often intense) and concentrated in a period of 70–80 days (Tappan and McGahuey 2007). Rainfall in the region is a result of the West African Monsoon and is prone to long-term cycling between relatively wet and dry periods. From the 1970s through the 1990s, the Sahel experienced a drought relative to 1925–2000 averages (Nicholson and Grist 2001; Le Barbé et al. 2002).

The geology of southern Mali consists predominantly of two rock types: the majority of the principal study region lies within the Birimian Precambrian weathered basement rock (largely comprised of metamorphosed sedimentary or igneous rocks). Lower Cambrian or Palaeozoic sedimentary rocks forming a "Sandstone Plateau" are present along the northern and eastern periphery of the Birimian (United Nations 1988; ARP Développement 2003). The cities of Bamako and Sikasso overlie this formation. Overlying the Birimian and the Sandstone Plateau rock are poorly sorted argillaceous sandstones. Clayey laterites (described in drilling records) are present near surface above the argillaceous sandstone; their age and thickness is variable, though the oldest are thought to be of Eocene age (United Nations 1988). Soils in the study region (at surface)

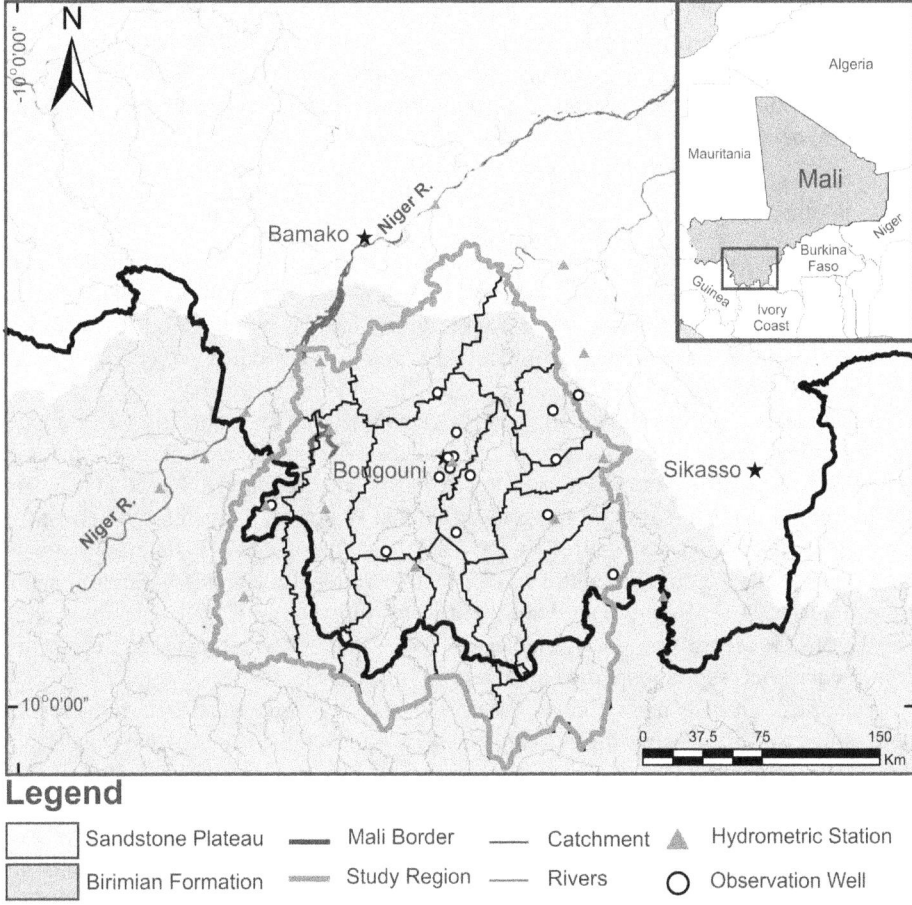

Figure 3.1. The study region in southern Mali (outlined in grey), showing the location of hydrometric stations, observation wells, and the major geological units. Inset map shows Mali in north-western Africa.

are mostly ferric luvisols; although gleyic luvisols, ferric acrisols, eutric nitosols and lithosols are also present (Food and Agricultural Organization 2003).

Both the Birimian and Sandstone Plateau units form fractured aquifer systems that may be unconfined or semi-confined. The aquifer systems generally consist of a shallow aquifer (formed in the laterite) that overlies a deeper fractured rock aquifer. The shallow aquifers typically have a high porosity yet limited permeability due to the clay content of the laterite. Groundwater here is extracted from traditional dug wells as the porosity is often too low to be exploited by drilling. The fractured units below conversely have a good permeability but a low porosity, and groundwater is obtained via hand or foot pumped drilled wells. The shallow aquifer is in hydraulic connection with the underlying unit via vertical or sub-vertical fractures, thus forming an unconfined system (which may be semi-confined if the units are not locally in contact by fractures; ARP Développement 2003). The majority of the groundwater is stored in the shallow system due to its higher porosity, and the drilled wells, screened only in the fractured zone, indirectly exploit these reserves because of the hydraulic connection between the two (ARP Développement 2003). Local groundwater flow is toward the valleys of the surface water courses to which they discharge, with groundwater divides beneath the rivers (ARP Développement 2003).

3.1.3   Methodology

The methodology employed for this study included: groundwater level–climate data analysis, GRACE satellite data analysis, and recharge modelling. In addition, climate change impacts on recharge were assessed in a preliminary fashion by adjusting the historical weather series used in the recharge models according to a median projection from an ensemble of global climate models.

Precipitation data (as daily and monthly accumulation) for Bamako and Bougouni were acquired from two sources: the Royal Netherlands Meteorological Institute (KMNI 2009) and the National Centre for Atmospheric Research (NCAR 2009). The KMNI rainfall data are based on the gridded Global Historical Climatology Network of the National Climatic Data Centre (NCDC-GHCN), while the NCAR data are obtained from the Climate Prediction Centre (CPC) of the National Centres for Environmental Protection (NCEP). Missing data were filled in by substituting KMNI data for months that the NCAR record was not available, and vice versa. Other climate data correction methods are discussed in more detail by Henry et al. (2011).

Historical groundwater storage anomalies and recharge were determined from observation well data using the water table fluctuation method (e.g., Healy and Cook 2002). The database (Government of Mali Sigma2), although incomplete, is an archive containing information for over 20,000 wells drilled throughout the country since the 1970s, including their lithologies, completion details, yield estimates, and water chemistry. The database also includes water level measurements for a small number of observation wells throughout Mali, including 15 within the study area (Fig. 3.1). The measurement frequencies for each monitoring well are variable, although generally more measurements were taken during the rainy season than the dry (Fig. 3.2). Observation time periods between monitoring wells was also inconsistent, and the records span variable lengths between the years 1982 and 2002 (Fig. 3.2). The equipment used to make the measurements is unknown; however, it is speculated that they were made manually rather than with a pressure transducer or limnograph, on account of the irregularity of the measurement regime.

The monthly storage variability time series was generated using a method similar to that of Rodell et al. (2007). First, the average water level over the period of record was computed for each well. Since the measurement frequencies were sporadic in some cases, mean annual hydrographs were generated by determining the average water level for each month of a specific year on record, and then averaging these monthly values for all years to establish a mean monthly level over the period of record for that well. The mean annual water level was determined by averaging the mean monthly levels for all 12 months, thereby providing a 'static' water level to which all measurements could be compared. The static level was then subtracted from each measurement to determine the hydraulic head anomalies, which were averaged over each month of the record to determine the monthly hydraulic head anomalies. Finally, monthly head anomalies were

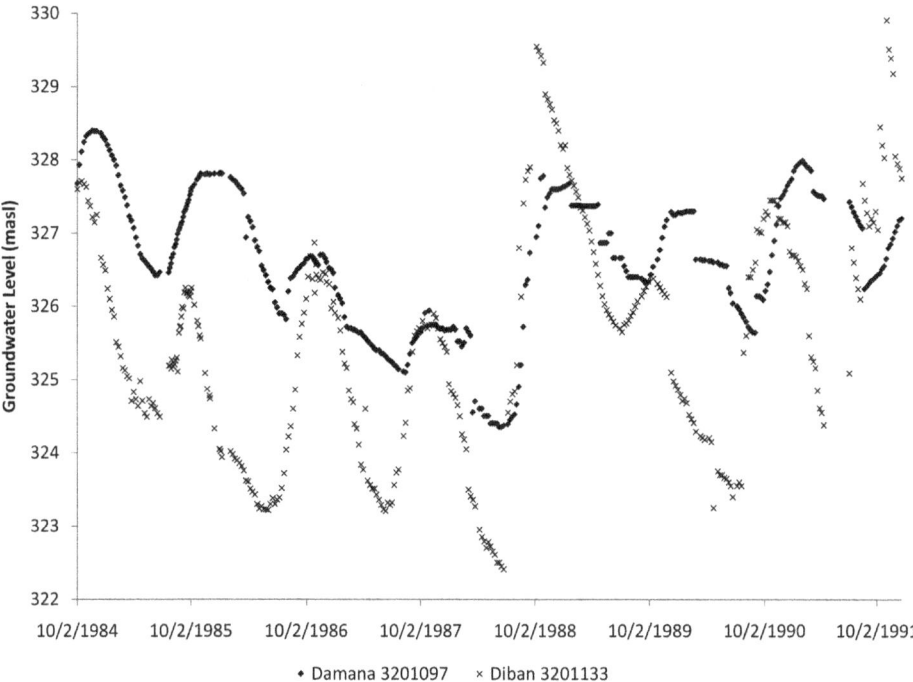

Figure 3.2. Examples of groundwater level hydrographs for two observation wells; 3201097 at Damana and 3201133 at Diban.

multiplied by an estimated aquifer specific yield (Sy) to give a time series of monthly storage anomalies. Estimates of Sy (often determined from the analysis of pumping test data) were not available for the study region. Thus, the Sy value was approximated by taking the mid-point of the range of reported values for the Birimian in the neighbouring Ivory Coast (range 0.03–0.07; ARP Développement 2003). A sensitivity analysis showed that for Sy values of 0.03 and 0.07, average annual recharge based on the historical observed water level data varies between 89.5 and 208.7 mm/a, respectively, a difference of up to ±40%. Finally, annual recharge was determined by subtracting the peak monthly storage anomaly from the low monthly storage anomaly for the year in consideration.

A regional monthly storage anomaly time series was generated by averaging the monthly anomalies for each well that had data for that month. The regional groundwater recharge was calculated by averaging the recharge estimated at each individual well, if an estimate was available for that year. Finally, the average annual recharge for the entire region over the time period 1982–2002 was calculated by averaging the regional estimate for each year, rather than using the average annual recharge of each well using its individual record.

Satellite data from the **G**ravity **R**ecovery **A**nd **C**limate **E**xperiment (GRACE) were used to estimate changes in monthly groundwater storage and annual recharge across a region in southern Mali (Henry et al. 2011). GRACE data were obtained and processed through Geomatics Canada. Monthly GRACE data were obtained for a 445 by 445 km area (9° to 5°W in longitude, 13° to 9°N in latitude) between August 2002 and

November 2008, encompassing the study region. This field is a running spatial average of Terrestrial Water Storage (TWS) expressed in water thickness equivalent (WTE) in mm, and each monthly field is a mean for the area over that month. The field was further discretised into 0.25° by 0.25° grids, so there is variability within the larger 445 by 445 km area. TWS anomalies from GRACE were corrected for soil moisture using the Global Land Data Assimilation System (GLDAS) operated by the Goddard Space Flight Centre (GSFC) to obtain monthly groundwater storage anomalies and annual recharge estimates. Because the GRACE TWS results are at a coarser resolution than the GLDAS soil moisture results, the GLDAS results were up-scaled using spherical harmonic analysis and filtered to provide a resolution equivalent to the GRACE data. Water storage fluctuations resulting from changes in biomass and surface water were not considered. A particular challenge was that the time series data for GRACE and observed groundwater level data do not overlap. To overcome this problem, GRACE time series data (2002–2008) were appended to the observed historical time series data (1982–2002), and the records were compared.

Recharge modelling was carried out using the **H**ydrologic **E**valuation of **L**andfill **P**erformance (HELP) code (Schroeder et al. 1994) using a stochastically-generated weather series representative of Bamako (100 years in length) as input to a series of representative unsaturated zone column models. The code accounts for the effects of surface storage, runoff, evapotranspiration, infiltration, vegetation, soil moisture storage, interflow, unsaturated vertical flow, and percolation. Recharge is simulated by setting the bottom of the column equal to the depth of the water table, where recharge is equal to the amount of percolation calculated at the column's base. The water table depth remains at a fixed depth throughout the simulation. HELP has been used in many recharge studies (e.g., Scanlon et al. 2002; Scibek and Allen 2006; Toews and Allen 2009; Liggett and Allen 2010). As yet, HELP has not been tested in tropical wet and dry regime, such as southern Mali, where rainfall is a result of the West African Monsoon, characterized by sporadic (yet often intense) and concentrated events spanning a period of 70–80 days.

Recharge was simulated through a series of 12 unsaturated zone columns. The soil layer in each model was 1 m thick and comprised of either clay or sandy loam based on the two dominant soil types in the region. The balance of the vadose zone (down to the water table) was comprised of either laterite or laterite overlying argillaceous sedimentary rock. In total, four Groups were considered. For each of these Groups, simulations were carried out for three different column depths; 5, 7.5 or 10 m depth (12 simulation columns in total). These depths are representative of the range of static water level depths measured in observation wells. Henry (2011) describes the simulation results in detail; for brevity, examples for two are given here (Groups 1 and 2; Fig. 3.3).

In Group 1, the top soil layer consists of ferric luvisol (sandy loam) overlying laterite. In Group 2, nitosol (clay) overlies laterite. In addition to the weather data series used to drive the recharge model (mean daily temperature, daily precipitation, daily solar radiation), several other parameters are needed for modelling recharge. These include parameters that control evapotranspiration (average annual wind speed, quarterly humidity, leaf area index, growing season, evaporative zone depth); the hydraulic properties of the media (saturated hydraulic conductivity, total porosity, wilting point, and field capacity); surface characteristics for runoff calculation; and initial moisture content. To ensure that initial moisture content had equilibrated (Liggett and Allen 2010), only the final 30 years of the 100 year simulation time series were used for analysis. The generated

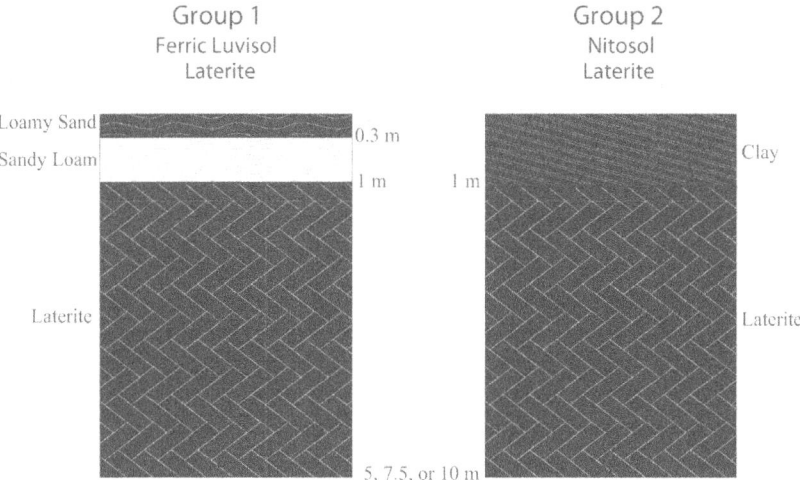

Figure 3.3.   Conceptual models of two unsaturated zone columns at three depth intervals thought to be most representative of the recharge environments of southern Mali based on the static water levels and stratigraphy of the observation wells in the region.

weather time series was verified against observed weather for Bamako by assessing daily temperature each month and monthly precipitation (Henry 2011).

The simulated recharge results were averaged over all columns and compared to the timing and magnitude of recharge calculated from the storage anomalies of the observed regional water level records and GRACE.

Climate change recharge simulations were also conducted for the 7.5 m depth of each Group 1 and 2 columns. Climate projections were taken from the United Nations Development Programme (UNDP) Climate Change Country Profiles (McSweeney et al. 2006a, 2006b). The global climate models (GCMs) used to make the projections are a sub-set of 15 of the 22-member ensemble used in the Intergovernmental Panel on Climate Change (IPCC) Fourth Assessment Report (Randall et al. 2007). A more detailed description of the generation of the climate projections is available in McSweeney et al. (2006b). For the recharge modelling, results for the SRES emissions scenario A2 were considered; this scenario predicts high population growth with slow economic development and technological advances, and is one of the worst-case scenarios in regards to high GHG emissions and consequent temperature increases. McSweeney et al. (2006a) depicted results in 2.5° × 2.5° grids cells for 10-year-average 'time-slices' for the 2030s, 2060s and 2090s. For each grid, seasonal minimum, maximum and median temperature and precipitation anomaly values were reported: Jan. Feb. Mar. (JFM), Apr. May Jun. (AMJ), Jul. Aug. Sep. (JAS), and Oct. Nov. Dec (OND). The climate change data within the grid at 10° to 12.5° latitude and −7.5° to −5° longitude (southern Mali) were used as the representative climate cell for the study region. Median climate from this 15-member ensemble was used for recharge modelling purposes. The seasonal data for JFM, AMJ, JAS and OND for magnitude of temperature change and percent change in precipitation were applied to the respective months. Adjusted monthly values for temperature and precipitation were input to the weather generator to shift the weather series (i.e., the delta approach). Shifts in variability were not considered; therefore, changes in precipitation intensity for example may not be accurately represented. This climate change assessment was very preliminary. Ideally, downscaled climate data from a range of GCMs and for different emission scenarios should be used for climate change impacts assessment in order to capture the range of uncertainty in projections (e.g. Toews and Allen 2009).

3.1.4   Relevance for GRAPHIC

This case study contributes to the GRAPHIC programme through analysis of several datasets that can be used for estimating recharge and understanding recharge processes in a tropical wet and dry regime region of Africa. These data include climate, groundwater level and GRACE satellite data. The integration of methods, as recommended by the GRAPHIC programme, is particularly suited to regions where groundwater information may be incomplete, such as in the case study area in southern Mali. The case study also demonstrates the use of modelling to estimate recharge, both for historical climate and future climate. The recharge modelling software, while used in numerous studies, had not yet been tested in a climate region characterized by sporadic and concentrated rainfall events. The recharge results for historical climate are shown to be consistent with groundwater level and GRACE results, lending credibility to the conceptual model for recharge and the magnitude and timing of recharge in this region.

The case study also highlights the value of research partnerships. Access to regional groundwater information for this research was provided through partnership with Global Aquifer Development Foundation (GADF) and the Mali Government. The database (Sigma2) included water well information, groundwater level records for observation wells, and a geochemical database (not discussed herein) and used by Henry (2011) along with water samples collected in the study region for geochemical and isotopic analysis of groundwater. In partnership with Geomatics Canada, GRACE (Gravity Recovery and Climate Experiment) satellite data were extracted and processed to generate monthly groundwater storage anomalies. In addition, at the outset of this project, GADF and Simon Fraser University coordinated with the Direction nationale de l'Hydraulique (DNH) in Mali, and offered training of groundwater technicians in data acquisition to augment groundwater management practices in Mali. This training included installation of limnographs for water level measurement in wells, data downloading and analysis techniques, and hydraulic testing (slug and bail) of wells for estimation of hydraulic properties of aquifers. It is hoped that this knowledge translation and local skill development will enable groundwater managers in Mali acquire much needed data for monitoring groundwater conditions and enabling sustainable management of the resource.

3.2   RESULTS AND DISCUSSION

3.2.1   Groundwater levels and storage anomalies

Analysis of water level measurements from 15 wells throughout the region (1982–2001) revealed that, on average, water levels are at their lowest in May or June, rise rapidly between July and September, and peak in September or October. Monthly storage anomaly time series generated for each well from the hydrographs (using the water table fluctuation method) suggest that average annual recharge ranged from 78.6 mm to 236.6 mm amongst the wells. Local differences in depth to the water table and unsaturated zone hydraulic conductivity are likely responsible for the differences in recharge timing between wells. Differences in magnitude could be related to local variations in recharge and/or to differences in the storage properties of the soils and aquifer. The basin average annual recharge was 149.1 mm based on a composite of all observation wells; recharge varied interannually between 98.4 mm in 1983 and 207.5 mm in 1994.

Mean monthly storage anomalies of GRACE TWS (2002–2008), modelled soil water storage (SM) using the GLDAS, interpreted groundwater storage (GRACE-SM), and observed groundwater storage (as the regional average) are shown in Figure 3.4, along with the mean monthly rainfall accumulation in Bamako and Bougouni, since 1982. The GRACE TWS shows lows and peaks in May and September, respectively. GRACE-SM reveals later peaks (November) than those of the regional observed average and GRACE TWS, but the recharge (149.7 mm) compares well. The discrepancy in timing is attributed to the GLDAS poorly predicting the timing of SM storage fluctuations (Henry et al. 2011). Notwithstanding, the soil moisture-corrected GRACE data (uncorrected) appear to accurately predict the timing and magnitude of groundwater storage changes, suggesting that GRACE data may invaluable for identifying recent regional changes in groundwater storage in areas with sparse hydrogeological data.

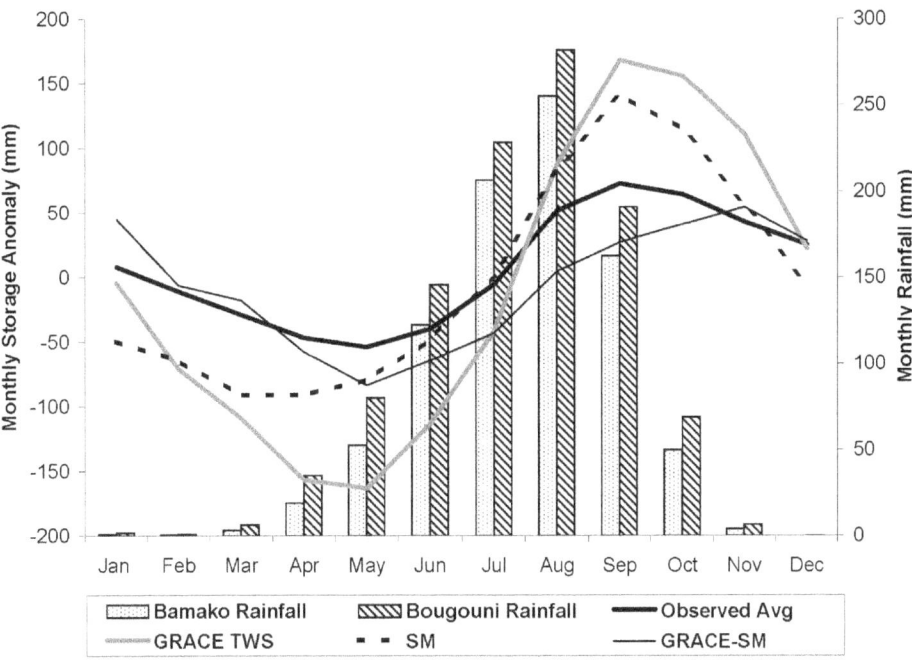

Figure 3.4.   Mean monthly storage anomaly (left y-axis) for regionally averaged groundwater storage (1982–2002), GRACE measured TWS (2002–2008), GLDAS predicted soil moisture (2002–2008), and interpreted groundwater storage (GRACE-SM) (2002–2008), with mean monthly rainfall accumulations (right y-axis) at Bamako and Bougouni, since 1982 (from Henry et al. 2011).

3.2.2   Recharge modelling

The above conceptual model was tested through recharge modelling. Figure 3.5 shows mean monthly simulated recharge (for all 12 simulations; Henry, 2011) along with the mean monthly storage anomaly curves for the entire region compiled from all observation wells and interpreted groundwater storage (GRACE-SM). The apparent flatline response of the simulated storage anomaly during the late recession (from March to June) is caused by the unsaturated soil column having no further stored water to continue to provide recharge to the deeper groundwater system. This is likely a realistic response given the relatively short period of intense rainfall in the region and the limited capacity of the shallow soils and rocks to hold water. One possible explanation for this flatline response is that the actual unsaturated zone is poorly represented in the simulations, such that in reality water is slower to drain the unsaturated zone, resulting in a prolonged recession. Another possibility, and perhaps the more likely one, is that the recharge columns for recharge simulation were not deep enough. The fractured rock aquifer is situated at roughly 40 m depth and well below the water table depth used for the simulations. Thus, there is potentially 30 m of rock available for storage that was not simulated. It has been customary in previous recharge simulation studies using HELP to set the base of the column to the water table depth (e.g., Liggett and Allen 2010); however, in this particular case, where there is a deeper aquifer the additional porosity offered by the overlying rock could provide the necessary storage to create the longer recession period observed in the storage anomaly hydrographs. Overall, however, there is generally good agreement between the timing and magnitude of the mean monthly simulated recharge and the regional mean monthly storage anomaly hydrograph generated from all monitoring wells.

As discussed above, the climate in southern Mali consists of eight dry or relatively dry months and a four month long rainy season (June–September). During the rainy season, events occur irregularly, yet they may be very heavy when they do. Due to the aridity for most of the year, as well as the relative impermeability of the clay-based

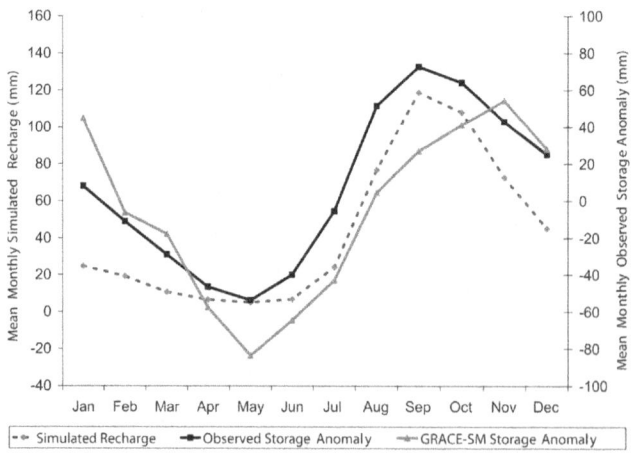

Figure 3.5.   Mean monthly simulated recharge for the average of all 12 simulations (left axis), with the mean monthly storage anomaly curves for the entire region compiled from all observation wells and interpreted groundwater storage GRACE-SM (right axis) (from Henry 2011).

laterite (though the permeability may be higher where the laterite is highly weathered), it might be expected that much of the precipitation occurring early in the rainy season will be evapotranspired by the water starved vegetation, will runoff to surface water courses, or be pooled in depressions of the relatively flat topography. As the rainy season continues, water will infiltrate the unsaturated zone directly from precipitation and locally from the pooled depressions. The clay-based laterite has a low hydraulic conductivity, is highly porous, and has a high storage capacity. These characteristics, coupled with the thickness of the unsaturated zone (up to 10 m deep), cause a time lag in the response of the water table from the onset of the rainy season. However, once the infiltrating water reaches the water table, a rapid rise is expected due to the nature of the fractured aquifer at depth, which is less porous, has a lower capacity to store water, but which is more permeable. The timing of the response may be more or less rapid depending on whether the shallow aquifer is in good hydraulic contact with the underlying fractured environment via vertical or sub-vertical fractures through the sedimentary formations dividing the two.

The recharge simulation results do not, however, compare well with the GRACE-SM storage anomalies. As noted above, the GLDAS used to estimate changes in soil moisture storage poorly resolved the timing of the soil moisture storage changes. Thus, the poor correlation between GRACE-SM with the simulated recharge results is not necessarily an indication of poor simulation performance, but rather is more likely due to the GLDAS estimation of SM variability.

The average annual recharge obtained from the simulations (132.2 mm/a, or 12.6% of simulated rainfall) is slightly less than the recharge calculated from the observation wells and GRACE-SM, which were 149.1 and 149.7 mm/a, respectively, or 16.4 and 14.8% of annual rainfall. These results are fairly close – differing by only 17 mm/a or only 2–4% of rainfall. One possible explanation for the modest discrepancy in results is the assigned storage properties for both the fractured rock aquifer for storage anomaly calculation and the unsaturated zone layers in HELP. The fact that there is only a relatively small discrepancy is an indication that the storage parameters used in the simulations and to calculate the storage anomalies are reasonable.

Under the SRES A2 scenario, the median change in annual mean daily temperature in southern Mali ranges from 1.4°C in the 2030s, to 2.8°C in the 2060s, to 4.6°C in the 2090s. Seasonal changes in temperature are relatively consistent with annual changes with up to ±0.7° variation between seasons about the annual change (McSweeney et al. 2006a). Under this scenario, precipitation amounts are also expected to change seasonally, both negatively (up to −10%) and positively (up to 8%) over the range of time periods. Figure 3.6 shows the magnitude changes in precipitation per season for the three time periods. Noticeably, by the 2090s, JAS (currently the rainiest season) will have less precipitation (by as much as 35 mm). Given that this is the main recharge period, this can be expected to significantly affect recharge.

Climate change recharge simulation results are shown in Figure 3.7. Both changes in evapotranspiration (as magnitude change) and percolation (recharge; as percent change) are shown for each column (Group 1 and Group 2). Evapotranspiration (ET) results are similar for each group, with AMJ experiencing less ET by the 2090s, and JAS and OND experiencing more ET by the 2090s. Both groups show little difference in ET in JFM. The lower ET in AMJ likely relates to the fact that precipitation is lower in that season despite a higher seasonal temperature. Actual evapotranspiration is simulated; therefore, water availability is a limiting factor. Given that this season marks the beginning of the

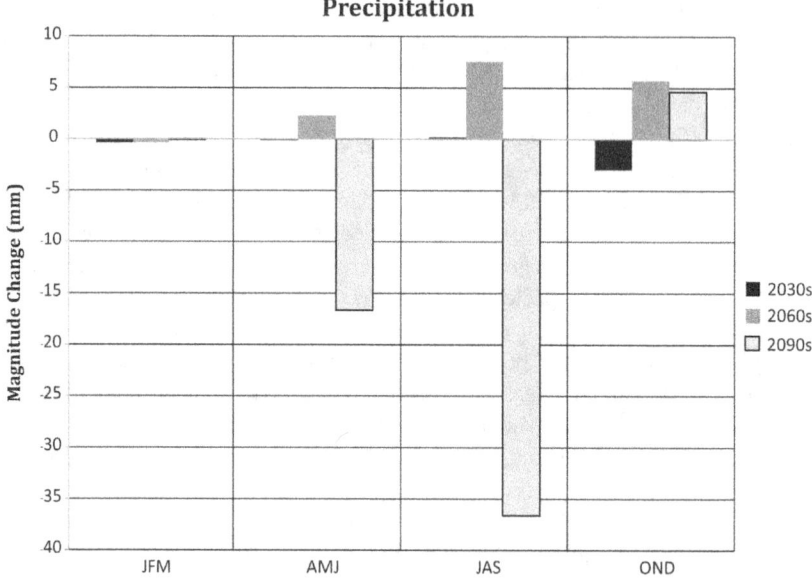

Figure 3.6.   Magnitude change in precipitation per season for the three time periods relative to historical values from the weather generator.

rainy season and the soil is very dry, antecedent moisture is likely very low. Higher ET in JAS is an interesting result given that precipitation is significantly lower in that season, but may relate to the moisture holding capacity of the soils and higher ET rates due to warmer seasonal temperatures. Finally, OND shows increased ET, likely due to higher temperatures and slightly greater precipitation in that season. Little change in ET is predicted for JFM.

Changes in percolation (or recharge) vary seasonally; and results are more strongly dependent on soil type. When examining these results, it is important to bear in mind the time delay for recharge. For Group 1, recharge in the 2090s in AMJ and JAS is expected to be lower, by as much as 15% compared to today. Lower recharge is also expected in OND (by ~4%), while higher recharge (by ~4%) is expected in JFM. Group 1 soil is luvosol, which has a high permeability and low moisture retention capacity. Thus, changes in precipitation and are almost directly translated into changes in percolation. The higher ET in JAS likely limits soil water storage, leading to less recharge in OND despite higher precipitation. The greater recharge in JFM is likely a delayed response to higher precipitation in OND.

Group 2 soil is nitosol, which has a lower permeability and greater moisture retention capacity compared to luvisol. In AMJ, recharge is higher (by ~2.5%) in the 2090s given higher precipitation and lower ET. In JAS, recharge is lower (by ~16%) given a combination of lower precipitation in that season and the previous season along with higher ET. In OND, recharge is lower (by ~10%) likely due to low moisture soil content and higher ET. Finally, in JFM, there is slightly less recharge. Overall, the average climate change recharge simulations for both groups indicate a slight decrease in annual ET by the 2090s. This decrease is due to moisture availability despite higher temperatures because annual precipitation is projected to decrease by ~4.8% in southern Mali by the

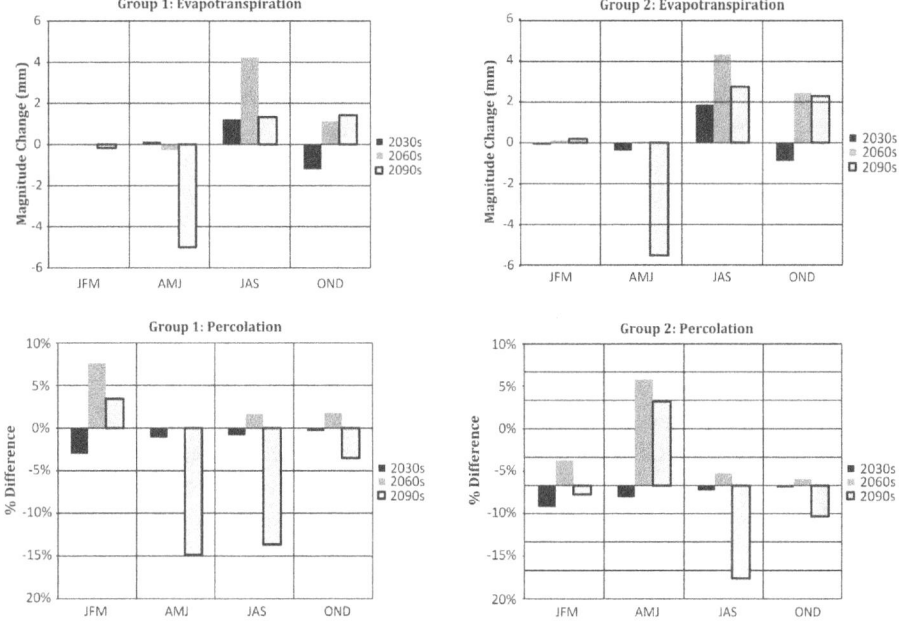

Figure 3.7. Results for Group 1 and Group 2 columns for the three future time periods. Magnitude change in evapotranspiration per season (top figures), and percent changes in percolation (recharge) per season (bottom figures).

2090s. Annual recharge is also projected to decrease by 8% for areas with luvisols and by 11% for areas with nitosols. Given this potential reduction in groundwater recharge, there may be added stress placed on an already stressed resource.

3.3   POLICY RECOMMENDATIONS

There is a significant need for aquifer characterization and resource assessment in Africa. However, there is a lack of basic observational data throughout much of Africa (Taylor et al. 2009) and many nations suffer from inadequate monitoring (including water levels, extraction rates, number of wells drilled, etc.). In Mali, there is no government source of historical meteorological or hydrological records akin to those offered freely online by many national agencies (e.g., Environment Canada). For this study, data were sought from outside sources, such as the Royal Netherlands Meteorological Institute (KMNI), National Centre for Atmospheric Research (NCAR), and the Global Runoff Data Centre; however, data are often incomplete and of questionable reliability (i.e., significant discrepancies in reported rainfall amounts for certain months between the KMNI and NCAR observed climate datasets). The national inventory for wells drilled in Mali – the Sigma2 database, which includes information concerning lithology, well completion, estimated yield, water chemistry, and water level data for a small number of monitoring wells – is disorganized, de-centralized, and incomplete. Currently, each region of the country is responsible for its own database; data are not shared amongst regions. A single centralized online well database is needed, with each region responsible for collecting and archiving data in that region.

Most hydrogeological research in Africa also tends to be highly localized and discontinuous (de Vries and Simmers 2002) due to the fact that internationally sponsored studies (which provide much of the data) are often funded for short time periods. In Mali, there is a lack of regional data synthesis that would allow for better resource assessment and prediction of future impacts climate change on groundwater. Also, from a water management perspective it is much more desirable for research to be of decadal timescales rather than of short period. Given the challenges of establishing and maintaining long term observing networks (climate, streamflow, groundwater level), it is important to synthesize information and draw on global data resources. A principal benefit of the GRACE satellite mission is the potential use of this technology as a groundwater management and monitoring tool in remote regions. However, most GRACE research published to date has focused on a validation of the technology in data rich regions. Few studies have assessed GRACE data without the benefit of concurrent ground observations as a constraint. This case study demonstrated that GRACE data show storage anomalies that appear to be consistent in both timing and magnitude with those derived from historical groundwater level variations even though the time series do not overlap, suggesting that in the absence of monitoring data, information on groundwater storage anomalies may be estimated from GRACE data. However, the research demonstrated that additional work is needed to correct for soil moisture storage at large scale.

3.4   FUTURE WORK

Swiss consultants working in the Bougouni district in the 1980s prepared a summary report of the well logs, hydraulic parameters estimated from pumping tests, and 15 water level records spanning the years 1981–2002; the region was also modelled using MODFLOW (ARP Développement 2003). While the report is useful as a compendium of the wells in the area, the hydrograph analysis was rudimentary, and the MODFLOW model assumed a uniform recharge. The current study has aimed to provide better constraints on recharge estimates and enhanced understanding of recharge processes in general. Future work in southern Mali could include modification of groundwater flow models to make use to these results.

The climate change assessment conducted as part of this study was a preliminary assessment. Uncertainty was not addressed by using a range of global climate models and different emission scenarios. Future work on climate change impacts on groundwater in southern Mali should examine changes in the groundwater system by applying different recharge scenarios to groundwater models. These models should also attempt to include changes in groundwater use and potentially land use change, which may far outweigh climate change impacts.

Finally, as part of this project, training was provided to groundwater technicians in Mali on the installation and operation of groundwater level recording instrumentation. It is our understanding the water level loggers have been installed in several monitoring wells in other regions of Mali. The data collected will be invaluable for monitoring groundwater level variations in areas not currently monitored.

ACKNOWLEDGEMENTS

The authors acknowledge Jianliang Huang from Geomatics Canada, Natural Resources Canada who provided the processed GRACE data used in this study. We wish to thank our collaborators in Mali from le Direction Nationale de l'hydraulique, who assisted with field work and provided the groundwater database. Funding for the project was from the Natural Sciences and Engineering Research Council (NSERC) of Canada and Global Aquifer Development Foundation.

REFERENCES

ARP Développement. (2003) Synthese des connaissances sur les ressources en eau souterraine dans la zone de l'ex-programme d'hydraulique villageoise Mali-Suisse [Summary of knowledge on groundwater resources in villages of southern Mali resulting from a Mali-Swiss partnership project]. Technical report for the Direction Nationale De L'Hydraulique. Ministere des Mines, de L'Energie et de L'Eau, Bamako, Mali, 85 pp.

Bear, J., Cheng, A.H.-D., Sorek, S., Ouazar, D. & Herrera, I. (eds.) (1999) *Seawater Intrusion in Coastal Aquifers: Concepts, Methods and Practices.* Kluwer Academic Publisher, Dordrecht/Boston/London, 625 pp.

Calow, R.C., Robins, N.S., Macdonald, A.M., Macdonald, D.M.J., Gibbs, B.R., Orpen, W.R., Mtwmbezeka, P., Andrews, A.J. & Appiah, S.O. (1997) Groundwater management in drought-prone areas of Africa. *International Journal of Water Resources Development*, 13, 241–261.

Christensen, J.H., Hewitson, B., Busuioc, A., Chen, A., Gao, X., Held, I., Jones, R., Kolli, R.K., Kwon, W.T., Laprise, R., Magaña Rueda, F., Mearns, L., Menéndez, C.G., Räisänen, J., Rinke, A., Sarr, A. & Whetton, P. (2007) Regional climate projections. In: Solomon, S., Qin, D., Manning, M., et al. (eds.) *Climate Change 2007: The Physical Science Basis.* Contribution of Working Group I to the Fourth Assessment Report of the Intergovernmental Panel on Climate Change. Cambridge, UK, Cambridge University Press.

de Vries, J.J. & Simmers, I. (2002) Groundwater recharge: An overview of processes and challenges. *Hydrogeology Journal*, 10, 5–17.

Food and Agricultural Organization of the United Nations. (2003) *The Digital Soil Map of the World Version 3.6.* [Online] Available from: http://www.fao.org/geonetwork/srv/en/metadata.show?id=14116 [Cited 1st June 2010].

Healy, R.W. & Cook, P.G. (2002) Using groundwater levels to estimate recharge. *Hydrogeology Journal*, 10, 91–109.

Henry, C. (2011) An integrated approach to estimating groundwater storage variability and recharge in southern Mali, Africa. *M.Sc Thesis*. Department of Earth Sciences, Simon Fraser University, Canada.

Henry, C.M., Allen, D.M. & Huang, J. (2011) Groundwater storage variability and annual recharge using well-hydrograph and GRACE satellite data. *Hydrogeology Journal*, 19(4), 741–755.

KMNI. (2009) *The Royal Netherlands Meteorological Institute Climate Explorer.* [Online] Available from: http://climexp.knmi.nl/register.shtml [Cited 1st June 2009].

Le Barbé, L., Lebel, T. & Tapsoba, D. (2002) Rainfall variability in west Africa during the years 1950–90. *Journal of Climate*, 15, 187–202

Liggett, J.E. & Allen, D.M. (2010) Comparing approaches for modeling spatially-distributed direct recharge in a semi-arid region. *Hydrogeology Journal*. 18(2), 339–357. Available from: doi:10.1007/s10040-009-0512-5.

McMurray, C. (2007) *Africa's Water Crises and the U.S. Response – Testimony before the Foreign Affairs Subcommittee on Africa and Global Health House of Representatives*. United States Department of State. [Online] Available from: http://www.globalsecurity.org/military/library/congress/2007_hr/070516-north.htm [Cited 20th April 2011].

McSweeney, C., New, M. & Lizcano, G. (2006a) *UNDP Climate Change Country Profiles: Mali*. [Online] Available from: http://country-profiles.geog.ox.ac.uk/ [Cited 15th January 2011].

McSweeney, C., New, M. & Lizcano, G. (2006b) *UNDP Climate Change Country Profiles: Documentation*. [Online] Available from: http://country-profiles.geog.ox.ac.uk/ [Cited 15th January 2011].

NCAR. (2009) *Research Data Archive of the Computational and Information Systems Laboratory at the National Center for Atmospheric Research*. [Online] Available from: http://dss.ucar.edu/datasets/ds512.0/ [Cited 1st June 2009].

Nicholson, S.E. & Grist, J.P. (2001) A conceptual model for understanding rainfall variability in the west African Sahel on interannual and interdecadal timescales. *International Journal of Climatology*, 21 1733–1757.

Randall, D.A., Wood, R.A., Bony, S., Colman, R., Fichefet, R., Fyfe, J., Kattsov, V., Pitman, A., Shukla, J., Srinivasan, J., Stouffer, R.J., Sumi, A. & Taylor, K.E. (2007) Climate models and their evaluation. In: Solomon, S., Qin, D., Manning, M., et al. (eds.) *Climate Change 2007: The Physical Science Basis*. Contribution of Working Group I to the Fourth Assessment Report of the Intergovernmental Panel on Climate Change. Cambridge, UK, Cambridge University Press.

Rodell, M., Chen, J., Kato, H., Famiglietti, J.S., Nigro, J. & Wilson, C.R. (2007) Estimating groundwater storage changes in the Mississippi River basin (USA) using GRACE. *Hydrogeology Journal*, 15, 159–166.

Rosen, J.E. & Conly, S.R. (1998) *Africa's Population Challenge: Accelerating Progress in Reproductive Health*. Washington, DC, Population Action International, 82 pp.

Scanlon, B.R., Christman, M., Reedy, R.C., Porro, I., Simunek, J. & Flerchinger, G.N. (2002) Intercode comparisons for simulating water balance of surficial sediments in semiarid regions. *Water Resources Research*. 38, 1323. Available from: doi:10.1029/2001WR001233.

Schroeder, P.R., Dozier, T.S., Zappi, P.A., McEnroe, B.M., Sjostrom, J.W. & Peyton, R.L. (1994) *The Hydrologic Evaluation of Landfill Performance (HELP) Model: Engineering Documentation for Version 3*. EPA/600/R-94/168b, Washington, DC, USEPA.

Scibek, J. & Allen, D.M. (2006) Modeled impacts of predicted climate change on recharge and groundwater levels. *Water Resources Research*. 42, W11405. Available from: doi:10.1029/2005WR004742.

Simmers, I. (1997) Groundwater recharge principles, problems, and developments. In: Simmers, I. (ed.) *Recharge of Phreatic Aquifers in (Semi-)Arid Areas*. Rotterdam, The Netherlands, AA Balkema. pp. 1–18.

Tappan, G. & McGahuey, M. (2007) Tracking environmental dynamics and agricultural intensification in southern Mali. *Agricultural Systems*, 94, 38–51.

Taylor, R.G., Koussis, A.D. & Tindimugaya, C. (2009) Groundwater and climate in Africa: A review. *Hydrological Sciences Journal*, 54, 655–664.

Toews, M.W. & Allen, D.M. (2009) Evaluating different GCMs for predicting spatial recharge in an irrigated arid region. *Journal of Hydrology*, 374, 265–281.

UN (United Nations). (1988) Mali. In: *Ground Water in North and West Africa*. Natural Resources/Water Series No. 18. New York, UN. pp. 247–264.

United States Geological Survey. (2003) *Global GIS Database: Digital Atlas of the World*. Digital Data Series DDS-62-H. US Geological Survey, Reston, VA.

CHAPTER 4

Groundwater discharge as affected by land use change in small catchments: a hydrologic and economic case study in Central Brazil

Henrique M.L. Chaves, Ana Paula S. Camelo & Rejane M. Mendes

ABSTRACT

The Pipiripau river basin, a $235\,km^2$ catchment in central Brazil, has experienced a substantial increase in land use intensity (mostly agriculture and pastureland) in the last 40 years. This has contributed to a significant decrease in the base flow discharge, responsible for the maintenance of the stream flow during the dry winter season. To assess the hydrological and economic benefits of three land conservation programs in the basin, an empirical relationship was obtained between the base flow index and the normalized basin curve-number, calibrated with observed stream flow and precipitation data. The results indicate that if reforestation and best management practices are implemented in the basin, up to $755 \times 10^6\,m^3/a$ of additional base flow discharge would result during the dry season, with additional revenues up to US$ 1.03 million per year for the water utility company. The simplicity of the presented methodology allows for its application in other data-scarce tropical basins.

4.1 INTRODUCTION

Base flow discharge is an important element of the hydrological cycle that describes the loss of water from the groundwater compartment to surface waters (GRAPHIC/UNESCO 2006). Base flow discharge, often equalled to groundwater discharge in river basins during dry seasons, helps to maintain aquatic ecosystems and provides for important hydrological services to human communities (Tomish et al. 2004).

Groundwater discharge is influenced by climate, watershed, and land use/management conditions. Forests are an important component in the stabilization of groundwater discharge and stream flow (Pereira 1989). When forest conversion is followed by land uses that interfere and alter natural bio-physical processes, hydrological consequences may follow (Pattanayak 2004). Although annual and storm flows typically increase with forest conversion to agriculture, base flows often decline owing to reduced infiltration and more episodic export of water (Poff et al. 1997; Siriwardena et al. 2005).

Several studies have highlighted the impacts of land use changes to groundwater discharge in river basins. Sloto (2008) reported that a decrease of up to 63% of base flow discharge occurred following full urban build-out of a small catchment in Pennsylvania. Klöcking and Haberlandt (2002), analyzing the effect of forestation and deforestation scenarios in base flow discharge at the outlet of five German catchments, reported that base flow discharge varied from +3% to −15%, with respect to the baseline conditions.

In general, conservation practices are known to improve base flow discharge in river basins. Schilling and Libra (2003) reported that the annual base flow index increased between 20 and 30% in two catchments in Iowa in a 60-year period, following soil and water conservation projects implemented during the 1940's.

Basin scale can also affect the response of groundwater discharge to changes in land use. In small basins, the influence of land use is more significant, whereas geologic and climatic factors become more dominant in large catchments (Pattanayak 2004).

Although data intensive, process-based mathematical models are capable of reproducing the surface and groundwater processes holistically and accurately in river basins, the lack of the necessary groundwater information often hinders their application in the tropics (Chaves 1996).

In recent years, economic valuation of environmental services and their extended benefits are becoming more and more effective for addressing water quality and quantity issues in river basins (Classen et al. 2001, Chaves et al. 2004). Pattanayak (2004) has shown that reforestation programs in small Indonesian catchments would increase the percentage of base flow discharge up to 24%, generating an equivalent amount of savings in water collection costs in small communities. This economic incentive can be a powerful motivator to resolve water management issues.

4.1.1 Purpose and scope

The purpose and scope of the present work was the analysis of the hydrological and economic impacts of land use change in a small basin in Brazil, particularly regarding base flow discharge responses. The study follows the UNESCO/GRAPHIC philosophy, namely the assessment of groundwater resources under the pressures of humanity and climate changes. In addition to the hydrologic response, economic valuation is important because its results are more easily comprehended by water users and decision-makers.

Additionally, economic assessments of hydrologic services allow for the establishment of incentive payment programs, which help the mitigation of existing hydrologic problems, as well as the adaptation to future impacts.

4.1.2 Description of the area: the Pipiripau river basin

The area studied under this research was the Pipiripau river basin, situated in the northeastern corner of the Federal District, in central Brazil (Figure 4.1).

The Pipiripau river basin has an area of $235\,km^2$, with central coordinates 15°27'14"S and 47°27'47"W. The river basin has a mean altitude of 950 m, gentle topography (average slope of 5.5%), and deep, well drained soils (red oxisols and ultisols), underlain by quartzites, phyllites, and rhythmites, the latter presenting medium to low permeabilities and transmissivities (Chaves and Piau 2008).

Presently, the main land uses in the catchment are agriculture, pastureland, and natural savannah. Figure 4.2 shows the main land uses in the basin at the present.

Mean annual precipitation in the basin is 1,300 mm, with wet summers and dry winters. Mean annual temperature is 22°C. Long-term mean annual stream flow, measured at the basin outlet, is $2.9\,m^3/s$, and base flow discharge represents 85% of the yearly mean stream flow (Chaves and Piau 2008).

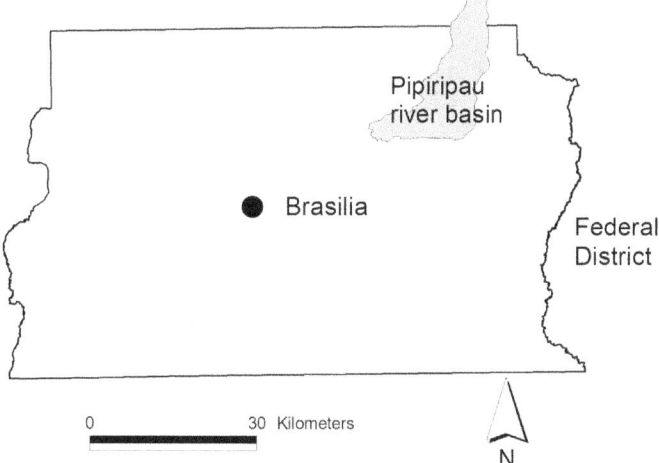

Figure 4.1.   Location of the Pipiripau river basin, with respect to the Federal District area (Brazil).

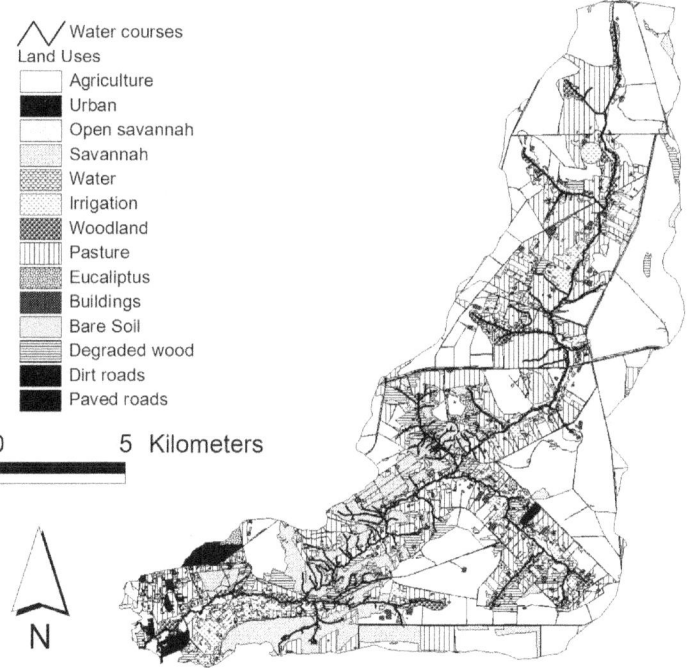

Figure 4.2.   Present land uses in the Pipiripau river basin. Source: Chaves and Piau (2008).

4.1.3   Relevance for GRAPHIC

In the last 20 years, the main water uses in the Pipiripau river basin (urban water supply and irrigation) have been competing for the seasonally available river water in the basin, particularly during the long dry winters. However, both users have to observe

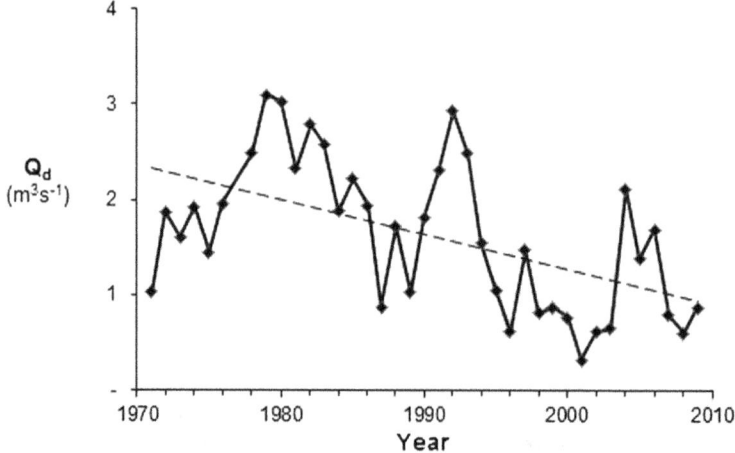

Figure 4.3.   Time series of the mean yearly dry season stream flow in the Pipiripau river.

their defined water allocation limits, in addition to the regulated minimum in-stream (environmental) flows, established by federal and state water agencies.

Since the Pipiripau river is not regulated by dams, the reduced water supply during the winter months often leads to water shortages, hindering the development of new businesses and other water-dependent activities.

Additionally, the decline in dry season stream flow indicates a reduction in groundwater discharge in the last 40 years, leading to further water-use conflicts in the basin. Figure 4.3 presents the mean dry-season stream flow values (mean stream flow in the driest 6 months) in the basin.

The dashed line in the time series of Figure 3 shows a significant (95% probability) decreasing linear trend in the dry season discharge in the Pipiripau river. Possible explanations for that behaviour are i) the decreasing trend of annual precipitation in the period, ii) reduced groundwater recharge in the basin as a consequence of increasing land development, iii) increasing water abstractions from the river, or iv) a combination of the three. Since groundwater pumping in the basin is insignificant, it was not considered a possible cause to the reduction of dry season stream flow.

The first process has to do with climate variability/change, and the latter two with human pressures in the basin, in the recent past. These are the key issues addressed by the UNESCO/GRAPHIC project. It is likely that the climate and human pressures are nested, making it difficult to pinpoint the leading cause for the base flow reduction, and therefore increasing the difficulty of establishing appropriate mitigation and adaptation policies (Tucci 2002).

Since the objectives of the present work were to investigate the possible relationships between the decreasing groundwater discharge trends and the basin land use intensity, to establish a relationship between base flow discharge and basin hydrologic parameters, and to devise an empirical method to estimate hydrological services resulting from land conservation practices in the basin, it is expected that the results could contribute to the GRAPHIC subjects, methods, and experiences.

Additionally, the methodology utilized in the present case study sheds some light on the relationship between climatic variability and groundwater discharge, particularly in data-scarce tropical basins.

4.2   METHODOLOGY

In order to achieve the objectives of the study, the methodology was divided in four steps: i) correlation of the annual base flow discharge with the basin land use intensity over the last 23 years; ii) determination of basin curve-number[1] and base flow discharge from existing precipitation and stream flow records; iii) establishment of an empirical relationship between the base flow index[2] basin curve-number and climate variability; and iv) estimation of the hydrological services and economic benefits resulting from three land conservation scenarios in the basin.

4.2.1   Correlating annual base flow discharge with basin land use intensity

In this step, basin land use dynamics in the last three decades was obtained from temporal analysis of land-use maps, generated from Landsat satellite images (1984–2006). Land-use designations for these maps were derived using a multi-band, maximum likelihood supervised classification routine (Mather 1999).

The proportions and frequencies of the land use classes in the basin were obtained for five periods: 1984, 1994, 1998, 2001, and 2006. Additionally, a land use intensity index was applied to the classes and frequencies of the resulting land use maps. The land-use intensity index used was (Ometo et al. 2000; Chaves and Santos 2009):

$$LUI = \sum_{i=1}^{5} f_i w_i \qquad (4.1)$$

where: LUI = land use intensity index, f_i = proportion of the land class i in the land-use map, and w_i = weight of the land-use class i. The weights in equation (4.1) increase as the land use intensity increases, i.e., w = 1 for natural conditions, w = 2 for little disturbances in the natural condition, w = 3 for agriculture and pasture, w = 4 for low density urban areas, and w = 5 for dense urban areas (Chaves and Santos 2009). The weights in equation (4.1) correlate well with NRCS (1972) curve-number, where impermeable urban areas has twice or more the runoff potential and recharge abatement of pervious natural areas.

In order to avoid climate interference in the analysis, mean annual stream flow of the driest 6 months (Q_d) of the Pipiripau river was normalized by annual basin precipitation. The normalized variable Q_d/P was then plotted against the corresponding LUI for the five years studied, and an empirical relationship was obtained.

4.2.2   Obtaining basin curve-number and base flow discharge from stream flow data

The next step was the estimation of the basin curve-number (CN2) from the daily rainfall and stream flow records for the 1991–2009 period. In order to separate storm runoff

---

[1]Runoff coefficient, developed by the NRCS (1972).
[2]Ratio of base flow to total stream flow.

from base flow, the daily stream flow data was passed through a digital filter, originally developed by Nathan and MacMahon (1990):

$$q_t = \alpha q_{t-1} + \frac{1+\alpha}{2} \left(Q_t - Q_{t-1} \right)$$
(4.2)

where: q_t $(m^3 s^{-1})$ = filtered stream flow; Q_t $(m^3 s^{-1})$ = original stream flow at day t; and α = filter parameter, taken as 0.925 (Nathan and MacMahon 1990). Daily base flow discharge (Q_b) is simply the difference between Q_t and q_t in equation (4.2).

After the hydrograph separation, significant storm flow hydrographs, resulting from precipitation events with a volume over 10 mm in three consecutive days, were integrated, and a direct runoff volume $(Q_r,$ in mm) was obtained. A corresponding CN2 value for each storm hydrograph was obtained through the iterative solution of the following equation (NRCS 1972):

$$Q_r = \frac{\left[P - 0.2 \left(\dfrac{25400}{CN2} - 254 \right) \right]^2}{\left[P + 0.8 \left(\dfrac{25400}{CN2} - 254 \right) \right]}$$
(4.3)

where: Q_r (mm) = runoff volume of the storm flow hydrograph, P (mm) = precipitation volume of the event, CN2 = curve number for medium antecedent moisture conditions. Appropriate corrections were made in the calculated CN2 whenever drier $(P_a < 3.56$ mm in the previous 5 days) or wetter $(P_a > 5.33$ mm) antecedent soil moisture conditions (AMC) occurred (Chow et al. 1988).

Annual means of calibrated and corrected CN2 values were obtained for each hydrologic year in the stream flow series, and a global mean of CN2 obtained for the basin. To verify the accuracy of the calibrated CN2 value, it was compared with the weighted average CN2 value obtained from the NRCS (1972) table (below), calculated by the GIS and using the basin soils and land-use map.

4.2.3   Empirical relationship between the base flow index and the normalized runoff coefficient

In this step, the dimensionless yearly mean base flow index (Q_b/Q) in the period 1992–2009 was plotted against the normalized yearly curve number (CN2/P). Outliers with too high or too low base flow indices were removed from the analysis. The obtained empirical relationship was later used in the estimation of the hydrologic services resulting from different land conservation scenarios, by corresponding changes in the basin CN2 coefficient.

4.2.4   Estimating and valuing hydrological services resulting from land conservation scenarios

Since reforestation and best management practices (BMPs) can increase base flow discharge (Klöcking and Haberlandt 2002; Schilling and Libra 2003), the mean annual base flow increment in the Pipiripau river basin was estimated under three hypothetical land use scenarios, starting from the present (baseline) land use conditions.

Target areas for land conservation were defined based on environmental liabilities previously identified in the basin, including forest cover deficits in riparian strips and property reserves (reforestation), and unprotected aquifer recharge zones (BMPs).

Table 4.1 below lists the three land conservation scenarios analyzed in the study. These scenarios considered three possible climate conditions: average (A, mean annual precipitation), dry (D, mean annual P minus one standard deviation), and wet (W, mean annual P plus one standard deviation) years.

According to Table 4.1, a total of nine combinations of land use/climate scenarios were analyzed. In order to estimate the mean annual base flow discharge increment resulting from the each of the nine scenarios above, and starting from the calibrated CN2 value for the basin (baseline), new weighted CN2 values were obtained for the basin with the aid of the GIS and Table 4.2 below.

The new basin curve-number (CN2′) under the different land conservation scenarios was the weighted mean of CN2 values of Table 4.2 and the new soils and land use/management frequencies:

$$CN2' = \frac{\sum_{i=1}^{11} CN2_i A_i}{A} \qquad (4.4)$$

Table 4.1 Land conservation scenarios and climate conditions used in the estimation of the hydrologic services in the Pipiripau river basin.

Scenario Number	Type of land conservation practice	% of Basin Area	Climate condition
1	Reforestation of riparian and legal reserve areas with native tree species	9.5%	D/A/W
2	BMPs in agricultural and pastureland in recharge areas (no-till agriculture, terraces)	44.1%	D/A/W
3	Both scenarios (1+2)	53.6%	D/A/W

Table 4.2 CN values for a combination of soil type and land use. Source: NRCS (1972).

Land Use	Land Management	Hydrologic Soil Group			
		A	B	C	D
Agriculture	Without conservation practices	72	81	88	91
	With conservation practices	62	71	78	81
Pastureland	Poor condition	68	79	86	89
	Good condition	39	61	74	80
Savannah	Good condition	30	58	71	78
Woodland	Regular cover	45	66	77	83
	Good cover	25	55	70	77

where: CN2′ = the new basin curve-number incorporating the land conservation scenario, $CN2_i$ = curve-number of the soil and land-use/management i, A_i = area corresponding to the soil and land-use combination i, and A (ha) = total basin area.

The base flow discharge under different land-use and climate scenarios was computed with an empirical linear regression, obtained from the observed data, with the form:

$$Q_b/Q = a(CN2/P) + b \qquad (4.5)$$

where: a = the angular coefficient of regression, b = the linear coefficient, Q and Q_b are the stream flow and base flow discharge (m^3/s), respectively, CN2 is the basin curve-number, and P (mm) is the annual precipitation.

After Q_b was calculated in the baseline (land use and climate) condition, a new base flow discharge (Q_{b1}) was computed by equation (4.5) for the desired land use and climate combination. P and Q values in equation (4.5) were set at three levels each (dry, average, and wet conditions). The increment in annual base flow discharge for a given soil and land conservation/climate scenario was therefore:

$$\Delta Q_b = Q_{b1} - Q_{b0} \qquad (4.6)$$

where: ΔQ_b (m^3/s) = mean annual increment in base flow discharge in the basin, Q_{b1} (m^3/s) = mean annual flow discharge under the new land conservation/climate scenario, and Q_{b0} (m^3/s) = mean annual flow discharge under the baseline scenario.

Once the base flow discharge increments were obtained for each soil and land conservation/climate combination, the resulting economic externality was estimated by the product of the annual volume of base flow increment and the average household water price in the basin, taken as US\$ 1.5 per m^3 (CAESB 2011).

4.3   RESULTS AND DISCUSSION

4.3.1   Correlation between the dry season discharge and basin land use intensity

Figure 4.4 presents the land use dynamics in the basin between 1984 and 2006. According to that Figure, savannah, woodland and grassland have been replaced with agriculture and other anthropic uses in a steady fashion during the period.

Figure 4.5 below shows the correlation between land use intensity index (LUI) and the normalized dry season discharge (Q_d/P) in the Pipiripau river basin, during the period 1984–2006.

LUI almost doubled in the period studied, from 1.88 in 1984 to 3.43 in 2006, with a corresponding decrease in Q_d/P. A good fit $(R^2 = 0.92)$ was obtained for the linear regression, with the advantage that climate interference was eliminated.

Mixed results regarding land use conversion and base flow discharge were found in the literature. While Krause (2002) reported a significant decrease of the base flow index (Q_b/Q) as a result of conversion of forests to agriculture in a German basin, Siriwardena

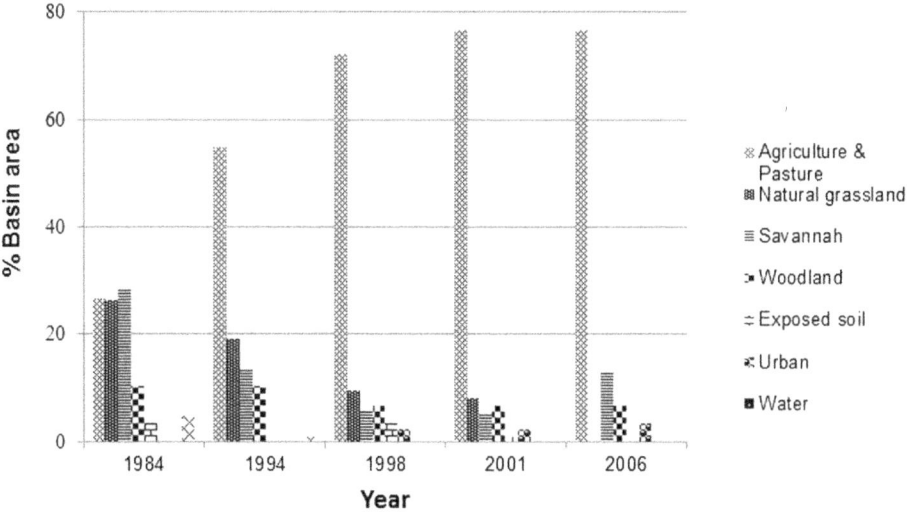

Figure 4.4. Land use dynamics in the Pipiripau river basin in the period 1984–2006.

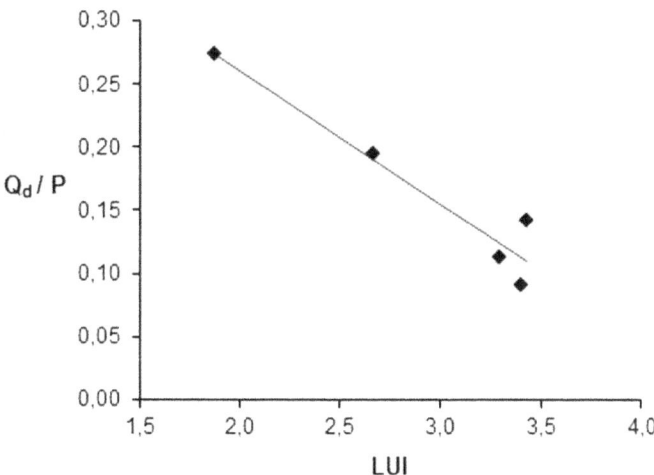

Figure 4.5. Relationship between the normalized dry-season stream flow and the land use intensity index in the Pipiripau river basin between 1984 and 2006.

et al., (2006) found the opposite in a basin in Queensland (Australia). Considering that other factors (rainfall variability, evapotranspiration) play a role in base flow discharge generation, it is reasonable to assume that results would be basin specific (Brooks et al. 2003).

Table 4.3   Yearly precipitation (P), mean stream flow (Q), mean base flow discharge (Q_b), base flow index (Q_b/Q), calibrated CN2, and normalized CN2 (CN2/P) for the Pipiripau river basin, in the period 1992–2009.

Year	P (mm)	Q (m^3/s)	Q_b (m^3/s)	Q_b/Q	CN2	CN2/P
1992	1,417	4.0	3.5	0.877	73.4	0.052
1993	1,383	4.6	4.1	0.887	73.4	0.053
1994	1,254	3.2	2.7	0.857	79.2	0.063
1995	1,431	2.1	1.9	0.908	66.4	0.046
1996	951	1.4	1.2	0.855	69.6	0.073
1997	1,479	2.0	1.6	0.807	73.3	0.050
1998	1,085	2.0	1.7	0.841	81.6	0.075
1999	1,340	1.8	1.5	0.840	72.0	0.054
2000	1,063	2.0	1.6	0.823	73.9	0.069
2001	912	1.2	1.0	0.809	74.1	0.081
2002	1,123	1.7	1.3	0.784	73.0	0.065
2003	1,213	1.6	1.3	0.851	68.2	0.056
2004	1,627	3.4	2.7	0.795	75.3	0.046
2005	1,316	2.7	2.3	0.862	70.4	0.054
2006	1,428	3.0	2.6	0.868	69.8	0.049
2007	1,228	2.5	2.1	0.859	69.0	0.056
2008	1,158	1.5	1.4	0.883	68.1	0.059
2009	1,147	1.7	1.5	0.882	68.2	0.059
Means	1,253	2.4	2.0	0.849	72.2	0.059

4.3.2   Base flow discharge hydrographs and basin curve-number (baseline condition)

Table 4.3 below presents the yearly precipitation, stream flow (Q), base flow discharge (Q_b), base flow index (Q_b/Q), and the calibrated CN2 values for the Pipiripau river basin, in the 1992–2009 period (baseline condition).

The mean calibrated CN2 for the catchment in the baseline condition, reflecting a combination of basin soils and land use characteristics, was 72.2, very close to the estimated (table) weighted average (CN2 = 70.7), indicating that the calibration and the hydrograph separation process were adequate.

Although Table 4.3 indicates that CN2 tended to decrease in the studied period, since it is strongly correlated with annual precipitation, the ratio CN2/P was a better indicator of basin condition.

Mean yearly precipitation in the period was 1,253 mm, with a standard deviation of 191 mm. Mean base flow was 2.0 m^3/s, and mean base flow index was 0.85. Figure 4.6 presents the correlation obtained between the base flow index and the normalized curve-number in the Pipiripau river basin.

Figure 4.6 shows that a relatively good fit was obtained for the linear regression between Q_b/Q and CN2/P. The dispersion in the data reflects the changes in watershed/aquifer conditions in the period studied.

Figure 4.6 also indicates that, for a given climate condition, an increase in CN2 would decrease the base flow index, i.e., the relative contribution of base flow discharge to total stream flow. The advantage of using the empirical function of Figure 4.6 is

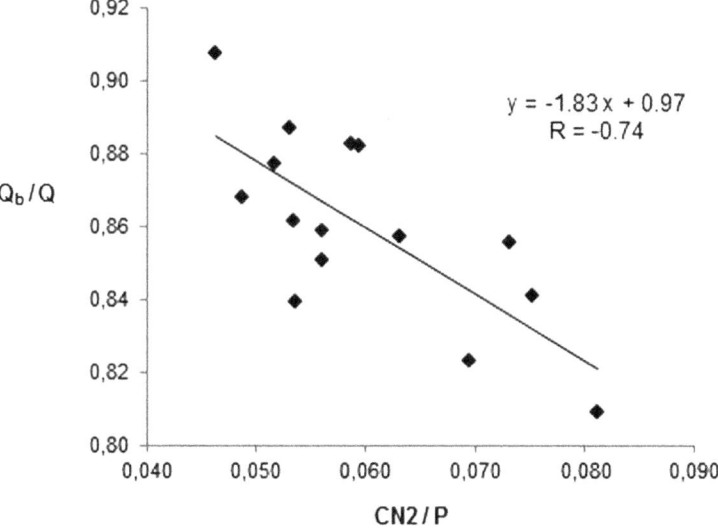

$$y = -1.83x + 0.97$$
$$R = -0.74$$

Figure 4.6. Correlation between the base flow index and the normalized curve-number for the Pipiripau river basin during studied period.

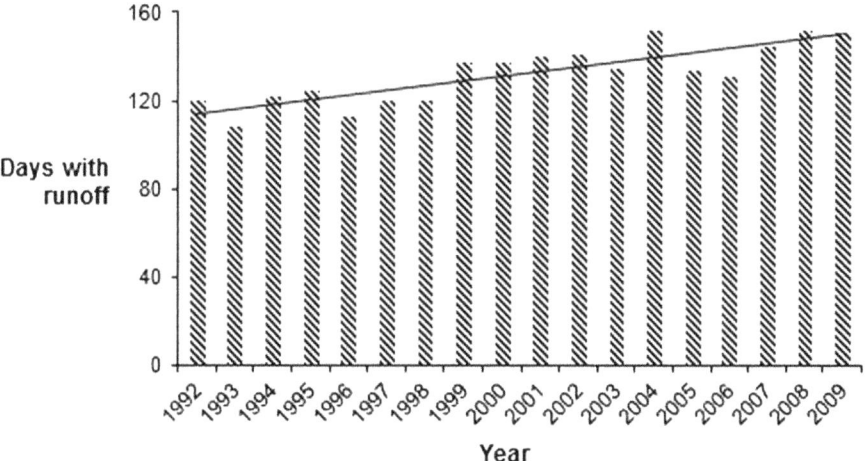

Figure 4.7. Number of days of the year with storm runoff in the Pipiripau river basin.

that the normalizations in Q_b and in CN2 eliminate intrinsic climatic variability, allowing for the isolation of the land use effects on base flow discharge.

Although there is a resemblance between Figures 4.5 and 4.6, the independent variable in the latter (CN2/P) has more hydrologic meaning, and therefore it is more useful for predictions of future basin behaviour.

Figure 4.7 below explains part of the behaviour of decreasing base flow index with increasing CN2/P, as seen from Figure 4.6. The former shows that the number of days with direct runoff in the Pipiripau basin has increased linearly in the period studied, contributing to the reduction of groundwater recharge and base flow discharge.

4.3.3   Hydrological services resulting from land conservation scenarios

The weighted averages of CN2 for the Pipiripau river basin under the baseline condition and land conservation scenarios are presented in Table 4.4.

According to Table 4.4, there would be a significant reduction in the basin CN2 after the introduction of the three land conservation scenarios, the highest decrease being under scenario 3, involving reforestation and the implementation of conservation practices in the agricultural areas.

Figure 4.8 shows the hydrological service with respect to base flow discharge (delta Q_b) and the additional revenue for the water utility company (delta R), as a function of the three land conservation scenarios, under the three climate conditions analyzed.

According to Figure 4.8, there would be a substantial increase in the dry season water availability in the basin, from 94,300 m^3 (0.4% of mean dry season flow) in a dry year under scenario 1, to 754,800 m^3 (3.0% of mean dry season flow) in a wet year under scenario 3. The additional revenues would range from US$ 123,900 per year to US$ 1,032,400 per year, respectively. The benefit/cost ratios were would be 0.08 and 2.93 for scenarios 1 and 3, respectively, indicating that the scenarios involving land conservation practices were more effective economically.

An increase of 25% in the base flow index of nine American basins was reported by Schilling and Libra (2003), as a result of improved conservation practices during the last 40 years, relative to previous conditions. Pattanayak (2004) reported that reforestation programs would increase the percentage of base flow discharge up to 24%, generating an equivalent amount of savings in water collection costs in small Indonesian communities.

Although land conservation practices such as reforestation can potentially increase basin evapotranspiration (ET) and decrease base flow discharge, the small proportion of

Table 4.4   Calibrated (baseline) and prospective (scenarios) CN2 values for the Pipiripau river basin.

Scenario	Baseline	Scenario 1	Scenario 2	Scenario 3
CN2	72.2	69.2	64.8	59.7

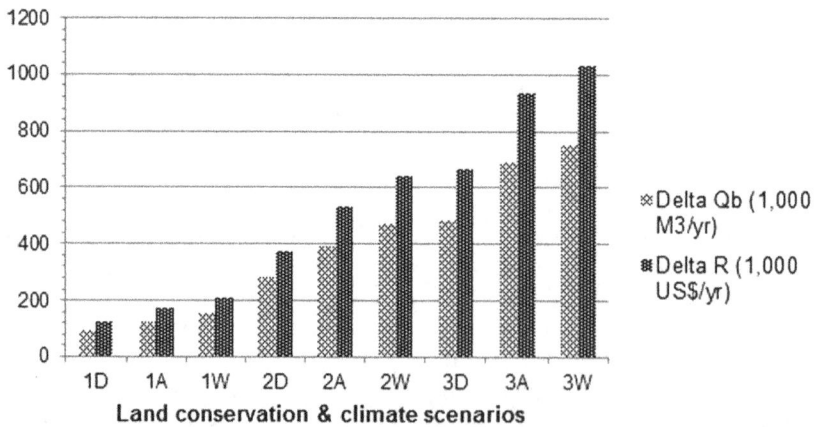

Figure 4.8.   Hydrological services and additional revenues resulting from the land conservation scenarios in the Pipiripau river basin, with respect to the baseline condition.

reforestation (<10% of the basin area) of scenarios 1 and 3 would not affect ET in a significant manner. In the case of scenario 2, the mulch cover of no-till agriculture would in fact reduce ET, with respect to the baseline condition.

4.4   POLICY RECOMMENDATIONS

The additional revenues generated by the increased base flow discharge (Figure 4.8) could be used to pay for the hydrologic services provided by participating farmers, creating a virtuous circle of land conservation and flow stabilization in the basin.

The hydrologic services and revenues would also demonstrate to the decision makers and water managers the economic feasibility of land conservation, which could lead to the design and implementation of land conservation projects in the future. In addition to the mitigation of the land use intensification in the basin, land conservation scenarios would also provide an adaptation strategy against future climate change.

Although there are supporting examples of the positive effect of land conservation on base flow discharge in the literature, the results of the present case-study reflect a specific situation, and discretion shall be used in transferring them to other river basins. However, the simplicity and robustness of the presented methodology facilitate its application to other data-scarce watersheds, where data intensive, process-based models could not be used.

4.5   FUTURE WORK

Considering that the Pipiripau river basin has been an object of hydrologic research in the last ten years, future activities in the Pipiripau river basin will involve:

- The survey and mapping of groundwater levels in the basin;
- The downscaling of GCM outputs to the basin scale and estimation of future climate change impacts in basin hydrology;
- The application of process-based models, such as SWAT;
- Continuous monitoring of stream flow and base flow in the basin, during and after the implementation of land conservation programs;
- The analysis of the vulnerability and risk of groundwater contamination by pesticides;
- Hydrologic and economic effectiveness of the implementation of agri-environmental (incentive payment) programs in the basin;
- The comparison of the results with other GRAPHIC case studies around the world.

REFERENCES

Brooks, D.N., Ffolliot, P.F., Gregersen, H.M. & DeBano, L.F. (2003) *Hydrology and the Management of Watersheds*. 3rd. edition. Ames, IA, Blackwell. 574 pp.

CAESB – Companhia de Saneamento Ambiental de Brasília. (2011) Personal communication by Suzana Alipaz, environmental scientist, on March 1, 2011.

Chaves, H.M.L. (1996) Mathematical modeling of water erosion: Past, present, and future. In: Alvarez, V.H. (ed.) *The Soil in the Large Morpho-Climatic Domains of Brazil*. Viçosa, UFV press. pp. 731–750 (in Portuguese).

Chaves, H.M.L., Braga, B., Domingues, A.F. & dos Santos, D.G. (2004) Estimating the environmental services and financial payments of ANA's Water Provider Program. *Revista Brasileira de Recursos Hídricos*, 9(3), 5–14 (in Portuguese).

Chaves, H.M.L. & Piau, L.P. (2008) Effects the rainfall variability and of the land-use and management on the direct runoff and sediment yield of a small catchment in the Federal District of Brazil. *Revista Brasileira de Ciencia do Solo*, 33, 333–343 (in Portuguese).

Chaves, H.M.L. & Santos, L.B. (2009) Land use, landscape fragmentation, and water quality in a small watershed. *Revista Brasileira de Engenharia Agrícola e Ambiental*, 13(Suppl.), 922–930.

Chow, V.T., Maidment, D.R. & Mays, L.W. (1988) *Applied Hydrology*. New York, NY, McGraw-Hill. 572 pp.

Claassen, R., Hansen, L., Peters, M., Breneman, V., Weinberg, M., Cattaneo, A., Feather, P., Gasby, D., Hellerstein, D., Hopkins, J., Johnson, P., Morehart, M. & Smith, M. (2001) *Agri-environmental policy at the crossroads: Guideposts on a changing landscape*. Washington, DC, USDA-ERS. Report number: 794. 67 pp.

GRAPHIC – UNESCO. (2006) *Groundwater Resources Assessment under the Pressures of Humanity and Climate Change*. International Hydrologic Program. Paris, UNESCO. 19 pp.

Klöcking, B. & Haberlandt, U. (2002) Impact of land use changes on water dynamics – a case study in temperate meso and macroscale river basins. *Physics and Chemistry of the Earth*, 27, 619–629.

Krause, P. (2002) Quantifying the impact of land use changes on the water balance of large catchments using the J2000 model. *Physics and Chemistry of the Earth*, 27, 663–673.

Mather, P.M. (1999) Computer processing of remotely-sensed images – an introduction. New York, NY, Wiley. 292 pp.

Nathan, R.J. & McMahon, T.A. (1990) Evaluation of automated techniques for base flow and recession analysis. *Water Resources Research*, 26(7), 1465–1473.

Natural Resource Conservation Service. (1972) *National Engineering Handbook – Chapter 4: Hydrology*. Washington, DC, Natural Resource Conservation Service.

Ometo, J., Martinelli, L.A., Ballerster, M.A., Gessner, A., Krusche, A.V., Victoria, R.L., & Williams, M. (2000) Effects of land use on water chemistry and macro-invertebrates in two streams of the Piracicaba river basin, Brazil. *Freshwater Biology*, 44, 327–337.

Pattanayak, S.K. (2004) Valuing watershed services: Concepts and empirics from southeast Asia. *Agriculture Ecosystems and Environment*, 104, 171–184.

Pereira, H. (1989) *Policy and Practice in the Management of Tropical Watersheds*. Boulder, CO, Westview Press. 125 pp.

Poff, N.L., Allan, D.J., Bain, M.B., Karr, J.R. & Prestegaard, K.L. (1997) The natural flow regime: A paradigm for conservation and restoration of riverine ecosystems. *Bioscience*, 47, 769–784.

Schilling, K.E. & Libra, R.D. (2003) Increased base flow in Iowa over the second half of the 20th century. *Journal of the American Water Resources Association*, 39(4), 851–860.

Sloto, R.A. (2008) *Effects of land-use changes and groundwater withdrawals on stream base flow, Pocono creek watershed, Monroe County, Pennsylvania*. Washington, DC, USGS. Scientific investigation report number: 2008-5030. 50 pp.

Tomich, T.P., Thomas, D.E. & van Noordwijk, M. (2004) Environmental services and land use change in southeast Asia: From recognition to regulation or reward? *Agriculture Ecosystems and Environment*, 104, 229–244.

Tucci, C.E.M. (2002) *Climate Variability and Land Use Impacts on Water Resources*. Brasilia, Brazilian Forum on Climate Change, National Water Agency. 150 pp.

CHAPTER 5

Effects of storm surges on groundwater resources, North Andros Island, Bahamas

John Bowleg & Diana M. Allen

ABSTRACT

Groundwater resources on islands may be particularly vulnerable to impacts of extreme events, sea-level rise and other effects related to climate change. This case study describes the consequences of a storm surge that contaminated the groundwater supply on North Andros Island, Bahamas during Hurricane Frances in 2004. Composite (depth averaged) chloride concentrations measured in the trench network increased from less than 400 mg/L (about 3 months before Frances), to over 13,000 mg/L in some wellfields. As part of corrective action, the trench and conduit systems were pumped down to encourage rapid recharge from the underlying resources.

5.1 INTRODUCTION

5.1.1 Purpose and scope

Groundwater resources on islands may be particularly vulnerable to impacts of extreme events, sea-level rise and other effects related to climate change. This case study describes the contamination of the groundwater supply on North Andros Island in The Bahamas (Fig. 5.1) due to a storm surge that impacted the island during Hurricane Frances in 2004. North Andros serves as the primary water supply for New Providence Island. While only preliminary research has been carried out to date, the case study presented here demonstrates the vulnerability of groundwater resources on the Caribbean Islands due to climate change related impacts.

5.1.2 Study area description: North Andros Island

The Bahamas, officially the Commonwealth of The Bahamas, consists of 700 islands and cays, with only a small percentage of the islands being inhabited (Meditz and Hanratty 1987). The population of The Bahamas is estimated at 356,658, concentrated in the cities of Nassau on New Providence Island and Freeport on Grand Bahama Island (Fig. 5.1). Population of The Bahamas is growing at a rate of 1.8 percent per annum and is projected to increase to 426,300 by 2030 (The Government of The Bahamas 2010). Tourism brings in about 4 million visitors a year placing a significant stress on the limited freshwater resources available.

The total land area encompasses 13,939 km^2. The Bahamas is divided into three geographical areas: New Providence Island, where the capital of Nassau is located; Grand

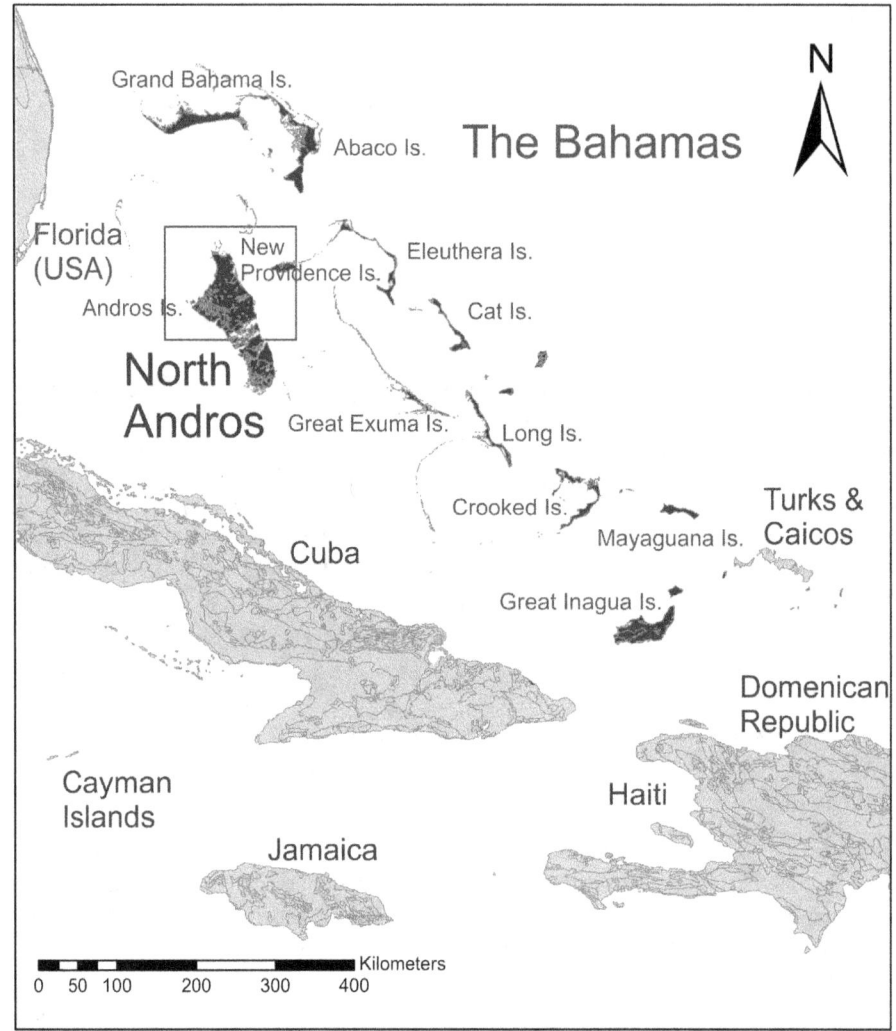

Figure 5.1.   The Bahamas location map. The study area on North Andros Island is shown in the box.

Bahama Island; and the Family Islands, consisting of all of the other islands and cays (Fig. 5.1). The chain of islands extends from 80 km east of Florida (USA) south-easterly to 80 km northeast of Cuba (Fig. 5.1). Haiti and the Turks & Caicos Islands lie to the southeast.

The islands are surface projections of two oceanic banks, the Little Bahama Bank and the Great Bahama Bank (Meditz and Hanratty 1987). Deepwater channels separate these banks. The highest elevation is only 63 metres above sea level (masl) on Cat Island; the island of New Providence reaches a maximum elevation of only 37 masl. The geography of the islands makes them susceptible to flooding and wind damage by tropical storms and hurricanes (ABC Country Book of The Bahamas 1997).

There are extensive coral reefs surrounding the islands. Patches of reefs and extensive sea grass beds are found on the interiors of the banks. In fact, The Bahamas provides

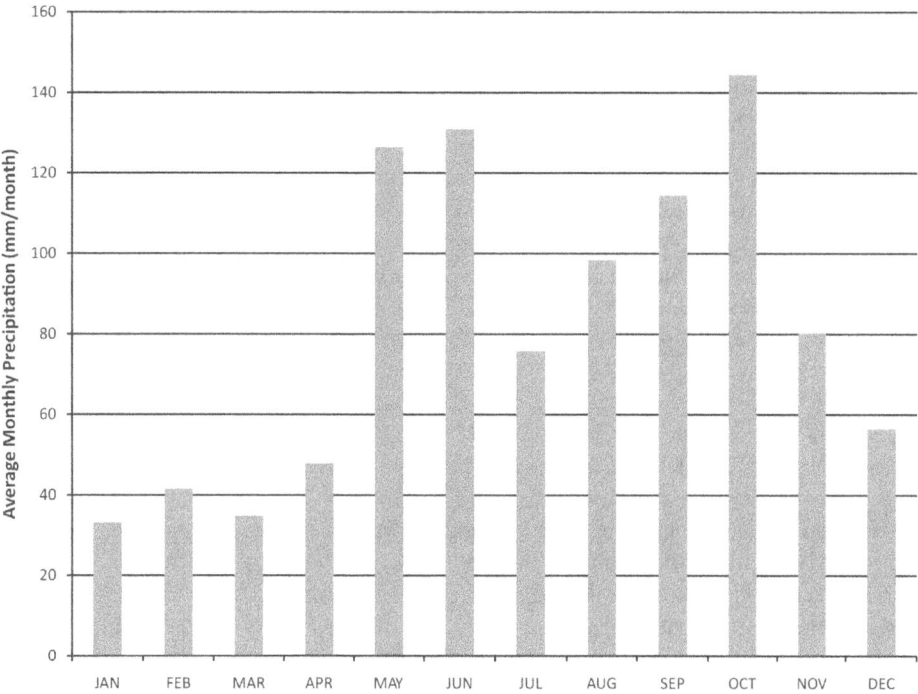

Figure 5.2.   Mean monthly precipitation for Kemp's Bay, Andros Island (station 78086) (data from 1971–2000).

the largest body of coral reef (as well as other marine organisms) in the Atlantic and Caribbean regions (The Government of The Bahamas 2002). The land has a foundation of fossil coral, but much of the rock is oolitic limestone. The land surface is either rocky or mangrove swamp. Low scrub covers much of the surface area. Timber is found in abundance on four of the northern islands: Grand Bahama, Great Abaco, New Providence, and Andros. On some of the southern islands, low-growing tropical hardwood flourishes. Although some soil is very fertile, it is also very thin. Freshwater in The Bahamas is primarily available as groundwater, which is naturally recharged by rainfall. The physical geology, hydrogeology, and water resources are directly linked, and there are no true rivers in The Bahamas.

The climate of the region is semitropical and has two seasons, summer and winter (Meditz and Hanratty 1987). During the summer, which extends from May through November, the climate is dominated by warm, moist tropical air masses moving north through the Caribbean. Midsummer temperatures range from 21°C to 34°C with a relative humidity of 60 to 100 percent. In winter months, extending from December through April, the climate is affected by the movement of cold polar masses from North America. Temperatures during the winter months range from 15°C to 24°C. Yearly rainfall averages 1,320 mm and is usually concentrated in the May–June and September–October periods (Fig. 5.2). Rainfall often occurs in short-lived, fairly intense showers accompanied by strong gusty winds, which are then followed by clear skies. Rainfall is unevenly distributed across the country, and the south-eastern islands receive as much as 40% less annual precipitation than the north and north-central islands (The Government of The Bahamas 2001). The islands in the south-eastern Bahamas are also prone to seasonal drought due to the limited rainfall. Winds are predominantly easterly throughout the year, but tend to become north-easterly from October to April and south-easterly from May to September. These winds seldom exceed twenty-four km per hour except during hurricane season; The Bahamas is located in the hurricane belt. Although the hurricane season officially lasts from June to November, most hurricanes in The Bahamas occur between July and October. More than fifty hurricanes have passed within 200 km of Nassau between 1886 and 1999. Bahama Islands Info (2008) provides a history of tropical storms and hurricanes in The Bahamas.

The Bahamas obtains the majority of its water supply from groundwater (US Army Corps of Engineers 2004). However, the availability of groundwater is highly variable. The only source of freshwater in the country is rainfall. The rainfall infiltrates the subsurface and forms freshwater lenses in the ground as well as wetlands, small pools at the surface, and seasonal ponds. Freshwater wetlands tend to be small, seasonal, and widely scattered. Groundwater is extracted from freshwater lenses contained within the Pleistocene and Holocene limestone and limesand aquifers using: (a) shallow hand-dug wells, (b) hand or electric pumps in uncased wells, and (c) trenches and pits. The freshwater lenses form when rainwater seeps through the porous surface material (Fig. 5.3). The freshwater is less dense than the saline seawater and floats on top of it within the pores of the rock. The lens also includes a zone of brackish water between the upper layer of freshwater and bottom layer of saltwater. The total estimated freshwater reserve is estimated at $7.7 \times 10^9 \, \text{m}^3$ and is scattered throughout the country in localized lenses of various sizes and quality (US Army Corps of Engineers 2004). Andros has the largest freshwater reserve, but Grand Bahama and Abaco also have large reserves.

The biggest threat to groundwater resources in The Bahamas is contamination by saltwater intrusion due to two causes: (1) storm surges resulting from hurricanes and

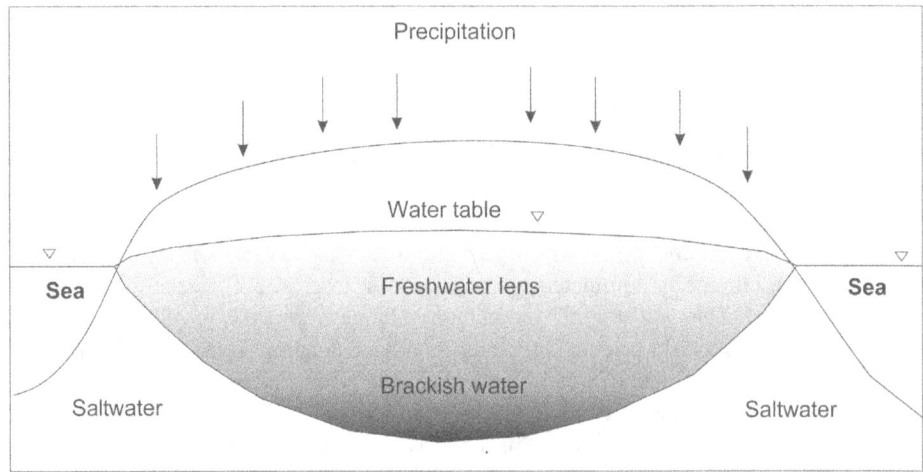

Figure 5.3.   Freshwater lens on an island. Brackish water is typically encountered at the base and edges of the freshwater lens where mixing with sea water occurs.

(2) over-pumping of freshwater aquifers (US Army Corps of Engineers 2004), although local contamination occurs from septic tanks, soakaways, and pit latrines and cess pits due to the low land elevations and a high water table; many septic tanks are located close to private wells. Salinity due to over-pumping from small freshwater lenses affects the water quality in many areas. Most of the water systems on Abaco have good quality (salinity below 400 mg/L). Individual well fields supply water for four systems in southern Eleuthera, where some well fields that are being over-pumped are yielding water with salinity levels above 1,200 mg/L and some above 2,000 mg/L. Salinity levels above 1,000 mg/L are generally considered unacceptable for domestic use. On some islands there are insufficient sources of freshwater, and others that had adequate water supplies are experiencing shortages due to growth of population, development and tourism. On islands with inadequate fresh groundwater resources, freshwater is often barged from islands with ample groundwater supplies, such as Andros. For example, New Providence previously transported over 50% of its water supply from North Andros Island (Fig. 5.1). On New Providence, which is home to over two thirds of the country's population, groundwater resources are threatened by pollution from anthropogenic activities.

In an effort to combat these vulnerabilities, the Government of The Bahamas utilizes reverse osmosis (RO) for desalinization on many islands to supplement or replace groundwater sources. RO is increasing in usage, and will most likely continue to increase, as fresh (ground) water availability continues to decline and water demands grow (US Army Corps of Engineers 2004). Rainwater catchment is rarely used. Due to quality issues and brackish water, a huge bottled water industry has developed. The estimated number of bottled water companies varies from eight on New Providence, up to a total of 27 (in 2004) for the entire country. According to the Central Statistical Office, approximately 85% of the population on New Providence buys bottled water for drinking and cooking (The Government of the Bahamas 2001). Responsibility for overseeing the water resources of the country is shared by several government agencies and institutions. The Bahamas Water and Sewerage Corporation (WSC) is the government body responsible for coordinating water resource management and development. Its mandate covers Water Supply, Waste Water Treatment, Water Sector Conservation, Development, Management and National Policy Development in the Water Sector. There are no other bodies at the sub-national level.

5.1.3 Methodology

Many of the islands of The Bahamas are low-lying, therefore storm surges, coupled with heavy rains, can cause widespread flooding. Ponding of seawater above freshwater aquifers results in increased and persistent levels of salinity until rainwater recharges and dilutes the groundwater. The Bahamas has a long history of tropical storms and hurricanes and is susceptible to storm damage from tidal surges and flooding from hurricanes, which can have dramatic effects on the groundwater supply. In 1992, Hurricane Andrew caused severe saltwater intrusion in a major farming area. Hurricane Lili caused flooding in 1996, which resulted in leaching of fertilizer and delay in replanting. In 1999, Hurricane Floyd caused a temporary rise in sea level and flooding in the Family Islands, resulting in water contamination, which required extensive pumping to restore water quality in some areas. Interruption of services was also experienced from a disruption of electricity at pumping stations and filtration system pumps, which lacked standby

Figure 5.4.   Areal view of the wellfield on North Andros Island.

generators. Also, broken pipelines and wind-damaged storage tanks caused some storage tanks to empty. The well fields on Grand Bahama and North Andros Islands were flooded with seawater and showed sharp increases in salinity after storm surges produced by Hurricane Floyd in 1999 and Hurricane Frances in 2004. Hurricane Jeanne in 2004 also caused significant damage to the aquifers on Grand Bahama and Abaco.

This particular case study focuses on the observed damage done to the groundwater lens aquifer on North Andros Island as a result of Hurricane Frances in 2004. The trenching network (Fig. 5.4 and 5.5) on North Andros caused the entire system to become inundated with saltwater. The trenches served as conduits for seawater passage during the hurricane, from the western trenches throughout the entire system.

5.1.4   Relevance for GRAPHIC

Freshwater resources of island nations are particularly vulnerable to climate change and human impacts. The Caribbean region, which relies on relatively shallow groundwater for much of its water use needs, is no exception. The Caribbean groundwater resources are sensitive because of the limited land areas and quantities of groundwater safe for drinking, growing demand over supply because of population growth, saltwater intrusion, and pollution from human and animal waste. Such problems may be exacerbated in response to climate variability and change, such as severe hurricanes and potential rising sea levels. Changes in the position of the distribution of fresh, brackish and saline groundwater might be anticipated due to increased storm surge activity, rising sea level, possible reductions in groundwater recharge, as well as increased abstraction.

Because of such concerns and because of the documented impact on the groundwater resource on North Andros Island in 2004 due to the storm surge produced during

Figure 5.5.   Photograph of one of the trenches.

Hurricane Frances, The Bahamas was proposed as a GRAPHIC case study area. The preliminary study reported herein is intended to serve as a representative example of potential consequences of storm surges on groundwater resources; similar events are expected to occur in other islands in the Caribbean which may threaten freshwater availability. Ongoing research in The Bahamas aims to identify the potential threats to groundwater resources due to climate change, with a particular focus on extreme weather events (storm surges), sea-level rise and changes in recharge, coupled with projections of increased demand for freshwater due to population growth.

5.2   RESULTS AND DISCUSSION

5.2.1   The well field on North Andros

The North Andros Water Resources, specifically the Barging Scheme Wellfields, comprises the "old" and "new" wellfields, which comprise a series of interconnected shallow trenches

(3 ft/0.9 m wide and a maximum of 10 ft/3 m deep) that form cruciforms (Fig. 5.4 and 5.5). In March 2004, production from the North Andros Water Resources was 4.27 million imperial gallons per day (MiG/day). Approximately 2.71 MiG/day was abstracted from the old Wellfields, and 1.86-Mig/Day from the new Wellfield. The target potable salinity for the area is 600 mg/L chloride, and corrective measures of trench isolation commence at 450 mg/L chloride.

Each cruciform in the "old" wellfield was designed to abstract 250,000 GPD (946.35 m^3/day), i.e., 43.4 GPM (0.16 m^3/min) per 1000 ft (304.8 m) of trench. The "old" wellfield area covers 6,000 acres (2,459 ha). The method of abstraction in the "old" wellfield is via 10 hp pumps at each cruciform site (total of 12 sites). Each cruciform is connected to a minimum of two and maximum of four trenches/limbs. Phases I and II of the "new" wellfield take the form of a network of parallel trenches from which water is permitted to flow by gravity into a conduit system, leading to a low level final sump. From this sump the water is pumped (total of four 700 GPM pumps each) directly to the reservoir storage. The criterion used for the abstraction rate for the design of Phases I & II of the "new" wellfield was 18.3 GPM (0.07 m^3/min) per foot (0.30 m) of trench. Each trench/limb in the "new' wellfield is approximately 1000 ft (304.8 m). Both sources ("old" and "new") are pumped to a final reservoir, prior to loading on the barge/tanker for transport to New Providence Island. Together, all the barging scheme wellfields are approximately 12,000 acres (4,858 ha), with an available freshwater supply of 6,000,000 GPD (22,712 m^3/day). The maximum abstraction from the area by WSC is approximately 4,800,000 GPD (18,170 m^3/day), based on 2003/2004 WSC Supply Records.

5.2.2   Hurricane Frances

Hurricane Frances was the sixth named storm, the fourth hurricane, and the third major hurricane of the 2004 Atlantic hurricane season (Wikipedia 2011). The system crossed the open Atlantic during mid to late August, strengthening while it moved northward. Its outer bands affected Puerto Rico and the British Virgin Islands while passing north of the Caribbean Sea. The storm achieved a Category 4 on the Saffir-Simpson Hurricane Scale. As the system slowed down its forward motion, the eye passed over San Salvador Island and very close to Cat Island in The Bahamas. Frances was the first hurricane to impact the entire Bahamian archipelago since 1866, and led to the nearly complete destruction of their agricultural economy (Wikipedia 2011).

5.2.3   Consequences of the storm surge in 2004

For North Andros, the storm surge associated with Hurricane Frances increased chloride concentrations in the trenches dramatically. A YSI temperature, water level and conductivity meter was used to profile electrical conductivity with depth in each cruciform and to determine composite chloride concentration. Composite (depth averaged) chloride concentrations increased from less than 400 mg/L (about 3 months before Frances), to over 13,000 mg/L in some wellfields (Table 1). The wetlands and low-lying lands to the west were the suspected source (about 1.5 km west of one trench in the "old" wellfield). The surge was not directly observed, primarily due to the sparse population to the west of the wellfield area. However, saltwater fish were observed inland in the forest. Residents had witnessed previous surges on both Grand Bahama and Abaco.

Table 5.1   Composite chloride concentrations (calculated from electrical conductivity) in May 2004 (prior to Frances), September 7, 2004 (immediately following Frances) and September 15–16, 2004 (after corrective action) in cruciforms in the new wellfield.

Cl (mg/L)	May 2004[1]	Sept 7, 2004	Sept 15–16, 2004
Minimum	165	3,740	900
Maximum	359	13,165	11,406
Average	258	9,042	4,227

[1]Values reported as ppm.

The storm surge did not appear to compromise the groundwater lens directly, based on measured concentrations of chloride in monitoring wells sampled following the surge (114 mg/L at 3.35 m depth). The "trenching" system of North Andros itself contributed to saltwater intrusion of the entire system; the trenches became flooded with sea water.

To correct the saltwater inundation from Hurricane Frances, the trench and conduit systems were pumped down, to encourage rapid recharge from the underlying resources. Table 5.1 also shows chloride concentrations measured in the trench network following corrective action.

5.3   POLICY RECOMMENDATIONS

The freshwater lens aquifers throughout most of The Bahamas (and other Caribbean islands) are vulnerable to storm surges, which may cause saltwater inundation of the aquifers. These damaging storm surges occur during hurricanes and other severe weather. Hurricanes Frances and Jeanne in 2004 caused significant damage in parts of the country. Water contamination from the storm surge on North Andros as described herein impacted about half of the freshwater supplies for New Providence. Open trench systems for water supply are only prevalent on Andros and New Providence. Both Grand Bahama and Abaco utilize boreholes for water supply. To prevent this occurrence of contamination of trenches during storm surges, in the future, the system should be isolated, with the trenches separated from each other (US Army Corps of Engineers 2004). Installed valves in each cruciform/chamber should be used to cut off the system. Ideally, all open trench systems should utilize perforated piping for abstraction and be backfilled to protect the underlying resources. Floodplain mapping is also needed to show areas vulnerable to flooding from storm surges (US Army Corps of Engineers 2004).

5.4   FUTURE WORK

This ongoing case study is a collaborative effort between regional and local groundwater experts of The Bahamas and international researchers. Partners include (from The Bahamas): the Water & Sewerage Corporation (WSC), the Meteorological Department, the BEST Commission of the Ministry of the Environment, College of the Bahamas (COB), and the Ministry of Foreign Affairs, and (international) UNESCO and Simon Fraser University in Canada. The objective of the extended case study is to identify and predict potential impacts of climate change on groundwater resources in the North

Figure 5.6.   Image of Hurricane Frances on August 31, 2004 at 17:55 UTC (1:55 PM EDT). At the time this image was taken Frances was located approximately 230 km north-northeast of San Juan, Puerto Rico and was moving towards the west near 26 km/hr. (Image captured from the MODIS instrument aboard NASA's Aqua satellite). (public domain; NASA 2004).

Andros Aquifer of The Bahamas, and to make recommendations for groundwater management schemes under such changing conditions. An integral component of the project will be to develop local and regional expertise through training of local water management personnel. The scope of work for the study will include:

- Compiling geologic, hydrologic, geochemical, soils, climate, water use, and other relevant geospatial data;
- Assessing the current status of groundwater resources (water use, current and past groundwater conditions, including water quantity and quality);
- Estimating groundwater recharge;
- Developing a calibrated density-dependent groundwater model to represent current groundwater conditions on North Andros, including distribution of fresh, brackish and saline groundwater;
- Using the model to simulate the response of the aquifer and trench network to the 2004 storm surge;

- Analyzing predicted shifts in climate, based on raw (or downscaled) Global Circulation Model (GCM) data and/or RCM simulations (several models);
- Using current and future predicted climate (and sea level) data as input to the numerical groundwater flow model;
- Generating scenarios for future land-use change, and changes in water use;
- Modelling potential future human impacts on groundwater resources;
- Evaluating model outputs for generating recommendations for groundwater monitoring and management under scenarios of human impact and climate change.
- Comparing major findings of The Bahamas case study to other GRAPHIC case studies that are assessing groundwater resources of other Small Island Developing States (SIDS).

ACKNOWLEDGEMENTS

The authors acknowledge UNESCO for providing funding to initiate this case study through a meeting held in the Bahamas in Nassau, March 17–19, 2008. The purpose of the meeting was to bring together groundwater and climate change experts from the Latin America-Caribbean (LAC) region, local researchers and partner agency representatives from the Bahamas, and GRAPHIC experts to discuss the conceptual project proposal as described above, and to establish a working plan for implementing the project. An additional planning meeting and site visit to North Andros by the U.S. Geological Survey (USGS) and U.S. Army Corps of Engineers' (USACE) International Centre for Integrated Water Resources Management (ICIWaRM) was conducted in the Bahamas during March 2 to 4, 2009.

REFERENCES

ABC Country Book of The Bahamas. (1997) *The Bahamas – Geography*. [Online] Available from: http://www.immigration-usa.com/wfb/the_bahamas_geography.html [Accessed 8th July 2011].

Bahama Islands Info. (2008) *Hurricane History in The Bahamas*. [Online] Available from: http://www.bahamaislandsinfo.com/index.php?option=com_content&view=article&id=935:hurricane-history-in-the-bahamas&catid=93:bahamas-weather&Itemid=212 [Accessed 8th July 2011].

Meditz, S.W. & Hanratty, D.M. (eds.) (1987) *Caribbean Islands: A Country Study*. Washington, DC, GPO for the Library of Congress. [Online] Available from: http://countrystudies.us/caribbean-islands/117.htm [Accessed 8th July 2011].

NASA. (2004) *Hurricane Frances (06L) Off Puerto Rico (Afternoon Overpass)*. [Online] Available from: http://visibleearth.nasa.gov/view_rec.php?id=6584 [Accessed 8th July 2004].

The Government of The Bahamas. (2001) *The Bahamas National Report: Integrating Management of Watersheds and Coastal Areas in Small Developing States of the Caribbean*. The Bahamas Environment, Science and Technology (BEST) Commission and ICF Consulting, March 2001. 8 pp.

The Government of The Bahamas (2002) *Bahamas Environmental Handbook*. Government of The Bahamas, Office of the Prime Minister, The Bahamas Environment, Science and Technology (BEST) Commission, Nassau, Bahamas. pp. 2–3.

The Government of The Bahamas. (2010) *Key Statistics 2010*. Department of Statistics, Government of The Bahamas. [Online] Available from: http://statistics.bahamas.gov.bs/key.php?cmd=view&id=203 [Accessed 8th July 2011].

US Army Corps of Engineers. (2004) *Water Resources Assessment of The Bahamas*. US Army Corps of Engineers, Mobile District & Topographic Engineering Center, December 2004. 114 pp.

Wikipedia. (2011) *Hurricane Frances*. [Online] Available from: http://en.wikipedia.org/wiki/Hurricane_Frances [Accessed 8th July 2011].

CHAPTER 6

Reducing groundwater vulnerability in Carbonate Island countries in the Pacific

Ian White & Tony Falkland

ABSTRACT

Many small island countries in the Pacific rely on shallow, fresh groundwater lenses in highly permeable coral sand and karst limestone aquifers, underlain and surrounded by seawater, as their main freshwater source. The delicate balance in small islands, between groundwater recharge and discharge into and mixing with seawater, as well as extraction of groundwater by communities, is threatened regularly by climate variability, storm surges, over-pumping, as well as by pollution from land uses and waste disposal. In low islands, predicted sea-level rises resulting from climate change are a further major threat. Together, the hydrogeology of fresh groundwater lenses and these natural climatic and anthropogenic threats make small coral island groundwater some of the most vulnerable systems in the world. This linkage between natural and human pressures means that small coral island groundwater systems are exemplars of the GRAPHIC approach to sustainable groundwater management. In this chapter, a simple steady-state approximation is used to provide insight into the key climatic, hydrogeological, physiographic, and management factors that influence the quantity of, and saline intrusion into freshwater lenses. Examples of the dynamic nature of freshwater lenses as they respond to these drivers are given. Natural and human-related threats to freshwater lenses are discussed. Strategies and policies processes to ensure groundwater sustainability and reduce its vulnerability in small islands are presented and suggested improvements to donor and aid programs in water are also advanced. It is argued that because of the extreme vulnerability of small island groundwater systems, management strategies and policy must be based on the very best technical information and using regional networks as exemplified by GRAPHIC.

6.1 INTRODUCTION

In the Pacific Ocean alone, there are about 1,000 populated, tropical and sub-tropical small islands. In many of these small island countries, because of the high permeability of their soils and regoliths, surface water is scarce or absent and groundwater is the main source of freshwater. The small land areas of these islands, typically of order 1 to $10\,km^2$, coupled with rising per capita water consumption mean that the amount of freshwater available to island communities is often limited, particularly in the frequent El Niño-Southern Oscillation (ENSO)-related droughts that occur in the central and western Pacific. In addition, increasing population growth means that urban densities are often high, with some population densities exceeding $12,000\,people/km^2$. Because of these densities and the added burden of domestic animals, particularly pigs, the quality of shallow groundwater is at risk of being polluted, as evidenced by high rates of deaths and diseases due to water-borne diseases in small island countries (WHO 2004).

Many small islands are less than 1 km in width (Dijon 1983) so their fresh groundwater systems are surrounded by seawater. In addition, the groundwater in many small islands occurs as fresh groundwater 'lenses' (Fig. 6.1). These lenses are relatively thin veneers of fresh groundwater overlying seawater in highly permeable, phreatic aquifers. These lenses, which are vital to small island communities, are extremely vulnerable to both natural variations and changes and human-caused perturbations. Both droughts and excessive pumping from inappropriate wells can cause seawater intrusion, salinising the lenses and making them unfit for consumption. Because of this, freshwater lenses are delicately balanced between additions from rainwater recharge and losses due to submarine discharge to the sea, tidal mixing with underlying seawater and extraction by communities. Such systems are vital to small island communities and require precise assessment of sustainable yields, vigilant monitoring of both quantity and quality and astute management (White et al. 1999b; van der Velde et al. 2007).

To compound problems, many Pacific island countries (PICs) have limited resources and few trained personnel with which to tackle these challenges. Many small island population centres are still in transition from subsistence cultures to urban settlements (Ward 1999; White et al. 1999b) so that their traditional adaptation strategies and modes of coping are often mismatched to present-day challenges of urban environments. Finally, the actual

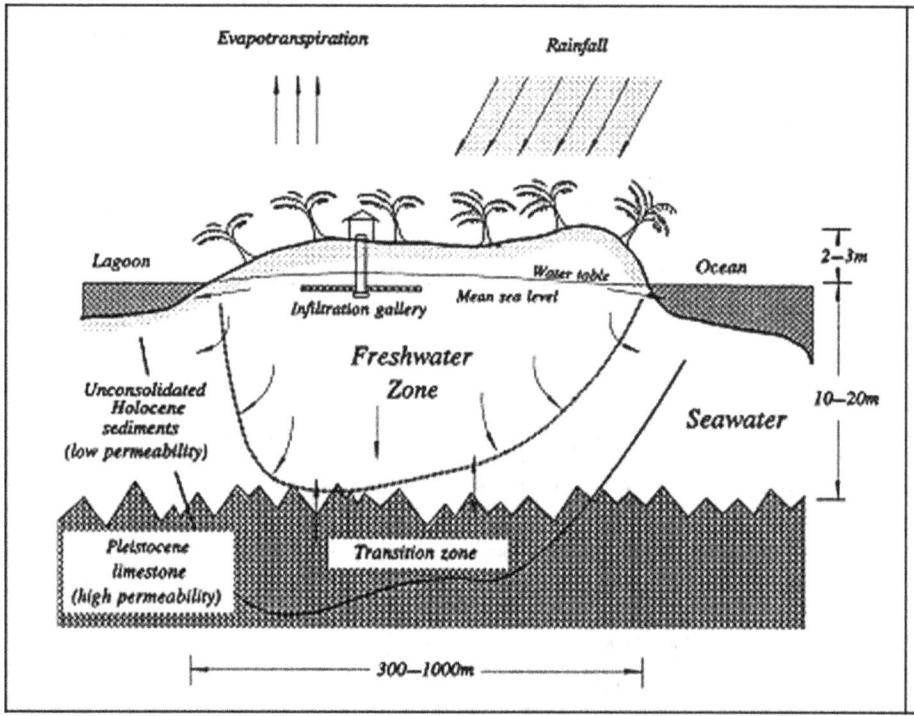

Figure 6.1.   Exaggerated vertical scale cross section through a small coral island showing the main features of a freshwater lens including the transition zone and the location of an infiltration gallery used for groundwater abstraction (White and Falkland 2010).

impacts of climate variability and the predicted effects of climate change, particularly sea-level rise add to the already complex challenges faced in small island nations.

6.1.1 Purpose and scope

This chapter firstly discusses the nature and characteristics of freshwater lenses in PICs and demonstrates the key factors which determine the quantity of fresh groundwater and its salinity. Then examples are given of how both natural and anthropogenic factors impact on island groundwater and of the main threats to small island groundwater systems. Finally, strategies to reduce the vulnerability of small island groundwater systems by carefully managing small island groundwater systems to ensure their sustainability are discussed.

6.1.2 Study area description: Pacific Island countries

This chapter will mainly draw on fresh groundwater systems in small islands in the Pacific (Fig. 6.2). Many of the issues discussed here, however, are germane to small islands throughout the world.

6.1.3 Methodology

In this chapter a simple, steady-state analysis based on the water balance is used to demonstrate the key drivers which determine the quantity of fresh groundwater and its salinity in small islands. More detailed modelling is used to determine safe groundwater yields. The impacts of these drivers are illustrated using groundwater monitoring results using piezometric surveys, water quality analyses and electromagnetic induction from a range of small islands.

Bredehoeft (1997) pointed out that commonly there are broader and often more weighty contributors that take precedence over technical and scientific issues in water resources management. In his hierarchy of factors contributing to decisions about water (Fig. 6.3), legal aspects come first followed by political, we have then added social-cultural issues followed by economic. He placed technical at the lowest level.

6.1.4 Relevance for GRAPHIC

Many of the past errors in water management have been due to placing technical issues as the lowest priority as in the hierarchy in Figure 6.3. A key thrust of GRAPHIC is to provide scientifically-based, recommendations to assist in improving health and laws, informing policies and social programs and in making wise economic decisions on water and particularly groundwater management in the face of increasing human pressures and the threat of climate change. As we shall to show in this chapter, this is exemplified in the management of PICs' groundwater systems. Another of the major benefits of GRAPHIC is that it provides a platform for the exchange of information and uses regional and global networks to improve the capacity to manage groundwater resources. For small island nations with few trained groundwater specialists, this exchange of information and the pooling of expertise through networks are vital (White et al. 2007a).

A primary goal of GRAPHIC is to establish case studies in different countries and regions of the world that exemplify the analysis of groundwater sustainability under

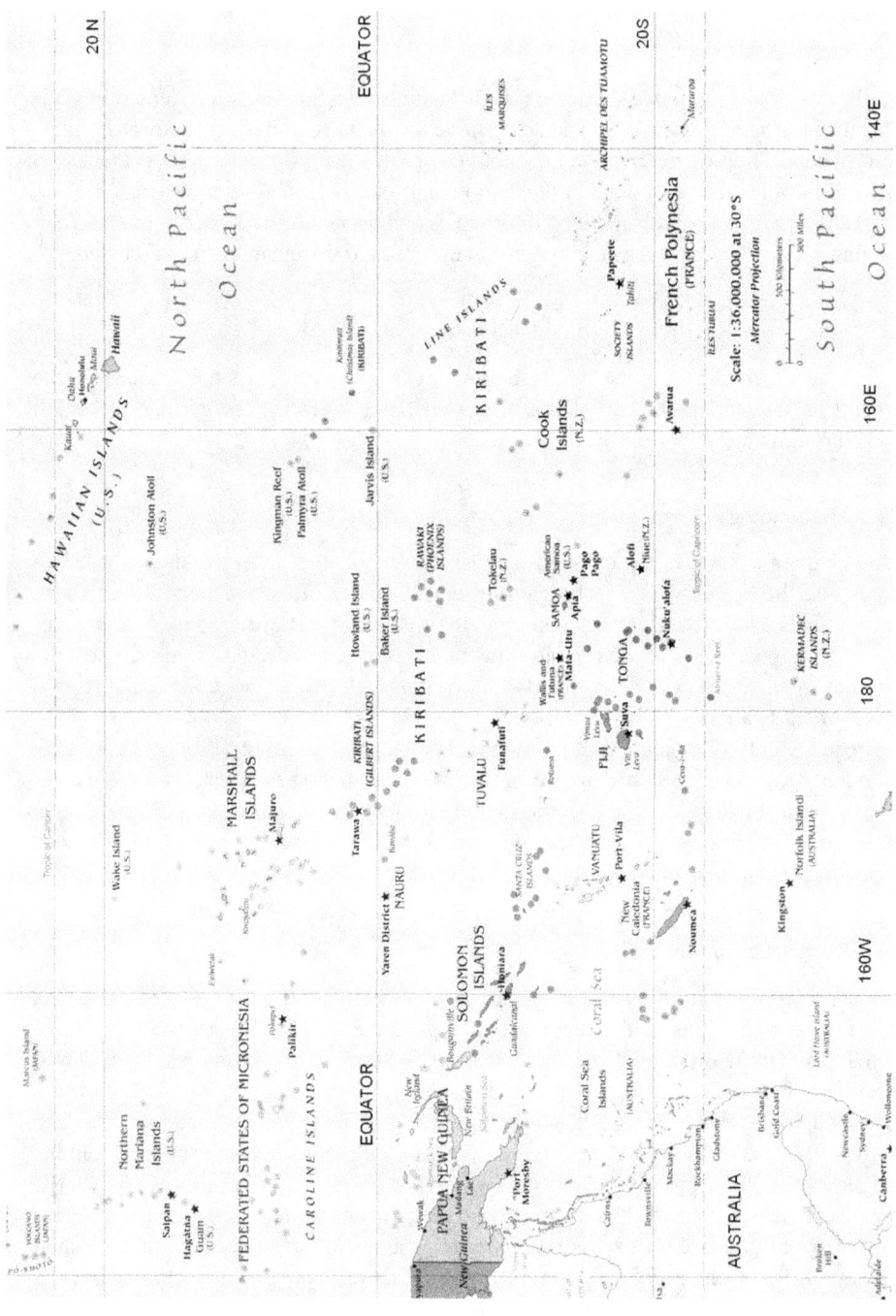

Figure 6.2. Small island countries in the central and southern Pacific Ocean [with permission from Pacific Islands Applied Geoscience and Technology Division (SOPAC)].

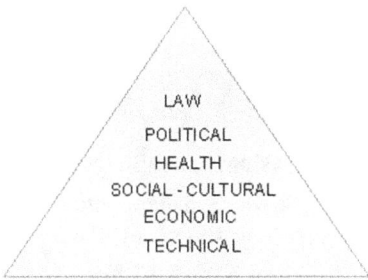

LAW
POLITICAL
HEALTH
SOCIAL - CULTURAL
ECONOMIC
TECHNICAL

Figure 6.3. The hierarchy of factors contributing to decisions about water (modified from Bredehoeft, 1997).

the impacts of human activities and climate change and identify suitable adaptation strategies. The intention is to provide insights and comparisons for similar countries/ regions, creating a global network of groundwater projects. Because groundwater in PICs is so vulnerable to climate and human influences and is at the cutting edge of climate change impacts, there are lessons that can be transferred not just to other small islands throughout the world but to larger countries.

Two of GRAPHIC's primary objectives are: understanding the impacts of human activities and climatic changes on groundwater availability and sustainability; and encouraging and coordinating consistent regional monitoring efforts. As has been briefly detailed above and described more fully elsewhere (White et al. 2007b; White and Falkland 2010), the impacts of human activities and current climate variability can have catastrophic impacts on small island groundwater systems. Their vulnerability to these influences means that systematic monitoring of groundwater and regional monitoring of the drivers of climate are crucial to the sustainable management of fresh groundwater lenses.

6.2 RESULTS, DISCUSSIONS, AND POLICY RECOMMENDATIONS

6.2.1 Characteristics of fresh groundwater lens

Freshwater lenses occur principally on low-lying carbonate islands where unconsolidated sands and gravels have been deposited unconformably on ancient karst limestone reefs as, for example, in Kiribati, Marshall Islands, Tuvalu, Tokelau and parts of the Cook Islands, French Polynesia, Hawaiian Islands, Papua New Guinea and Tonga (Fig. 6.2). Others are on limestone islands of uplifted reef deposits as in the island states of Nauru and Niue and a number of islands in Tonga. An excellent overview of the geology and hydrogeology of carbonate islands is provided by Vacher and Quinn (1997). While detailed studies of the nature and dynamics of many freshwater lenses have been published (UNESCO 1991), many more remain to be investigated.

The lower boundary between freshwater and underlying seawater in the thin lenticular groundwater body is not sharp as envisaged by the classical 'Ghyben-Herzberg' model (Badon Ghijben 1889; and Herzberg 1901). Rather the lower boundary occurs as a wide transition or mixing zone where groundwater salinity increases with depth from freshwater to seawater (Fig. 6.1) due to tidal mechanical mixing and dispersion. The practical limit of the base of the freshwater zone of the lens is defined by some measure of groundwater salinity acceptable for human use such as the World Health Organization drinking water guideline for chloride ion concentration of 250 mg/L (WHO 2004) or an equivalent electrical conductivity, EC, used routinely to measure groundwater salinity in the field.

Climate, hydrogeological, and physiographic factors affecting freshwater lenses

A simple, steady-state analysis provides insight into the factors which influence the quantity and salinity of freshwater lenses in small islands. The maximum thickness of a freshwater lens to an assumed sharp interface between fresh and saltwater, H_u (m) in the centre of a circular uniform island is approximated by (Volker et al. 1985):

$$H_u = \frac{W}{2}\left[(1+\alpha)\frac{R}{2K_0}\right]^{1/2} \tag{6.1}$$

where W (m) is the width of the island, $\alpha = (\rho_s - \rho_0)/\rho_0$ with ρ_s, ρ_0 the densities (t/m^3) of sea and freshwater respectively, R (m/a) the mean annual groundwater recharge rate and K_0 (m/a) the assumed uniform, horizontal saturated hydraulic conductivity of the phreatic aquifer.

Eq. 6.1 predicts that: wider islands should generally have thicker freshwater lenses than narrower islands; and islands with higher groundwater recharge rates should also have thicker lenses than those with lower recharge. Eq. 6.1 also shows that islands whose phreatic aquifers have high saturated hydraulic conductivities should have thinner freshwater lenses than those with lower K_0. For a typical island 1 km wide with a recharge rate of 1 m/a and a K_0 of 1,300 m/a ($\sim 4 \times 10^{-5}$ m/s), Eq. 6.1 predicts the average thickness of the freshwater lens to the sharp interface should be about 10 m.

In the sharp interface model, the elevation of the water table above mean sea level (msl), h_0 (m) is governed by the density difference between fresh and seawater:

$$h_0 = \frac{\alpha}{\alpha+1}.H_u \tag{6.2}$$

Water table elevations are typically of order 0.2–0.5 m above msl and vary with recharge rate.

In practice, islands are neither circular nor uniform. Islands lying within the cyclone belt generally have very coarse unconsolidated sediments and large hydraulic conductivities, such as those in Tuvalu (Fig. 6.2). These islands often have very limited or no viable freshwater lenses, consistent with Eq. 6.1. Freshwater lenses on atoll islands are often asymmetric with the deepest portion often displaced towards the lagoon side of the island, due to cross-island differences in permeability. Figure 6.4 shows an example of the measured asymmetric freshwater lens on the island of Bonriki on Tarawa atoll, Kiribati.

Islands in the Pacific are often subject to prolonged ENSO-related droughts of up to 44 months duration. To a first approximation, Eq. 6.1 can be used to estimate the impact of long droughts on the thickness of freshwater lenses. If H_d (m) is the thickness under a prolonged period of drought with recharge R_d (m/a), then: $H_d/H_u = (R_d/R)^{1/2}$. In a prolonged drought, if recharge is decreased to 25% of the mean recharge, the thickness of the freshwater lens should be reduced by about 50%.

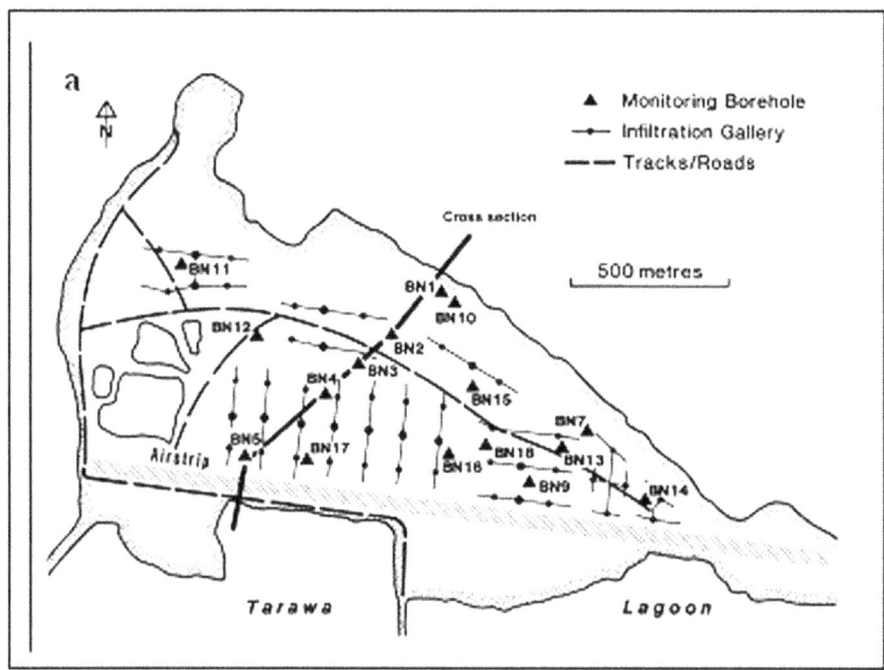

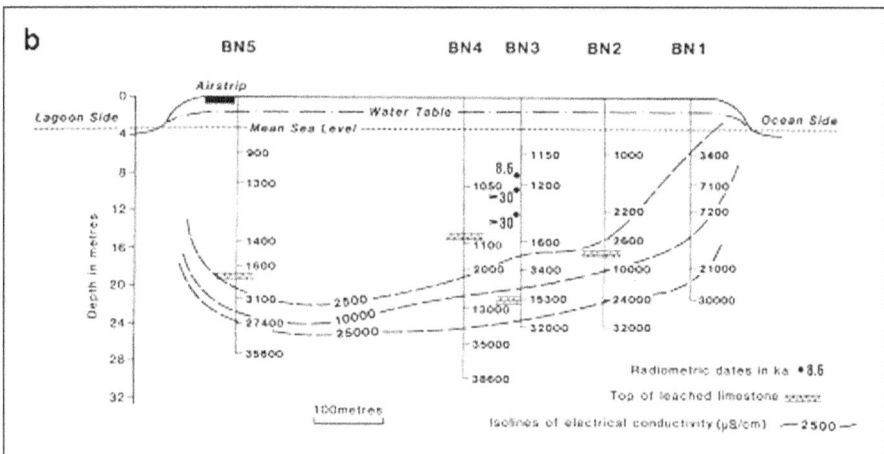

Figure 6.4. Asymmetric lens on Bonriki island, Tarawa atoll, Republic of Kiribati; (a) plan of Bonriki showing location of cross section and (b) cross section through the Bonriki freshwater lens showing depths to freshwater limit (taken as 2,500 μS/cm) and mid-point of transition zone (25,000 μS/cm) [From Falkland and Woodroffe (1997) with permission from Elsevier].

Volker et al. (1985) provide an approximate, steady-state approach to predicting the width of the salinity transition zone under a freshwater lens based on the work of Wooding (1963, 1964) who examined two-dimensional flow in freshwater overlying salt-water. The ratio of the mean width, δ_u (m), of the transition zone to the mean maximum freshwater lens thickness at the centre of a lens in a low coral island in the absence of pumping can be written as:

$$\frac{\delta_u}{H_u} = \frac{K_0}{R}\left(\frac{D}{\alpha WK_0}\right)^{1/2} \tag{6.3}$$

Here D (m^2/a), is the dispersion coefficient. Eq. 3 predicts that the relative thickness of the transition zone will increase as K_0 increases and decrease with increasing island width and increasing recharge rate. At low recharge rates, the width of the transition zone may equal the thickness of freshwater. This is important if the lens is used as freshwater source since the practical thickness of usable freshwater, H_{wu} is approximately $H_{wu} = H_u - \delta_u/2$. Useable freshwater lenses exist when $\delta_u < 2H_u$ or $R/K_0 > (1/2)(D/[\alpha WK_0])^{1/2}$. These are typically about 5–20 m thick, with transition zones of similar thickness.

Eqs 6.1 and 6.3 provide insight into the impact of long droughts on the width of the transition zone, δ_d (m). The ratio of the width in the drought to that under mean conditions is given by $\delta_d/\delta_u = (R/R_d)^{1/2}$. If recharge during a long drought is reduced to 25% of mean recharge then the width of the transition zone could be doubled. The practical thickness of useable freshwater during a drought, H_{wd} (m) is reduced by $H_{wd} = (R_d/R)^{1/2}(H_u - \delta_u[R/\{2R_d\}])$ and it is expected that the salinity of abstracted water should also increase during droughts. A useable freshwater lens will still exist provided $\delta_u < (2R_d/R)H_u$.

Impacts of pumping on freshwater lenses

When water is being pumped at a constant rate, Q (m^3/s), from a freshwater lens of land area A (m^2), the steady-state analysis predicts that the maximum lens thickness to an assumed sharp interface between fresh and saltwater, H_p (m), at the centre of an island is given by (Volker et al. (1985):

$$H_p = \frac{(1-q)^{1/2}W}{2}\left[(1+\alpha)\frac{R}{2K_0}\right]^{1/2} = (1-q)^{1/2}H_u \tag{6.4}$$

where q is the ratio of the specific pumping rate to recharge rate, $q = (Q/A)R$. Eq. 6.4 suggests when the specific pumping rate is 50% of the mean recharge rate, the maximum thickness of the freshwater lens will be about 71% of the mean thickness of the unpumped lens. During pumping, the increase in the width of the transition zone, δ_p (m), is given by:

$$\frac{\delta_p}{H_p} = \frac{1}{1-q}\left(\frac{K_0}{R}\right)\left(\frac{D}{\alpha WK_0}\right)^{1/2} = \frac{\delta_u}{(1-q)H_u} \tag{6.5}$$

or $\delta_p = \delta_u/(1-q)^{1/2}$

So it can be seen that, as with drought, the width of the transition zone should increase, as should the salinity of the lens, as the pumping rate increases. The thickness of useable freshwater from a pumped lens, H_{wp} follows from Eqs. 6.4 and 6.5:

$$H_{wp} = H_p - \delta_p/2 = (1-q)^{1/2}\left(H_u - \delta_u/\{2[1-q]\}\right) \tag{6}$$

Eq. 6.5 indicates that when the specific pumping rate is 50% of the mean recharge rate, the transition zone will be 41% wider than that in an unpumped lens. So the effect of pumping is to decrease the lens thickness and to increase the width of the transition zone. This means that the practical depth of usable water under pumping is further reduced as in Eq. 6.6. A useable freshwater lens will continue to exist provided $\delta_u < 2(1-q)H_u$ and $q < 1$.

The approximate steady-state analysis above demonstrates how the freshwater lens thickness and the width of the transition zone at the base of the lens varies with: the effect of climate, through recharge, R; island physiography through the characteristic island width, W; aquifer hydrogeology through the saturated hydraulic conductivity, K_0; pumping through the non-dimensional specific pumping rate, q, seawater-freshwater interaction through the density ratio parameter α; and mixing through the dispersion coefficient D.

Several important dimensionless groups besides α emerge including the dimension-less recharge rate R/K_0, the dimensionless specific pumping rate $q = (Q/A)/R$ and an atoll Peclet number, $\alpha W K_0/D$. These three factors determine the rate of inputs of water to the freshwater lenses, the rates of pumping losses and the rate of mixing with sea-water. The balance between these factors determines the viability of a freshwater lens as a source of potable water and illustrate why in small island hydrology it is essential to assess accurately recharge rates, hydraulic conductivity and monitor pumping rates. While this analysis provides insights into the impacts of human influences and climate changes on freshwater lenses, it cannot describe the detailed dynamics of freshwater lenses in heterogeneous small islands. For those situations dynamic models involving numerical solutions are required (see e.g. Underwood et al. 1992).

Freshwater lenses in raised limestone islands

Freshwater lenses in raised limestone islands, unlike low islands, are predominantly contained in highly permeable karst limestone aquifers. Lenses vary from extensive and relatively thick on some islands, such as Niue and Tongatapu in Tonga, to very thin or absent on others, such as Banaba in Kiribati and Nauru (Fig. 6.2). Groundwater investigations on Niue (Jacobson and Hill 1980; Wheeler and Aharon 1997; GWP 2006) and Tongatapu in Tonga (Furness 1997; White et al. 2009a) show signifi-cant fresh groundwater resources available for pumping for urban and village water supplies.

In contrast, groundwater investigations in 1987 (Jacobson and Hill 1988) after a heavy rainfall period found Nauru had a relatively small freshwater zone (Fig. 6.5) which was then simulated by numerical models which suggested a viable lens (Jacobson et al. 1997). Drilling, piezometric and salinity investigations in 2008 during a drought, however, found no freshwater lens in the centre of the island at all and only a few small pockets of fresh to brackish groundwater at one end of the island most of which have now become saline. This demonstrates the dynamic nature of freshwater lenses, their vulnerability to climate variability, the fundamental importance of long-term monitoring in small islands and the sensitivity of groundwater models to input or calibration data.

The simple steady-state model above assumes that the aquifers are uniform and deep. In practice, in low coral atolls and islands in the Pacific, where Holocene sands and gravels are deposited unconformably over karst Pleistocene limestones, the depth of the unconformity frequently determines the freshwater lens thickness because of the extremely high permeability of the karst limestone as further discussed below.

Natural influences on freshwater lenses

As the steady-state approximation above illustrates, the main natural influences on the occurrence of viable freshwater lenses on small islands are rainfall recharge to the ground-water system the storage capacity of the freshwater lens and the losses from the aquifer: evapotranspiration by phreatophytes, hydraulic losses at the edges of the island and dispersive mixing with underlying seawater. These natural influences are discussed below.

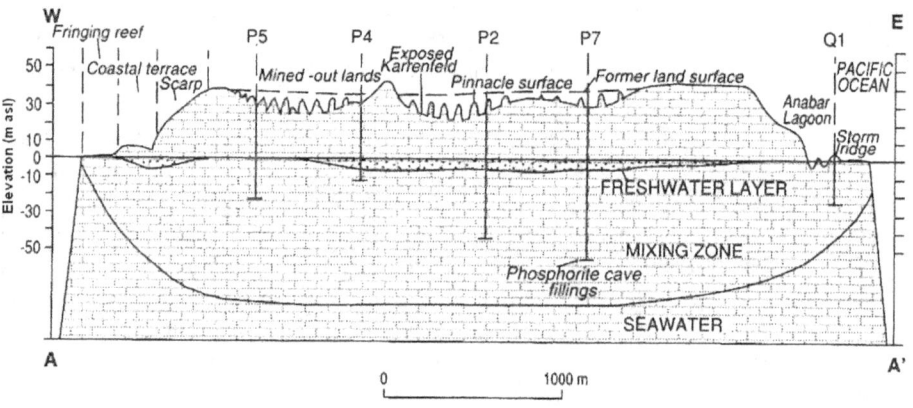

Figure 6.5.   Cross section through Nauru showing freshwater and mixing (transition) zones after heavy rainfalls in 1987 [from Jacobson et al. (1997) with permission from Elsevier]. Recent investigations in a dry period found no major viable freshwater lens.

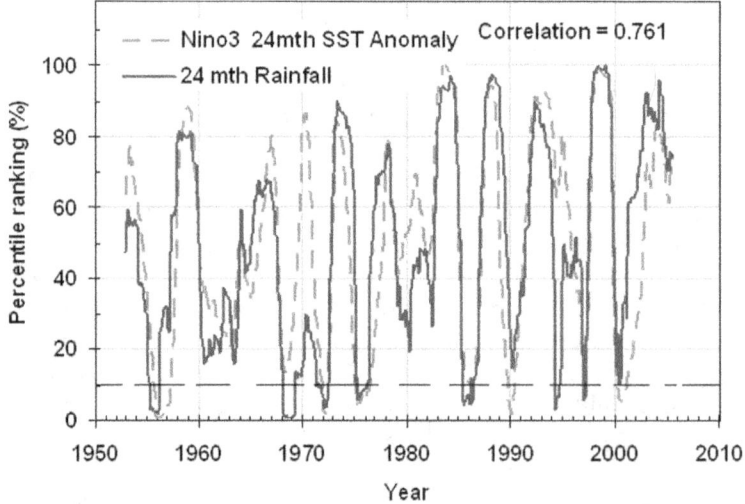

Figure 6.6.   The strong correlation between rainfall (solid line) in Kiritimati atoll, Kiribati and the Niño Region 3 SST anomaly (dashed line). Rainfall percentiles dropping below the 10 percentile ranking (dashed horizontal line) are severe droughts.

Climate variability

Climate, particularly rainfall and evapotranspiration are key drivers of groundwater recharge. Average annual rainfall and rainfall variability varies considerably throughout the Pacific. On Kiritimati atoll in the dry equatorial zone of eastern Kiribati (Fig. 6.2), average annual rainfall is less than 1,000 mm and the coefficient of variation (CV) is greater than 0.7. In the western Pacific, near Funafuti atoll in Tuvalu (Fig. 6.2), the average annual rainfall is over 3,500 mm and the CV is much lower, near 0.2. Variations in monthly and annual rainfalls in the Pacific are influenced strongly by inter-annual El Niño and La Niña cycles as the Pacific warm pools migrates from the eastern to western equatorial Pacific. On many Pacific islands, there is a strong correlation between sea surface temperatures (SST) or the Southern Oscillation Index (SOI) and rainfall patterns (e.g. White et al. 1999a, Van der Velde et al. 2006; White et al. 2007b)). Figure 6.6 shows the strong correlation between percentile rankings of 24 month rainfall in Kiritimati atoll, Kiribati (Fig. 6.2) and the percentile ranking of the 24 month Niño Region 3 SST anomaly. This correlation carries through to recharge. Groundwater salinity and thickness of freshwater lenses are also correlated with SST or SOI (e.g. van der Velde et al. 2006; White et al. 2007a).

Evapotranspiration is a very important part of the hydrological cycle for small islands and can exceed more than half of the rainfall on an annual basis. It often exceeds rainfall for individual months or consecutive months during dry seasons or drought periods but the variability of evapotranspiration is much lower than that of rainfall. Typical annual potential evapotranspiration rates in the tropical areas of the Pacific are between 1,600 and 1,800 mm (Nullet 1987). Measurements of evapotranspiration on Bonriki island, Tarawa atoll suggested a lower annual rate closer to the equilibrium rate of about 1,420 mm due to the rapidly draining coral sands (White et al. 2002).

Island physiography

Island size, shape and topography, particularly width and height of the island above mean sea level play a critical role in island water resources. Larger, higher and wider islands are more likely to have either surface and groundwater resources or groundwater in greater quantities than smaller and narrower islands (Eqs. 1 and 3).

Raised limestone islands are likely to have higher groundwater recharge for the same rainfall and vegetation conditions than low islands, as the roots of deep rooted trees such as coconuts are unable to reach the water table and transpire water directly from the freshwater lens, as happens in low-lying atoll and reef islands (White et al. 2002). Where islands have narrow necks and peninsulas, the potential for seawater mixing and intrusion there is increased. Figure 6.7 illustrates the groundwater salinity distribution in Tongatapu, Kingdom of Tonga (Fig. 6.2) and shows narrow areas where seawater intrusion increases groundwater salinity to the extent that the water is not fit for use.

Hydrogeological properties

As demonstrated in the simple steady-state model, hydraulic conductivity of the aquifer material has a direct bearing on the size, salinity and sustainability of freshwater lenses on small islands. Small limestone islands are generally karst Pleistocene limestone, which has weathered from alternate periods of submergence and exposure due to fluctuating sea levels. Caves and solution cavities are often found along the shoreline and within the island. The hydraulic conductivity of the limestone is often greater than 1,000 m/d and, consequently, freshwater lenses are generally no more than about 10 m thick, even in wide islands such as Tongatapu (Fig. 6.7).

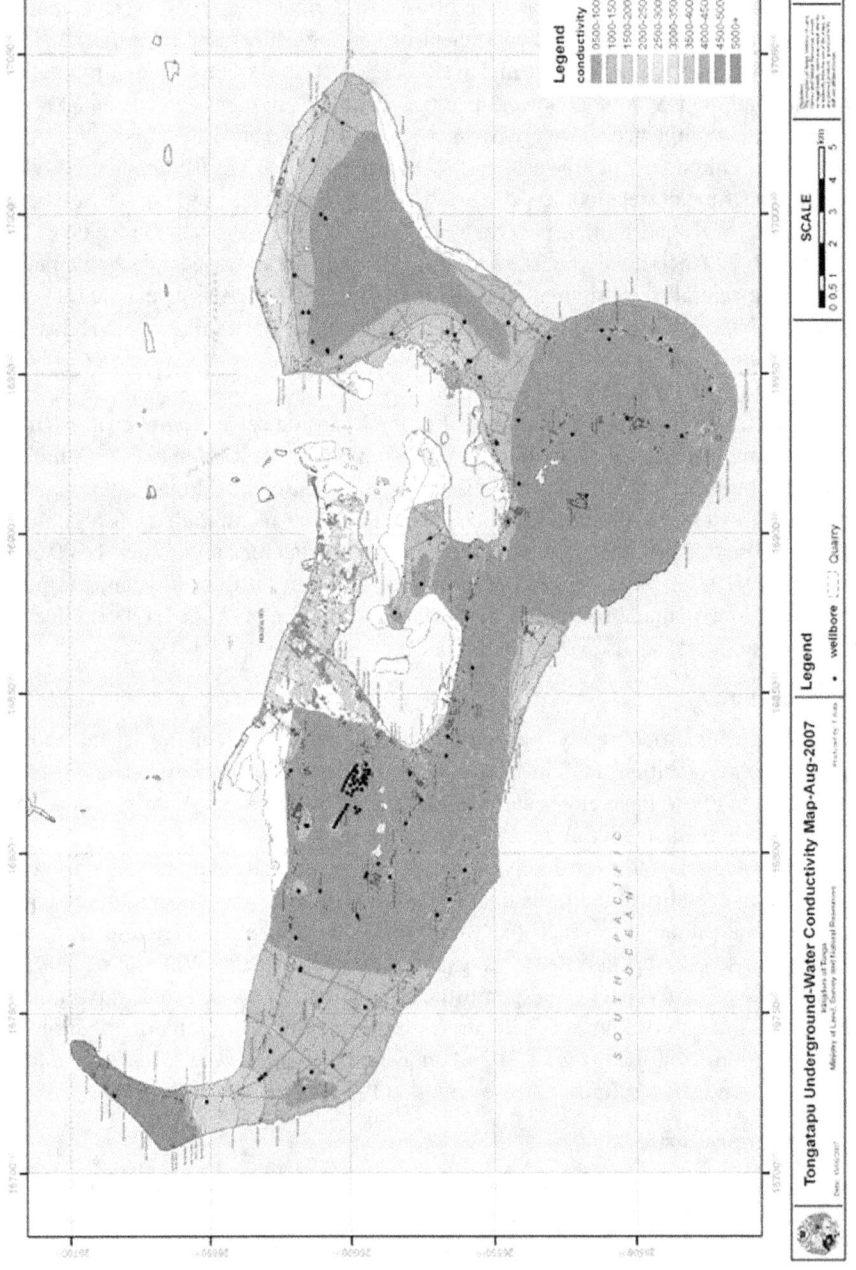

Figure 6.7. Salinity (EC in µS/cm) distribution in the top 2 m of the groundwater sampled in vertical wells and bores (black points) in Tongatapu, Kingdom of Tonga, in August 2007 (White et al. 2009).

On atoll and reef islands the aquifer material consists of two significant layers. The upper layer consisting of recent Holocene sediments, mainly coral sands and fragments of coral, lies unconformably over an older Pleistocene karst limestone deposit (Woodroffe 2008). The unconformity, typically at depths of 10–15 m below mean sea level (Fig. 6.1), is one of the main controls to freshwater lens thickness (e.g. Hunt and Peterson 1980; Wheatcraft and Buddemeier 1981; Jacobson and Taylor 1981). Uranium-series dating of the older limestone in Tarawa atoll indicates that it was formed 125,000 years ago (Jacobson and Taylor 1981). The upper unconsolidated sediments have been laid down over about the last 8,000 years with vertical accretion rates of order 5–8 mm/a. The freshwater zone is generally contained in the relatively low permeability coral sediments (with typical hydraulic conductivities of 5–20 m/day) as mixing of freshwater and seawater is rapid in the high permeability karst limestone. Figure 6.8 shows the wide range of saturated hydraulic conductivities of the shallow, phreatic aquifers in the unconsolidated Holocene sediments on Bonriki and Buota islands, Tarawa atoll, Kiribati estimated from pump drawdown tests in horizontal skimming wells or infiltration galleries (White et al. 2007b).

There are marked differences in hydrogeological properties of the upper sediments between atoll and reef islands located in areas which are prone to cyclonic activity and others in the central Pacific Ocean where cyclones do not occur. Islands within the cyclone belt such as Funafuti in Tuvalu and Pukapuka, Manihiki, Rakahanga and Penrhyn in the Northern Cook Islands (Fig. 6.2) are characterised, particularly on ocean sides, by coral rubble sediments with a significant proportion of boulders embedded in sands and gravels. By comparison, islands in Kiribati and in the Maldives in the Indian Ocean which are not within cyclone regions have finer sediments with a much higher

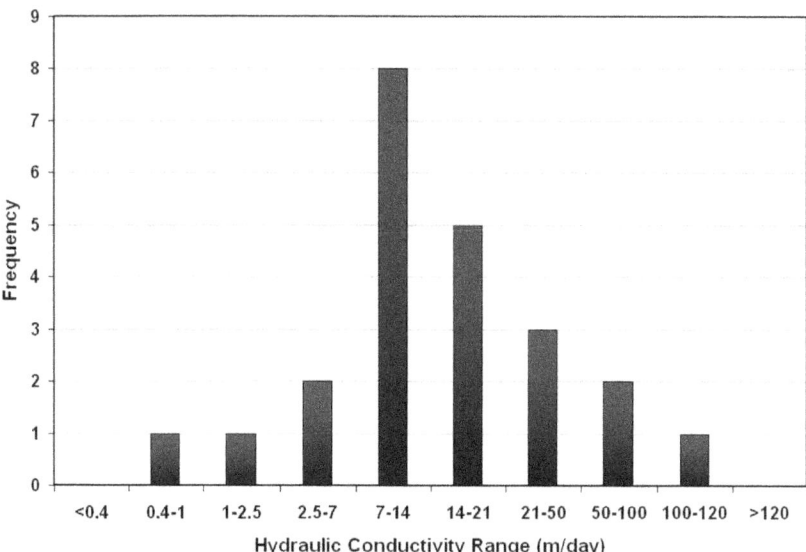

Figure 6.8. Distribution of saturated hydraulic conductivity in the phreatic, unconsolidated Holocene aquifers, Bonriki and Buota islands, Tarawa atoll, Kiribati measured by pump drawdown tests (White and Falkland 2010).

proportion of sand and gravel. These finer sediments have lower hydraulic conductivities and, as predicted by Eq. 1, have thicker freshwater lenses for similar width islands. There are thicker freshwater lenses in the Gilbert chain of islands in the non-cyclonic region of western Kiribati (Fig. 6.2) than in the nearby cyclone-prone islands of Tuvalu, despite the higher and less variable annual rainfall in Tuvalu.

Tidal effects

On small islands, daily fluctuations in sea level, primarily due to tides, cause movement of the freshwater lens and promote mixing of fresh and seawater, increasing the transition zone thickness (Hunt and Peterson 1980; Wheatcraft and Buddemeier 1981, Oberdorfer et al. 1990). The classical theory of tidal signal propagation in continental coastal aquifers predicts that the ratio of the tidal amplitude in the groundwater to that in the sea, called the tidal efficiency, should decrease with distance from the coast. Correspondingly, the lag between the response of the groundwater to the tidal forcing should increase with distance from the coast. That is not the case in atoll islands where tidal lags and efficiencies in wells and boreholes in the unconsolidated sediments are independent of horizontal distance from the shore (e.g. Hunt and Peterson 1980; Wheatcraft and Buddemeier 1981; Ayers and Vacher 1986). Tidal lags and efficiencies on atolls are, however, greatly influenced by the depth of the boreholes. The reason for this apparent anomaly is the rapid transmission of the tidal pressure signal in the underlying high permeability karst Pleistocene limestone. Vertical propagation of tidal signals tends to be dominant in the middle of the island whereas both horizontal and vertical propagation are significant near the seawater margins. This aspect is important in developing conceptual models of groundwater flow in freshwater lenses and in their management, as described later.

The average tidal efficiency and lag on islands with finer textured sediments are typically around 5% and 2.5 hours, respectively (Peterson 1997; White et al. 2002) while those with coarser sediments can be approximately 45% and 2 hours (Falkland 1999). In karst limestone islands, such as Nauru (Fig. 6.2), the tidal efficiency is generally higher, nearly 50% and the tidal lags shorter, around 1.5 hours, than on atoll islands owing to the higher permeabilities.

Soils and vegetation

Soils and vegetation are important in groundwater recharge through their influence on evapotranspiration and infiltration. The high permeability soils of atoll, reef and limestone islands promote rapid infiltration and lead to negligible surface runoff. Atoll and reef islands have only a thin surface soil covering above coral sands, which are generally deficient in organic matter and nutrients. They have low water retention capacity and offer very little protection to underlying freshwater lenses from surface pollution. Soils on limestone islands are generally similar. Some have moderately thick to thick volcanic soils, such as the limestone islands of Tonga, due to past volcanic eruptions and these offer better protection from surface pollutants (van der Velde 2006).

The vegetation on small carbonate islands generally consists of a variety of trees, particularly coconut trees, and a limited range of bushes and grasses. The coconut tree is remarkably salt tolerant and can grow in water with relatively high salinity levels (Foale 2003). Deeper rooted tree species such as coconuts can act as phreatophytes, transpiring directly from shallow groundwater. Both interception and transpiration decrease recharge and hence the amount of groundwater available for use. In dry periods, direct transpiration from groundwater significantly reduces the available groundwater. Measurements of individual coconut tree transpiration showed daily rates of 150 litres per tree and above on Tarawa atoll (White et al. 2002). The high transpiration rate of coconut and other trees has management implications for freshwater-scarce areas where demand is high and groundwater abstraction needs to be maximised. Selective clearing can increase both recharge and the sustainable yield and decrease groundwater salinity. Many subsistence island communities, however, depend on these trees.

6.2.2   Threats to fresh groundwater

Freshwater lenses and coastal aquifers on small islands are vulnerable to threats from both natural events and human activities

Natural threats

Current priorities focus on the predicted threats of climate change. Small island communities in the Pacific, however, have had to live with natural threats for at least the last 4,000 years. The main natural threat to fresh groundwater lenses on small islands are extended droughts (Scott et al. 2003) and, for low lying islands, partial or complete overwash from storm waves or storm surge particularly those associated with major tropical cyclones (Terry 2007; Spennemann 2006). Tropical cyclones are a major problem for many small island communities (Terry 2007). They often cause widespread damage and can generate storm surges with overwash of parts or all of some islands, resulting in seawater intrusion into freshwater lenses. Climate variability associated with inter-annual El Niño and La Niña cycles has significant impacts on groundwater availability in small Pacific islands. There are major concerns that climate change may increase the severity and frequency of these threats in small islands with increased frequency of drought, enhanced cyclone activity, rising mean sea levels and increased risk of island overtopping (Ali et al. 2001).

Droughts

Droughts in the Pacific are closely associated with El Niño and La Niña episodes (White et al. 1999a; Scott et al. 2003). Islands in the southern and northern part of the Pacific are largely drought-affected during El Niño events such as Tonga (van der Velde et al. 2006) while those in the central Pacific, particularly Kiribati, are impacted by droughts during La Niña events (White et al. 2007b). As discussed above, during droughts the fresh groundwater lens contracts. Figure 6.9 shows the relationship between the depth of the freshwater lens at the edge of a coral atoll island in the central western Pacific and La Niña and El Niño events identified by the SOI.

The effect of droughts on recharge results in a non-linear behaviour where a threshold of rainfall is required before any recharge occurs as illustrated for the raised island of Tongatapu, Tonga in Figure 6.10.

In small islands, where mean rainfall is relatively low and where annual rainfall has a high CV, such as in Kiritimati Island, Kiribati (Fig. 6.2), only large freshwater lenses remain viable at the end of major droughts. Some severe droughts have forced the abandonment of several very small islands when fresh groundwater was exhausted.

Sea overtopping

Overtopping of low islands from storm waves or by seawater inundation due to storm surges, sometimes associated with high sea levels has salinised fresh groundwater on low-lying islands (Richards 1991; Oberdorfer and Buddemeier 1984). Six months after

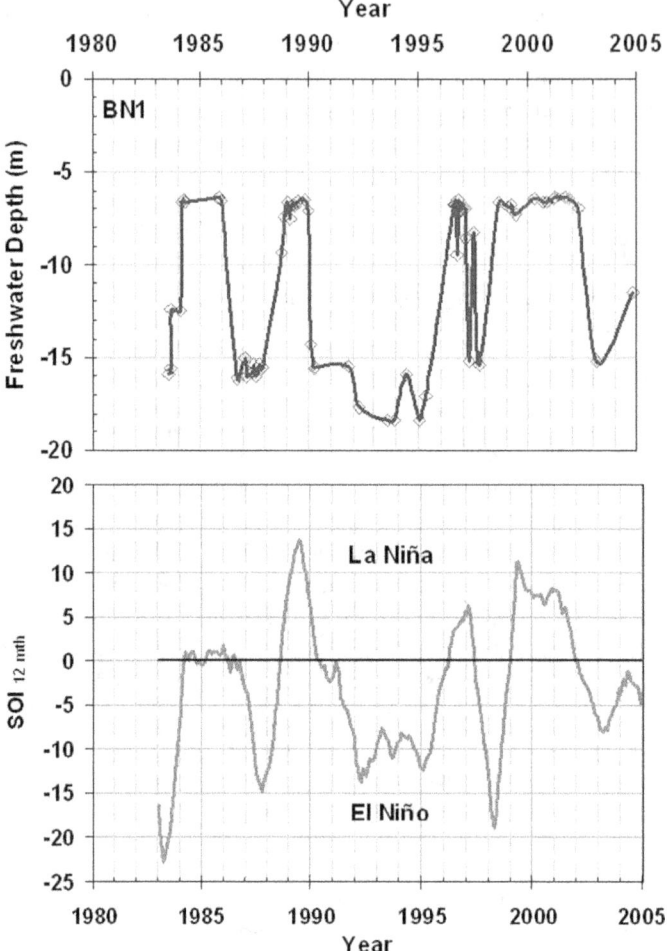

Figure 6.9.   Relationship between the depth of the freshwater lens below the land surface at the ocean edge of Bonriki island, Kiribati and La Niña and El Niño events identified by the average SOI over the previous 12 months (White and Falkland, 2010).

a storm surge which sent waves across part of Enewetak Island, Enewetak atoll, Marshall islands the salinity of the groundwater dropped sharply to 15–25% of immediate post-storm values a period during of negligible rainfall (Oberdorfer and Buddemeier 1984), indicating that recharge was not the factor which decreased salinity. More recently, saline intrusion into freshwater lenses as result of cyclone-generated waves and storm surge on the three islands on Pukapuka atoll, northern Cook Islands in 2005 was found to have dissipated within 12 months due to density-driven downward migration of the seawater (Terry and Falkland 2010).

Sea-level rise

The impacts of droughts and overwash on freshwater lenses are temporary, as illustrated in Figure 6.9. Freshwater lenses recover over periods of months or years after droughts

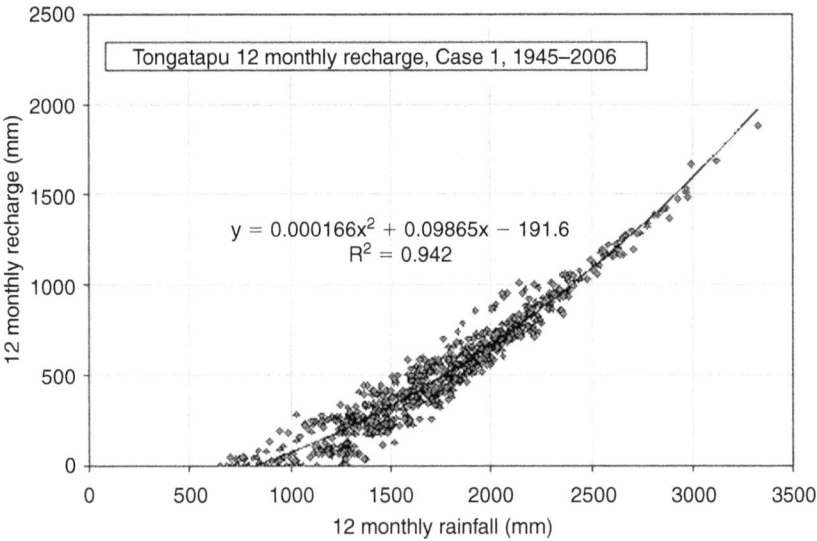

Figure 6.10. Non-linear relationship between estimated 12 month recharge and rainfall for Tongatapu, Tonga for every 12 month period between 1945 and 2006. The relationship suggests that 12 month rainfall needs to exceed 820 mm for significant groundwater recharge to occur (White et al. 2009).

or seawater inundation following recharge from significant rainfalls. More permanent changes occur with rising sea level and this is a major concern for the communities on low-lying PICs as a consequence of climate change-induced sea-level rise (Ali et al. 2001; Burns 2002). Figure 6.11 shows the predicted impacts of sea-level rise on an islet in Tarawa atoll, Kiribati.

The impacts on freshwater lenses from projected mean sea level rises and possible changes in recharge have been studied on several atoll islands using groundwater models. The two-dimensional model SUTRA (Voss 1984; Voss et al. 1997) has been used to analyse impacts for Enjebi Island, Enewetak atoll (Oberdorfer and Buddemeier 1988) and Bonriki island, Tarawa atoll (Alam and Falkland 1997; World Bank 2000; Alam et al. 2002). Sea level rises of up to 1 m would have little impact on freshwater lenses provided that land was not lost at the edges of the island. Indeed the freshwater zone was predicted to slightly increase in thickness and volume as more of the freshwater lens will be within the upper, lower-permeability, Holocene sediments. However, when land is lost due to erosion at the edges of an island (Fig. 6.10), then the island area is reduced decreasing the volumes of freshwater lenses. Potential changes in recharge resulting from changes in rainfall were found to be more likely to have a larger impact on freshwater lenses. Further work is required to assess the relative vulnerability of different shorelines to sea-level rise (Woodroffe 2007).

In addition to the impacts of droughts, overwash and potential sea-level rise, extreme events such as tsunamis and earthquakes can also impact on small islands and cause disruption to groundwater resources. Groundwater in many low-lying lands in the Indian Ocean was impacted following the Boxing Day tsunami in 2004. Tsunamis, which have devastated many islands and continents around the Pacific rim, are not normally a major

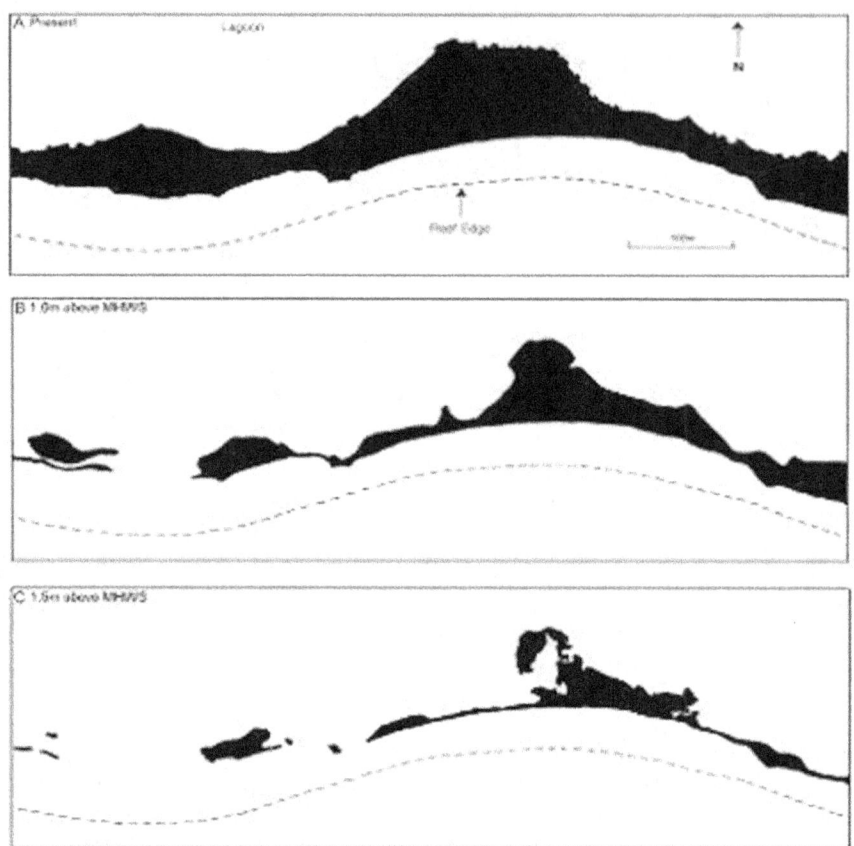

Figure 6.11.   Projected inundation of Bikenibeu island, Tarawa, under the Worst-Case Emissions Scenario. a: Present status, b: Residual island under a worst case scenario in 2100; c: Worst case scenario and storm surge in 2100 (World Bank 2000).

threat to mid-oceanic islands except where islands are close to tectonically active areas, such as Tonga. Submarine landslides can also cause catastrophic changes to atolls with the loss of whole or parts of islands as occurred, for example, on the atolls in the northern Cook Islands (Hein et al. 1997).

Climate change, rainfall, and groundwater recharge

The impacts of climate change are a predominant concern in PICs. As can be seen from the above, the impact of climate-change induced sea-level rise is the highest concern, particularly to low-lying atolls. The impacts of climate change on future weather and particularly rainfall recharge of groundwater is also of major concern in nations with already stressed water supply systems.

The problem of predicting the impacts of climate change on future rainfall in PICs is complex. This is because simulations using ocean-atmosphere general circulation models (GCMs) are not presently carried out at sufficiently fine horizontal scale resolution. Their ability to generate climate change scenarios for Small Island States are therefore limited

(Ali et al. 2001). *"GCMs used to predict the impacts of green-house gas emission scenarios on future climates) are not good at simulating changes to the hydrological cycle and are notoriously bad on rainfall, especially in the tropics. There are two basic reasons for this: (i) they generally don't simulate tropical convection very well, and (ii) they can't repro-duce some of the major modes of current climate variability, including El Niño- Southern Oscillation (ENSO)."* (Will Steffen, private communication, 23 February 2009).

In PICs, the increase in average annual temperature has been less than 0.5°C since 1900. Rainfall records across the Pacific for 1900 to 1995 reveal no clear general trend (Ali et al. 2001). The records do, however, show decadal fluctuations of mean annual rainfall linked to ENSO fluctuations or sea surface temperatures (Fig. 6.6). The surrounding oceans, none-the-less, have a strong influence on the climate of Pacific islands. The Pacific Ocean is predicted, with a doubling of atmospheric carbon dioxide (CO_2) concentration, to warm in the future by 1 to 2°C and mean rainfall intensity in small islands may increase by about 20–30% across the tropical oceans (Ali et al. 2001).

Unfortunately, the rainfall record in most small island countries is too short to reveal any major long-term trend but mid-western Pacific country records do show a sustained drying period followed by a wetter period possibly linked to the Pacific decadal oscilla-tion. Any trend in rainfall with increasing global temperatures is difficult to discern.

One of the major concerns with global warming is that increasing sea surface temperatures will have significant impacts on the frequency of ENSO events with corre-sponding major impacts on periods of intense rainfall and droughts in the Pacific. Some GCMs predict increasing rainfall variability in the Pacific as a result of rising atmos-pheric temperatures. Since, however, ENSO is a major driver of rainfall variability in Tarawa and since current GCMs cannot reproduce ENSO events; these predictions have to be treated with extreme caution. A recent study using 14 GCM's in attempt to predict the impacts of climate change on rainfall, evaporation and groundwater recharge in the island of Tongatapu, Tonga, concluded that the GMCs used in the study are not capable of predicting future changes in drought frequency, severity or duration or possible changes in groundwater recharge and that the observed trend in evaporation was opposite to that predicted (White et al. 2009a).

Threats from human activities

The main human threats to freshwater lenses are over-abstraction of groundwater and pol-lution from surface sources, particularly human, animal and industrial wastes and spillages. Other threats include mining of sand and gravel for building materials from groundwater source areas and shoreline works which induce erosion. Rapidly expanding populations due to both natural growth and inward migration to urban centres are placing increasing demands on water supply systems which abstract groundwater from freshwater lenses. This is especially noticeable in population centres as, for example, on Tarawa and Kiritimati atolls in Kiribati and, to a lesser extent, on Tongatapu in Tonga (van der Velde 2006). Significant losses, sometimes up to 70%, place additional stress on the limited groundwater resources.

Over-pumping

Over-abstraction can be island-wide or localised. Localised over-abstraction is generally caused by inappropriate methods of pumping from vertical boreholes, which increases salinity through up-coning of the transition zone (Fig. 6.1). Island-wide abstraction at

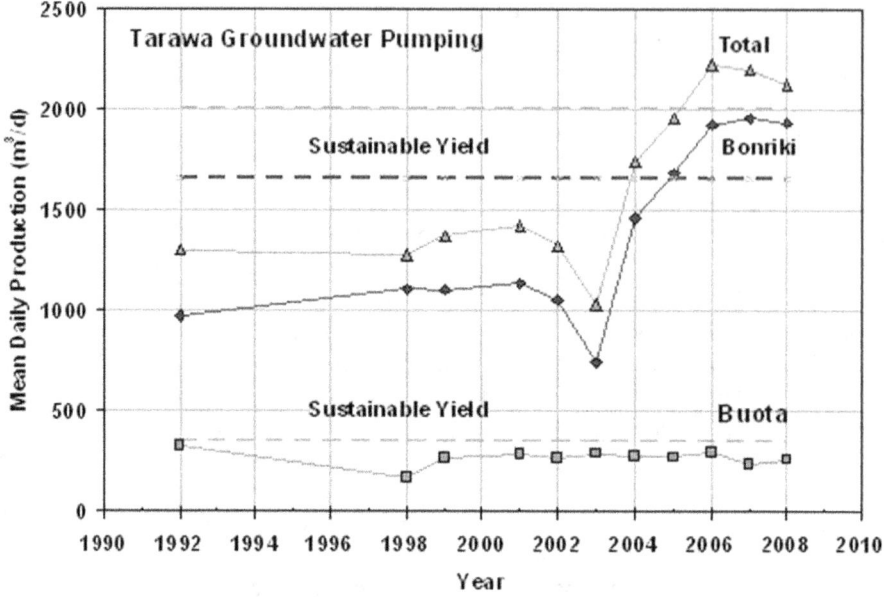

Figure 6.12.   Annually averaged mean daily groundwater pumping rates using infiltration galleries from freshwater lenses in the islands of Bonriki and Buota, Tarawa atoll, Kiribati (solid lines and points) compared with estimated sustainable yields (dashed lines) [White and Falkland, 2010].

greater than the sustainable yield of the island can be due to poor understanding, insufficient information or lack of regular monitoring, use of inappropriate pumping systems or demand pressures forcing management to pump at higher than sustainable rates. Figure 6.12 shows the change in pumping rate since 2004 compared to the estimated sustainable rate at Tarawa atoll's groundwater sources when pumps at existing infiltration galleries were incorrectly refurbished.

One good indicator of over-pumping is an increase in salinity of the produced groundwater (discussed below). In some cases salinisation due to over-pumping can be so severe that the produced water is unfit for any use except flushing toilets.

Contamination

On Tarawa, population pressures and the limited land area for urban development, places additional pressure on the water reserve areas of Bonriki and Buota islands used for groundwater abstraction. Illegal settlements and inappropriate land uses such as on the designated water reserves is a major problem (White et al. 1999b, White et al. 2007a). Figure 6.13 shows the impact of encroachment by people on the quality of water produced from infiltration galleries in the Bonriki water reserve.

Groundwater contamination in PICs, caused by a variety of biological and chemical sources including sanitation systems, particularly pit toilets and septic tanks, animal wastes, rubbish disposal areas, cemeteries, leaking fuel tanks, fertilisers and agricultural chemicals (van der Velde et al. 2007), poses significant health risks. In some countries the practice is for burials within the home block of ground often adjacent to shallow groundwater wells. Detay et al. (1989) comprehensively reviewed pollution

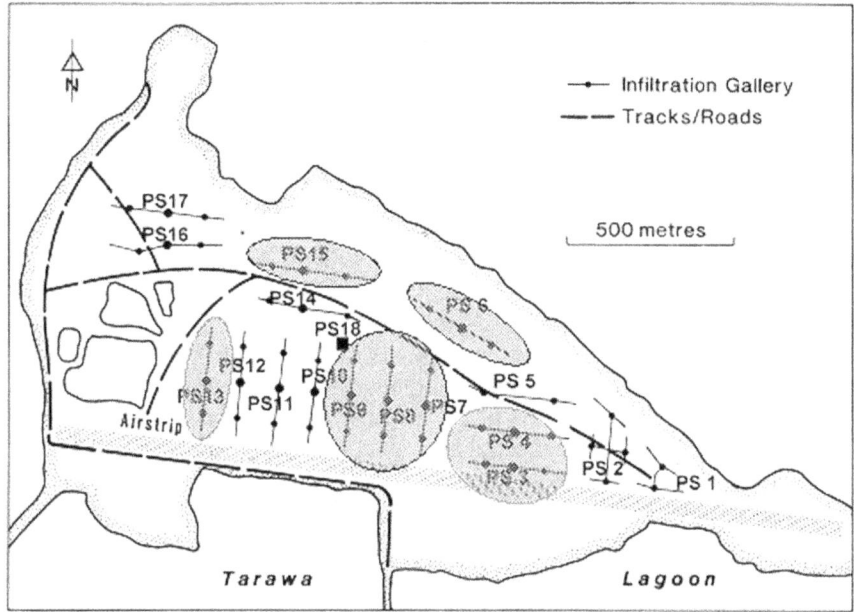

Figure 6.13.   Distribution of positive E. Coli water samples from pumping galleries on Bonriki (White et al. 2005).

problems in small islands in the Federated States of Micronesia, the Marshall Islands and Belau (Fig. 6.2), many of which are atoll islands.

Human settlements over freshwater lenses are of major concern because of the potential for rapid pollution due to the shallow permeable soils and short travel times to the water table. This has caused the contamination of large areas of urban Tarawa, Kiritimati and Nauru so that groundwater is only fit for non-potable purposes. Mining of sand and gravel from freshwater lens areas creates additional risks by decreasing the soil depth over the water table and increasing vulnerability to pollution and, in some cases, increasing evaporation losses directly from exposed groundwater, even in higher islands such as Tongatapu in Tonga (White et al. 2009). The problems caused by human settlements are exacerbated by high population growth rates and inward migration to urban centres.

Land ownership

Land ownership is also a central problem for governments in many PICs with virtually all land owned by traditional owners. This means the creation of water reserves on privately owned land to protect groundwater sources creates conflicts between governments and landowners sometimes resulting in vandalism of water infrastructure (White et al. 1999b) or blockades. Land ownership is fundamentally culturally important as it is the primary source of wealth conferring subsistence rights, including fishing rights, on land owners and also providing a social security system for parents through the prospect of inheritance by their children who are therefore obliged to care for their parents.

Customary law in many PICs assigns ownership of groundwater to land owners. Because of that governments are reluctant to enact water legislation specifying that water belongs to all people (the government) or banning polluting land uses for fear of infringing property rights. As a consequence, in some PICs there is no legal protection of groundwater from over pumping or from contamination or misuse. Also in some cases, more than adequate groundwater exists in some regions of an atoll but its transfer to other parts in severe deficit is virtually impossible of land ownership issues.

Inadequate governance

In addition to these direct threats, fresh groundwater can also be especially vulnerable because of inadequate legislation and regulations to protect water sources, inappropriate national water policies providing no clear priorities or directions to government ministries and agencies or assignment of responsibilities and very limited financial and human resources to manage water resources and water supply systems. Many PICs have no national water policies or water laws or regulations. Indeed in some, there is no systematic policy and planning process at all and poorly resolved mechanisms for policy and plan implementation save ministerial fiat.

Despite their small size, government ministries and agencies in PICs usually act in isolation with no or limited collaboration between organisations with responsibilities in water even in single island nations. In addition, there is frequently no participation of communities in water resource management and planning because of a disconnect between government ministries and communities. Frequently there are only one or two water specialists in small island nations so the provision of appropriate training to water resource personnel is a critical need in PICs as is succession planning so that acquired knowledge can be passed on.

Inadequate knowledge

Despite their vulnerability, many PICs do not know the full extent and quality of their water resources or their sustainable yields. In order to conserve and manage their fresh groundwater sustainably and to protect the security of vulnerable groundwater supply systems a clear understanding of the location, extent, sustainable yield and the impact of both climate variability and human activities on the groundwater system is required.

The extreme fragility of small island groundwater systems and the increasing demands on them need thorough resource assessment, a commitment to ongoing monitoring and analysis. The importance of monitoring was exemplified in one small island nation where, during the widespread 1998–2001 drought a state of disaster was declared despite the fact that the main groundwater supply lenses had more than adequate freshwater. The lead water agency had stopped monitoring groundwater during the drought so the actual availability of water was unknown. The problem they faced was inadequate water distribution not declining supply.

Knowledge is also required to develop groundwater appropriately to minimise salinity and for effective management of groundwater source areas, targeted demand management, public participation, training and mentoring of staff and attentive management of the water abstraction and supply systems. The limited human and financial resources in many island counties mean that these are major challenges (van der Velde et al. 2007).

6.2.3   Reducing the vulnerability of groundwater systems

While the natural and human threats to groundwater described above may seem daunting, there are a range of strategies that can be implemented to reduce the vulnerability of groundwater and water supply systems in PICs (Falkland 2002).

Severe droughts

The length and severity of droughts in many PICs (Fig. 6.6) mean that it is vitally important to develop strategies to survive droughts. Several that can be identified include:

- Develop or use reliable early warning systems for the onset of droughts
- Early report of the risk of drought to Governments and their Ministries and agencies
- Early warning to the community of the risk of drought
- Develop a drought contingency plan
- Control demand and leakage during droughts
- Conjunctive use of groundwater, rainwater and desalinated water during droughts
- Monitor rainfall, groundwater pumping, water use, leakage and groundwater salinity.

The very strong relation between sea surface temperature, SOI and rainfall and the amount of groundwater available in many PICs (Figs 6.6 and 6.9) and SOI provide a basis for predicting the probability of below average rainfall several months in advance. The SCOPIC program developed by the Australian Bureau of Meteorology for PICs (http://www.bom.gov.au/climate/pi-cpp/scopic.shtml) is designed to provide seasonal climate forecasts in the Pacific, three months in advance. The decrease of rainfall percentiles below the 40 percentile (Fig. 6.6) also provides an early warning system for the onset of severe droughts with a mean lead time of 6 months but with an accuracy of only 50% (White et al. 1999b).

The use of SCOPIC or any other early warning method, however, has to be accompanied by a plan of action which alerts the government and appropriate authorities and agencies, as well as the community and needs to be coupled to strategies for increasing conservation of water and decreasing demand. Reducing the continuing large water losses from the domestic reticulation systems would be a major contribution to conserving water.

The certainty of drought in the central Pacific makes drought contingency plans essential so that droughts are not treated as extraordinary, emergencies. Such a plan should include:

- An effective early drought warning system
- An effective public drought risk communication and water conservation strategy
- A list of highest priority consumers such as hospitals and schools and the disabled and elderly
- A strategy to isolate excessively leaking parts of reticulation systems and to supply water from fixed distribution points or by tanker
- Ensuring that there are sufficient, well maintained water tankers to deliver water to community or household tanks where they are installed
- Increased frequency of monitoring of groundwater sources
- Ensure there is adequate legislative or regulatory basis for emergency interventions

Some Pacific small island nations have no means of controlling demand other than by supplying water intermittently. The inequity in these systems is apparent. Several have unmetered reticulation or supply systems so there is either a bulk charge or no charge at all for water. In many cases the charge is only a small fraction of the cost of water production so the systems are financially unsustainable. Even in systems with metered systems unaccounted for water (water losses) can be as high as 90%. In severe droughts it may be necessary to disconnect the leaky domestic reticulation system and supply water either from fixed location distribution centres, such as the village tanks or by tanker delivery. This would reduce the large losses from the domestic reticulation pipes and would allow revenue collection from distribution centres, as is often done for bulk water tanker deliveries. This, of course would disadvantage the poor.

Rainwater harvesting has the potential to supplement groundwater drawdown. In general however, domestic household rainwater harvesting systems have too small rainwater tank capacities, too small roof collecting areas and too high a demand on them in PICs to last through significant droughts. Larger rainwater storages connected to bigger public buildings such as churches, schools, meeting halls and government buildings have the potential to act as emergency supplies during drought, as occurs in Tuvalu (Fig. 6.2).

In some small island countries such as Nauru (Fig. 6.2), there are insufficient groundwater (Fig. 6.5) and rainwater storages to last through severe droughts. Because of this desalination of seawater is used to supplement these sources. This is an expensive but necessary option.

The principal way of storing freshwater in coral atolls is in groundwater lenses. While groundwater lenses in narrower islands and some high islands, such as Nauru, tend to become brackish or saline during droughts, those on larger low islands can survive long droughts well. The Bonriki and Buota groundwater lenses in Tarawa atoll are prime examples. The 1998 to 2001 drought is the worst on record for Tarawa for 30 month rainfalls.

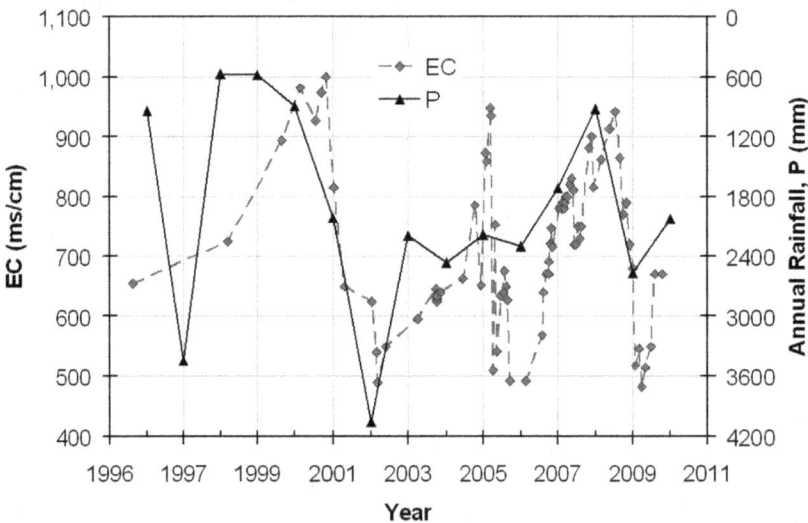

Figure 6.14.   Salinity (EC) of combined groundwater pumped from infiltration galleries on Bonriki and Buota, Tarawa (dashed line), compared with annual rainfall (solid line). The 1998 to 2001 and the 2008 droughts and 1996 and 2002 wet periods are evident as is installation of higher capacity pumps in 2004 (White and Falkland 2010).

During this drought, the thickness of the freshwater lenses had halved at the end of the drought and the salinity (EC) of the water pumped from the combined lenses increased, but not to problem levels (Fig. 6.14). These groundwater lenses were able to supply water to South Tarawa throughout the drought which underlines how important it is to monitor.

Monitoring is a major problem in many PICs because of the lack of trained personnel and shortages of equipment and resources, such as access to transport. A failure to appreciate the importance of adequate water resources monitoring pervades government agencies responsible for water supply and regulation. In some cases, lack of clear definition of roles between government agencies and competition for limited resources are further impediments. Some of these problems are being addressed through projects such as the Pacific HYCOS project (Pacific HYCOS 2009). Ongoing capacity building and training and mentoring of staff in PICs is, however, required (van der Velde et al. 2007). These are poorly handled in many aid projects which are too short term to have lasting impacts (White et al. 2008).

Over-pumping and sustainable groundwater yield

The large natural variations in rainfall and groundwater recharge in PICs (Figs 6.6 and 6.10) mean that determination of the sustainable groundwater yields require careful estimation. Knowledge of groundwater recharge is fundamental to estimating sustainable yields from freshwater lenses. A range of techniques are available for this from using empirical curves relating annual recharge to annual rainfall (Falkland and Brunel 1993) through chloride ion balance approach (Ayers 1981; Chapman 1985; White et al. 2002), measuring the components of the water balance (Falkland 1993; White et al. 2002), continuous watertable and rainfall measurements (Furness and Gingerich 1993), measurements of the drainage fluxes beneath the root zone (van der Velde et al. 2005), and groundwater models, most commonly two-dimensional variable-density models such as the SUTRA model (Voss 1984; Voss et al. 1997). These techniques, however, are beyond the current skills of many PICs water specialists are point to the need for long-term training.

Preliminary estimates of sustainable groundwater yield are usually taken to be a fraction of mean annual recharge. This recognises that only a proportion of recharge (of order 25% to 50%) can be abstracted leaving a significant amount for maintaining the integrity of the lens.

Groundwater modelling to estimate sustainable yield relies on the availability of good quality medium to long-term climatic and groundwater data. Using models with limited data can lead to optimistic estimates of groundwater viability (Fig. 6.5). Table 6.1 shows the progressive estimates of sustainable groundwater yields for the major water supply lenses in Tarawa atoll over the past 40 years. These used a variety of modelling and field investigation techniques. The current estimate is equivalent to just over 43 L/person/day of treated freshwater and demonstrates the importance of accurate estimation of sustainable yield.

Minimising salinity of pumped groundwater

The delicate hydrostatic balance between freshwater and the surrounding and underlying seawater in small islands is easily disturbed by inappropriate groundwater extraction particularly during droughts. The most common method of obtaining groundwater on low-lying, coral islands is from hand-dug dug wells typically two to three meters deep and approximately one metre below groundwater level. Groundwater is abstracted by buckets, hand pumps or small electric pumps. Such household systems work provided pumping rates are low.

For public water supply pumping systems, single or multiple dug wells or drilled boreholes have been used on some coral islands. These vertical abstraction systems cause upconing of the transition zone and increases salinity of the abstracted water. In some cases, the groundwater can becomes too saline for potable use. Pumping from horizontal infiltration galleries or skimming wells has proved a far better abstraction method, particularly in islands with thin freshwater lenses. Infiltration galleries, consisting of up to 300 m long horizontal slotted pipes buried a short distance below the water table (Figs 6.1 and 6.15), skim the fresh groundwater from the surface of the lens, and thus distribute the pumping drawdown over a wide area. In so doing, they avoid excessive local drawdown and upconing of saline water that occurs in pumping from vertical boreholes.

Freshwater lenses contract during major droughts, so it is important that the local impacts on the freshwater lenses caused by pumping are minimised. By maintaining a small drawdown at each gallery pump well, the impact of pumping on the freshwater lens and on the salinity of abstracted water is minimised.

Measurements in twenty five Bonriki and Buota gallery pump wells in Tarawa atoll, Kiribati have shown that the average groundwater drawdown of all approximately 300 m long galleries is 33 mm when pumped at mean rates of 88 m^3/day (White et al. 2007b). This drawdown is less than the magnitude of daily groundwater fluctuations due to tidal

Table 6.1   Successive estimates of the daily sustainable groundwater yield from Bonriki and Buota groundwater reserves, Tarawa atoll, Kiribati.

Year	Estimates of Sustainable Yield (m^3/day)			
	Bonriki	Buota	Combined	Reference
1973	110			Mather (1973)
1978	<85	<85	<170	Richards and Dumbleton (1978)
1982	750	250	1,000	DHC (1982)
1992	1,000	300	1,300	Falkland (1992)
2002	1,350	350	1,700	Alam et al. (2002)
2004	1,660	350	2,010	Falkland (2004)

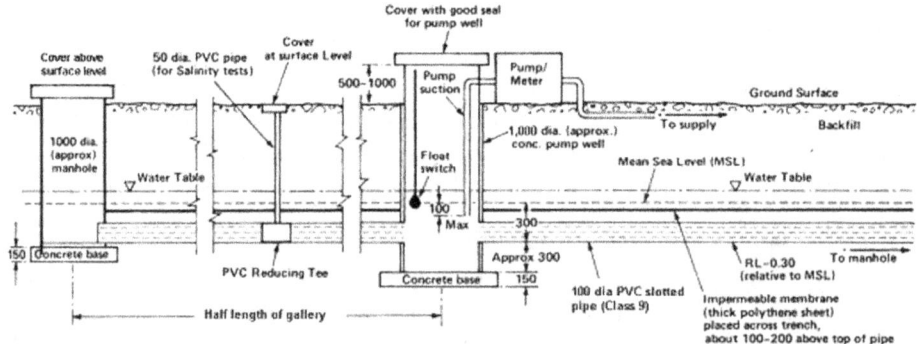

Figure 6.15.   Cross section through a typical infiltration gallery or skimming well designed to minimise drawdown and upconing of saline water [modified from Falkland and Brunel (1993) with permission from Cambridge University Press].

influence there of typically 70–80 mm). It is also much less than the longer term ground-water level fluctuations of about 450 mm between very wet and very dry periods.

For limestone islands, where depths to the groundwater table are greater than 10 m and up to 50 m or more, abstraction using vertical drilled boreholes is currently the most practical method of developing fresh groundwater lenses. In the future, directional drilling from the surface may be an option for installing horizontal infiltration galleries on these islands.

Reducing contamination

Effective land use planning and management is essential to protect shallow groundwater resources from contamination on low-lying carbonate islands. 'Groundwater protection zones' or 'water reserves' coupled with regulation of land uses has been trialled in some PICs. Human settlements, agriculture involving raising of livestock or intense, the use of chemicals and fertilisers and mining of sand and gravel all increase the risk of groundwater contamination. Such reserves, however, are often difficult to manage owing to pressures on limited land areas in small islands and the problem of ownership of land and water. The resolution of management problems requires appropriate administrative and financial provisions, such as land rental agreements, the involvement of the local community in managing the reserve, or even the water supply system as in Tongatapu, and the provision of social services such as sports fields on the reserves. Rather than land rental payments it has been proposed that land owners be contracted to be reserve managers (White et al. 1999).

Sanitation systems using septic tanks and pit latrines are a major source of biological groundwater contamination, particularly of household wells, and health risks on many islands. In parts of some heavily populated atolls such as Majuro, Marshall Islands and Tarawa, Kiribati, piped sewerage systems using seawater for flushing, to conserve limited freshwater supplies, have been installed to overcome this problem. Compost toilets protect freshwater lenses as well as also conserving scarce water resources (Crennan and Berry 2002). While these toilets have many advantages and have been accepted in some communities, cultural attitudes have so far limited their wide spread use in others. Other technical solutions are available including improved septic tanks and relatively simple effluent disposal and treatment systems (UNEP 2002; Bower et al. 2005; and WHO 2008).

The most appropriate strategy is to ensure that human settlements including their sanitation systems are placed well away from groundwater sources used for public water supply. A study using bromide tracer in Lifuka, Tonga concluded there was no safe distance between pit latrines or septic tanks and water supply wells in an urban area because of the density of sanitation facilities (Crennan et al. 1998).

Improving governance

There are no easy prescriptions for the rapid translocation of relatively recent water governance reforms and water management frameworks from developed countries to small island developing countries. For example, in developed countries there are frequently hundreds of people engaged in the planning, management and use of water. For them, some of the major challenges are addressing the environmental impacts of water supply and water effluent treatment systems and planning for the predicted impacts of global and climate change. In many small island countries in the Pacific region there are often only one or two water professionals whose tasks may range from replacing washers in domestic taps, replacing groundwater pumps, and unblocking clogged sewers to advising on national water policy and representing the country at international climate change meetings. The major daily challenges in water governance they face are maintaining supplies of adequate quantities of safe freshwater to growing populations in dispersed and isolated communities with very limited resources, no economies of scale, as well as coping with the complex cultural and institutional changes necessary in the transition from subsistence to urban communities.

Quick, developed-world formulaic solutions that take no account of island priorities, traditions and practices developed over millennia, are often politely ignored (Dray et al. 2007). Transformation of the water sector in the Pacific involves behavioural change (SOPAC and ADB 2002), generally a long-term process, which requires appreciation of the different nature of freshwater in small island countries and the prevailing culture, traditions and practices.

Policy, legislation and regulation and their implementation are essential over-arching components of integrated water resource management which encompasses health, environmental, economic, social, cultural, infrastructural and technical issues (Fig. 6.3). Because of this, it is essential that all Ministries, agencies, community organisations and businesses with broad interests in freshwater participate in deliberations on policy and plans. The establishment of a whole-of-government and community National Water and Sanitation Coordination Committee (NWSCC) reporting to Cabinet is a key first task in small island countries with Ministries that act as 'silos'. Because many are unfamiliar with policy processes a simplified model of the policy process can be adopted, in which the NWSCC plays a key driving role (Fig. 6.16) (White et al. 2009b).

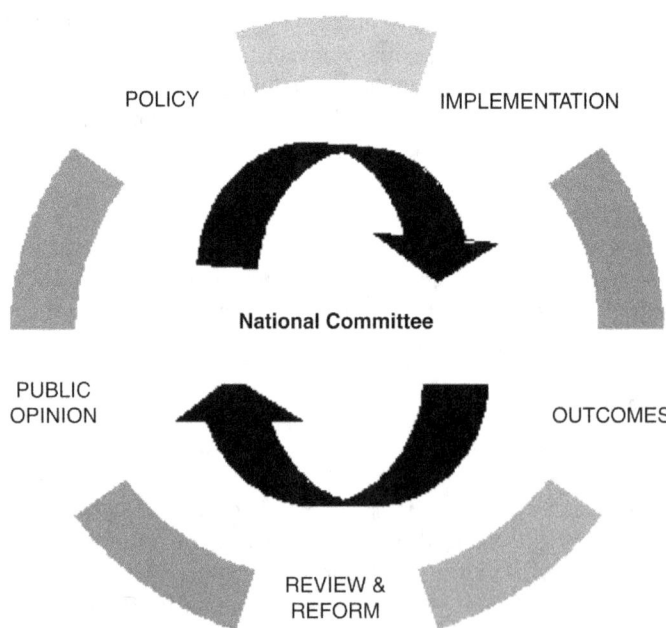

Figure 6.16.   Simplified model of the policy cycle showing the central role of the NWSCC in considering public opinion, implementing, reviewing and recommending revisions to policy.

A five step adaptive planning process based on Ackoff (1999) has been used to develop policy and their accompanying implementation plan. To simplify the process, phases I and II are incorporated into a policy document and phases III, IV and V into an implementation plan. Phase I of the process identifies any needs for national water legislation or regulation. This process was used in assisting Kiribati in developing its National Water Resources Policy and 10 year Implementation Plan which was endorsed by Cabinet in February 2009 (White et al. 2009b) and is being trialled in Nauru.

In small island countries the time required for the policy and planning process can be longer than expected. The pressing day-to-day problems faced by Ministry staff and the limited number of staff meant that reading lengthy reports is of secondary priority. In the islands, oral transmission is the customary way of transmitting information. It is therefore necessary to have personal discussions with staff together with more frequent and widely based workshops to explain the process and the contents.

Finally, attracting, training and retaining skilled staff in sufficient number in the water sector in Pacific small island states remains a problem. Fortunately, there are regional organisations such as the Applied Geoscience and Technology Division (SOPAC 2005; 2006) of the Secretariat of Pacific Community that provide a resource and expertise base for Pacific Island Counties. Where aid donors can assist is in providing long-term training opportunities for small island students in water resource and supply management.

6.3 FUTURE WORK

This chapter has attempted to show that the climatic, hydrogeological and physiographic factors compounded by human activities make fresh groundwater used for water supply on small islands in the Pacific some of the most vulnerable groundwater systems in the world. Some small island population centres are already close to the limit of sustainable groundwater abstraction. Faced with climate change, rising sea levels, increasing frequency of extreme events, growing populations, restricted land areas, limited resources and capacity, their prospects appear bleak. Yet island populations have proved remarkably resilient and communication technologies now mean that their isolation is lesser of a problem when it comes to knowledge about groundwater management.

It has been argued here that, because small island groundwater systems are so vulnerable, knowledge is a key to successful and sustainable management both now and in a climate change-impacted future. Already the probability of below average rainfall in many Pacific small island states can be predicted with a good degree of success three months in advance, particularly using remotely sensed sea surface temperature data. Further developments there are likely to improve contingency planning. These coupled to groundwater recharge models should improve the ability to small island states to manage groundwater more precisely when coupled with careful climate, groundwater and land use monitoring. There is potential for higher resolution remote sensing to assist in some aspects of monitoring.

We have also argued here that despite the lack of established policy processes in PICs, a process can be used to develop a water policy and planning framework which has been successfully trialled in one PIC. Because of their small population numbers, many PICs have the potential to form integrative government-community national water and sanitation committees which cut across all the sectors in Fig. 6.3. The involvement

Table 6.2   The five phases of interactive planning. Adapted from Ackoff (1999).

Phase	Objective	Components	Principle Outputs
I. Formulation of the Issues	Determine problems and opportunities	Problems, opportunities, and their interactions; constraints to effective management	Issues to be addressed by policy, plans, legislation
II. Ends Planning	Determine where you want to be and the gaps between that and now	Extract vision, principles, goals, and objectives matched to issues to achieve the desired ends.	Policy principles, Policy goals and objectives
III. Means Planning	Choosing mechanisms to achieve goals and objectives	Develop and select actions for achieving goals and objectives and indicators for completion of actions	Implementation Plan Actions
IV. Resource Planning	Determine resources required for planned actions	Define resource needs and identify if resources are available or how they will be generated or acquired	Implementation Plan Resources needs
V. Implementation and Control	Determine responsibilities and schedules for implementation	Identify who is responsible for actions, when they are to be implemented and how implementation is to be monitored	Implementation Plan Schedule and Responsibilities for implementation. Ministerial Operations Plans

of community representatives in such committees is essential. They have the potential to drive and oversee policy initiation, implementation planning and legislative and regulatory reform using a simplified model of the policy process (Fig. 6.16) and the adaptive planning process outlined in Table 6.2. If successful these might present a process for more developed countries.

Many of the pressing future problems in PICs can be addressed through six policy objectives:

- improve understanding, assessment and monitoring of water resources and their use;
- increase access to safe and reliable water supplies and appropriate sanitation;
- achieve financially, socially and environmentally sustainable water resource management;
- increase community participation in water management and conservation;
- improve governance in the water and sanitation sector with regular reviews of policy outcomes and planning milestones; and
- provide training opportunities for and mentoring of staff in the sector.

External donors and lenders have assisted and continue to assist PICs to develop and manage their groundwater resources. A number of aid and loan projects in the water sector, however, have had only limited success. Almost all projects have been relatively short-term and narrowly focused on infrastructure, and some have been driven by the agendas of international agencies rather than the priorities of local populations. Many have also assumed that developed world solutions, concepts and "tool boxes" are universally transportable and applicable. Donor and aid programs that are sensitive to cultural nuances, recognise local priorities, value effective community participation, mentor staff in the water sector and appreciate the long time-scale for behavioural change are more likely to be successful (White et al. 2008).

Village-level water committees have proved successful in rural areas in Tonga and Samoa and appear to offer a model for other PICs. They are appropriate for the cultural contexts in many PICs and would help return control of the protection and management of water resources at the local level in rural areas and outer islands. Such committees need, however, to be nurtured and resourced adequately.

The increasing complexity of water management as islands approach the limit of sustainable yield and the future impacts of climate change will pose difficulties for small islands, where sometimes only one or two people are responsible for water management. Regional organisations which pool expertise and local experience and provide training opportunities have and should continue to have a key role in supporting PICs in water management. It is important that these regional organisations are incorporated as partners into aid and donor programs in the water and sanitation sector in the Pacific.

ACKNOWLEDGEMENTS

Parts of this work were supported by the Kiribati Adaptation Program Phase II supported by AusAID, NZaid and the World Bank, by the European Union, by the Applied Geoscience and Technology Division (SOPAC) Pacific Programme for Water Governance, by the European Development Fund 8 Reducing the Vulnerability of Pacific APC States administered by SOPAC, by the Australian Centre for International Agricultural Research Project LWR1/2001/050 and by the Australian Research Council's Grant LX0214943. This work was initiated under the Humid Tropics Programme of UNESCO IHP V. We are grateful to our colleagues throughout the Pacific who have generously worked and shared with us and have taught us a great deal.

REFERENCES

Ackoff, R.L. (1999) *Ackoff's Best. His Classic Writings on Management.* New York, John Wiley and Sons.

Alam, K. & Falkland, A. (1997) *Vulnerability to Climate Change of the Bonriki Freshwater Lens, Tarawa.* Prepared for Ministry of Environment and Social Development, Tarawa, Republic of Kiribati, September 1997.

Alam, K., Falkland, A. & Mueller, N. (2002) *Sustainable Yield of Bonriki and Buota Freshwater Lenses.* SAPHE Project, Hydrogeology Component. Tarawa, Republic of Kiribati, February 2002.

Ali, M., Hay, J., Maul, G. & Sem, G. (2001) *Chapter 9: Small island states.* In: Watson, R.T., Zinyowera, M.C., Moss, R.H. & Dokken, D.J. (eds.) *IPCC Special Report on the Regional Impacts of Climate Change. An Assessment of Vulnerability.* UNEP and WMO. [Online] Available from: www.grida.no/publications/other/ipcc_sr/?src=/climate/ipcc/regional/index.htm [Accessed 9th October 2011].

106   *Ian White & Tony Falkland*

Ayers, J.F. (1981) *Estimate of recharge to the freshwater lens of northern Guam*. Technical report number: 21. Water Resources Research Center, University of Guam, Mangilao, Guam.

Ayers, J.F. & Vacher, H.L. (1986) Hydrogeology of an atoll island, a conceptual model from a detailed study of a Micronesian example. *Groundwater*, 24(2), 185–198.

Badon Ghijben, W. (1889) Nota in verband met de voorgenomen putboring nabij Amsterdam [Notes on the probable results of the proposed well drilling near Amsterdam]. Tijdschrift het Koninklijk Instituut voor Ingenieurs, The Hague. pp. 8–22.

Bower, R., Crennan, L. & Navatoga, A. (2005) *The sanitation park project Fiji*. Suva, Fiji, SOPAC. SOPAC Technical report number: 386.

Bredehoeft, J.D. (1997) Water Management in the United States – A Democratic Process (Who are the Managers?). In: Burras, N. (ed.) *Reflections on Hydrology: Science and Practice*. Washington, DC, American Geophysical Union, 35–62, doi:10.1029/SP048p0035

Burns, W.C.G. (2002) Pacific Island developing country water resources and climate change. In: Gleick, P.H., Burns, W.C.G., Chalecki, E.L. & Cohen, M. (eds.) *World's Water 2002–2003: The Biennial Report on Freshwater Resources*. Washington, DC, Island Press. pp. 113–131.

Chapman, T.G. (1985) The use of water balances for water resource estimation with special reference to small islands. Bulletin number: 4. Pacific Regional Team. Canberra, Australia, Australian Development Assistance Bureau.

Crennan, L. & Berry, G. (2002) Review of community-based issues and activities in waste management, pollution prevention and improved sanitation in the Pacific Islands region. In: Wright, A. & Stacey, N. (eds.) *Issues for Community-based Sustainable Resource Management and Conservation: Considerations for the Strategic Action Plan for the International Waters of the Pacific Small Island Developing States*. IWP technical report 2002/03. Apia, Samoa, The International Waters Programme, South Pacific Regional Environment Programme.

Crennan, L., Fatai, T. & Fakatava, T. (1998) *Groundwater pollution study*, Completion of Phase Two, Lifuka, Kingdom of Tonga. Report submitted to UNESCO Office of Pacific States, Apia, Samoa, December 1998.

Detay, M., Alessandrello, E., Come, P. & Groom, I. (1989) Groundwater contamination and pollution in Micronesia. *Journal of Hydrology*, 112, 149–170.

DHC. (1982) *Kiribati – Tarawa water resources, Pre-design study*. Prepared for the Australian Development Assistance Bureau by the Australian Government Department of Housing and Construction.

Dijon, R. (1983) Some aspects of water resources planning and management in smaller islands. *Natural Resources Forum*, 7(2), 137–144.

Dray, A., Perez, P., LePage, C., D'Aquino, P. & White, I. (2007). Who wants to terminate the Game? The role of vested interests and meta-players in the Atoll Game experience. *Simulation and Gaming*.: doi:10.1177/1046878107300673

Falkland, A. (1992) *Review of Tarawa Freshwater Lenses, Republic of Kiribati*. Report HWR92/681. Hydrology and Water Resources Branch, ACT Electricity and Water. Prepared for Australian International Development Assistance Bureau.

Falkland, A. (1993) Hydrology and water management on small tropical islands. In: Gladwell, J.S. (ed.) *Proceedings of the Symposium on Hydrology of Warm Humid Regions*. International Association of Hydrological Sciences Publ. No. 216. pp. 263–303.

Falkland, A. (1999) *Water Management for Funafuti, Tuvalu*. Prepared for the Australian Agency for International Development, Canberra, ACT, August 1999.

Falkland, A. (2002) Tropical island hydrology and water resources: Current knowledge and future needs. In: Gladwell, J.S. (ed.) *Proceedings of the Second International Colloquium on Hydrology and Water Resources Management in the Humid Tropics*, 22–25 March 1999. Panama City, Panama, UNESCO-IHP, CATHALAC. pp. 237–298.

Falkland, A. (2004) *Preliminary Design Report for Four Infiltration Galleries at Bonriki, Tarawa, Kiribati*. SAPHE Project, Hydrogeology Component. Tarawa, Republic of Kiribati, August 2004.

Falkland, A. & Brunel, J.P. (1993) Review of hydrology and water resources of humid tropical islands. In: Bonell, M., Hufschmidt, M.M. & Gladwell. J.S. (eds.) *Hydrology and Water Management in the Humid Tropics.* UNESCO, International Hydrology Series, Cambridge, Cambridge University Press.

Falkland, A. & Woodroffe, C.D. (1997) Geology and hydrogeology of Tarawa and Christmas Island, Kiribati. Chapter 19. In: Vacher, H.L. & Quinn, T.M. (eds.) *Geology and Hydrogeology of Carbonate Islands, Developments in Sedimentology 54.* Amsterdam, The Netherlands, Elsevier. pp. 577–610.

Foale, M. (2003) *The coconut odyssey: The bounteous possibilities of the tree of life.* Australian Centre for International Agricultural Research, Canberra, Australia.

Furness, L.J. (1997) Hydrogeology of carbonate islands of the Kingdom of Tonga. Chapter 18. In: Vacher, H.L. & Quinn, T.M. (eds.) *Geology and Hydrogeology of Carbonate Islands, Developments in Sedimentology 54.* Amsterdam, The Netherlands, Elsevier. pp.565–576.

Furness, L.J. & Gingerich, S. (1993) Estimation of recharge to the fresh water lens of Tongatapu, Kingdom of Tonga. In: Gladwell, J.S. (ed.) *Proceedings of the Symposium on Hydrology of Warm Humid Regions.* International Association of Hydrological Sciences, Publication number: 216. pp. 317–322.

GWP. (2006) *Groundwater resources investigations on Niue Island.* Prepared for the Government and South Pacific Applied Geoscience Commission (SOPAC), GWP Consultants, Charlbury, Oxford, UK, March 2006.

Hein, J.R., Gray, S.C. & Richmond, B.R. (1997) Geology and hydrogeology of the Cook Islands. Chapter 16. In: Vacher, H.L. & Quinn, T.M. (eds.) *Geology and Hydrogeology of Carbonate Islands, Developments in Sedimentology 54.* Amsterdam, The Netherlands, Elsevier. pp. 503–535.

Herzberg, A. (1901) Die Wasserversorgung einiger Nordseebäder [The water supply on parts of the North Sea coast]. *Journal für Gasbeleuchtung und Wasserversorgung*, 44, 815–819; 45, 842–844.

Hunt, C.D. & Peterson, F.L. (1980) *Groundwater resources of Kwajalein island, Marshall Islands.* Water Resources Research Centre, University of Hawaii, Honolulu, Hawaii, USA. Technical report number: 126.

Jacobson, G. & Hill, P.J. (1980) *Groundwater resources of Niue Island.* Canberra, ACT, Bureau of Mineral Resources. Record number: 1980/14.

Jacobson, G. & Hill, P.J. (1988) *Hydrogeology and groundwater resources of Nauru Island, Central Pacific Ocean.* Canberra, ACT, Bureau of Mineral Resources. Record number: 1988/12.

Jacobson, G., Hill, P.J. & Ghassemi, F. (1997) Geology and hydrogeology of Nauru Island. Chapter 24. In: Vacher, H.L. & Quinn, T.M. (eds.) *Geology and Hydrogeology of Carbonate Islands, Developments in Sedimentology 54.* Amsterdam, The Netherlands, Elsevier. pp. 707–742.

Jacobson, G. & Taylor, F.J. (1981) *Hydrogeology of Tarawa atoll, Kiribati.* Canberra, ACT, Bureau of Mineral Resources. Record number: 1981/31.

Mather, J.D. (1973) *The Groundwater Resources of Southern Tarawa, Gilbert and Ellice Islands.* UK, Hydrogeological Department, Institute of Geological Sciences. London. 54 pp.

Nullet, D. (1987) Water balance of Pacific atolls. *Water Resources Bulletin*, 23(6), 1125–1132.

Oberdorfer, J.A. & Buddemeier, R.W. (1984) *Atoll Island groundwater contamination: Rapid recovery from saltwater intrusion.* Annual Meeting of the Association of Engineering Geologists. Boston, MA, 9–11 October 1984.

Oberdorfer, J.A. & Buddemeier, R.W. (1988) Climate change, effects on reef island resources. *Sixth International Coral Reef Symposium.* Vol. 3. Townsville, QLD. pp. 523–527.

Oberdorfer, J.A., Hogan, P.J. & Buddemeier, R.W. (1990) Atoll island hydrogeology, flow and fresh water occurrence in a tidally dominated system. *Journal of Hydrology*, 120, 327–340.

Pacific HYCOS. (2009) *Pacific Hydrological Cycle Observing System.* [online] Available from: http://pacific-hycos.org/ [Cited 23rd July 2009].

Peterson, F.L. (1997) Hydrogeology of the Marshall Islands. Chapter 20. In: Vacher, H.L. & Quinn, T.M. (eds.) *Geology and Hydrogeology of Carbonate Islands, Developments in Sedimentology 54*. Amsterdam, The Netherlands, Elsevier. pp. 611–636.

Richards, R. (1991) Atoll vulnerability: the storm waves on Tokelau on 28 February 1987. In: Hay, J. (ed.) *South Pacific Environments: Interactions with Weather and Climate*. Environmental Science, University of Auckland, Auckland, New Zealand. pp. 155–156.

Richardson and Dumbleton International. (1978) *Water resources, Tarawa, report on feasibility study*. Prepared for the UK Ministry of Overseas Development on behalf of the Government of the Gilbert Islands.

Scott, D., Overmars, M., Falkland, A. & Carpenter, C. (2003) *Pacific dialogue on climate and water synthesis report*. Suva, Fiji, SOPAC, February 2003. 36 pp.

SOPAC. (2005) *First groundwater training course*. Hydrological training on surface water and groundwater, SOPAC Secretariat, 4–22 April 2005, sponsored by NZAid, NIWA, UNESCO and SOPAC.

SOPAC. (2006) *Second groundwater training course*. Hydrological training on surface water and groundwater, SOPAC Secretariat, 12–30 June 2006, sponsored by NZAid, NIWA and SOPAC.

SOPAC & ADB. (2003) *Pacific Regional Action Plan on Sustainable Water Management*. Suva, Fiji, South Pacific Applied Geoscience Commission and Asian Development Bank. 53 pp.

Spennemann, D.H.R. (2006) Freshwater lens, settlement patterns, resource use and connectivity in the Marshall Islands. *Transforming Cultures*. 1(2), 44–63. Available from: http://express.lib.uts.edu.au/journals/Tfc [Accessed October 9, 2011]

Terry, J.P. (2007) *Tropical Cyclones, Climatology and Impacts in the South Pacific*. New York, Springer.

Terry, J.P. & Falkland, A.C. (2010) Response of coral island freshwater lenses to storm surge in the Northern Cook Islands. *Hydrogeology Journal*, 18, 749–759.

Underwood, M.R., Peterson, F.L. & Voss, C.I. (1992) Groundwater lens dynamics of atoll islands. *Water Resources Research*, 28(11), 2889–2902.

UNEP. (2002) A directory of environmentally sound technologies for the integrated management of solid, liquid and hazardous waste for Small Island Developing States (SIDS) in the Pacific Region. Compiled by OPUS International in conjunction with South Pacific Regional Environment Programme (SPREP) and South Pacific Applied Geoscience Commission (SOPAC), July 2002.

UNESCO. (1991) Hydrology and water resources of small islands: A practical guide. In: Falkland, A. & Custodio, E. (eds.) *Studies and Reports on Hydrology*. Paris, France, UNESCO. Report number: 49.

Vacher, H.L. & Quinn, T.M. (eds.) (1997) *Geology and Hydrogeology of Carbonate Islands, Developments in Sedimentology 54*. Amsterdam, The Netherlands, Elsevier.

van der Velde, M. (2006) Agricultural and climatic impacts on the groundwater resources of a small island. Measuring and modeling water and solute transport in soil and groundwater on Tongatapu. *PhD Thesis*. Faculty of Biological, Agricultural and Environmental Engineering, Catholic University of Louvain, Brussels, Belgium. 263 pp.

van der Velde, M., Green, S.R., Gee, G.W., Vanclooster, M. & Clothier, B.E. (2005) Evaluation of drainage from passive suction and nonsuction flux meters in a volcanic clay soil under tropical conditions. *Vadose Zone Journal*, 4, 1201–1209.

van der Velde, M., Green, S.R., Vanclooster, M. & Clothier, B.E. (2007) Sustainable development in small island developing states: Agricultural intensification, economic development, and freshwater resources management on the coral atoll of Tongatapu. *Ecological Economics*, 61, 456–468.

van der Velde, M., Javaux, M., Vanclooster, M. & Clothier, B.E. (2006) El Niño-Southern Oscillation determines the salinity of the freshwater lens under a coral atoll in the Pacific

Ocean. *Geophysical Research Letters.* [online] 33, L21403. Available from: doi:10.1029/2006GL027748 [Accessed October 9th 2011]

Volker, R.E., Mariño, M.A. & Rolston, D.E. (1985) Transition zone width in ground water on ocean atolls. *Journal of Hydraulic Engineering*, 111(4), 659–676.

Voss, C.I. (1984) SUTRA, A finite-element simulation model for saturated-unsaturated, fluid-density-dependent ground-water flow with energy transport or chemically-reactive single-species solute transport. USGS Water Resources Investigation Report 84-4389. 409 pp.

Voss, C.I., Boldt, D. & Sharpiro, A.M. (1997) A graphical-user interface for the US Geological Survey's SUTRA code using Argus ONE (for Simulation of Variable-Density Saturated-Unsaturated Ground-Water Flow with Solute or Energy Transport). Reston, VA, US Geological Survey. Open-File report number: 97-421.

Ward, R.G. (1999) *Widening worlds, shrinking worlds, the reshaping of Oceania* [Pacific Distinguished Lecture 1999]. Centre for the Contemporary Pacific, Australian National University, Canberra, ACT.

Wheatcraft, S.W. & Buddemeier, R.W. (1981) Atoll island hydrology. *Ground Water*, 19(3), 311–320.

Wheeler, C. & Aharon, P. (1997) Geology and hydrogeology of Niue. Chapter 17. In: Vacher, H.L. & Quinn, T.M. (eds.) *Geology and Hydrogeology of Carbonate Islands, Developments in Sedimentology*. Volume 54. Amsterdam, The Netherlands, Elsevier. pp. 537–564.

White, I. & Falkland, A. (2010) Management of freshwater lenses on small islands in the Pacific. *Hydrogeology Journal*, 18, 227–246.

White, I., Falkland, A., Crennan, L., Jones, P., Metutera, T., Etuati, B. & Metai, E. (1999a) Groundwater recharge in low coral islands Bonriki, South Tarawa, Kiribati. Issues, traditions and conflicts in groundwater use and management. *UNESCO IHP-V, Technical Documents in Hydrology*, Number: 25. Paris, France, UNESCO. 39 pp.

White, I., Falkland, A., Etuati, B., Metai, E. & Metutera, T. (2002) Recharge of fresh groundwater lenses: Field study, Tarawa Atoll, Kiribati. In: Gladwell, J.S. (ed.) *Proceedings of the Second International Colloquium on Hydrology and Water Resources Management in the Humid Tropics*, 22–25 March 1999. Panama City, Panama, UNESCO-IHP, CATHALAC.

White, I., Falkland, A. & Fatai, T. (2009a) *Vulnerability of groundwater in Tongatapu, Kingdom of Tonga: Groundwater evaluation and monitoring assessment.* Report to Pacific Islands Applied Geoscience Commission, EU EDF8, February 2009, ANU, Canberra.

White, I., Falkland, A., Metutera, T., Katatia, M., Abete-Reema, T., Overmars, M., Perez, P. & Dray, A. (2008) Safe water for people in low, small island Pacific nations: The rural–urban dilemma. *Development*, 51, 282–287.

White, I., Falkland, A., Metutera, T. & Metai, E. (2005) Effects of landuse on groundwater quality in a low coral atoll: Coliforms, nutrients and metals. ACIAR Project LWR1/2001/050. *Equitable Groundwater Management for the Development of Atolls and Small Islands.* Canberra, ACT, Australian Centre for International Agricultural Research, May 2005.

White, I., Falkland, A., Metutera, T., Metai, E., Overmars, M., Perez, P. & Dray, A. (2007a) Climatic and human influences on groundwater in low atolls. *Vadose Zone Journal*, 6, 581–590.

White, I., Falkland, A., Perez, P., Dray, A., Metutera, T., Metai, E. & Overmars M. (2007b) Challenges in freshwater management in low coral atolls. *Journal of Cleaner Production*, 15, 1522–1528.

White, I., Falkland, A., Rebgetz, M., Overmars, M., Katatia, M. & Metutera, T. (2009b) Development of National Water Resources Policy and Implementation Plans for the Kiribati Adaptation Program. In: *Proceedings 32nd Hydrology and Water Resources Symposium, 30 November–3 December 2009*, Newcastle, Australia. ISBN 978-08258259461. Engineers Australia. pp. 1603–1618.

White, I., Falkland, A. & Scott, D. (1999b) Droughts in small coral islands: Case study, South Tarawa, Kiribati. UNESCO IHP-V, Technical Documents in Hydrology, Number 26. Paris, France, UNESCO. 55 pp.

WHO. (2004) *Guidelines for Drinking-water Quality*. 3rd edition (including addenda in 2006 and 2008). Geneva, Switzerland, World Health Organization.

WHO. (2008) *Sanitation, Hygiene and Drinking-Water in the Pacific Island Countries, Converting Commitment into Action*. Geneva, Switzerland, World Health Organization and SOPAC.

Wooding, R.A. (1963) Convection in a saturated porous media at large Rayleigh number or Peclet Number. *Journal of Fluid Mechanics*, 15, 527–544.

Wooding, R.A. (1964) Mixing-layer flows in a saturated porous media. *Journal of Fluid Mechanics*, 19, 103–112.

Woodroffe, C.D. (2008) Reef-island topography and the vulnerability of atolls to sea-level rise. *Global and Planetary Change*, 62, 7–96.

World Bank. (2000) *Cities, Seas and Storms, Managing Change in Pacific Island Economies. Volume IV, Adapting to Climate Change*. Papua New Guinea and Pacific Island Country Unit, World Bank, November 2000.

Dry (Arid and Semiarid) Climates

CHAPTER 7

Groundwater resources increase in the Iullemmeden Basin, West Africa

Guillaume Favreau, Yahaya Nazoumou, Marc Leblanc,
Abdou Guéro & Ibrahim Baba Goni

ABSTRACT

The Iullemmeden Basin, ~620,000 km^2 in area, is the fourth largest transboundary sedimentary aquifer in sub-Saharan Africa. In its southern part, the unconfined aquifers of the Continental Terminal and Continental Hamadien are recharged by focused infiltration through ponds and gullies. A long-term rise in the water table of ~4 m (1963–2007) was shown to have occurred in response to land clearing over ~10,000 km^2 in the south-western part of the basin. This paper presents new evidence that a long-term rise of up to 0.4 m/a have also occurred over ≥100,000 km^2 during various time periods through the 1930s to the 2000s in response to land clearing. This represents one of the longest periods (~80 years) and the largest surface area at the global scale showing a rise in the water table. Increased pumping would be needed to avoid soil salinization in valleys where the water table is now increasingly outcropping. Irrigation using groundwater ("blue" water) will also help to mitigate the impacts of rainfall variability on rain-fed crop production.

7.1 INTRODUCTION

7.1.1 Purpose and scope

The groundwater response to environmental change is a complex function that depends on climate, the land surface and vegetation dynamics, as well as aquifer characteristics and human influences. Groundwater pumping for irrigation has, in recent years, often masked or outweighed other factors, and there is significant evidence of a decrease in groundwater reserves at a global scale (Wada et al. 2010).

In Sub-Saharan Africa (SSA), irrigation is limited to <4% of cultivated lands, whereas at a global scale, irrigation accounts for ~19% (Siebert et al. 2010). In western SSA, most irrigated areas are located in valleys of large rivers, where surface water is used for irrigation. Meanwhile, groundwater use in large sedimentary basins remains limited to domestic purposes and/or livestock breeding activities. Despite a threefold increase of the population in the region over the 1950s–2010s, groundwater is largely under-used and there is significant evidence that water table fluctuations are still governed by climatic or environmental changes (Carter and Parker 2009).

In semiarid regions, land use has been shown to often have a greater impact than climate to explain changes in recharge and discharge (Scanlon et al. 2007). In western Africa, a very detailed and documented example of the indirect impact of land clearing on groundwater recharge is the Continental Terminal (CT) aquifer of the Iullemmeden

Basin (IB) in southwest Niger (Leduc et al. 2001). At the scale of ~10,000 km², several independent methods were used to show that despite a long lasting rainfall deficit since the early 1970s (Lebel and Ali 2009), the water table has risen in response to land clearing that has increased runoff through sandy gullies to endoreic ponds, where groundwater recharge was shown to occur (Favreau et al. 2009).

The aim of this paper is to expand temporal and spatial observations of the rise in the water table for unconfined aquifers of the IB, in Niger and Nigeria. Direct and indirect evidence from various sources are summarized to show that the rise in the water table occurred over at least 100,000 km² at different time periods as a function of diachronic land clearing dynamics. Recommendations to increase irrigated surface in valleys using available fresh groundwater should be made to help mitigate impacts of soil salinization in valleys, where the water table is locally outcropping. Increased crop production in valleys will help to reduce erosion on slopes where extensive rainfed cropping has resulted in soil losses through Hortonian runoff and uncontrolled gullying. Increased irrigation could also help mitigate projected changes in the Sahelian climate as a consequence of global warming, which will increase crop transpiration demand. Further studies should take advantage of remote sensing satellite data (including gravity surveys) to quantify hydrological changes at the basin scale.

7.1.2   Description of the study area: the Iullemmeden Basin

The IB, located in central West Africa is a sedimentary basin shared over about 95% of its surface area between Niger, Mali and Nigeria (Fig. 7.1). With a surface area of ~620,000 km², it is the fourth largest sedimentary basin in SSA. A 2,000 m thick series of sedimentary aquifer formations, dating back mainly to the Secondary (Continental Intercalaire (CI), Continental Hamadien (CH)) and Tertiary (CT, known as Gwandu formation in Nigeria), represent the main natural resource (Moody 1997). The CT/Gwandu aquifer outcrops over ~150,000 km² of the basin and the CI/CH over 250,000 km², most of it being covered by aeolian sands in its northern part. Quaternary aquifers comprised of coarse alluvial sands are found in up to 10 km-wide paleo-valleys originating in the north of the basin and account for about 10% of the total surface area of the basin (Fig. 7.1).

The IB is located both in the Sahara and Sahel zones, where arid to semiarid conditions prevail (Fig. 7.1). Rainfall gradients are typically 1 mm/a/km with monsoon rainfall decreasing to the north. Potential is about 2,500 mm/a in the south-western part of the IB (Niamey) and increases northward. The natural vegetation (prior to removal) was a dry savannah with ligneous species a few metres in height, but under increasing land clearing, is now almost completely dominated by rainfed cropping (millet, sorghum). Irrigation is very limited spatially and occurs near villages in paleo-valleys where the water table is mostly shallow (<10 m). Population density decreases from south to north and east to west across the basin, with extreme values of about 150 inhab./km² (projection for 2010; Guengant and Banoin 2003; CIESIN and CIAT 2005) in the Maradi and Sokoto regions and <2 inhab./km² to the desertic north in Mali and Niger. Population growth was about 1.5%/a in the 1950s and about 3%/a during the 1990s (UNPP 2005). Most of the population lives in small villages with a few hundreds of inhabitants in the southern part of the basin, where rainfed agriculture is dominant. To the north, livestock breeding is a main economic activity (Guengant and Banoin 2003).

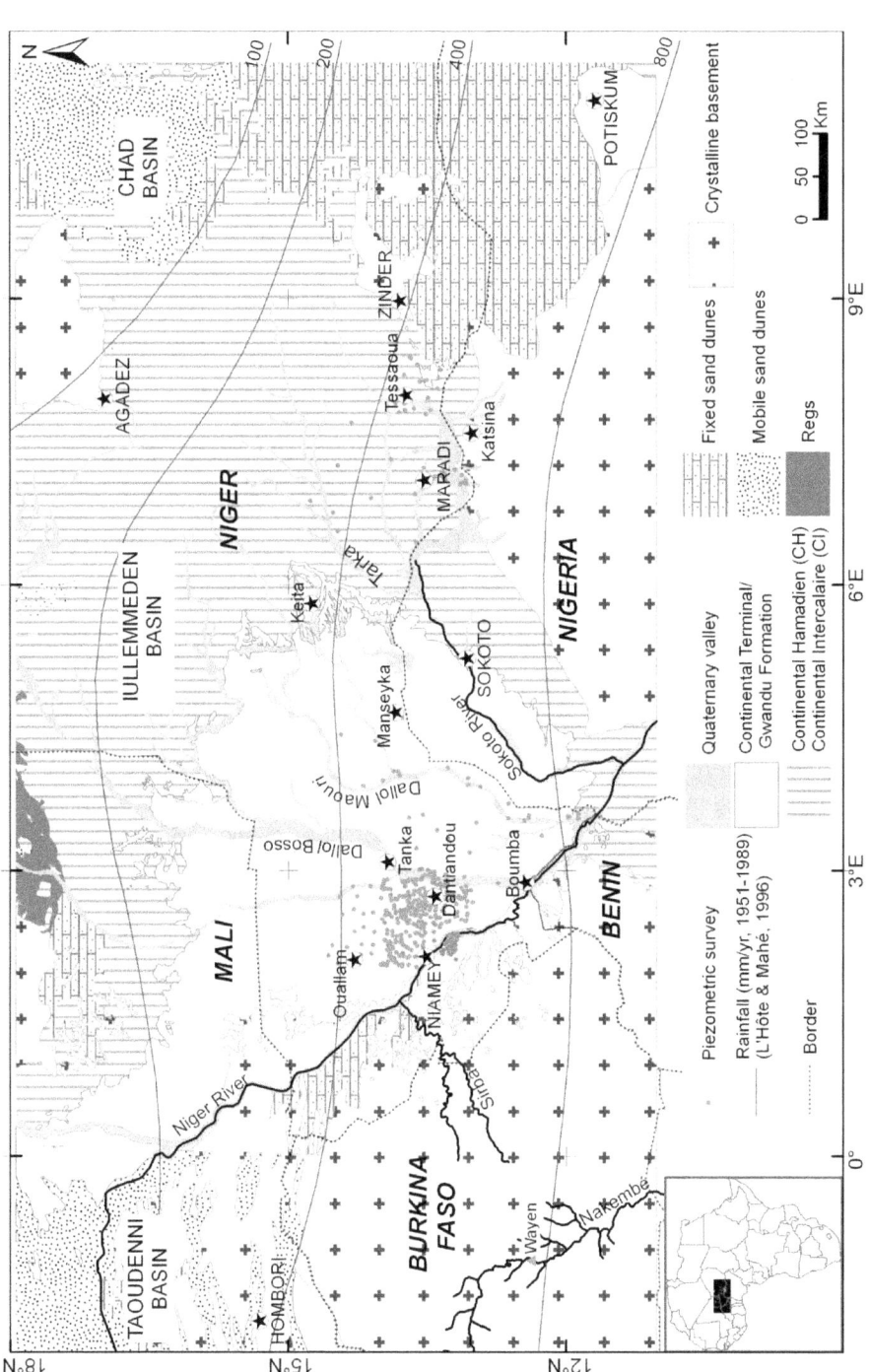

Figure 7.1. Hydrological and geological characteristics of the southern part of the Iullemmeden Basin, West Africa, with contours of main valleys and aquifers (Geological mapping after Greigert 1961). Location of long-term piezometric (1990s to 2000s), hydrological and/or land clearing surveys are also reported.

The large volumes of fresh groundwater resources stored in the IB has attracted scientific interest during recent years and several collaborative projects have used complementary approaches to better estimate the groundwater flux, using either environmental isotope tracers (Wallin et al. 2005), remote sensing (Saradeth and Weissman 2008), or numerical groundwater modeling (OSS 2008). More detailed information was obtained through specific case studies based on groundwater chemistry and/or hydrodynamic investigations in the unconfined parts of the CT/Gwandu aquifer (Guéro 2003; Adelana et al. 2008; Favreau et al. 2009) and in the confined parts of the basin (Geyh and Wirth 1980; Andrews et al. 1993; Le Gal La Salle et al. 2001). Main aquifer characteristics are the following:

1. The water table is a continuous, smooth surface, with hydraulic gradients <0.1%, and little seasonal variation. The water table depth varies from ~80 m below surface in the lateritic plateau to positive values where the water table is outcropping in valley bottoms.
2. Several confined aquifers do exist at depth, with basin scale length paths. Most of these confined groundwaters were shown to have infiltrated during more humid periods of the Quaternary (Beyerle et al. 2003).
3. Groundwater recharge is mainly indirect, occurring through ponds and gullies where surface runoff concentrates. Average recharge is estimated at few tens of mm/a depending on study area and methods used. Increased recharge was shown to have occurred in response to land clearing and gullying in the southern Niger part of the basin (Favreau 2000; Guéro 2003).
4. Groundwater is mainly fresh, with mean total dissolved solids (TDS) of ~100 mg/L (Greigert and Bernert 1979). Higher values of up to 500 mg/L were measured in the downstream part of the large paleo-valleys, where direct fluxes from the water table and solute recycling occur (Barbiero and Valles 1992). Nitrate content values were above the WHO limit of 10 mg/L NO_3-N in some boreholes and this was shown to be related to deforestation and nitrogen leaching to the water table (Favreau et al. 2009).
5. The Niger River erodes the Precambrian basement and is a natural outlet of the aquifers through springs or downstream the Dallol valleys (Chaperon 1970). In the southern part of the basin, the Sokoto River and its tributaries had perennial flows of up to a few tens of m^3/s in the 1960 and 1970s (Adelana et al. 2003), but under increased water use and damming, are no longer exoreic in the upper part of the catchment.

7.1.3 Methodology

A wide range of methods was used to estimate the dynamics and renewal rate of groundwater in the IB. These include remote sensing (aerial photographs, satellite data), subsurface geophysics, environmental geochemistry (isotopes, inorganic tracers) hydrodynamic measurements (limnimetry, piezometric) and numerical modelling at various scales. Detailed information is provided in the literature that reviews current knowledge on the IB aquifers in Niger and Nigeria (Adelana et al. 2008; Favreau et al. 2009).

In this paper, we summarize direct and indirect evidences of the large spatial and temporal extent of the rise in the water table. Direct evidence is provided by piezometric surveys. More than 30,000 manual measurements were made at about 200 sites during the 1990s and 2000s in the Niger part of the basin. Details on these surveys can be found in Leduc et al. (2001) and Guéro (2003). These measurements were mostly made in shallow (<25 m deep) piezometers or much deeper (up to 75 m) hand dug wells used for domestic purposes. We are unaware of regular, pluri-annual measurements in the Mali or Nigeria parts of the basin.

Several reports document previous rises in the water table in the IB. These include unpublished theses (e.g. Guéro 2003), technical reports (e.g. Boeckh 1992) and papers published in journals of limited distribution (e.g. Jones 1960).

Indirect evidence of the rise in the water table can be inferred from the change in land cover and subsequent increase in gullying. Because the IB is mostly endoreic (Descroix et al. 2009) and considering that the discontinuous drainage network is the main source of indirect recharge to the aquifer, any increase in gullying and pond surface area would contribute to an increase in aquifer recharge. A critical review of grey literature and published papers on land use change, erosion and hydrological changes was made at the IB scale and its surroundings to show that land clearing and increased runoff is a widespread phenomenon in the basin. This review expands the spatial scale considered in Leblanc et al. (2008) by new examples from the IB and its surroundings.

7.1.4   Relevance to GRAPHIC

Groundwater represents a widespread perennial fresh water resource available in the IB and most of the population relies on aquifers for domestic use. In the northern part of the basin, access to groundwater is a key factor for successful cattle breeding activities. In the southern, more humid part of the basin, irrigation using groundwater during the dry season, although still limited $< 0.5\%$ of the land surface (USGS 2008), was shown to increase in response to the rapid ($+3\%/a$) population growth and urbanization.

Land clearing for expanding rainfed cropping has been both a rapid and recent (20th century) process in the Sahel region (Guyer et al. 2007). This has modified soil surface characteristics and, in turn, has had an impact on aquifer recharge. In the IB, this has resulted in increased runoff and enhanced focused recharge to the unconfined aquifer (Leblanc et al. 2008). Very little is known about changes in recharge rates in SSA (Scanlon et al. 2007); in western Africa, where the population is rapidly growing, a better estimate of the impact of land use – land cover (LULC) change on groundwater resources will help to better manage shared aquifer resources. With higher crop water demand resulting from increased air temperature ($\sim +3$–$4°C$ in western Africa by 2100; IPCC 2007), sustainable irrigation rates for a "green revolution" (Rockström et al. 2007) require a detailed knowledge of renewable aquifer resources.

7.2   RESULTS AND DISCUSSION

7.2.1   Land use and land cover change

Dramatic changes in LULC have occurred in the Sahel region during the 20th century in response to demand for land to meet the population growth. In the IB, changes that occurred in the Niamey region are summarized in Leblanc et al. (2008). Using historical aerial photographs (e.g. Fig. 7.2), it was shown that most of the land clearing

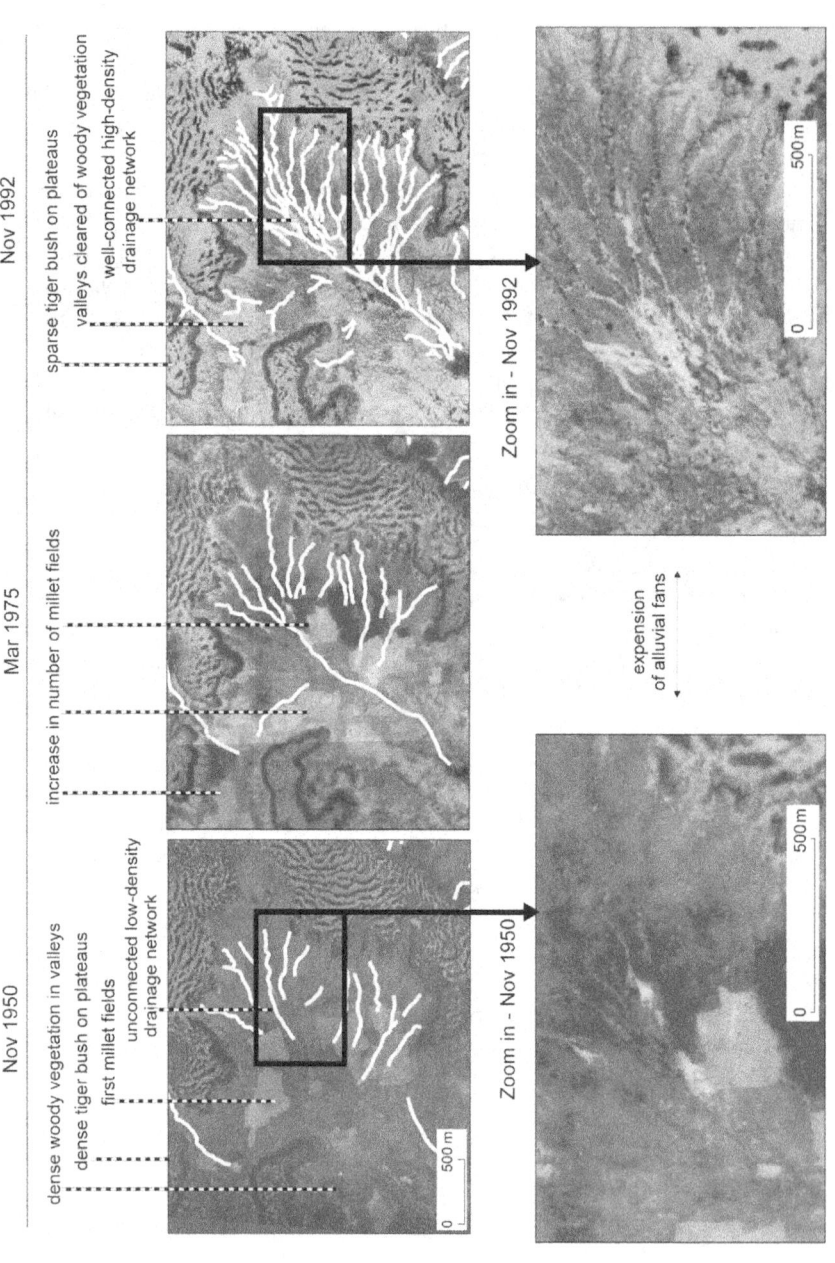

Figure 7.2. Historical aerial photographs (IGN, Niger) showing long-term changes in LULC and in the drainage network over the Continental Terminal aquifer in southwest Niger (Kolo Bossey, 2°35'E, 13°44'N). An increase in the number of millet fields is visible from 1950 to 1992 and is accompanied by the disappearance of woody vegetation on the valleys and hillslopes. On the plateaus, a significant clearing of the 'tiger bush' is observed during the study period. Concurrently, a strong increase in the density and connectivity of the drainage network is evident in the aerial photographs. The two lower panels show the expansion of a system of alluvial fans between 1950 (lower left) and 1992 (lower right).

occurred during the 1970s and 1980s, with almost 80% of the natural vegetation cover disappearing between 1950 and the 1992.

Using historical aerial photographs, Gentizon (1998) showed that the loss in forested areas near Tanka (Fig. 7.1), over a surface area of $246\,km^2$ and over the period 1950–1992, was 52% (77% for the plateaus). In the Keita region (Fig. 7.1), vegetation change, as observed on historical aerial photographs (1956–1975) and through inquiries made to local inhabitants, showed similar decreases, with vegetation cover decreasing from 25–40% down to 2–5% on the plateaus, and from 10–15% down to 2–5% in valleys (Bouzou Moussa, 1988).

Land use changes from the 1972–1975 time period to the years 1999–2000 were estimated in the southern part of Niger using Landsat MSS, TM and ETM+ images downscaled to a $2\,km$ pixel resolution (USGS 2008). Images and analyses were obtained from the Agrhymet research centre (Niamey, Niger) and the IB part of the country was isolated to quantify LULC changes south of 18°N. Re-analysis of the data showed that natural vegetation cover was 75% in the mid-1970s but 65% in 2000. For the same time period, cultivated areas increased by $\sim$50%. These estimates have lower resolution and accuracy than the ones performed using historical aerial photographs but nonetheless confirm the general trend of land clearing and encroachment of rainfed cropping on the landscape at the whole IB scale.

7.2.2 Increased runoff and erosion

Increased Hortonian runoff, erosion and gullying are the main hydrological consequences of land clearing in the basin. Over a surface area of $500\,km^2$, Leblanc et al. (2008) showed a $\times2.5$ increase in drainage density between 1950 and 1992 using aerial photographs. In a $2\,km^2$ small watershed, numerical modelling showed a $\times2.8$ increase in runoff capacity for the same time period (Séguis et al 2004). Increased runoff to ponds, new ponds and increased silting were observed on the landscape in response to this increase in runoff capacity (Fig. 7.2).

Further north, an increase in runoff was shown to have occurred in the Keita valley (length is $140\,km$, surface area of $3,020\,km^2$; Fig. 7.1) in response to land clearing, where the length of gullies was observed to grow annually by several meters (Bouzou Moussa 1988; FAO 1994). Close to Nigeria, the Mountséka valley (Fig. 7.1) experienced a return of seasonal river flow conditions over $70\,km$ in the 1990s.

This abrupt hydrological change occurred in response to land clearing and higher runoff to the valley: Remote sensing data (1986–2000) and field surveys showed a significant increase in surface areas with high runoff capacity over the $5,500\,km^2$ of the watershed (Bouzou Moussa et al. 2009).

Near Maradi, the upper Sokoto River (Goulbi) valley is supporting intensive farming activities. Damming of the Goulbi in the 1990s induced a decrease in ephemeral river flow and, consequently, a decrease in the seasonal recharge to the downstream alluvial aquifer. Under "no dam" conditions, erosion was shown to be the main factor explaining the rapid silting up of the economically important Madarounfa lake (100–850 ha), with rapid changes in ephemeral rivers feeding the lake caused by land clearing of five classified forests (46,000 ha). The maximum depth of the lake was $2.5\,m$ in 2000 (Ibrahim 2001) compared to $5\,m$ in the early 1970s (Carré and Robin 1973).

Westwards of the IB, increasing river discharge in response to a decreased vegetation cover was shown to have occurred during the 1970s–1990s decades over the $39,000\,km^2$ of the Sirba watershed (Amani and Nguetora 2002) and in the $21,000\,km^2$ of the Nakambé watershed (Mahé et al. 2005). Increased ponding during the 1950s to the 2000s was also shown using remote sensing data, over a ~20,000 km² endoreic region near Hombori in Mali (Gardelle et al. 2010). However, evidence of rises in the water table was not reported in these regions, probably because of the discontinuous nature of the underlying Precambrian fractured aquifers (Fig. 7.1).

7.2.3   Long-term changes in the water table

Boreholes and hand dug wells are widespread in the IB but because irrigation is negligible, groundwater pumping was limited to ~180 Mm³/a in 2004 (OSS 2008). Groundwater used for irrigation accounts for much less than the estimated 0.5% of irrigated cultivated lands in the Niger part of the basin, with possible significant impact limited to a few Quaternary valleys in Niger and "fadamas" in Nigeria (Siebert et al. 2010). Groundwater pumping from the unconfined CT aquifer for domestic water use was reported to be of the order of 10 to 30 L/inhab./day (Favreau et al. 2009). Given the rural population densities from 10 to 150 inhab./km² in the southern IB (Guengant and Banoin 2003; CIESIN and CIAT 2005), this implies domestic groundwater use is < 2 mm/a. Because groundwater use remains low, changes in aquifer recharge can be monitored by water table surveys.

Piezometric surveys in the Niger part of the CT aquifer were reviewed by Guéro (2003) and Favreau et al. (2009). All of the ~150 surveys performed to date showed a rise in the CT water table, irrespective of the time period considered. Present day (2011) water table levels are the highest ever recorded (Fig 7.3a). Measured rise intensities range from <0.1 m/a to up to 0.4 m/a (Leduc et al. 2001). New, unpublished evidence from the eastern part of the CT aquifer showed rises of up to 0.4 m/a in valleys during the 1990s and further extends the surface area of the rise at the aquifer scale (Fig. 7.3b).

In the south-eastern part of the basin, the CH aquifer outcrops over ~80,000 km². Water table surveys performed during the 1990s–2000s show no trend in the water table. However, several papers or reports suggest that a water table rise from the 1930s to the

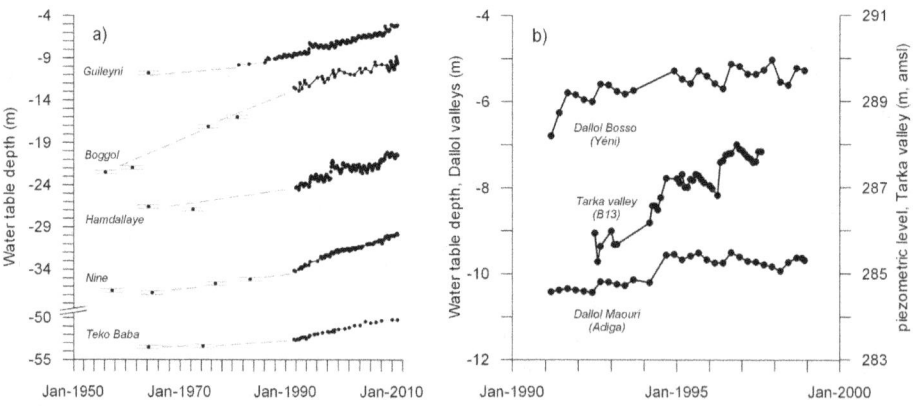

Figure 7.3.   Examples of (a) long term rises in the water table in the Continental Terminal aquifer, (b) stable or rising water table levels in the Dallols and Tarka Quaternary valleys located in the centre of the Basin.

Table 7.1. Characteristics and changes in unconfined aquifers of the IB during the last decades.

Aquifers	Surface area (10³ km²)	Observed land clearing periods	Pop. density, 2000–2001 (inhab. km⁻²)	Gullying, increased Hortonian runoff	Periods with rise in the water table	Surface area with rise in the water table (10³ km²)
CH	80	1930s–1957 [a] 1975–2000 [b] 1964–1992 [c]	60–120 [d,e]	1957 [a] 1990s [f]	1930s–1957 [a] 1964–1992 [c]	≥ 2 [a] 0.6 [c]
CT	150	1950–1975 [g,h,i] 1975–1992 [g,h] 1975–2000 [b]	10–70 [d,e]	1950–2005 [i] 1980s [i,k] 1986–2007 [l]	1955–2011 [j,m,n]	≥100*
Dallols**	~10% of the total surface area	< 20th century [m]	20–150 [d]	not reported	1990s–2011 [m,o]	south of ~13.5°N

* The rise has been observed on all the piezometric chronicles in the Niger part of the IB (no data from Mali and Nigeria).
** Large Quaternary valleys coming from the north of the IB (cf. Fig. 1)

(a) Jones, 1960 ; (b) Boeckh, 1992 ; (c) Boeckh, 1992 ; (d) Guengant & Banoin, 2003; (e) CIESIN/CIAT, 2005; (f) Ibrahim, 2001; (g) Leblanc et al, 2008 [and references therein]; (h) Gentizon, 1998; (i) Bouzou Moussa, 1988; (j) Favreau et al, 2009; (k) FAO, 1994; (l) Bouzou Moussa et al, 2009; (m) Guéro, 2003; (n) this study; (o) Niger local authorities, Pers. Com., 2011.

1950s occurred in the CH aquifer following clearance of the natural vegetation. Jones (1960) states: *"During the last 20 years, a rise in the water table has been observed in parts of the Daura and Katsina Emirates (...). It is believed to be due to an increasing proportion of the rainfall percolating to the water table, as a result of reduced transpiration losses, through land clearance of the natural vegetation"*. This study was performed over ~2000 km^2 over 80 representative wells and lead to the conclusion that for the 1937–1957 time period, a rise in the water table had occurred. Similar explanations were reported by Boeckh (1992) when comparing measurements made in 1964 and 1992 in the nearby Tessaoua region (Fig. 7.1). Earlier consequences of the change in land use on the water table were also reported to the east in the Potiskum region (Fig. 7.1) where the water table was reported to have risen by up to 20 m during the 1930s to the 1950s in response to land clearing (Barber and Dousse 1965).

Direct and indirect evidence of the rise in the water table suggest that the total surface area of the IB that has been subject to a rise in the water table following land use change during the 20th century is at least 100,000 km^2 (Table 7.1).

7.2.4   Impacts of climate and land use changes on groundwater resources

Water table surveys have shown that from a long term perspective, land use change is a stronger factor influencing changes in groundwater resources than climate variability in the IB. Quantifying this influence would require a coupled surface water – groundwater model, integrating a representative part of the unconfined aquifer. A first analysis of the respective influences of land cover and climate on runoff was performed for the 1950–1992 time period over a small watershed of southwest Niger (Séguis et al. 2004). Numerical, 2D physically based modelling showed that for a given rainfall, land clearing had increased runoff by $\times 2.8$, whereas for a given land use situation, rainfall variability would decrease runoff by at most 40%.

The impact of land use change is expected to be even greater for groundwater resources. Groundwater surveys near Niamey showed, for instance, that half of the rise in the water table ("excess recharge") during the 1992–2005 time period occurred during only three rainy seasons characterized by frequent high rainfall events (Fig. 7.4). Because focused recharge is highly dependent upon heavy rainstorms (Séguis et al. 2004), increased frequency of intense rainfall events predicted by climate models (IPCC 2007) as a consequence of global warming would likely increase recharge in the IB.

7.3   POLICY-RELEVANT RECOMMENDATIONS

The long term change in the water balance is the result of lower water use efficiency by natural vegetation during intense monsoon rainfall events. Over the past decades, this loss of efficiency has resulted in Hortonian runoff, soil loss, gullying and increasing ponding in the lower parts of the landscape (Fig. 7.5). Because natural groundwater outflow is low in the IB, most of the excess in groundwater recharge has been stored in the unconfined aquifers and is available for pumping.

Increasing irrigation using groundwater has been observed during recent years in most populated regions of the southern parts of the basin; this study demonstrates that a sustainable use of groundwater can be achieved by using more groundwater where the water table is rising. In the downstream parts of the landscape, near ponds, pumping of

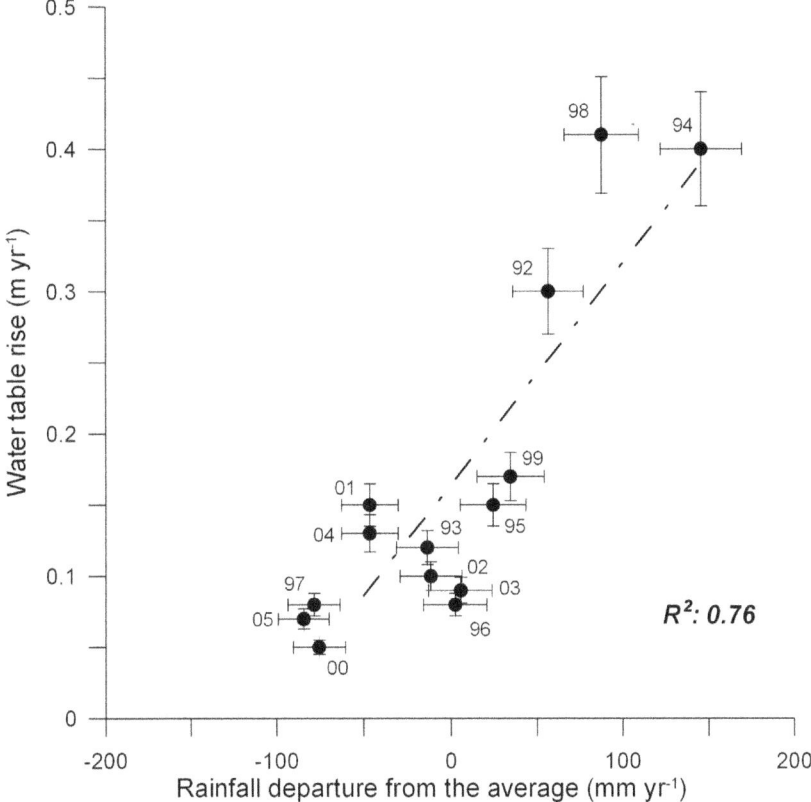

Figure 7.4. Mean computed water table rise as a function of departure from the mean annual rainfall, 1992–2005. Years are reported as labels for each year (e.g. 92 for 1992). Modified after Favreau et al. (2009).

groundwater enriched in nitrogen by land clearing (Fig. 7.5) should be encouraged and could help to reduce the cost of fertilizers.

Increased pumping would be needed to avoid soil salinization in valleys where the water table is now outcropping (Fig. 7.6). This salinization process is different from the one resulting from leaching of the vadose zone in other semiarid regions (e.g. SE Australia; Leaney et al. 2003) and would remain limited in space in lowlands where the water table is outcropping. Irrigation in these valleys using groundwater ("blue" water; Rockström et al. 2007) will also help to mitigate impacts of rainfall variability on rain-fed crop production and help reduce hunger vulnerability (Bassett 2010). Increased pumping would also help to reduce flood risks in large valleys (eg, Dallols) where the water table has been observed to getting close to the soil surface and is contributing to seasonal floods (Table 7.1, Fig. 7.6).

In the upper part of the landscape, near plateaus, recent initiatives of soil water harvesting and runoff control have been successfully performed during the 1980s–1990s (FAO 1994). This costly land management is the most promising in the long term to restore soil fertility and increase "green water" use, and should be seen complementary to sustainable aquifer management.

7.4   FUTURE WORK

Groundwater surveys showed that unconfined aquifers of the IB are very sensitive to changes in environmental land conditions, with little time lag between land clearing, focused recharge and consequent water table rise. This is mainly due to indirect recharge processes being dominant in the basin, whereas at a global scale, diffuse recharge is often

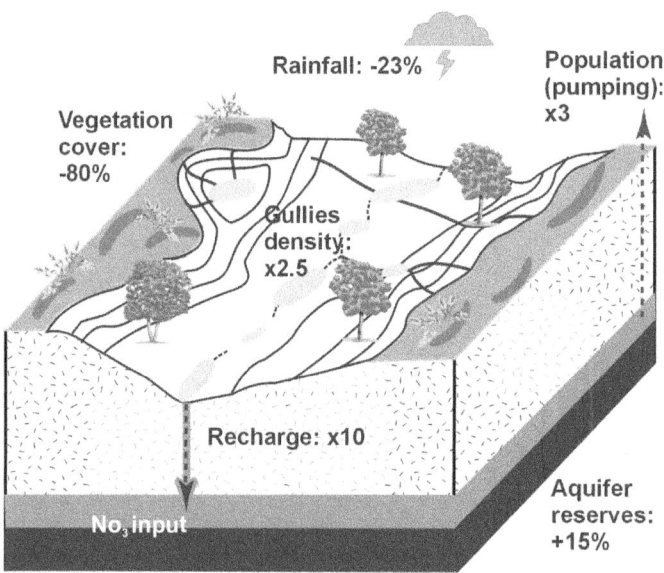

Figure 7.5.   Synopsis of relative changes in groundwater recharge conditions and fluxes in the southwestern part of the Continental Terminal aquifer for the 1963–2007 time period. (Data from Table 1 in Favreau et al. 2009).

Figure 7.6.   (a) Outflow of the aquifer to the Niger River south of the dallol Bosso valley at Boumba (Sept. 4, 2009, 12°24'39"N, 2°50'01"E, estimated groundwater outflow of ~1 m^3/s) and (b) salinization processes in play at the Tinga pond near Ouallam where the water table outcrops over ~50 ha in the valley (June 12, 2010, 14°18'16"N, 2°04'02"E).

more significant (Scanlon et al. 2007). In addition, most of the aquifer can be considered as endoreic, with little southward outflow to the Niger River (Fig. 7.6). This makes it possible to record local changes in the hydrological regime by monitoring the unconfined aquifers of the basin.

Basin-scale remote sensing data analyses would be needed to estimate increased ponding surface areas and gullying through time. This task remains challenging for most satellite imagery because of the fine scale required in time and space; ponds being limited to a few ha in surface area and sometimes do not last more than a few weeks during the (cloudy) rainy season. Accurate census of ponds and gullies is necessary for coupled surface water – groundwater modelling in the complex environment of the IB (Massuel et al. 2011).

Groundwater surveys are very limited temporally and spatially in the IB. Long-term surveys (Fig. 7.3a) exist only because of meticulous data mining and rigorous field surveys and are limited in number to a few tens of sites (Favreau 2000; Guéro 2003). Most piezometric surveys started during the 1990s at a monthly to quarterly frequency but remain discontinuous (Fig. 7.3b). The averaged piezometric survey density can be estimated to $\sim$3 surveys/10,000 km^2 for the whole basin, with great disparities (Fig. 7.1). More numerous piezometric surveys are needed for transient modelling and better management of the shared aquifer resources. Results from the GRACE satellite mission (Gravity Recovery And Climate Experiment, launched in 2002) could be used to supplement groundwater surveys at the aquifer scale (Swenson and Famiglietti 2011). However, the CT surface area is on the low side of present-day capacity for estimating terrestrial changes in the water balance using GRACE and the groundwater component need to be constrained by regional piezometric measurements (Leblanc et al. 2011) and/or local field gravity surveys (e.g. Pfeffer et al. 2011) to provide more elaborated results.

ACKNOWLEDGEMENTS

The Ministry of Hydraulics of Niger and its Regional Hydraulics Directions made available most of the water table surveys discussed in this paper. Additional funds for water table surveys were provided by the IRD-supported AMMA-Catch research program (2001–2013, http://www.amma-catch.org). The Danish, German and Swiss cooperation agencies in Niger were also long-term supporters, through different development programs, of the water table surveys carried out by local authorities in the CT and CH aquifers.

REFERENCES

Adelana, S.M.A., Olasehinde, P.I., Bale, R.B., Vrbka, P., Edet, A.E., & Goni, I.B. (2008) An overview of the geology and hydrogeology of Nigeria. In: Adelana, S. & McDonald, A. (eds.) *Applied Groundwater Studies in Africa*, IAH book. SP 13, chapter 11. Taylor & Francis, London, UK. pp. 171–197.

Adelana, S.M.A., Olasehinde, P.I. & Vrbka, P. (2003) Isotope and geochemical characterization of surface and subsurface waters in the semi-arid Sokoto Basin, Nigeria. *African Journal of Science and Technology*, 4(2), 76–89.

Amani, A. & Nguetora, M. (2002) Evidence d'une modification du régime hydrologique du fleuve Niger à Niamey. In: Van Lannen, H. and Demuth, S. (eds.) *FRIEND 2002, Regional Hydrology: Bridging the Gap between Research and Practice, Proceedings of the Conference*, Cape Town, South Africa. IAHS Publication Number: 274. pp. 449–456.

Andrews, J.N., Fontes, J.C., Aranyossy, J.F., Dodo, A., Edmunds, W.M., Joseph, A. & Travi, Y. (1993) The evolution of alkaline groundwaters in the continental intercalaire aquifer of the Irhazer Plain, Niger. *Water Resources Research*, 30, 45–61.

Barber, W. & Dousse, B. (1965) Rise in the water-table in parts of Potiskum division, Bornu province. Further observations. *Records of the Geological Survey of Nigeria*, 9, 41–57.

Barbiero, L. & Valles, V. (1992) Geochemical aspect of soil alcalinisation in the dallol Bosso valley (Niger). Cahier ORSTOM, série pédologie, XXVII(2), 143–152.

Bassett, T. (2010) Reducing hunger vulnerability through sustainable development. PNAS, 107(13), 5697–5698.

Beyerle, U., Rueedi, J., Leuenberger, M., Aeschbach-Hertig, W., Peeters, F., Kipfer, R. & Dodo, A. (2003) Evidence for periods of wetter and cooler climate in the Sahel between 6 and 40 kyr BP derived from groundwater. *Geophysical Research Letters* [Online] 30, 4. Available from: doi:10.1029/2002GL016310

Boeckh, E. (1992) *Mise à disposition d'un expert hydrogéologue auprès du projet hydraulique rurale dans l'arrondissement de Tessaoua (PHRT). Partie II: rapport technique*. Technical report. Germany, GTZ, Tessaoua/Hanovre.

Bouzou Moussa, I. (1988) L'érosion dans la vallée de Keita (Adar, Niger). Contribution géomorphologique. *PhD Thesis*. Université Joseph Fourier, Grenoble 1, France. 248 pp.

Bouzou Moussa, I., Faran Maiga, O., Ambouta-Karimou, J.M., Sarr, B., Descroix, L. & Moustapha Adamou, M. (2009) Geomorphological consequences of land use and climate change in a sahelian rural catchment basin. *Sécheresse*, 20(1), 145–152.

Carré, P. & Robin, J. (1973) Le Goulbi de Maradi et le Lac de Madarounfa. Données hydrologiques de base. Technical report. Paris, France, Banque Africaine de Développement, Orstom. 88 pp.

Carter, R.C. & Parker, A. (2009) Climate change, population trends and groundwater in Africa. *Hydrological Sciences Journal* (Special issue: Groundwater and climate in Africa), 54(4), 676–689.

Center for International Earth Science Information Network (CIESIN), Columbia University and Centro Internacional de Agricultura Tropical (CIAT). (2005). *Gridded Population of the World Version 3 (GPWv3): Population Density Grids*. [Online] Palisades, NY, Socioeconomic Data and Applications Center (SEDAC), Columbia University. Available from: http://sedac.ciesin.columbia.edu/gpw [Accessed 18th January 2011].

Chaperon, P. (1970) *Projet de mise en valeur du Dallol Maouri, Etude hydrologique, Rapport terminal*. ORSTOM/FAO publ., Paris, France, 172 pp.

Descroix, L., Mahé, G., Lebel, T., Favreau, G., Galle, S., Gautier, E., Olivry, J.C., Albergel, J., Amogu, O., Cappelaere, B., Dessouassi, R., Diedhiou, A., LeBreton, E., Mamadou, I. & Sighomnou, D. (2009) Spatio-temporal variability of hydrological regimes around the boundaries between sahelian and sudanian areas of west Africa: A synthesis. *Journal of Hydrology*, 375, 90–102.

FAO. (1994) *Le projet de développement rural intégré de Keita*. [Online] Rome, Italie. Available from: http://www.fao.org/docrep/x5306f/x5306f00.HTM [Accessed September 29, 2011]

Favreau, G. (2000) Caractérisation et modélisation d'une nappe phréatique en hausse au Sahel : dynamique et géochimie de la dépression piézométrique naturelle du kori de Dantiandou (sud-ouest du Niger) [Characterization and modelling of a rising water table in the Sahel: dynamic and geochemistry of the Dantiandou kori natural hollow aquifer (southwest Niger)]. *PhD Thesis*. University of Paris-XI, Orsay, France.

Favreau, G., Cappelaere, B., Massuel, S., Leblanc, M., Boucher, M., Boulain, N. & Leduc, C. (2009) Land clearing, climate variability and water resources increase in semiarid southwest Niger: A review. *Water Resources Research*. [Online] 45, W00A16. Available from: doi:10.1029/2007WR006785

Gardelle, J., Hiernaux, P., Kergoat, L. & Grippa, M. (2010) Less rain, more water in ponds: A remote sensing study of the dynamics of surface waters from 1950 to present in pastoral Sahel (Gourma region, Mali). *Hydrology and Earth System Sciences*, 14, 309–324.

Gentizon, B. (1998) *Evolution des surfaces boisées dans la region de Tanka (Niger) entre 1950 et 1992*. Travail de Diplôme (Master 2). Switzerland, Institut de Botanique Systématique et Géobotanique, Université de Lausanne. 54 pp.

Geyh, M.A. & Wirth, K. (1980) ^{14}C ages of confined groundwater from the Gwandu aquifer, Sokoto basin, Northern Nigeria. *Journal of Hydrology*, 48, 281–288.

Greigert, J. (1961) *Carte géologique de reconnaissance du bassin des Iullemeden, échelle 1/1 000 000ᵉ*. France, BRGM publ.

Greigert, J. & Bernert, G. (1979) Atlas des eaux souterraines du Niger; état des connaissances (mai 1978). Rapport BRGM 79/AGE/Q01, Orléans, France.

Guengant, J.P. & Banoin, M. (2003) Dynamique des populations, disponibilités en terre et adaptation des régimes fonciers: Le Niger. [Online] FAO/CICRED. 144 pp. Available from: http://www.cicred.org/Eng/Publications/pdf/MonoNiger.pdf [Accessed September 29, 2011]

Guéro, A. (2003) Etude des relations hydrauliques entre les différentes nappes du complexe sédimentaire de la bordure sud-ouest du basin des Iullemmeden (Niger): approches géochimique et hydrodynamique. *PhD Thesis*. University of Paris-Sud, France. 250 pp.

Guyer, J.I., Lambin, E.F., Cliggett, L., Walker, P., Amanor, K., Bassett, T., Colson, E., Hay, R., Homewood, K., Linares, O., Pabi, O., Peters, P., Scudder, T., Turner, M. & Unruh, J. (2007) Temporal heterogeneity in the study of African land use. *Human Ecology*, 35, 3–17.

Ibrahim, M. (2001) Dégradation des bas-fonds et stratégies adaptatives paysannes: cas du Lac de Madarounfa et de son bassin d'alimentation (région de Maradi). *Mémoire de Maîtrise*. Département de géographie, Université Abdou Moumouni, Niamey. 114 pp.

Intergovernmental Panel on Climate Change (IPCC). (2007) Summary for policy makers. In: Solomon, S., Qin, D., Manning, M., Chen, Z., Marquis, M., Averyt, K.B., Tignor, M. and Miller, H.L. (eds.) *Climate Change 2007: The Physical Science Basis*. Contribution of Working Group I to the Fourth Assessment Report of the Intergovernmental Panel on Climate Change [Online] Cambridge, UK, Cambridge University Press. Available from: http://www.ipcc.ch/pdf/assessment-report/ar4/wg1/ar4-wg1-spm.pdf (last accessed 29 Sept 2011).

Jones, D.G. (1960) The rise in the water table in parts of Daura and Katsina Emirates, Katsina province. *Records of the Geological Survey of Nigeria*, 1957, 24–28.

Leaney, F.W., Herczeg, A.L. & Walker, G.R. (2003) Salinization of a fresh palaeo-groundwater resource by enhanced recharge. *Ground Water*, 41(1), 84–92.

Lebel, T. & Ali, A. (2009) Recent trends in the central and Western Sahel rainfall regime (1990–2007). *Journal of Hydrology*, 375, 52–64.

Leblanc, M., Favreau, G., Massuel, S., Tweed, S., Loireau, M. & Cappelaere, B. (2008) Land clearance and hydrological change in the Sahel: SW Niger. *Global and Planetary Change*, 61, 135–150.

Leblanc, M., Tweed, S., Ramillien, G., Tregoning, P., Frappart, F., Fakes, A. & Cartwright, I. (2011) Groundwater change in the Murray Basin from long-term in-situ monitoring and GRACE estimates. In: Treidel, H., Martin-Bordes, J. L., Gurdak, J.J. (Eds.), Climate Change Effects on Groundwater Resources: A Global Synthesis of Findings and Recommendations. IAH - International Contributions to Hydrogeology (27), CRC Press, London.

Le Gal La Salle, C., Marlin, C., Leduc, C., Taupin, J.D., Massault, M. & Favreau, G. (2001) Renewal rate estimation of groundwater based on radioactive tracers (^{3}H, ^{14}C) in an unconfined aquifer in a semiarid area, Iullemmeden Basin, Niger. *Journal of Hydrology*, 254, 145–156.

Leduc, C., Favreau, G. & Schroeter, P. (2001) Long-term rise in a Sahelian water table: The Continental Terminal in South-West Niger. *Journal of Hydrology*, 243, 43–54.

L'hôte, Y. & Mahé, G. (1996) *Afrique de l'ouest et centrale: carte des précipitations moyennes annuelles au 1/6 000 000ème (période 1954–1989)*. Collection des cartes ORSTOM, ORSTOM éd., Paris, France.

Mahé, G., Paturel, J.E., Servat, E., Conway, D. & Dezetter, A. (2005) The impact of land use change on soil water holding capacity and river flow modeling in the Nakamnbé River, Burkina-Faso. *Journal of Hydrology*, 300, 33–43.

Massuel, S., Cappelaere, B., Favreau, G., Leduc, C., Lebel, T. & Vischel, T. (2011) Integrated surface water-groundwater modeling in the context of increasing water reserves of a regional Sahelian aquifer. *Hydrological Sciences Journal*, 56(7), doi: 10.1080/02626667.2011.609171.

Moody, R.T.J. (1997) The Iullemmeden Basin. In: Selley, R.C. (ed.) *African Basins, Sedimentary Basins of the World*. Vol. 3. Elsevier, Amsterdam, The Netherlands, pp. 89–103.

OSS. (2008) *Système aquifère d'Iullemeden (Mali, Niger, Nigeria): Gestion concertée des ressources en eau partagées d'un aquifère transfrontalier sahélien*. Organisation du Sahara et du Sahel, Collection Synthèse n°2. Tunis, Tunisie, OSS Publ. 33 pp.

Pfeffer, J., Boucher, M., Hinderer, J., Favreau, G., Boy, J.P., de Linage, C., Cappelaere, B., Luck, B., Oï, M. & Le Moigne, N. (2011) Local and global hydrological contributions to time-variable gravity in Southwest Niger. *Geophysical Journal International*, 184, 661–672.

Rockström, J., Lannerstad, M. & Falkenmark, M. (2007) Assessing the water challenge of a new green revolution in developing countries. *PNAS*, 104(15), 6253–6260.

Saradeth, S. & Weissmann, T. (2008) AQUIFER – remote sensing as support for the management of internationally shared transboundary aquifers in Africa. In: Lee, C. & Schaaf, T. (eds.) *The Future of Drylands*, Proc. International Scientific Conference on Desertification and Drylands Research – Tunis, Tunisia, 19-21 June 2006 *UNESCO Publishing, Paris*, pp. 217–228.

Scanlon, B.R., Jolly, I., Sophocleous, M. & Zhang, L. (2007) Global impacts of conversions from natural to agricultural ecosystems on water resources: Quantity versus quality. [Online] *Water Resources Research*, 43, W03437. Available from: doi:10.1029/2006WR005486

Séguis, L., Cappelaere, B., Milési, G., Peugeot, C., Massuel, S. & Favreau, G. (2004) Simulated impacts of climate change and land-clearing on runoff from a small Sahelian catchment. *Hydrological Processes*, 18, 3401–3413.

Siebert, S., Burke, J., Faures, J.M., Frenken, K., Hoogeveen, J., Doll, P. & Portmann, F.T. (2010) Groundwater use for irrigation – a global inventory. *Hydrology and Earth System Sciences*, 14, 1863–1880.

Swenson, S. & Famiglietti, J. (2011) Sustainable groundwater management for large aquifer systems: Tracking depletion rates from space. In: Treidel, H., Martin-Bordes, J. L., Gurdak, J.J. (Eds.), Climate Change Effects on Groundwater Resources: A Global Synthesis of Findings and Recommendations. IAH – International Contributions to Hydrogeology (27), CRC Press, London.

UNPD. (2005) World population prospects: The 2004 revision. [Online] Available from: http://www.un.org/esa/population/publications/WPP2004/wpp2004.htm [Accessed February 23, 2011]

USGS. (2008) *Land cover changes in Niger: A synthesis*. Technical report. IGNN/USAID, Niamey, Niger. 8 pp.

Wada, Y., van Beek, L.P.H., van Kempen, C.M., Reckman, J.W.T.M., Vasak, S., & Bierkens, M.P.F. (2010) Global depletion of groundwater resources. *Geophysical Research Letters*, 37, L20402. doi:10.1029/2010GL044571,

Wallin, B., Gaye, C., Gourcy, L. & Aggarwal, P. (2005) Isotope methods for management of shared aquifers in Northern Africa. *Ground Water*, 43(5), 744–749.

CHAPTER 8

Climate change and its impacts on groundwater resources in Morocco: the case of the Souss-Massa basin

Lhoussaine Bouchaou, Tarik Tagma, Said Boutaleb,
Mohamed Hssaisoune & Zine El Abidine El Morjani

ABSTRACT

The study reviews the present tools and results which deal with the identification and prediction of the impacts of climate change (CC) on groundwater resources in Morocco. The paper describes the available data, which could be used to indicate the CC effects on groundwater in the Souss-Massa basin in south-western Morocco. The average of rain ranges from 100 mm/a in the plains to 600 mm/a in the Atlas Mountains. Precipitation data indicate an overall decrease during the three last decades. Water resources show an important depletion in surface and subsurface. The recurrent droughts and decreases in recharge directly affect the groundwater level. This is coupled with increased groundwater abstraction and explains the water crisis witnessed in the area, which is predicted to be particularly affected by CC in the future. Chemical and isotopic tracers indicate a degradation of water quality with increasing salinity originating from seawater intrusion, evaporates and anthropogenic pollution (fertilizers, waste water). By this result, these natural tracers confirm the fact that CC impact directly on groundwater in the Souss-Massa basin. It was argued that the heavily exploited aquifer along the coastal areas is more vulnerable to marine intrusion given the relatively longer residence time of the water and salinization processes in this part of the aquifer. The data compiled in this study provide the framework for a comprehensive management plan in which water exploitation should shift toward the eastern part of the basin where current recharge occurs with young and high quality groundwater. Any variation in the natural recharge can affect immediately the capacity of groundwater to meet the demands of humans and ecosystems in this area. These results should be taken in consideration for the future water management in the country.

8.1 INTRODUCTION

8.1.1 Purpose and scope

In spite of its geographical situation on the Atlantic and Mediterranean coast, Morocco is one of the most arid areas of the world; it experiences highly variable rainfall and recurrent droughts. The limited water resources are threatened by increasing demands and accelerated quality degradation. In addition, the Intergovernmental Panel on Climate Change (IPCC) predicts in its 4th Assessment Report that "*Annual rainfall is likely to decrease in much of Mediterranean Africa and northern Sahara, with the likelihood of a decrease in rainfall increasing as the Mediterranean coast is approached*" (IPCC 2007) Chapter 11, p. 866). The fact that climate is changing has become increasingly clear over the past decade. This prediction is based on several climate models that simulate global and regional mean precipitation. While the multi-model predictions show conflicting results for some areas, in the Mediterranean and Northern Africa the multi-models show consistent results of significant reduction in precipitation (IPCC 2007). The Intergovernmental Panel on Climate Change reports that heat waves have been increasing towards the end of the 20th century and are projected to continue to increase in frequency, intensity and duration worldwide (WHO 2009; Baccini et al. 2008; IPCC 2007; Meehl et al. 2007; Robinson 2001). Climate change is caused by the combined impact of growing human population and economic activities (WHO 2003b). The IPCC has developed a series of 40 scenarios of plausible future trajectories for population growth, economic and technological development (IPCC 2000). Each scenario gives estimates for greenhouse gas emission levels, and predicts the changes in the temperature (Campbell-Lendrum and Woodruff 2007). For example, temperature may be estimated to increase by 0.54°C (scenario B2 – low emission scenario), 0.84°C (scenario A1B – middle emission scenario) or 1.02°C (scenario A2 – high emissions scenario) in 2030, relative to the baseline period (WHO 2009).

Consequently, projections for future renewable water resources in Morocco are bleak, and climate change coupled with increasing water demands are likely to exacerbate the water crisis in Morocco.

The surface water represents two-thirds of the hydro reserves of the country. From 1967 to 2010 more than 120 large dams were built to increase the storage capacity from 2.3 billion cubic meters (BCM) in 1967 to 18 BCM in 2004. During the last four decades the Moroccan water resources development planning has been focusing on increasing the storage of the country's surface water resources for an optimal use in irrigated agriculture, drinking water, industrial supplies, and hydroelectric power production. Huge state funds have been invested in the essential infrastructure to control surface water flows. Further capturing and utilization of about two-thirds of the surface water potential is projected, and a number of major infrastructure projects are in advanced stages of planning and/or construction.

Due to its large geological diversity, a broad range of aquifers from almost all the geological periods are present in Morocco. Overall, 32 deep (200 to 1,000 m) and 48 shallow aquifers are tapped in Morocco. The deep aquifers are often not accessible due to the high economic cost of drilling, whereas the shallow aquifers are more accessible, but also more vulnerable to climate change, pollution and evaporation. Unfortunately, a comprehensive model for evaluation of the full potential and yield of these aquifers has not been completed and several unresolved questions for the potential yield of these aquifers remain. Overall, in the 80 already identified aquifers, the potential sustainable groundwater yield under feasible economical and technological constraints is estimated at 4 BCM/a (Bzioui 2004).

The present study focuses on the Souss-Massa basin which is intensively studied. This paper reviews literature and the observed aspects showing the current state water resources in Morocco through the case of the Souss-Massa basin under the pressures of demographic growth and climate change.

The objectives of the study are to examine: (i) whether any trends can be identified from observed rainfall and temperature data in the region and how can be related to CC; and (ii) what are the impacts of the projected changes in temperature and precipitation on groundwater, more specifically, what impacts these changes will have on recharge, renewal and water quality of the reserves.

8.1.2   Description of the study area: the Souss-Massa basin

The Souss-Massa basin, located in south-western Morocco, is one of the country's most important hydrological catchments with an area of 27,000 km² (Fig. 8.1). Elevations in the catchment range from 0 m (Atlantic Ocean) to 4,168 m (Toubkal peak in the High Atlas Mountains). The plain area lying between 0 and 700 m above sea level (m.a.s.l.) covers about 5,700 km² and contains the groundwater reservoir, while the remaining part is mountainous.

The Plio-quaternary formations of the Souss-Massa plain situated between the High Atlas Mountain in the north and Anti-Atlas in the South (Fig. 8.1) represent the most important aquifer in the southern High Atlas Mountain in Morocco. The economy of the Souss-Massa is primarily based on agriculture, sea fishing and tourism. Surface water and groundwater resources are used both, intensively and extensively throughout the area. Overexploitation, aridity of the climate and the various sources of pollution in the area threaten both the quantity and quality of the hydrous reserves. In order to improve the management of these precious resources, several studies have been carried out during the last years to reach a better understanding of the hydrological functioning of the aquifer system (Boutaleb et al. 2000; Ekwurzel et al. 2001; Hsissou et al. 2002, Ahkouk et al. 2003; Dindane et al. 2003; Krimissa et al. 2004; Bouchaou et al. 2005) using different approaches and tools. The shallow aquifer in the plain is sensitive to climate change and several sources of pollution (wastewater, solid discharges, and agricultural fertilizers).

The rivers of the region, locally called "Oued", have an intermittent flow regime, because the dry season is typically very long (6 to 8 months. The main oueds in this basin are Souss and Massa, which receive important inflow, in particular from the rain-laden High Atlas Mountains in the North and the Anti Atlas Mountains in the South. This inflow coming from a high altitude is infiltrated in the piedmont area and in the beds of rivers which consist of high permeable conglomerates. The shallow aquifer of the Souss-Massa plain is the main resource for drinking, irrigation and industrial water in the region.

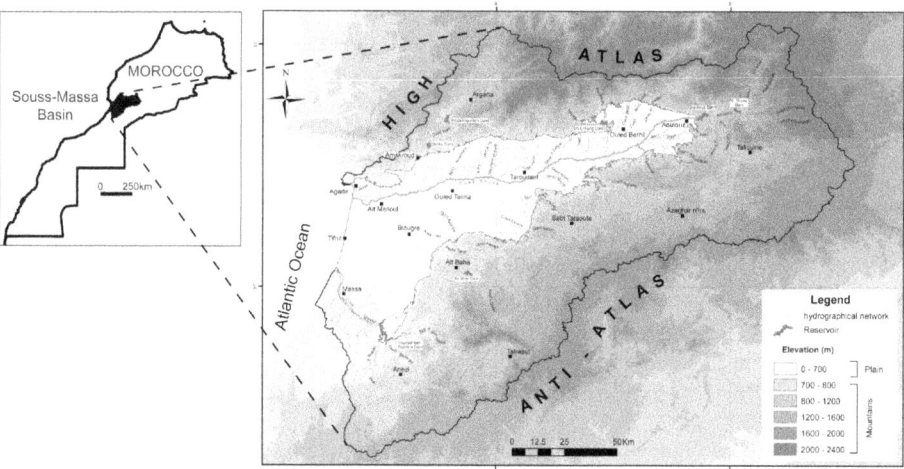

Figure 8.1.   Runoff network and location of dams in the Souss-Massa basin.

8.1.3   Methodology

The paper describes the available meteorological time series data, which could be used to predict CC impacts in a hydrological basin under arid climate in Morocco. The paper reviews the existing results and literature on the impacts of climate change (CC) on water resources through the quantity and quality degradation in the study area.

8.1.4   Relevance to GRAPHIC

The current estimated actual water consumption in Morocco is 11 BCM per year including 3.5 BCM per year from groundwater, of which approximately 90 percent is used for agriculture. The ratio of renewable water resource to the population in Morocco is less than $1000\,m^3$ per person per year, and thus Morocco is defined as "water stressed" country. Due to population growth it is estimated that by 2030, 35% of the population will have reduced access to water and will be below the level of severe stress conditions of $500\,m^3$ per person per year. Thus, growing scarcity is anticipated due to rising demands resulting from expansion of irrigated areas and urban development coupled with diminishing water resources. Morocco is already ranked on the 155th position on a scale of 180 countries, in terms of available fresh water resources. The current overexploitation of the aquifers results in a serious decrease of groundwater levels and degradation of water quality, in particular increasing salinity in all groundwater basins, which reduces potable water availability. In some of the more intensively exploited coastal aquifers groundwater levels have declined and saltwater intrusion migrates inland. In other basins high salinity of groundwater originating from human activities (agriculture, wastewater...) has become a limiting factor for sustainable management of water resources in Morocco. Water deficiency and water quality degradation have important implications for future economical development and social and political stability in Morocco, as water authorities are already struggling to distribute and provide potable water to the domestic and agricultural sectors.

Sustainable water supply is an essential element of Morocco's economy and prosperity. In addition, tourism, which is heavily dependent on water supply, is an important economic component in southern Morocco. Degradation of water quality in the region might reduce agricultural production and the ability to sustain the growing tourism industry in the region. Inadequate water supply might also increase the competition between the rural and urban societies over diminishing water resources. Mitigation of a water crisis and salinization phenomena is therefore a critical factor for economical growth and political stability in Morocco.

The Souss-Massa basin is a good example in Morocco showing quantitative and qualitative aspects of the combined impacts of CC and human activities on groundwater resources. This local case study compares the climate change impacts in a small scale basin with global climate change models. As such, this case study will contribute to the development of scenarios within GRAPHIC.

8.2   RESULTS AND DISCUSSION

An attempt has been made to examine whether the climate is actually changing throughout the study area. This has been done by trend analysis of the available time series of rain and changes affecting the quantity and quality of groundwater. The groundwater level in the Souss-Massa aquifer is closely related to fluctuations in precipitation and shows an inter-annual variation according to the seasonal regime.

8.2.1 Rainfall variation

In general, precipitation in Morocco decreases from the north to the south and from the west to the east. The north-western section receives the greatest quantities of precipitation. The annual average rainfall reaches more than 1,000 mm in the northern Rif Mountains and less than 300 mm in most other parts of Morocco. The rainwater potential shows an annual average of 150 BCM in which 15% of the land surface receives more than 50% of precipitation. The rainfall amount largely fluctuates from 50 BCM in a dry year to 400 BCM in an exceptionally rainy year. In an average year, the potential renewable water is about only 20% of total precipitation; that is 29 BCM including 19 BCM of surface water. Taking into account potential storage sites and groundwater development possibilities, only 22 BCM are annually potentially accessible, 18 BCM from surface water and 4 BCM from groundwater (Agoumi 2003; Bzioui 2004; Agoussine and Bouchaou 2004). Figure 8.2 shows the spatial variation of rainfall in Morocco, with a decrease from the north to the south and indicating the low precipitation in major part of the country.

The climate in the Souss-Massa basin is semi-arid to arid, the rainy season extending from November to March and the dry season from April to October. Locally, the rainfall varies in time and space (Fig. 8.2 and Fig. 8.3), ranging from 200 mm/a in the plain (mean altitude: 460 m.a.s.l.) to 600 mm/a in the mountains (altitude >700 m.a.s.l.). The long-term mean annual precipitation decreased in 20 years from 343 mm in Aoulouz (700 m.a.s.l.) to 232 mm in Taroudant (500 m.a.s.l.). A very clear decrease is observed in the High Atlas Mountains in the north of the area (Tamri) to the south (Massa station) where the monthly rainfall exceeds rarely 100 mm/a (Fig. 8.2).

The variation of rainfall is very important in time and space showing a clear decrease from the mountains to the plains and towards the desert zone in the south. The monthly values indicate a decrease over the last three decades after the most important intensity during the 1960's. The variation shows a clear seasonal irregularity (Fig. 8.3).

8.2.2 Temperature and heat waves

The spatial distribution of the intensity levels of heat wave hazard for Morocco within the Mediterranean context indicates that a large part of the country shows a high temperature, mainly in the south (Fig. 8.4).

According to the seasonal variation using 12-month moving average, the monthly values of the temperature indicate an increase during the last decades since the 1970's (Fig. 8.5). The inter-annual evolution shows the same trend with strong increase in temperature. Mean annual temperature is plotted versus time. The solid black line represents a locally-weighted polynomial regression using kernel smoothing (Fig. 8.6). The bandwidth was selected using the Sheather-Jones method (see the package SiZer for R at http://www.r-project.org for more detail). The grey band represents the 95% confidence intervals.

In conclusion, the area experiences high inter-annual variability of the scarce rainfall distribution making it most vulnerable to climatic change. Consequences of climate

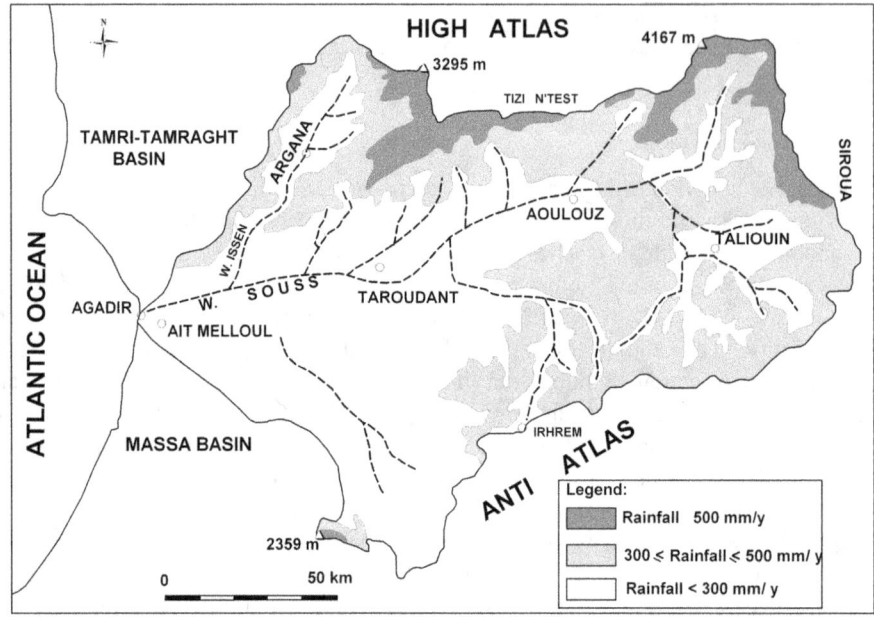

Figure 8.2.   Altitude and rainfall distribution (Elmouden et al. 2005).

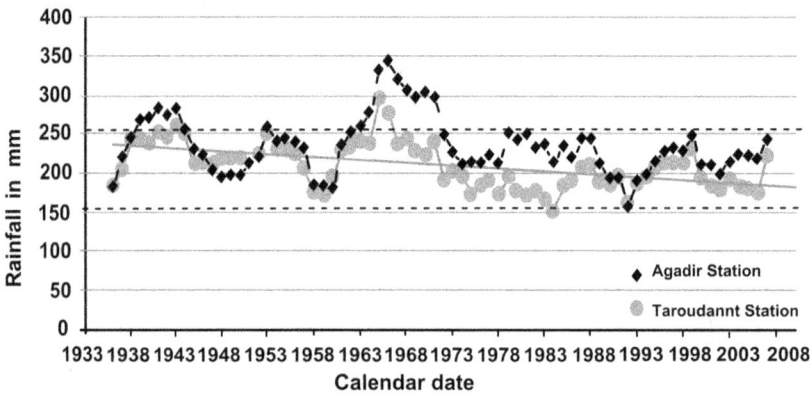

Figure 8.3.   Monthly variation of precipitation in two main stations (Agadir and Taroudant) in the Souss-Massa basin.

change are expected to have major implications on agricultural production and tourism in the area. In this context, increased variability of rainfall and temperature in the region, associated with climate variability and change, implies increased vulnerability. This is likewise true for changes in land use, agricultural production and other climate variability hat affects and responds to the water resources. These changes are often amplified by an increase in population numbers or density or decreasing water supply.

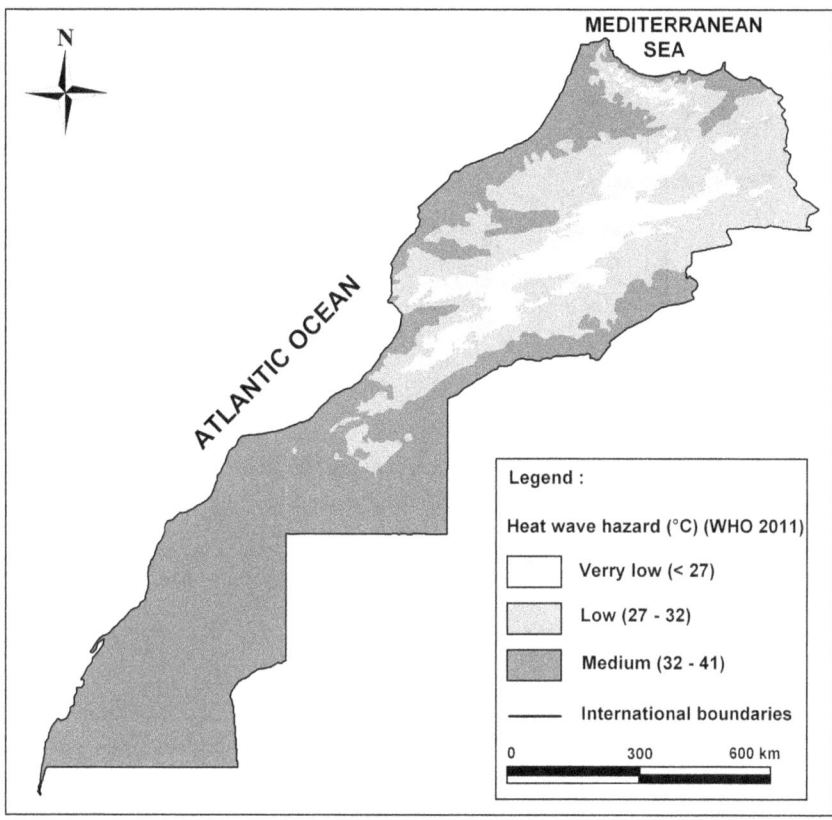

Figure 8.4.   Spatial distribution of the intensity levels of heat wave hazard for Morocco (WHO 2011).

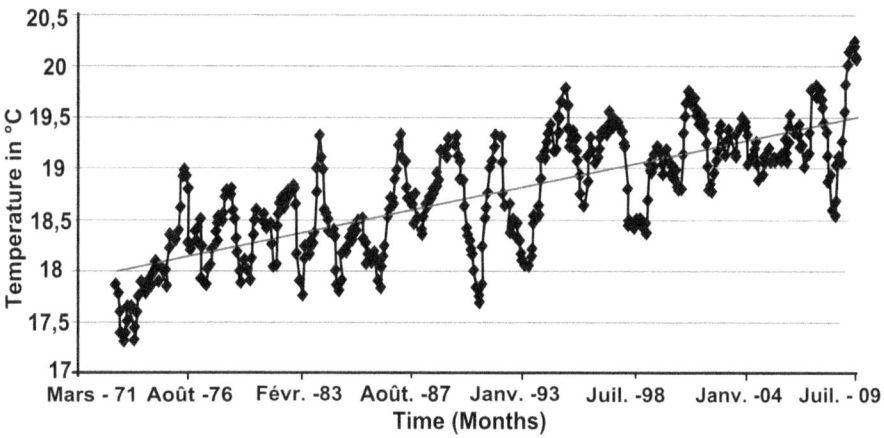

Figure 8.5.   Seasonal variation of the monthly temperature in Agadir station: moving average.

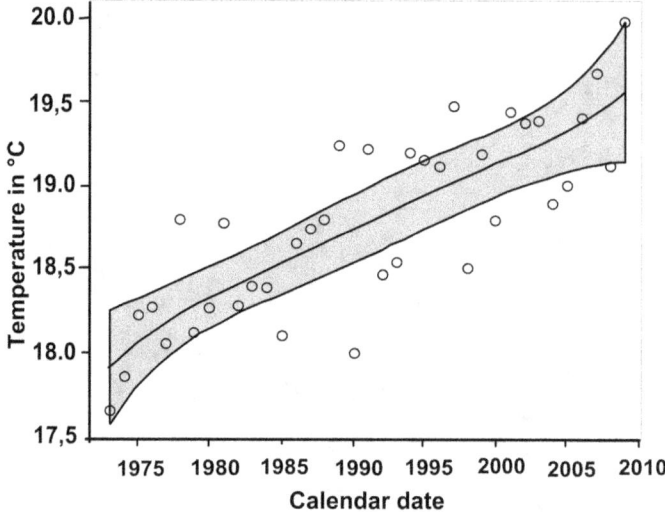

Figure 8.6.   Seasonal variation of the temperature in Agadir station: inter-annual trend evolution. The solid black line represents a locally-weighted polynomial regression using kernel smoothing.

8.2.3   Impacts on groundwater level

As a consequence of decrease in precipitation and demographic pressures, groundwater pumping has been accelerated and the exploitation of the aquifers has intensified. Subsequently, wells were deepened and equipped with more powerful pumps, allowing for much deeper resources like the Turonian aquifer to be exploited. As a result, groundwater tables declined significantly since 1990 (ABHSM 2004).

A number of problems associated with increasing water scarcity and recurrent and extended droughts have been noticed in the Souss area particularly in the past few decades (Bouchaou et al. 2008; Tagma et al. 2009). The extended drought periods are often thought to be related to climate variability arising and can be indicators of CC. The studies supporting these observations in the area are being carried out. Manifestations of water scarcity include, among others, an alarming decline of both surface and groundwater resources, which resulted in the introduction of strict water control by the hydraulic agency. An increasing number of private wells were developed to circumvent these restrictions (Fig. 8.7), which in turn resulted in a further adverse impact on the groundwater table by increasing the extractions (Fig. 8.7). Increased water shortage during the last 20 years, often perceived to be caused by recurrent droughts led to the cancellation of the plans to extend the area irrigated by the main dams in the region. There are concerns that the recurrent droughts will further deplete the available water resources.

Figure 8.8 indicates the impact of rainfall variability in the basin on groundwater quantity. The reaction of the water table to the rainfall recharge is very variable in space and time within the basin. The major trend indicates an overall decrease in water resources, due to the combination of the natural decreased recharge and human activities (extractions). This depletion affects directly the water availability per capita in the

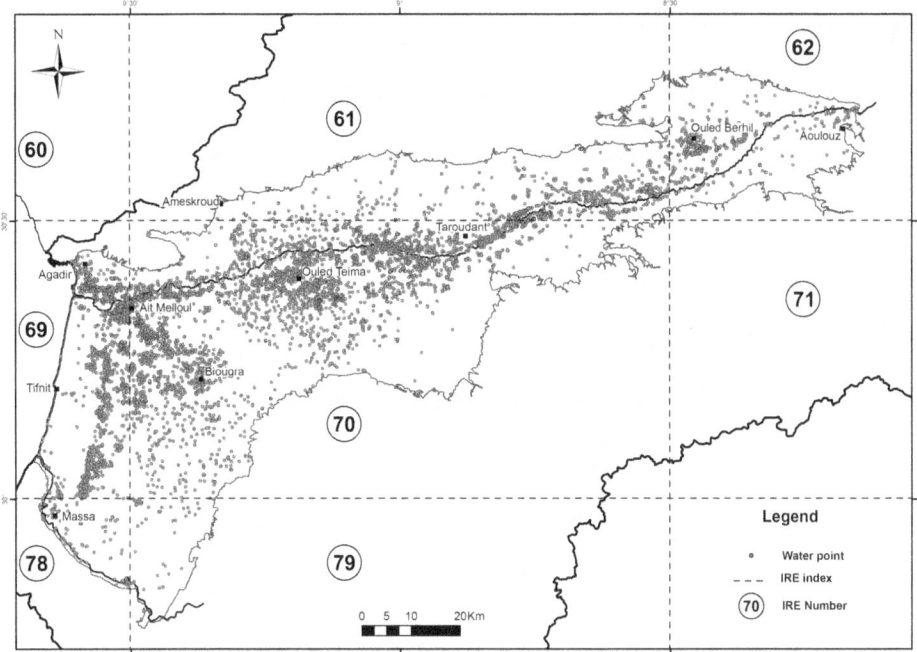

Figure 8.7.   Wells and boreholes in Souss-Massa plain aquifer (more than 25,000 wells).

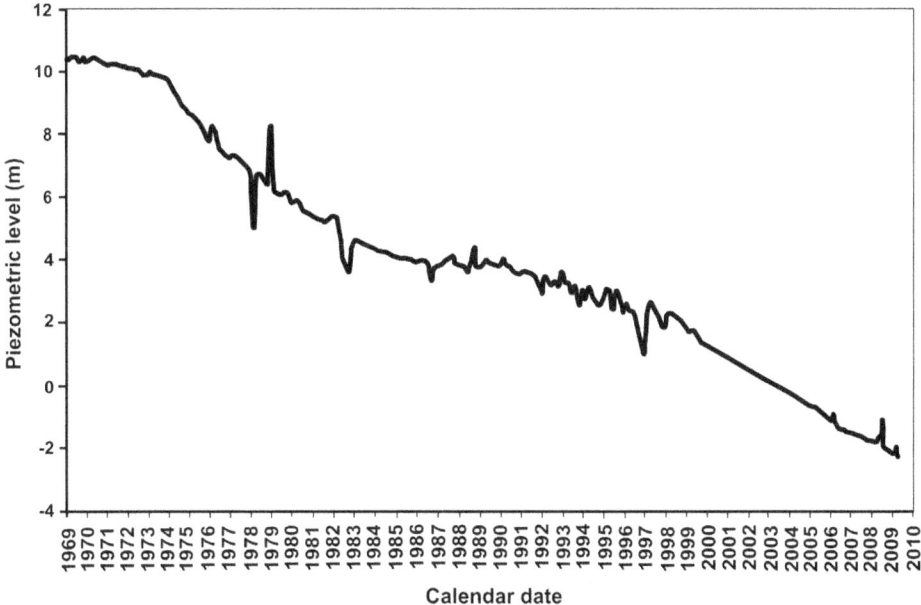

Figure 8.8.   Decrease in water table during the last decades obtained from a monitoring of a representative well located in the middle of the Souss-aquifer.

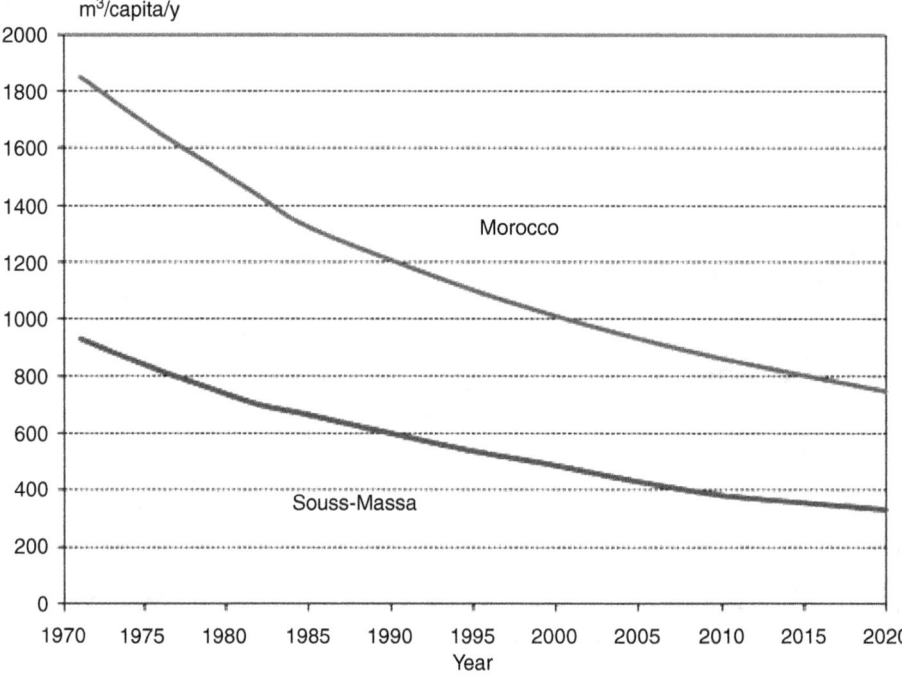

Figure 8.9.   Scenarios evolution of water availability in Morocco and in Souss-Massa region (ABHSM 2008).

country and in particular in the Souss-Massa basin. According to water monitoring and simulation established by hydraulic department, the scenario for the 2020 predicts a possible water crisis in the area (Fig. 8.9).

8.2.4   Impacts on groundwater quality

The depletion of groundwater level induced by limited recharge and overexploitation has induced degradation of water quality in the Souss-Massa plain aquifer. The differ-ent chemical and isotopic data used in some studies carried out in the area (Boutaleb et al. 2000; Hsissou et al. 2002; Bouchaou et al. 2008; Tagma et al. 2009) indicate that salinity of groundwater originates from multiple sources and not only from seawater intrusion as previously thought. These studies have revealed a complex hydrogeological system in which several sources of salinity have been identified, including seawater intrusion, entrapped saline groundwater within the aquifer, saline water derived from salt dissolution, and infiltration of agricultural return flows. The stable isotope data also indicate that the Souss-Massa shallow aquifer is highly influenced by the contribution of recharge water from the High Atlas Mountain, which has a high rainfall, particularly in its upstream part along the Upper Souss plain. The results of age estimates based on ^{3}H and ^{14}C data suggested that relatively old groundwater is practically mined at some wells. This indicates that the Souss-Massa basin is very vulnerable to contamination processes, and the rates of salinization (e.g. seawater-intrusion) or anthropogenic contamination

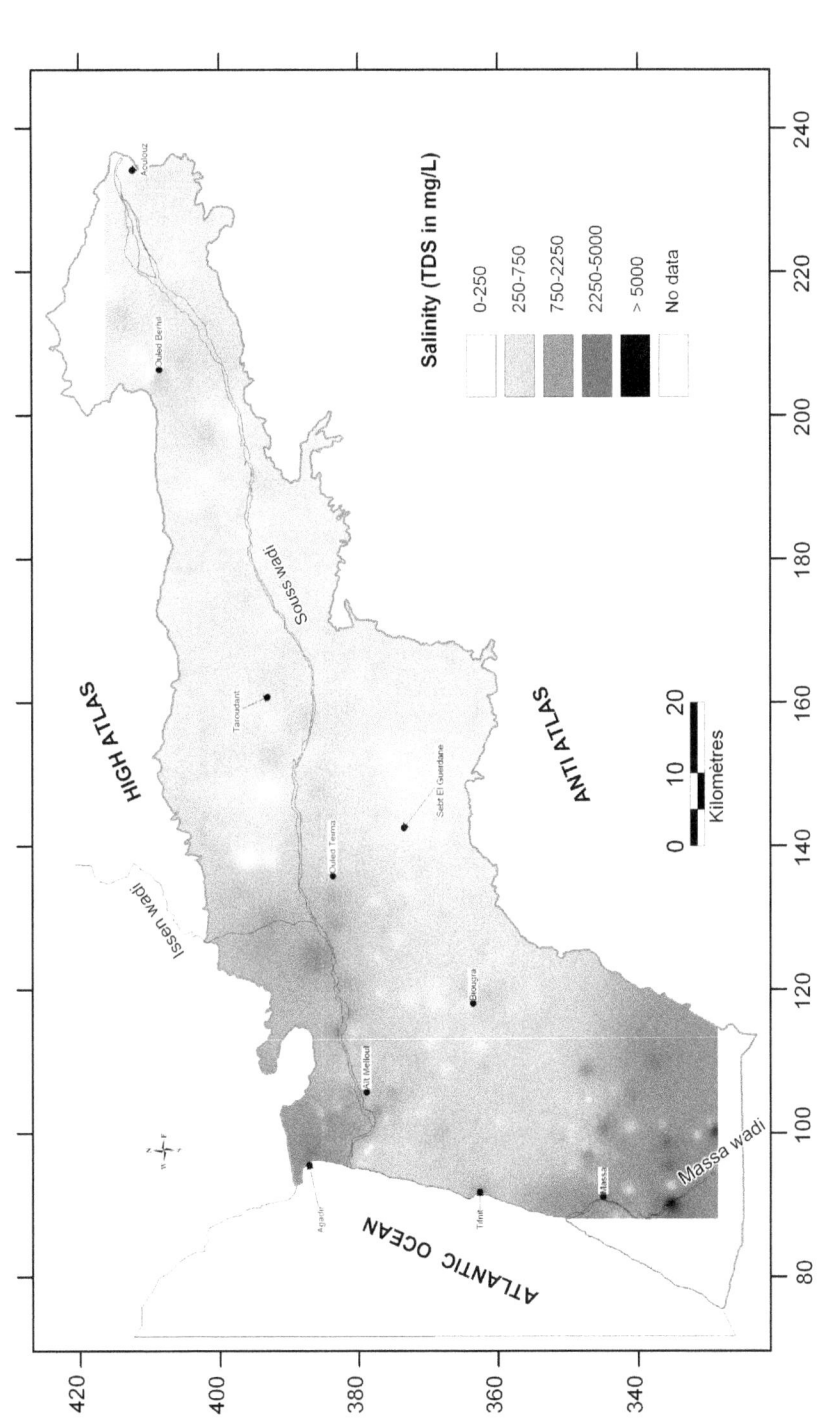

Figure 8.10. Spatial variation of salinity (2004) expressed in mg/L as the total dissolved solids (TDS) in groundwater of the Souss-Massa basin (Tagma et al. 2009).

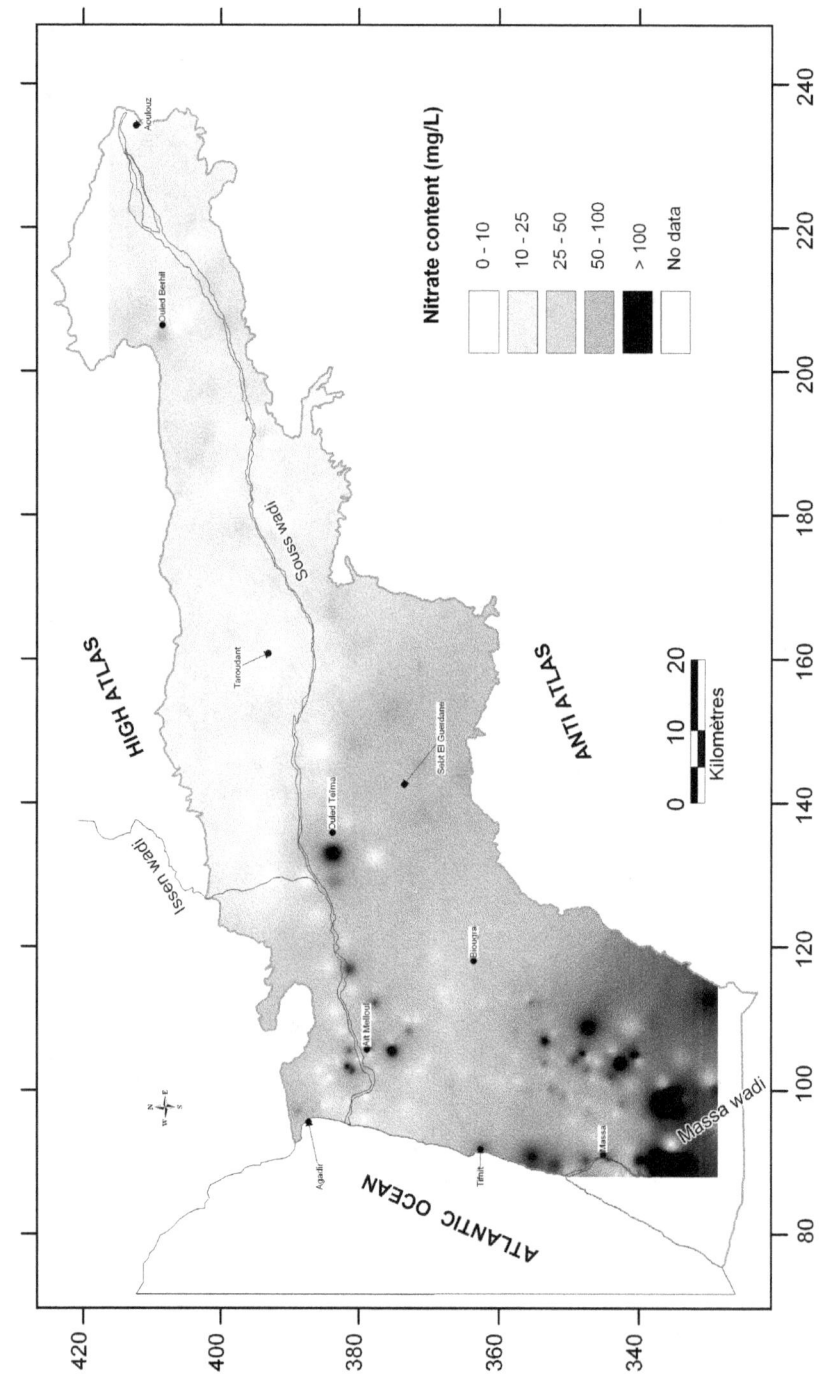

Figure 8.11. Spatial variation of nitrate contents (2004) in groundwater of the Souss-Massa basin (Tagma et al. 2009).

agricultural return flow) are faster than natural replenishment of the aquifer. According to the decrease of the aquifer recharge, the renewal of the groundwater in the reservoir is very low which minimises dilution in the aquifer. Two major areas were distinguished in the basin (Fig. 8.10 and Fig. 8.11): (i) The Souss upstream part with a high water quality and modern recharge, and (ii) downstream in coastal areas with low water quality derived from different salinity sources and long residence time of groundwater, increased by the low replenishment. This is a reflection of climatic change and pressure of human activities. Intense agricultural activities using fertilizers highly affect the water quality with high nitrate contents (Fig. 8.11). The contamination is highlighted mainly in irrigated perimeter. This result means that we cannot focus only on the quantitative understanding of CC impacts that can be also cause deterioration of the groundwater quality in the country. This hides the potential impacts of CC in small scale on the individual river basins or irrigation schemes. Without serious precautions, the degradation of the quality can induce a water crisis in the future. The hydrochemical and isotopic tracers indicating the possible future climatic change in the Souss-Massa basin can be used to quantitatively assess the impacts of CC on water availability and, consequently, to formulate adaptation strategies.

8.3 POLICY RECOMMENDATIONS

Precipitation in the Souss-Massa basin has high variability, as typical for semi-arid climates. Several global-scale studies have concluded that the Mediterranean will be one of the regions most significantly affected CC in the future, in terms of decreasing precipitation and increasing temperatures. This will eventually lead to decreasing availability of water resources. The case of the Souss-Massa basin confirms these global-scale findings at the local level. The rain stations indicate growing rainfall deficits, possibly related to CC, especially over the High Atlas Mountains, which are the main source of water for the entire hydro-system in southern Morocco. The hydrology of the upstream basin and the High Atlas Mountains supplying the main dams and the aquifer in the system has not been assessed sufficiently under present-day conditions, neither for future scenarios of CC. The implications of future water availability scenarios for irrigation system management (diversifying cropping pattern, expanding sprinkler irrigation, improving institutional aspects, conjunctive use of surface and groundwater) are not yet well examined. Obviously, continued water utilization in the coastal and near-coastal areas would further increase the depletion of water resources and degradation of groundwater quality.

More detailed studies on CC impact on groundwater resources in this region will be required. The findings of current and future studies should be taken into consideration for the preparation of water management plans. Groundwater abstraction should be shifted from the heavily populated areas along the coast, which are more vulnerable to contamination, towards the high quality and renewable water resources along the upper zone of the basin. This can affect the recharge area in the upstream and limits the contribution supplying the downstream part. Therefore, the application of artificial recharge in the area and improving of the irrigation schemes may help substantially to anticipate further aggravation of the water resources situation, both quantitatively and qualitatively. Desalinisation of brackish or sea water can be an option in the cases of crisis. Convincing the farmers to use the treated wastewater is as important as challenging in the country.

8.4   FUTURE WORK

The data generated in this study provide the framework for a comprehensive management plan in which water exploitation should shift toward the eastern part of the basin where current recharge occurs with young and high quality groundwater. In contrast, it was argued that the heavily exploited aquifer along the coastal areas is more vulnerable given the relatively longer residence time of the water and salinization processes in this part of the aquifer. What management measures can be implemented in the area in order to administer the possible impacts of CC and pressures of humanity (overexploitation and contamination)? The impacts of CC change on snow cover in the Atlas Mountains that constitute the main recharge area of the basin need to be further assessed. What is the contribution of snow melt and stream flow to groundwater recharge? Options to develop and operate the water-related infrastructure in the basin under CC conditions need to be examined. The use of the isotopic tracers is a good tool to estimate the recharge and the contribution of different component to the groundwater.

A combination of adaptive models should form a good basis to examine how the future water use in the Souss-Massa area can be optimized. The development of this assessment methodology will form an integral part of future studies in many basins in Morocco.

ACKNOWLEDGEMENTS

We thank a lot all the personnel of the Hydraulic Agency Basin of Souss-Massa (ABHSM) for their help and their cooperation in providing the climatic data base.

REFERENCES

ABHSM. (2004) Plan Directeur du Souss-Massa, 2004. Agadir, Maroc, Agence du Bassin Hydraulique du Souss-Massa (ABHSM).

ABHSM. (2008) Plan Directeur du Souss-Massa, 2008. Agadir, Maroc, Agence du Bassin Hydraulique du Souss-Massa (ABHSM).

Agoumi, A. (2003) Vulnerability of North African countries to climatic changes: adaptation and implementation strategies for climatic change. Developing Perspectives on Climate Change: Issues and Analysis from Developing Countries and Countries with Economies in Transition. IISD/Climate Change Knowledge Network. 14 pp. http://www.iisd.org http://www.cckn.net.

Agoussine, M. & Bouchaou, L. (2004) Les problèmes majeurs de la gestion de l'eau au Maroc. *Secheresse*, 15, 187–194.

Ahkouk, S., Hsissou, Y., Bouchaou, L., Krimissa, M. & Mania, J. (2003) Impact des fertilisants agricoles et du mode d'irrigation sur la qualite des eaux souterraines: cas de la nappe libre des Chtouka (bassin du Souss-Massa, Maroc). *Africa Geoscience Review*, 10, 355–364.

Baccini, M., Biggeri, A., Accetta, G., Kosatsky, T., Katsouyanni, K., Analitis, A., Anderson, H.R., Bisanti, L., D'Ippoliti, D., Danova, J., Forsberg, B., Medina, S., Paldy, A., Rabczenko, D., Schindler, C. & Michelozzi, P. (2008) Heat effects on mortality in 15 European cities. *Epidemiology*, 9(5), 711–719.

Bouchaou L., Hsissou Y., Krimissa M., Krimissa S., Mudry J. (2005) – ^{2}H and ^{18}O isotopic study of ground waters under a semi-arid climate. Environmental Geochemistry, pp. 57–64, Green

Chemistry and Pollutants in Ecosystems, E. Lichtfouse, J. Schwarzbauer, D. Robert (Eds.) XXVI, chapter 6, 780 p. 289 illus. ISBN: 3-540-22860-8 Springer.

Bouchaou, L., Michelot, J.L., Vengosh, A., Hsissou, Y., Qurtobi, M., Gaye, C.B., Bullen, T.D. & Zuppi, G.M. (2008) Application of multiple isotopic and geochemical tracers for investigation of recharge, salinization, and residence time of water in the Souss-Massa aquifer, southwest of Morocco. *Journal of Hydrology*, 352, 267–287.

Boutaleb, S., Bouchaou, L., Mudry, J., Hsissou, Y. & Chauve, P. (2000) Effects of lithology on quality of water resources. The case of oued Issen (Western Upper Atlas, Morocco). *Hydrogeology Journal*, 8, 230–238.

Bzioui, M. (2004) Politique et stratégies de gestion des ressources en eau au Maroc. Académie du Royaume du Maroc. Session "La politique et la sécurité alimentaire du Maroc à l'aube du xxie siècle" 20-21-22 novembre 2000, Rabat, 2000, 60 pp.

Campbell-Lendrum, D. & Woodruff, R. (2007) Climate change: Quantifying the health impact at national and local levels. *WHO Environmental Burden of Disease* Series No. 14. Editors: Annette Prüss-Üstün & Carlos Corvalán. Public Health and the Environment, Geneva 2007. http://whqlibdoc.who.int/publications/2007/9789241595674_eng.pdf

Chaponniere, A. & Smakhtin, V. (2006). *A review of climate change scenarios and preliminary rainfall trend analysis in the Oum er Rbia Basin*, Morocco. Working Paper 110 (Drought Series: Paper 8). Colombo, Sri Lanka, International Water Management Institute (IWMI).

Dindane, K., Bouchaou, L., Hsissou, Y. & Krimissa, M. (2003) Hydrochemical and isotopic characteristics of groundwater in the Souss Upstream Basin, southwestern Morocco. *Journal of African Earth Sciences*, 36, 315–327.

Ekwurzel, B., Moran, J.E., Hudson, G.B., Bissani, M., Blake, R., Krimissa, M., Mosleh, N., Marah, H., Safsaf, N., Hsissou, Y. & Bouchaou, L. (2001) An Isotopic Investigation of Salinity and Water Sources in the Souss-Massa Basin, Morocco. First international conférence on Salt Water and Coastal Aquifers (SWICA)-Monitoring, Modeling and Management. Essaouira, Morocco, April 23–25, 2001.

Elmouden, A., Bouchaou, L., Snoussi, M. & Wildi, W. (2005) Comportement des métaux et fonctionnement d'un estuaire en zone sub aride: Cas de l'estuaire du Souss (Côte Atlantique Marocaine). Revista Estudios Geologicos, volume 61 (1–2). Madrid-Espana, Museo Nacional de Ciencias Naturales.

Hsissou, Y., Mudry, J., Bouchaou, L., Chauve, P. & Mania, J. (2002) Use of chemical tracy to study acquisition modality of mineralization and behaviour of unconfined groundwater under semi-arid climate: The case study of the Souss plain (Morocco). *Environmental Geology*, 42, 672–680.

IPCC. (2000) *Emissions scenarios*. A Special Report of IPCC Working Group III for the Intergovernmental Panel on Climate Change. ISBN: 92-9169-113-5. IPCC, 2000 – Nebojsa Nakicenovic and Rob Swart (Eds.) Cambridge University Press, UK. pp 570. www.ipcc.ch/pdf/special-reports/spm/sres-en.pdf

IPCC. (2007) *Climate Change 2007: The Physical Science Basis*. Contribution of working group I to the Fourth Assessment Report of the Intergovernmental Panel on Climate Change. Cambridge, UK and New York, NY, Cambridge University Press.

Krimissa, S., Michelot, J.-L., Bouchaou, L., Mudry, J. & Hsissou, Y. (2004) Sur l'origine par altération du substratum schisteux de la minéralisation des eaux d'une nappe côtière sous climat semi-aride (Chtouka-Massa, Maroc). *Comptes Rendus Geoscience*, 336, 1363–1369.

Meehl, G.A., Stocker, T.F., Collins, W.D., Friedlingstein, P., Gaye, A.T., Gregory, J.M., Kitoh, A., Knutti, R., Murphy, J.M., Noda, A., Raper, S.C.B., Watterson, I.G., Weaver, A.J. & Zhao, Z.C. (2007). Global climate projections. In: Solomon, S., Qin, D., Manning, M., Chen, Z., Marquis, M., Averyt, K.B., Tignor, M. & Miller, H.L. (eds.) *Climate Change 2007: The Physical Science Basis*. Contribution of working Group I to the Fourth Assessment Report of the Intergovernmental Panel on Climate Change. Cambridge, UK and New York, NY, Cambridge University Press.

Robinson, P.J. (2001) On the definition of a heat wave. *Journal of Applied Meteorology*, 40(4), 762–775.

Tagma, T., Hsissou, Y., Bouchaou, L., Bouragba, L. & Boutaleb, S. (2009) Groundwater nitrate pollution in Souss-Massa basin (south-west Morocco). *AJEST*, 3, 301–309.

Willmott, C.J. & Matsuura, K. (1995) Smart interpolation of annually averaged air temperature in the United States. *Journal of Applied Meteorology*, 34(12), 2577–2586.

World Health Organization (WHO), (2003a). The health impacts of 2003 summer heat waves. Briefing note for the delegations of the fifty-third session of the WHO Regional Committee for Europe. Vienna, Austria, 8–11 September 2003. 12pp. At (http://www.irisknet.cn/riskdb/global-risk/juzai/10.6.pdf;Accessed December 15, 2010)

World Health Organization (WHO), (2003b). Methods of assessing human health vulnerability and public health adaptation to climate change. World Health Organization, Geneva (http://www.euro.who.int/document/e81923.pdf; Accessed December 15, 2010)

World Health Organization (WHO). (2009) *Improving public health responses to extreme weather/heat-waves: EuroHEAT*. Technical Summary. Copenhagen, Denmark, World Health Organization Regional Office for Europe.

World Health Organization (WHO). (2011) *Methodology document for the WHO e-atlas of disaster risk. Volume 1. Exposure to natural hazards* Version 2.0: Heat wave hazard modelling. Zine El Abidine El Morjani & Soufiane Idbraim. Ibn Zohr University editions.

CHAPTER 9

Vulnerability of groundwater quality to human activity and climate change and variability, High Plains aquifer, USA

Jason J. Gurdak, Peter B. McMahon & Breton W. Bruce

ABSTRACT

Groundwater resources are affected by human activities and climate change and variability in many complex ways. Although many such studies to date have focused on the quantity of groundwater resources, the quality of groundwater may be a limiting factor for some intended uses such as drinking- or irrigation-water supply and to the long-term sustainability of many groundwater resources worldwide. This chapter characterizes the vulnerability of groundwater quality to coupled human activities and climate change and variability by focusing on important findings and policy implications of a recent (1999 to 2009) groundwater-quality study of the High Plains aquifer (450,000 km^2). In the United States, the High Plains aquifer is the most heavily used aquifer and currently supports one of the largest agricultural economies, but groundwater from this important aquifer is being used at unsustainable rates. Results indicate that contamination, especially elevated nitrate concentrations primarily from agricultural sources, is typically found in shallow groundwater and may be a limiting factor for some intended uses such as drinking-water supply at local and some sub-regional scales. The quality of groundwater from deeper in the High Plains aquifer, where most private, public-supply, and irrigation wells are screened, is generally suitable for drinking and as irrigation water. However, during the past 50 to 60 years, small but measurable increases in concentrations of contaminants in water-supply wells have occurred because of differences in spatial climate patterns, the conversion of rangeland to irrigated cropland, mobilization of natural and anthropogenic solutes in the unsaturated zone, chemical transport to the water table along fast or preferential pathways, mixing of shallower (younger) and deeper (older) groundwater by high-capacity and multiple-screened wells, a limited ability within the aquifer to naturally attenuate some contaminants such as nitrate, and the influence of interannual to multidecadal climate variability on recharge rates and chemical mobilization to the water table. Findings and policy implications of this study may help managers of the High Plains aquifer and other similar aquifers worldwide to prioritize areas that are most sensitive to coupled human activities and climate change and variability, and implement best management practices in the context of adaptability and long-term sustainability of groundwater resources.

9.1 INTRODUCTION

Groundwater resources supply fresh drinking water to more than 1.5 billion people and support streams, lakes, wetlands, aquatic communities, economic development and growth, and agriculture worldwide (Alley et al. 2002). However, the sustainability of global groundwater resources is in question because of coupled socioeconomic and climatic stresses (Taniguchi and Holman 2010; Green et al. 2011). Alley et al. (1999) define *groundwater sustainability* as development and use of groundwater in such a manner that can be maintained for an indefinite time without causing unacceptable environmental, economic, or social consequences. Although *unacceptable consequences* are subjective and specific to intended uses (i.e., drinking-water supply, irrigation supply, ecosystem needs, etc.) (Alley et al. 1999), a universal threat to sustainability is groundwater vulnerability to human activities and climate change and variability (Holman 2006; Bates et al. 2008; Green et al. 2011). We adopt the definition of Farley et al. (2011) that conceptualizes *vulnerability* as a function of sensitivity and adaptability to stresses. Here, sensitivity is the relative degree of response in the groundwater system to human activities and climate change and variability, and adaptability is the relative degree of possible adjustments in groundwater management to human activities and projected or actual climate change and variability (Farley et al. 2011).

Although often overlooked, the quality of groundwater is often as critically important as the quantity of groundwater in terms of groundwater sustainability (McMahon et al. 2007). Groundwater quality that has declined may become a limiting factor for some intended uses such as drinking- or irrigation-water supply. Having adequate water supplies of appropriate quality is vital to the continued economic health and well being of many aquifers worldwide. Furthermore, relatively few studies have evaluated the effects of climate change and variability on groundwater quality (Green et al. 2007, 2011). In one of the few studies, Stuart et al. (2011) conclude that the implications for nitrate leaching to groundwater of the United Kingdom as a result of climate change is not yet well enough understood to make useful predictions without additional monitoring data. Studies that have addressed climate change effects on natural soil and agricultural processes in the United Kingdom report a range of nitrate leaching rates from a slight increase to possibly doubling nitrate concentrations in groundwater by 2100 (Stuart et al. 2011). It is critical to better understand the coupled effect of human and climate stresses on groundwater quality in other aquifers to best predict and manage future groundwater sustainability.

9.1.1   Purpose and scope

This chapter summarizes many of the major scientific findings and management implications of an assessment of water quality in the High Plains aquifer (450,000 km²) of the central United States (U.S.) (Fig. 9.1) by the U.S. Geological Survey (USGS) National Water-Quality Assessment (NAWQA) Program. Serving as a case study for the United Nations Educational, Scientific, and Cultural Organization (UNESCO) study Groundwater Resources Assessment under the Pressure of Humanity and Climate Change (GRAPHIC) (UNESCO 2008), results and implications from the High Plains aquifer NAWQA study are presented in the context of sensitivity and adaptability to help inform resource managers and policy makers responsible for managing groundwater vulnerability in the face of coupled human activities and climate change and variability. Detailed technical information, data and analyses, collection and analytical methodology, models, graphs, maps, and publications that support the following findings and implications are summarized by McMahon et al. (2007) and Gurdak et al. (2009b).

9.1.2   Study area description: High Plains aquifer

The High Plains aquifer in the Great Plains physiographic province underlies about 450,000 km² in parts of eight States (Colorado, Kansas, Nebraska, New Mexico,

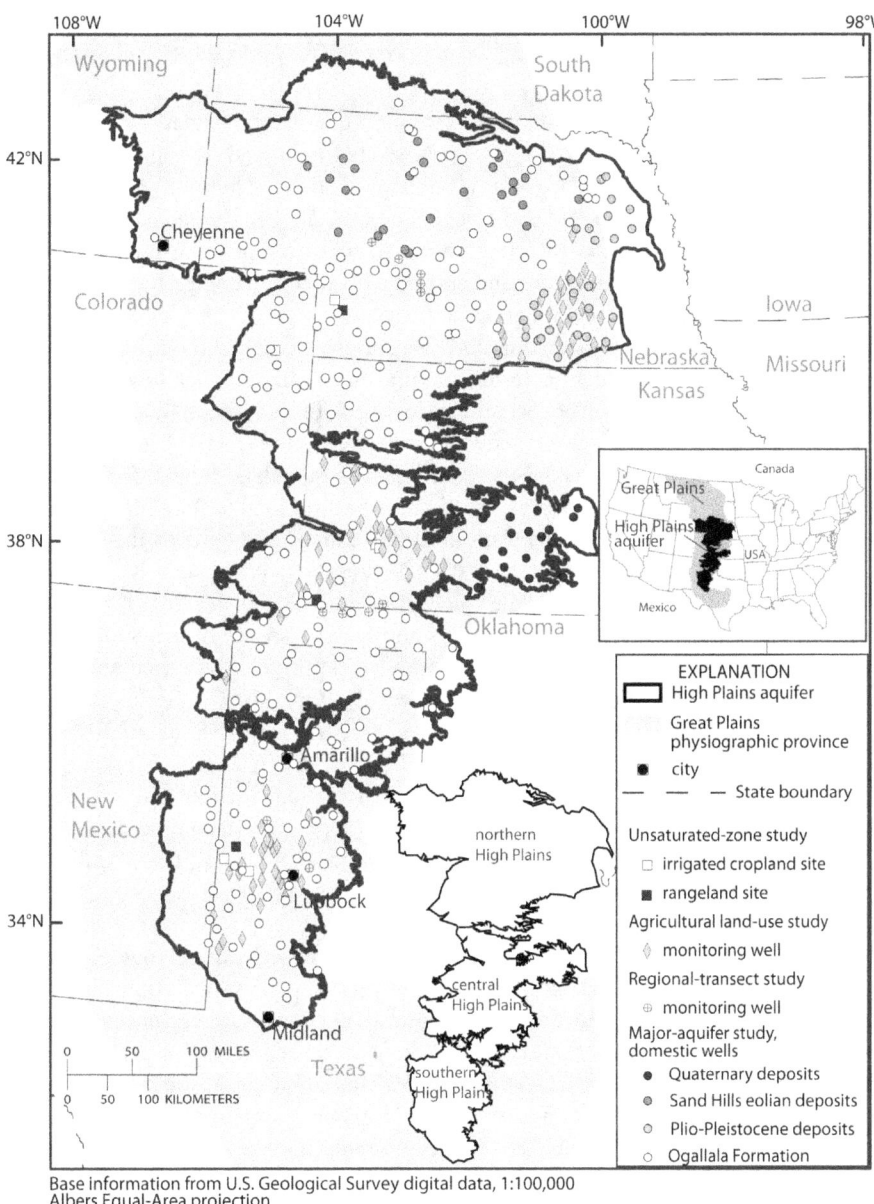

Figure 9.1. The High Plains aquifer underlies approximately 450,000 square kilometres (km²) beneath eight States in the central United States (modified from Gurdak et al. 2009b). The locations are shown for the study components (unsaturated-zone study, land-use study, regional-transect study, and major aquifer study).

Oklahoma, South Dakota, Texas, and Wyoming) of the U.S. (Fig. 9.1). In 2000, the High Plains aquifer had an estimated 3.67×10^{12} cubic meters (m³) of drainable water in storage, making it one of the largest aquifers in the world (McGuire, 2007). Elevations of the

High Plains range from about 300 m along the eastern boundary to about 2,400 m along the north-western boundary (McMahon et al. 2007). The topography is characterized by flat to gently rolling terrain.

The High Plains has a middle-latitude, semi-arid continental climate (average annual air temperature ranges from 4 to 18°C), characterized by abundant sunshine, moderate average annual precipitation (30 to 84 cm), frequent winds, low humidity, and a high rate of annual evaporation (152 to 277 cm) (Dennehy et al. 2002). The large areal extent of the High Plains results in relatively large north-to-south temperature gradients (cooler in the north and warmer in the south) and east-to-west precipitation gradients (more precipitation in the east and less precipitation in the west).

The population of the study area in 2000 was approximately 2.3 million people. The majority (77%) of this population reside in rural areas and smaller towns and cities, while only 23% live in the four largest cities: Lubbock, Texas (199,564), Amarillo, Texas (173,627), Midland, Texas (94,996), and Cheyenne, Wyoming (53,011) (McMahon et al. 2007) (Fig. 9.1). The dominant land use/land cover is rangeland (about 56%) and agriculture (about 38%, in cultivated crops, small grains, fallow, and pasture/hay); urban land use accounts for only 3% (McMahon et al. 2007). About one-third of the agricultural land is irrigated, mostly concentrated in eastern Nebraska, south-western Kansas, and the west-central part of the Texas Panhandle (McMahon et al. 2007).

Use of groundwater from the High Plains aquifer as a source of irrigation water has transformed the study area into one of largest and most productive agricultural regions, earning it the nickname "breadbasket of the world" (Opie 2000). Groundwater withdrawals from the High Plains aquifer in the year 2000 accounted for about 20% of total groundwater withdrawn in the U.S. (Maupin and Barber 2005). Most (97%) of the water withdrawn from the High Plains aquifer is used for irrigation (Maupin and Barber 2005). Although withdrawals for drinking water account for a relatively small percentage of the total groundwater use, they provide drinking water for about 82% of the 2.3 million people who live within the study area boundary (McMahon et al. 2007).

The High Plains aquifer consists of sedimentary deposits that form six hydraulically connected hydrogeologic units. The most extensive of these hydrogeologic units is the Ogallala Formation, which makes up about three-fourths of the total High Plains study area (McMahon et al. 2007). The depth to water below land surface (unsaturated-zone thickness) ranges from 0 to approximately 152 m, averages about 30.5 m, and is generally greatest in the central and southern High Plains (McMahon et al. 2007). The saturated thickness of the High Plains aquifer ranges from less than 1 to more than 300 m and averages about 61 m (McMahon et al. 2007). The saturated thickness varies geographically and is greatest in the northern High Plains.

Evaporation rates exceed precipitation rates across much of the High Plains, so little water is available to recharge the aquifer. Recharge to the High Plains aquifer occurs by infiltration of irrigation water, aerially diffuse infiltration from precipitation, focused infiltration of storm- and irrigation-water runoff through streambeds and other topographic depressions (Gurdak et al. 2008), and upward movement of water from underlying aquifers (McMahon et al. 2007). Discharge from the High Plains aquifer is primarily to irrigation well pumping, streams and underlying aquifers, groundwater flow across the eastern boundary of the aquifer, and evapotranspiration. Regional groundwater flow is generally from west to east; however, local variability in hydraulic gradients can result in different directions of groundwater flow, particularly near high-capacity pumping wells and major rivers.

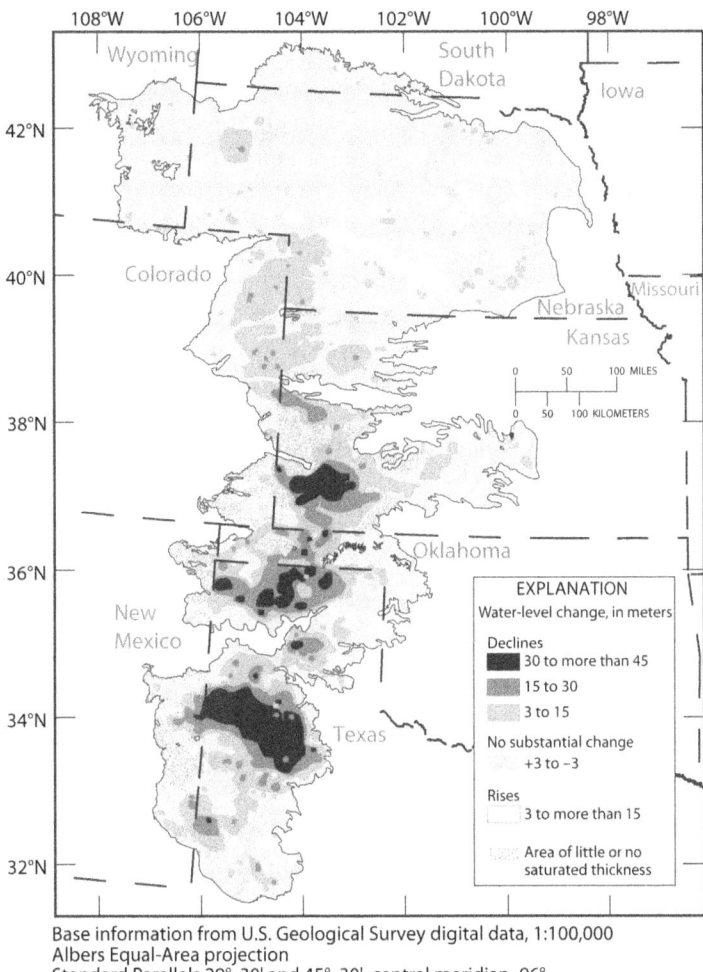

Figure 9.2.   The most substantial declines in water levels since predevelopment (approximately the mid-1950s) are in parts of Texas, New Mexico, Oklahoma, and Kansas (modified from McGuire (2007) and the High Plains water-level monitoring study (http://ne.water.usgs.gov/ogw/hpwlms/, accessed January 16, 2009).

Water levels have declined (Fig. 9.2) substantially since predevelopment times (approximately the mid-1950s) because groundwater withdrawals have greatly exceeded recharge across much of the aquifer (McGuire et al. 2003). The largest water-level declines range from 15 to more than 45 m, primarily across parts of Kansas, Oklahoma, New Mexico, and Texas (Fig. 9.2). The saturated thickness of the aquifer has decreased by more than 50% in some parts of Kansas and Texas (McGuire 2007). This groundwater depletion has led to increased pumping costs and a reduction of water discharging to streams, among other things. Ecosystems along riparian corridors that rely on groundwater discharge are adversely affected by even small volume changes in the groundwater system (Alley 2006).

9.1.3   Methodology

The High Plains aquifer study used a Source-Transport-Receptor model to assess the linkages between the quality of the water recharging the aquifer (Source), the effects of transport through the hydrologic system on water quality (Transport), and the quality of the used resource as represented by water pumped from private, public-supply, and irrigation wells (Receptors). The study was divided into four study components: unsaturated-zone studies, land-use studies, major-aquifer studies, and regional-transect studies (Figs. 9.1 and 9.3). The objectives of the unsaturated-zone studies were to measure the mass of chemicals stored in the subsoil above the water table and to determine their transit times from land surface to the water table in rangeland and irrigated settings across regional climate gradients. The objective of the land-use studies was to measure and explain the quality of young or recently recharged groundwater (defined here as approximately less than 50 years old based on tritium-age dating) in irrigated cropland and urban settings by sampling water-table monitoring wells installed as part of the NAWQA Program. The objective of the major-aquifer studies was to broadly assess water-quality conditions in the aquifer by sampling networks of randomly selected, aerially distributed existing private wells. The objectives of the regional-transect studies

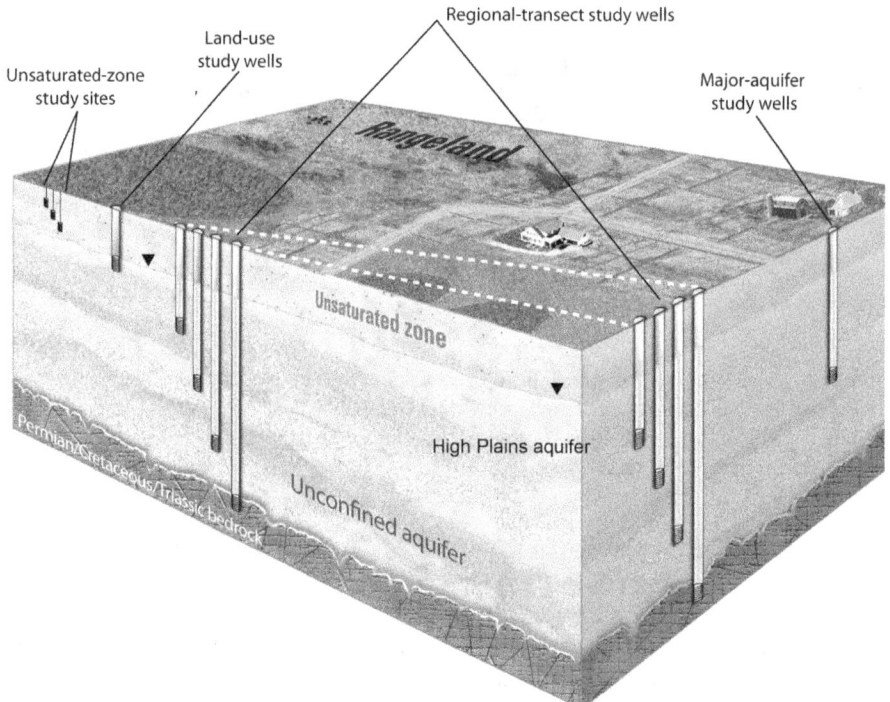

Figure 9.3.   The four distinct study components enabled a systematic understanding of the quality of water as it moves from land surface, through the unsaturated zone, to the water table, and finally to deeper depths in the aquifer that are commonly used to supply water for drinking and irrigation (modified from Gurdak et al. 2009b).

were to characterize vertical gradients in groundwater chemistry and apparent age in the thickest areas of the aquifer and to identify major biogeochemical reactions affecting the quality of water along groundwater flow paths leading from recharge areas to down-gradient wells. An important concept of the High Plains aquifer assessment design was the nesting of the study components (Figs. 9.1 and 9.3), which resulted in improved understanding of assessment results and supported extrapolation of results to unmoni-tored areas of the aquifer.

Monitoring water and chemical movement from the land surface to the water table and from the water table to the base of the aquifer along flow paths leading from recharge to discharge areas resulted in an improved understanding of the timescales at which that movement occurred. This enabled interpretation of local-scale data over a broad range of variability contained within the regional aquifer. Thus, a beneficial aspect of the High Plains aquifer study design is that it bridges the interpretation gap between local-scale studies and the aggregation to regional and national-scale synthesis. Additional methodology details about study design, sampling locations, sampling frequency, and data collected is presented by McMahon et al. (2007) and Gurdak et al. (2009b).

9.1.4   Relevance for GRAPHIC

Many themes and long-term goals of the High Plains aquifer study support the mission of GRAPHIC (UNESCO 2008). The High Plains aquifer is a globally prominent example of a regionally important groundwater resource that is being used at unsustainable rates (Alley et al. 2002), vulnerable to climate change and variability (Gurdak et al. 2007a; Gurdak 2008; Gurdak and Roe 2010), and has groundwater contamination from natural and anthropogenic sources (McMahon et al. 2007; Gurdak et al. 2009b). The High Plains aquifer is the most intensively used groundwater resource in the U.S., producing almost twice the volume of water than any other U.S. aquifer (Maupin and Barber 2005). Groundwater from this aquifer supports domestic agricultural production (Dennehy et al. 2002) and is an important *virtual water* source for grains and other agricultural products that are exported globally (Konikow and Kendy 2005). Since widespread development in the 1940s and '50s, more than half the groundwater in storage has been used in some areas of the High Plains aquifer (McGuire et al. 2003) and elevated concentrations of nitrate and other contaminants have increased (see this chapter). Similar to other trans-boundary aquifers separated by geopolitical boundaries, sub-regions of the High Plains aquifer have differing degrees of groundwater-level declines and groundwater contami-nation because the aquifer underlies parts of 8 States that each have different water laws, regulations, and groundwater-management practices and policies. The following **Results, Discussion, and Policy Implications** section details many lessons learned that may have specific relevance to groundwater science and management in aquifers worldwide that are in semi-arid climates and have substantial agricultural land-use activities and climate sensitivity. Additionally, the nested design of the High Plains study is an example that could be used to help bridge the interpretation gap between regional-scale GRAPHIC case studies and the aggregation to a global-scale synthesis. Some of the long-term research goals of the High Plains study as related to GRAPHIC are detailed below in the **Future Work** section.

9.2   RESULTS, DISCUSSION, AND POLICY RECOMMENDATIONS

The results from the High Plains NAWQA assessment show where, when, why, and how specific water-quality conditions occur in groundwater across the High Plains aquifer and yield science-based implications for assessing and managing the quality of this water resource as affected by human activities and climate change and variability. Understanding groundwater quality conditions and the natural and human factors that control water quality in this aquifer is important because of the implications to human health, the sustainability of rural agricultural economies, and the substantial costs associated with land and water management, conservation, policy, and regulation. McMahon et al. (2007) and Gurdak et al. (2009b) present the details that support the following major findings and implications, which are framed within the previously mentioned conceptualization of groundwater *vulnerability* as a function of sensitivity and adaptability to human and climate stresses (Farley et al. 2011).

9.2.1   Groundwater availability and sustainability are a function of quantity and quality

Groundwater is the primary source of water used in the High Plains; thus, knowledge about groundwater availability and sustainability are essential for the informed management of this limited resource. Groundwater availability and sustainability are sensitive to many factors, one of which is water quality. Understanding groundwater quality is important because it directly affects how water can be used now and in the future. Water quality generally has been overlooked in the High Plains because the primary focus has been on obtaining a sufficient water supply, and it has been broadly assumed that the aquifer contains high-quality water. For the most part, results from the NAWQA study support that assumption.

Implications for Adaptability

- Groundwater quality, particularly regarding elevated nitrate or dissolved-solids concentrations, may be a limiting factor for some intended uses such as drinking- or irrigation-water supply at local and, in some instances, sub-regional scales.
- Groundwater quality, particularly regarding elevated nitrate concentrations in recently recharge groundwater, is changing over time and, because of the slow rates of water movement in the aquifer and lack of nitrate attenuation capacity, could affect groundwater availability for decades, centuries, or longer.

9.2.2   Conversion of rangeland to irrigated cropland affects water quality

Nitrogen and pesticide applications on irrigated cropland result in substantially more nitrate and pesticides being transported to the water table at irrigated cropland settings than at rangeland settings (McMahon et al. 2007). In some more arid climate regions of the High Plains, however, the change in recharge chemistry following conversion of rangeland to cropland results from the mobilization by irrigation water of large natural nitrate and chloride deposits in the unsaturated zone (above the water table).

Conversion of rangeland to irrigated cropland could result in increasing groundwater nitrate and chloride concentrations over time in the southern High Plains aquifer as natural subsoil salt deposits are mobilized and transported to the water table by irrigation return flow. Mobilization of natural salt deposits has occurred in some areas of the High Plains, but slow chemical transit times indicate that many of those salts generally have not reached the water table. The amount of salts (chemical mass) entering the aquifer could increase in the future as the stored chemicals, which currently reside in the unsaturated zone, are mobilized, transported downward in the unsaturated zone, and reach the water table. Thus, groundwater quality may be sensitive to future climate change and variability that increases unsaturated zone water fluxes and chemical transport and mobilization to the water table.

Generally, less nitrate is stored in the subsoil beneath rangeland than beneath irrigated cropland in the High Plains (Fig. 9.4), which is consistent with the much larger nitrogen application rates on cropland compared to rangeland. Large natural nitrate deposits, however, have been detected in the subsoil beneath rangeland in some locations (Fig. 9.5). Some nitrate could be mobilized in the unsaturated zone by irrigation return flow if the rangeland is converted to cropland, which could lead to increased nitrate concentrations in the aquifer. Nitrogen isotopic data indicate that groundwater quality is sensitive to nitrate from both fertilizer and natural sources that are mobilized under the high recharge conditions that occur beneath irrigated fields (Fig. 9.5).

Implications for adaptability

- Reducing the acreage of rangeland that is converted to irrigated cropland in areas that contain large natural subsurface nitrate and chloride deposits (currently not delineated across the entire High Plains aquifer) is likely to be an effective way to reduce the dissolution and transport of natural chemical constituents in recharge water and resulting adverse effects on water quality in the aquifer.
- Implementing efficient agricultural-chemical application and irrigation technologies in areas not currently implemented to reduce deep percolation of irrigation water and

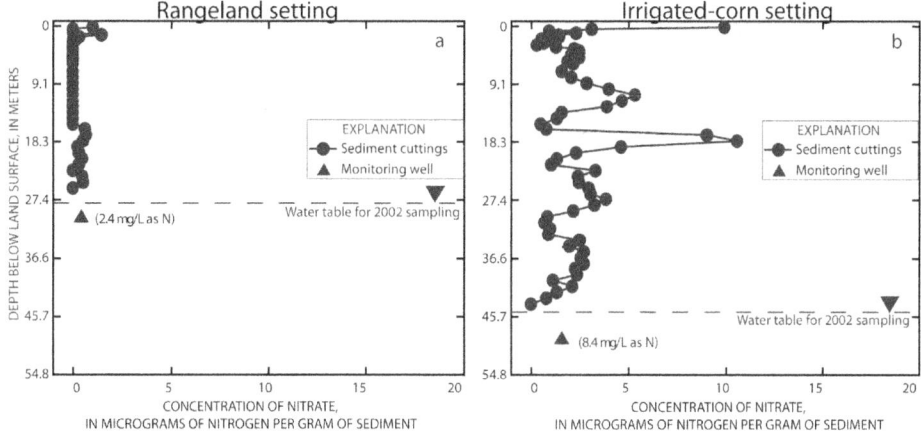

Figure 9.4.   Generally, less nitrate is stored in the subsoil beneath (**a**) natural rangeland than beneath (**b**) irrigated cropland, such as corn fields, in the High Plains. The subsoil nitrate can be derived from natural and agricultural sources (modified from Gurdak et al. 2009b).

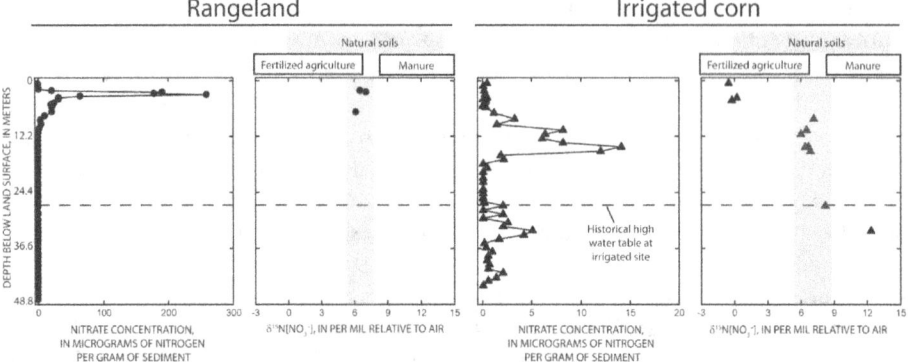

Figure 9.5.   Chemical and isotopic data indicate that natural nitrate deposits exist in relatively large concentrations in the unsaturated zone beneath rangeland in some parts of the High Plains (McMahon et al. 2006). When that rangeland is converted to irrigated cropland, some of the nitrate can be mobilized by irrigation return flow, thereby increasing the amount of nitrate moving to the water table. In many locations, these large nitrate concentrations in unsaturated-zone pore water have not yet reached the water table (McMahon et al. 2006) (modified from Gurdak et al. 2009b).

chemicals is likely to be an effective way to reduce the adverse effects on the quality of the recently recharged groundwater.

- Improving estimates of climate change and variability effects on subsurface water flux under different land uses is likely to lead to better planning and management of future chemical loading to the water table.

9.2.3   Chemical transport to the water table follows fast and slow paths

Groundwater quality is sensitive to chemical transport to the water table by fast and slow paths through the unsaturated zone (McMahon et al. 2007). Estimated transit times of water and chemicals through the unsaturated zone beneath irrigated cropland at the unsaturated-zone study sites are highly variable, ranging from 49 to 373 years (McMahon et al. 2006). A major finding is that although the estimated chemical transit times are only approximations, the times exceeded the irrigation period at all of the study sites (28 to 46 years), implying that agricultural chemicals should not have reached the water table, when in fact they have (McMahon et al. 2006). Agricultural chemicals are commonly detected at the water table beneath irrigated cropland. The discordant finding of long chemical transit times beneath irrigated cropland and the presence of agricultural chemicals at the water table cannot be explained as a single flow process in the unsaturated zone and indicates that both fast and slow paths for water and chemical movement are present in the unsaturated zone (Fig. 9.6) (McMahon et al. 2006; Gurdak et al. 2007a; Everett 2011).

Fast paths (with transit times of months to decades) through the unsaturated zone are most likely to occur beneath topographic depressions in the land surface in which surface runoff from irrigation or precipitation collects (such as seasonal ponds or in playas in the southern High Plains (Gurdak and Roe 2010) because of the increased hydraulic head by the ponded water (Fig. 9.6) (McMahon et al. 2006). Fast paths also

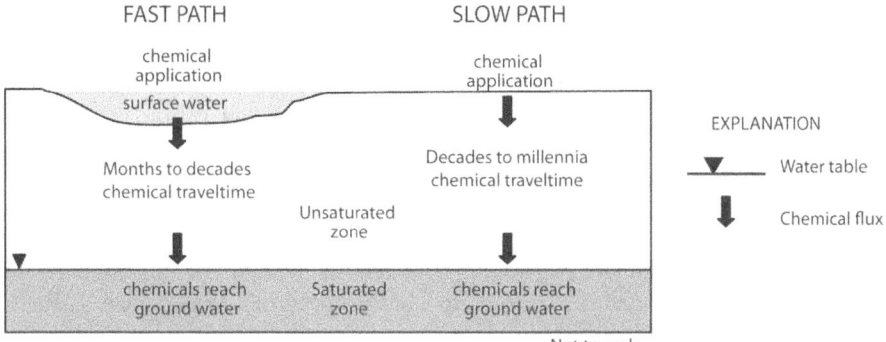

Figure 9.6.   Water and chemical movement from land surface to the water table may be slow (decades to millennia) or fast (months to decades) depending on the natural hydrogeological conditions in the unsaturated zone, land-management practices, and climate variability (McMahon and others, 2006; Gurdak et al. 2007a; Gurdak et al. 2008). Therefore, it is important to understand the rate of water and chemical movement and the time scales in which best-management practices may affect groundwater quality for determining the implications on the sustainability of a high-quality groundwater supply in the High Plains aquifer (modified from Gurdak et al. 2009b).

could occur through wellbore leakage and through cracks, burrows, fissures, and other soil structural features. Climate variability that occurs on interannual to multidecadal timescales also has been shown to be an important control on transit times beneath fast paths (Gurdak et al. 2007a). Slow paths (with transit times of decades to centuries) may occur in fine grained sediments or beneath flat terrain, and possibly in other environments. Evapotranspiration can restrict water movement in both fast and slow path settings (Fig. 9.6) (McMahon et al. 2006).

Implications for adaptability

*   Reducing the amount of untreated agricultural and urban runoff to topographic depressions or other fast-path zones is likely to reduce the rapid transport of contaminants to the water table and be an effective management strategy toward minimizing groundwater contamination.
*   Improving current understanding of climate change and variability effects on fast path contaminant transport is critical for future management of the High Plains groundwater resource.

9.2.4   The quality of shallow and deep groundwater are substantially different

Changes in water quality have occurred over time that may affect the sustainability of the groundwater resource in the High Plains aquifer. Important spatial differences in the concentrations of dissolved solids, nitrate, pesticides, arsenic, and other constituents were observed between the quality of shallow and relatively young (recently recharged) groundwater and the quality of deep and relatively old groundwater (McMahon et al. 2007). Dissolved-solids concentrations in shallow groundwater generally increased from north to south in the High Plains aquifer and were significantly greater in the southern part (median of 800 milligrams per litre (mg/L)) than in the northern and central parts (medians of about 450 to 500 mg/L). Nitrate above local background concentrations (defined here as 4 mg/L as nitrogen (N) (McMahon et al. 2004)) in shallow groundwater was detected throughout the High Plains aquifer. Nitrate concentrations in shallow groundwater were greater than the background concentration in 90% of samples from the northern, 60% of samples from the central, and 55% of samples from the southern High Plains. Although widely detected, pesticide concentrations in shallow groundwater were less than U.S. Environmental Protection Agency (USEPA) drinking-water standards in all but 2 of 119 samples. Atrazine and its degradate deethylatrazine were the most commonly detected pesticide compounds. Concentrations of arsenic in shallow groundwater were significantly greater in samples from the southern High Plains than in samples from the northern High Plains. About 40% of the samples in the southern High Plains exceeded the USEPA arsenic drinking-water standard of 10 µg/L, whereas none exceeded the standard in the northern High Plains. A major control on the spatial differences in groundwater quality is the strong regional climate gradient from north to south in the High Plains that affects recharge rates and contaminant fluxes to the water table (McMahon et al. 2007). Therefore, the future quality of shallow and recently recharged groundwater may be sensitive to climate change and variability that substantially affects the regional patterns in recharge rates and contaminant fluxes.

Understanding the connection between chemical use at the land surface and the quality of relatively young and recently recharged groundwater is important but costly to measure across the entire High Plains aquifer. NAWQA study design provides a basis for extrapolating results to unmonitored areas of the aquifer. Using the monitoring data, a probability map of elevated nitrate concentrations (greater than background concentrations of 4 mg/L as N) at the water table (Fig. 9.7) was developed to evaluate the chemistry of recharge in unmonitored areas of the High Plains aquifer (Gurdak and Qi 2006). The analysis showed that the features of the aquifer affecting vulnerability to nitrate concentrations greater than background include nonirrigated agriculture, irrigated agriculture, organic matter in the soil, depth to the water table, and average percentage of clay in the unsaturated zone (Gurdak 2008; Gurdak et al. 2007b). Output from these types of predictive models can be used to identify potentially vulnerable areas for enhanced monitoring and protection under current climate conditions. The vulnerability map (Fig. 9.7) was created using data collected from 1990 to 2004 and therefore might not be appropriate for forecasting future conditions in the aquifer (Gurdak and Qi 2006). Therefore, additional process-based models would be necessary to evaluate the effects of future climate change and variability on groundwater vulnerability to elevated nitrate concentrations.

The quality of deeper groundwater that is used by many private, public-supply, and irrigation wells in the High Plains generally is of suitable quality for most uses, although some spatial differences in water quality are observed across the region (McMahon et al. 2007). Based on USEPA drinking-water standards, the quality of deeper groundwater decreases from the northern to the southern High Plains where the water contains greater concentrations of dissolved solids, chloride, nitrate, fluoride, manganese, arsenic, and uranium. These elevated constituent concentrations in deeper groundwater from the southern High Plains are the result of natural processes, such as water/rock interactions, and human activities, such as the mixing of high-quality groundwater with lower-quality water from underlying hydrogeologic formations, induced by pumping of high-capacity wells (McMahon et al. 2007).

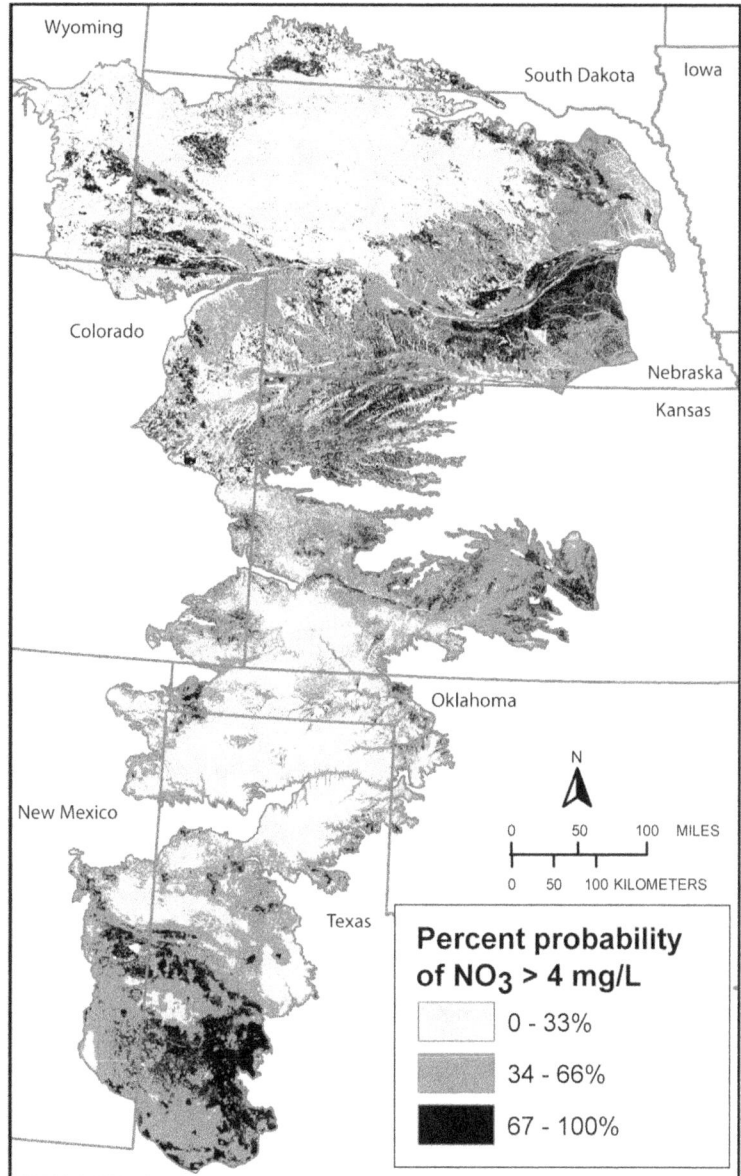

Figure 9.7.   Spatial distribution of the probability of detecting nitrate greater than 4 mg/L as nitrogen (N) in recently recharge groundwater (defined here as less than 50 years) (modified from Gurdak and Qi, 2006).

Dissolved-solids concentrations in the aquifer in rangeland or undeveloped areas typically increase gradually with depth below the water table as the residence time of groundwater increases and water/rock interactions progress (Fig. 9.8a). However, concentrations of dissolved solids generally decreased with depth below the water table near agricultural land and indicate increased concentrations of dissolved solids in recharge in agricultural areas. Agricultural activity at land surface can increase concentrations of dissolved solids in recharge above what would be expected under natural conditions, resulting in a reversed gradient in dissolved-solids concentration near the water table (Fig. 9.8b).

Natural upwelling of saline groundwater from underlying geologic units at the base of the aquifer in regional discharge zones can cause large increases in dissolved-solids concentrations with depth below the water table (Fig. 9.8c). The upwelling could be enhanced by pumping in the aquifer that lowers hydraulic heads and increases the upward-hydraulic gradient across the base of the aquifer (McMahon et al. 2007). These results indicate that agricultural inputs, natural upwelling, and pumping-enhanced upwelling all have the potential to alter groundwater quality at different depths in the High Plains aquifer.

Implications for adaptability

- The quality of groundwater near the water table may not be currently suitable for human consumption in some locations of the aquifer because of elevated concentrations of salts, arsenic, nitrate, and (or) pesticides. However, the quality of groundwater in deeper parts of the aquifer, with some exceptions, currently is generally suitable for human consumption.
- The quality of groundwater that is influenced by mixing with brackish surface water or water from underlying hydrogeologic formations may not be suitable for irrigation or drinking-water supply in some locations of the aquifer because of the

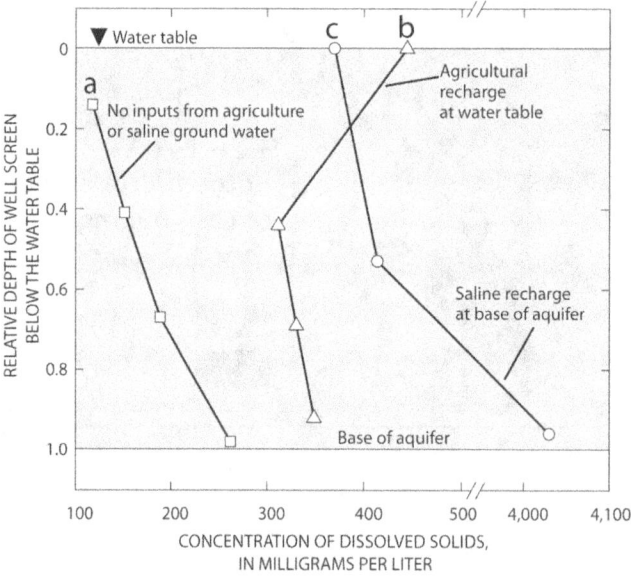

Figure 9.8. Vertical gradients of dissolved-solids concentrations at (**a**) an undisturbed location generally increase with depth, but can be (**b**) reversed from what would be expected in undisturbed areas because of agricultural recharge. Natural and pumping-enhanced upwelling of brackish water from underlying units results in (**c**) substantial increases in dissolved-solids concentrations near the base of the aquifer (modified from Gurdak et al. 2009b).

elevated concentrations of dissolved solids. Future climate change and variability that decrease precipitation during the growing season and (or) cause drought may, in turn, lead to increased pumping rates from high-capacity wells, which could increase the upward flow of saline groundwater from underlying geologic units and further reduce groundwater quality of some deep groundwater that is primarily used by many private, public-supply, and irrigation wells.

• Groundwater remediation is expensive, slow, and impractical across regional-scale aquifers. Therefore, irrigation, agriculture, and other land-use practices that prevent groundwater contamination are likely to be a more effective way to maintain the availability and sustainability of groundwater in the High Plains aquifer for the intended uses for human consumption and agricultural practices.

• Long-term monitoring is likely to be an effective way to detect gradual temporal trends and to provide early warning of water-quality problems from human activities and climate change and variability.

9.2.5   Mixing of groundwater by high-capacity wells adversely affects water quality

The quality of deeper groundwater from private, public-supply, and irrigation wells, which typically are completed using one long or multiple screens across much of the saturated thickness of the High Plains aquifer, is sometimes affected by mixing with poorer-quality water from the water table. Mixing is also caused by leakage through long or multiple well screens and by long-term pumping of high-capacity public-supply and irrigation wells. Mixing caused by leakage (or short-circuiting) and pumping is a major process for moving chemical constituents from near the water table to deeper zones more rapidly than would occur otherwise under natural hydraulic gradients. Pumping by wells may also accelerate the upward movement of saline water from underlying formations. In the central High Plains, halite-dissolution brines with dissolved-chloride concentrations up to 35,000 mg/L are present in geologic units underlying the aquifer and contribute about 5% of the water at the base of the High Plains aquifer in the regional-discharge area of the Cimarron River valley (McMahon et al. 2004). Water level declines in the aquifer caused by overpumping could enhance upwelling of the saline water. Thus, human activity is not only a source of chemical constituents to the water table, but it could enhance the transport of natural constituents into the aquifer.

Implications for adaptability

• The following management strategies are likely to help reduce the adverse effects on the quality of water in the deep zones of the aquifer caused by mixing processes:

• reduce the screen length of public-supply and irrigation wells, and optimize the spatial and temporal distribution of pumping to minimize mixing and water-table drawdown;

• eliminate the practice of screening wells across confining layers in the aquifer; and

• practice screening wells below confining layers in the aquifer where contaminant sources at the land surface are a concern, and practice screening wells above confining layers in the aquifer where contaminant sources in underlying bedrock are a concern.

• Proper construction of wells to limit the movement of low-quality groundwater above and (or) below a well screen will result in maintaining a higher-quality groundwater supply for a variety of uses.

9.2.6   Limited ability to naturally attenuate some contaminants

The High Plains aquifer is limited in its ability to naturally attenuate some contaminants, such as nitrate. Denitrification, the primary natural process for removing nitrate in groundwater, generally occurs very slowly in the aquifer because of oxic conditions and relatively low dissolved carbon concentrations in the groundwater, and would require hundreds to thousands of years to lower nitrate concentrations by just 1 mg/L as nitrogen (N) in many locations (McMahon et al. 2007). Additionally, because water residence times in the aquifer are long (several thousand years in some locations), simply flushing nitrate from the aquifer could take many years. These results highlight the importance of managing land use in the High Plains to minimize the amount of nitrate entering the groundwater.

Implications for adaptability

- In the indefinite future, the aquifer will continue to be subject to the effects of mobilized agricultural chemicals on groundwater quality from ongoing irrigation, conversion of rangeland to cropland, chemicals already stored in the unsaturated zone, or climate change and variability.
- The long transit times of nitrate through the unsaturated zone will undoubtedly delay future improvements in water quality from implementation of best management land-use practices.

9.2.7   Interannual to multidecadal climate variability affects recharge and groundwater quality

The influence of interannual to interdecadal climate variability on groundwater levels, deep infiltration events, and downward displacement of subsoil chloride and nitrate pore-water reservoirs in the High Plains may have a substantial effect on groundwater resources (Gurdak et al. 2007a). Variability in precipitation and groundwater-level time-series data was partly coincident with known climate cycles, such as the Pacific Decadal Oscillation (PDO) (10- to 25-year recurrence interval), North American Monsoon System (NAMS) (6 to 10 years), and the El Niño/Southern Oscillation (ENSO) (2 to 6 years) (Fig. 9.9). Most of the variance in the precipitation time-series was attributed to >PDO (greater than 25-year recurrence interval) and PDO periods, capturing 51 to 64% and 23 to 78% of the variance, respectively (Gurdak et al. 2007a) (Fig. 9.9). These findings indicate the importance of lower-frequency climatic forcings, with lesser controls by NAMS and ENSO, on precipitation variability across the High Plains. The strength of correlation between >PDO and PDO periods and precipitation variability is consistent with the findings from McCabe et al. (2004) that indicate that most of the variance in drought frequency across the conterminous U.S. is attributed to lower frequency climatic forcings. These findings from the High Plains aquifer are consistent with other recent studies (Hanson et al. 2004; Holman et al. 2009; Holman et al. 2011; Kuss 2011) that have identified groundwater-level fluctuations in other aquifers in the United States and United Kingdom that are affected by low frequency (interannual to multidecadal) atmospheric and ocean circulation systems.

Although water-level fluctuations in the High Plains aquifer are dominated by seasonal groundwater abstraction for irrigation, climatic variability within groundwater

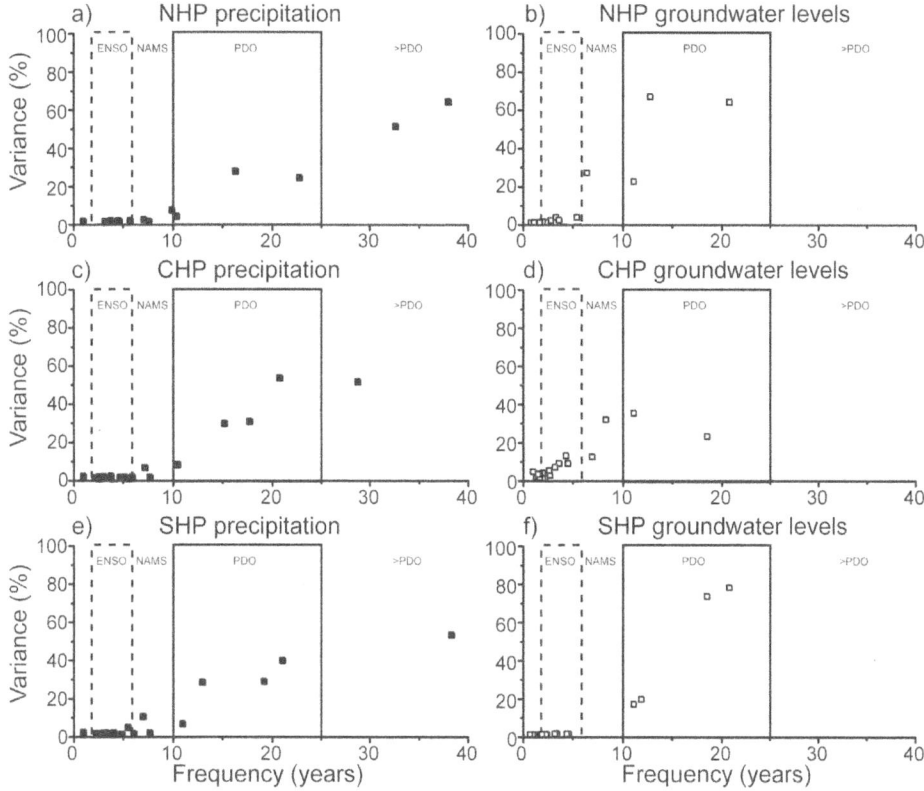

Figure 9.9.   Reconstructed components from the singular spectrum analysis of precipitation and groundwater level time series from the northern High Plains (NHP) (**a, b**), central High Plains (CHP) (**c, d**), and southern High Plains (SHP) (**e, f**) have frequencies (yeas) that are partially consistent with known climate cycles, including the El Niño/Southern Oscillation (ENSO) (2- to 6-year periodicity), North American Monsoon System (NAMS) (possible 6- to 10-year periodicity), Pacific Decadal Oscillation (PDO) (10- to 25-year periodicity), and a greater than 25-year periodicity (>PDO) that may be associated with the Atlantic Multidecadal Oscillation (AMO) (see Gurdak et al. 2007a for additional details). Total climate-related variance is the sum of the variance from individual reconstructed components within a given frequency range.

level hydrographs also was identified (Gurdak et al. 2007a). PDO-like variations in groundwater levels accounted for most of the total climate-related variance at the northern (67 to 87%) and southern (94 to 96%) High Plains sites, while accounting for only 23 to 35% of the variance at the central High Plains sites, similar to results for the precipitation time-series (Fig. 9.9). The precipitation and water-level time-series were lag correlated, which could be an indication of the importance of interannual to interdecadal climate variability on recharge in the High Plains aquifer (Gurdak et al. 2007a).

Even though lower-frequency climate forcings such as PDO may exert major controls on unsaturated-zone water fluxes and aquifer recharge in the High Plains, annual- and event-scale variability also can be important. Total potential profiles measured at one of the southern High Plains unsaturated-zone sites in 2001–05 indicate that a relatively large wetting front moved to a depth of 7 m below land surface during an unusually wet period in 2004 (Gurdak et al. 2007a). Such wetting-front movement had not been observed at the 7-m depth during the 4 years of monitoring prior to that time. The onset of deep infiltration appears to have coincided with particularly large total monthly precipitation from June to December 2004, especially the near record-setting precipitation during November 2004. Total precipitation recorded at the rangeland unsaturated-zone site was 3.5 times greater than the average for that month. The National Oceanic and Atmospheric Administration (2006) reported similar record precipitation across much of the southern High Plains for November 2004: Lubbock, Texas, 9.4 times greater than the average; and Amarillo, Texas, 6 times greater than the average. More recent observations (2006–10) at unsaturated-zone sites in the central and northern High Plains indicate that at least four wetting fronts rapidly propagated to previously unobserved depths (7.3 to 23.0 m below land surface) (Everett 2011). Corresponding water fluxes are orders of magnitude greater than previous estimates using tracer-based techniques and are likely the result of pore-scale dual-domain flow (Everett 2011). Such observations indicate that at any particular location within the aquifer, unsaturated zone water fluxes and recharge rates may have considerable temporal variability and are particularly sensitive to extreme climate events.

Implications for adaptability

- Interannual to multidecadal climate variability have important implications for groundwater resource sustainability, yet are poorly documented and not well understood in most aquifer systems.
- Findings from this study illustrate the importance of longer frequency climatic variations on long-term groundwater-level fluctuations and aquifer sustainability but also indicate that shorter frequency climatic events could influence water and chemical fluxes in the unsaturated zone.
- Wetting fronts such as those observed at the unsaturated-zone sites could be important for moving chemicals in the shallow subsurface into deeper parts of the unsaturated zone, where hydraulic gradients are consistently downward to the water table.

9.2.8   The quality of most water produced by private, public-supply, and irrigation wells is suitable for the intended uses

The quality of groundwater from deeper in the High Plains aquifer, where most private, public-supply, and irrigation wells are screened, is generally suitable for drinking and as irrigation water. Comparison of private-well water quality to USEPA national primary and secondary drinking-water standards indicates that water from the Ogallala Formation in the northern and central High Plains had the best water quality, whereas water from the Ogallala Formation in the southern High Plains had the poorest quality. Most exceedances of primary and secondary drinking-water standards were those for dissolved solids, nitrate, arsenic, fluoride, iron, manganese, and nitrate. The most frequently detected pesticide compounds were atrazine and deethylatrazine, and the most frequently detected volatile organic compound was chloroform. None of the pesticide compounds or volatile organic compounds detected in private wells exceeded a primary drinking-water standard.

The changing quality of groundwater recharge during the past 50 to 60 years and transport processes in the High Plains aquifer have resulted in small but measurable increases in concentrations of anthropogenic contamination in water-supply wells. This is illustrated by comparing nitrate concentrations in High Plains groundwater since the 1930s. Median nitrate concentrations in groundwater were similar from the 1930s to the 1960s, somewhat larger in the 1970s, and significantly larger yet in the 1980s and 1990s (Fig. 9.10) (Litke 2001; McMahon et al. 2007). This type of long-term gradual increase in groundwater nitrate concentrations is an example of the concept called "creeping normalcy" (Fogg and LaBolle 2006) because the gradual increase generally is hard to separate from the background when viewed in the context of short timeframes.

Implications for adaptability

- If contaminated, the deep zones in the aquifer, in which production wells are screened, are not likely to be remediated quickly because of slow recharge rates, long water residence times in the aquifer, and slow rates of contaminant degradation.
- An implication of the trend data observed in water from supply wells of the High Plains aquifer is that some constituent concentrations in the used resource will continue to increase in the future as the fraction of recent recharge captured by supply wells increases. The upward nitrate trend could limit groundwater sustainability given current uses and therefore is an important factor for future adaptability planning and management.

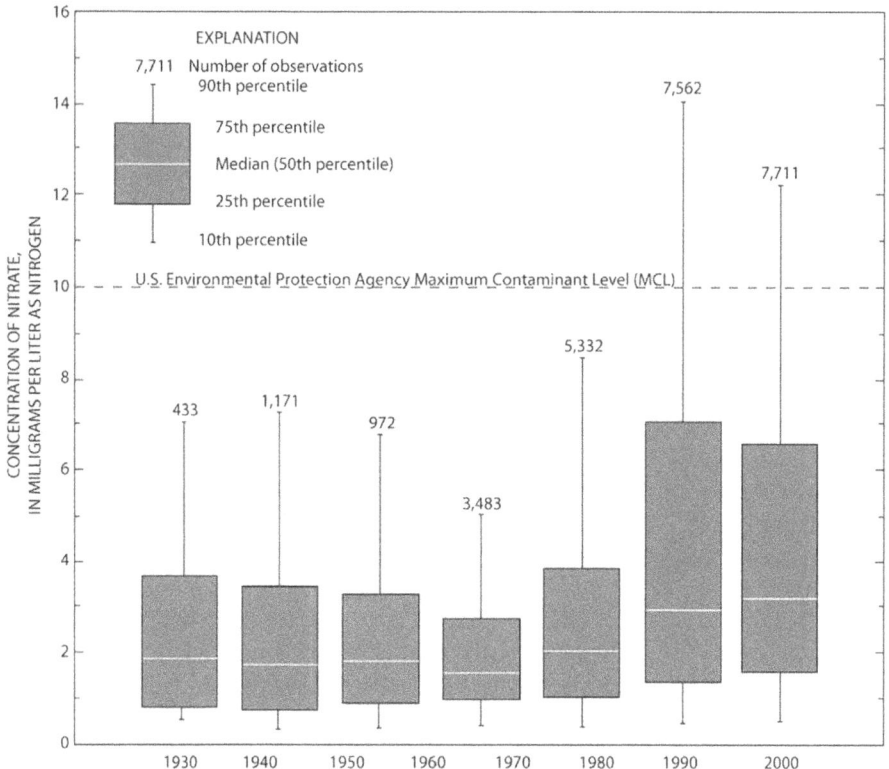

Figure 9.10. A "creeping normalcy" over time of nitrate concentrations in water from supply wells may limit the sustainability of the aquifer as a source of drinking water (modified from Litke 2001).

9.3   FUTURE WORK

Understanding coupled human activities and climate change and variability can be integral to successful management of groundwater resources (Hanson et al. 2004; Clark et al. 2011). Thus, there is an urgent and ongoing need to evaluate and understand climate change and variability over the long term to better plan and manage groundwater resources well into the future, while taking into consideration the increasing stresses on groundwater resources from population growth and industrial, agricultural, and ecological needs (Warner 2007). Several critical questions remain regarding human activities and climate change and variability on groundwater resources of the High Plains aquifer and other similar aquifers worldwide. Remaining critical questions that require additional research are adapted from Gurdak et al. (2009a) and include, but are not limited to:

1. How do recharge, discharge, and change in storage in the High Plains aquifer respond to natural climate cycles on interannual to multidecadal timescales and anthropogenic climate change?
2. Considering future climate change, what fraction of hydrologic response is due to natural variability versus human activities?
3. Are certain time periods of climate variability more cost-effective for artificial storage and recovery efforts, and so necessitate time-varying groundwater management policies?
4. Are certain regions of the High Plains aquifer more/less vulnerable to changes in storage and water quality due to climate change and variability?
5. During the next 30–100 years, the Great Plains physiographic province (Fig. 9.1) may receive less snowfall in winter, the snow will begin falling later and melt earlier, and more winter precipitation will be rain rather than snow (Intergovernmental Panel on Climate Change 2007). Under such a climate scenario of more winter rainfall, winter recharge and associated chemical mobilization and transport may increase because of the relative lack of water loss due to evapotranspiration during winter as compared with summer evapotranspiration loss (Gurdak and Roe 2010).
6. The Intergovernmental Panel on Climate Change (2007) reported that annual precipitation across parts of the Great Plains is likely to decrease; the largest decreases are predicted in the southern High Plains, especially New Mexico and Texas. Under such climate scenarios of less annual precipitation, recharge and associated chemical mobilization and transport may decrease.
7. In contrast to the southern High Plains, the northern High Plains, especially parts of Nebraska, may have substantial increases in precipitation during the summer during the next 30–100 years (Intergovernmental Panel on Climate Change 2007). Precipitation across the Great Plains region, however, is likely to continue to be highly variable with great local variation in amounts and intensity (Nippert et al. 2006), which could result in local droughts and regional flooding (Covich et al. 1997). Under increased summer precipitation in the northern High Plains, recharge and associated chemical mobilization and transport may be increased. However, higher rates and intensity of precipitation may increase overland flow and not have a significant effect on recharge rates or water quality.

9.4   ADDITIONAL INFORMATION

Additional information about the High Plains aquifer study can be found at the USGS NAWQA High Plains Regional Groundwater study homepage at http://co.water.usgs. gov/nawqa/hpgw/HPGW_home.html.

ACKNOWLEDGEMENTS

The High Plains aquifer groundwater-quality study outlined in this chapter was funded by the USGS National Water-Quality Assessment (NAWQA) Program. We gratefully acknowledge landowners that agreed to the installation and long-term operation of monitoring equipment on their land. USGS scientists Kevin Dennehy, Pixie Hamilton, Tracy Hancock, Wayne Lapham, Suzanne Paschke, Gary Rowe, Sharon Qi, and Mike Woodside provided constructive suggestions on earlier versions of the findings and implication presented here.

REFERENCES

Alley, W.M. (2006) Tracking U.S. ground water-reserves for the future? *Environment*, 49, 11–29.
Alley, W.M., Healey, R.W., LaBaugh, J.W. & Reilly, T.E. (2002) Flow and storage in groundwater systems. *Science*, 296, 1985–1990.
Alley, W.M., Reilly, T.E. & Franke, O.L. (1999) *Sustainability of ground-water resources. US Geological Survey Circular 1186*. 79 pp.
Bates, B., Kundzewicz, Z.W., Wu, S., Palutikof, J.P. (2008) *Climate change and water*. Technical paper VI of the Intergovernmental Panel on Climate Change. Intergovernmental Panel on Climate Change Secretariat, Geneva, 210 pp.
Clark, B.R., Hart R.M., & Gurdak, J.J. (2011) *Groundwater availability of the Mississippi Embayment. US Geological Survey Professional Paper* 1785, 86 p.
Covich, A.P., Fritz, S.C., Lamb, P.J., Marzolf, R.D., Matthews, W.J., Poiani, K.A., Prepas, E.E., Richman, M.B., & Winter TC (1997) Potential effects of climate change on aquatic ecosystems of the Great Plains of North America. *Hydrol Process*, 11, 993–1021.
Dennehy, K.F., Litke, D.W. & McMahon, P.B. (2002) The High Plains aquifer; USA – groundwater development and sustainability. In: Hiscock, K.M., Rivett, M.O. & Davison, R.M. (eds.) Sustainable Ground-Water Development. *Geological Society Special Publication (London)*, 193, 99–119.
Everett, B.C. (2011) Pore-scale dual-domain flow and temporal variability in recharge, High Plains aquifer, USA. *MS Thesis*. Department of Geosciences, San Francisco State University, San Francisco, CA. 226 pp.
Farley, K.A., Tague, C. & Grant, G.E. (2011) Vulnerability of water supply from the Oregon Cascades to changing climate: Linking science to users and policy. *Global Environmental Change*, 21, 110–122.
Fogg, G.E. & LaBolle, E.M. (2006) Motivation of synthesis, with an example on groundwater quality Sustainability. *Water Resources Research*. 42, W03S05. : doi:10.1029/2005WR004372
Green, T.R., Taniguchi, M. & Kooi, H. (2007) Potential impacts of climate change and human activity on subsurface water resources. *Vadose Zone Journal*. 6(3), 531–532.: doi:10.2136/vzj2007.0098
Green, T.R., Taniguchi, M., Kooi, H., Gurdak, J.J., Allen, D.M., Hiscock, K.M., Treidel, H. & Aureli, A. (2011) Beneath the surface of global change: Impacts of climate change on groundwater. *Journal of Hydrology*. : doi:10.1016/j.jhydrol.2011.05.002

Gurdak, J.J. (2008) *Ground-water vulnerability: Nonpoint-source contamination, climate variability, and the high plains aquifer.* Saarbrucken, Germany, VDM. 223 pp.

Gurdak, J.J. & Qi, S.L. (2006) *Vulnerability of recently recharged groundwater in the High Plains aquifer to nitrate contamination. US Geological Survey Scientific Investigations Report,* 2006–5050. 39 pp.

Gurdak, J.J., Hanson, R.T. & Green, R.T. (2009a) *Effects of climate variability and change on groundwater resources of the United States. US Geological Survey Fact Sheet,* 2009–3074. 4 pp.

Gurdak, J.J., Hanson, R.T., McMahon, P.B., Bruce, B.W., McCray, J.E., Thyne, G.D. & Reedy, R.C. (2007a) Climate variability controls on unsaturated water and chemical movement, High Plains aquifer, USA. *Vadose Zone Journal.* : doi:10.2136/vzj/2006.0087

Gurdak, J.J., McCray, J.E., Thyne, G.D. & Qi, S.L. (2007b) Latin hypercube approach to estimate uncertainty in ground water vulnerability. *Ground Water,* 45(3), 348–361.

Gurdak, J.J., McMahon, P.B., Dennehy, K.F. & Qi, S.L. (2009b) Water quality in the High Plains Aquifer, Colorado, Kansas, Nebraska, New Mexico, Oklahoma, South Dakota, Texas, and Wyoming,1999–2004. *US Geological Survey Circular 1337.* 63 pp.

Gurdak, J.J. & Roe, C.D. (2010) Review: Recharge rates and chemistry beneath playas of the High Plains aquifer, USA. *Hydrogeology Journal,* 18(18), 1747–1772. : doi:10.1007/s10040-010-0672-3

Gurdak, J.J., Walvoord, M.A. & McMahon, P.B. (2008) Susceptibility to enhanced chemical migration from depression-focused preferential flow, High Plains aquifer. *Vadose Zone Journal.* 7(4), 1–13. : doi:10.2136/vzj2007.0145

Hanson, R.T., Newhouse, M.W. & Dettinger, M.D. (2004) A methodology to assess relations between climatic variability and variations in hydrologic time series in the southwestern United States. *Journal of Hydrology,* 287, 252–269.

Holman, I.P. (2006) Climate change impacts on groundwater recharge-uncertainty, shortcomings, and the way forward? *Hydrogeology Journal,* 14, 637–647.

Holman, I.P., Rivas-Casado, M., Bloomfield, J.P. & Gurdak, J.J. (2011) Identifying non-stationary groundwater level response to North Atlantic ocean-atmosphere teleconnection patterns using wavelet coherence. *Hydrogeology Journal.* : doi:10.1007/s10040-011-0755-9

Intergovernmental Panel on Climate Change. (2007) *Climate change 2007: Impacts, adaptation and vulnerability.* Contribution of Working Group II to the Fourth Assessment Report of the Intergovernmental Panel on Climate Change. Cambridge University Press, Cambridge, UK. 976 pp.

Konikow, L.F. & Kendy, E. (2005) Groundwater depletion: A global problem. *Hydrogeology Journal,* 13, 317–320.

Kuss, A.J.M. (2011) Effects of climate variability on recharge in regional aquifers of the United States. *MS Thesis.* Department of Geosciences, San Francisco State University, San Francisco, CA. 227 pp.

Litke, D.W. (2001) Historical water-quality data for the High Plains Regional Ground-Water study area in Colorado, Kansas, Nebraska, New Mexico, Oklahoma, South Dakota, Texas, and Wyoming, 1930–98. *Water Resources Investigations Report (US Geological Survey),* 00–4254, 65 pp.

Maupin, M.A. & Barber, N.L. (2005) Estimated withdrawals from principal aquifers in the United States, 2000. *US Geological Survey Circular 1279.* 46 pp.

McCabe, G.J., Palecki, M.A. & Betancourt, J.L. (2004) Pacific and Atlantic Ocean influence on multidecadal drought frequency in the United States. *Proceedings of the National Academy Sciences of the United States of America,* 101, 4136–4141.

McGuire, V.L. (2007) Changes in water levels and storage in the High Plains aquifer, predevelopment to 2005. *US Geological Survey Fact Sheet,* 2007–3029. 2 pp.

McGuire, V.L., Johnson, M.R., Schieffer, R.L., Stanton, J.S., Sebree, S.K. & Verstraeten, I.M. (2003) Water in storage and approaches to ground-water management, High Plains aquifer, 2000. *US Geological Survey Circular 1243.* 51 pp.

McMahon, P.B., Böhlke, J.K. & Christenson, S.C. (2004) Geochemistry, radiocarbon ages, and paleorecharge conditions along a transect in the central High Plains aquifer, southwestern Kansas, USA. *Applied Geochemistry*, 19, 1655–1686.

McMahon, P.B., Dennehy, K.F., Bruce, B.W., Böhlke, J.K., Michel, R.L., Gurdak, J.J. & Hurlbut, D.B. (2006) Storage and transit time of chemicals in thick unsaturated zones under rangeland and irrigated cropland, High Plains, United States. *Water Resources Research*, 42, W03413. : doi:10.1029/2005WR004417

McMahon, P.B., Dennehy, K.F., Bruce, B.W., Gurdak, J.J. & Qi, S.L. (2007) Water-quality assessment of the High Plains aquifer, 1999–2004. *US Geological Survey Professional Paper*, 1749. 136 pp.

National Oceanic and Atmospheric Administration. (2006) National Oceanic and Atmospheric Administration National Climatic Data Center, *Climate of 2004, Annual review*. [Online] Available from: *http://www.ncdc.noaa.gov/oa/climate/research/2004/ann/drought-summary.html* [Accessed 25th January 2006].

Nippert, J.B., Knapp, A.K. & Briggs, J.M. (2006) Intra-annual rainfall variability and grassland productivity: Can the past predict the future? *Plant Ecology*, 184(1), 65–74.

Opie, J. (2000) *Ogallala: Water for a Dry Land*. Lincoln, University of Nebraska Press. 475 pp.

Stuart, M.E., Gooddy, D.C., Bloomfield, J.P. & Williams, A.T. (2011) A review of the impact of climate change on future nitrate concentrations in groundwater of the UK. *Science of the Total Environment*. [Online] 409(15), 2859–2873. : doi:10.1016/j.scitotenv.2011.04.016

Taniguchi, M. & Holman, I.P. (2010) *Groundwater Response to Changing Climate*. CRC Press. Boca Raton, London, New York, Leiden, 232 pp.

United Nations Educational, Scientific and Cultural Organization (UNESCO). (2008) Groundwater resources assessment under the pressures of humanity and climate change (GRAPHIC) – A framework document, GRAPHIC series number: 2. Paris, France, UNESCO Pub. 31 pp.

Warner, S.D. (2007) Climate change, sustainability, and ground water remediation – The connection. *Ground Water Monitoring and Remediation*, 27(4), 50–52.

CHAPTER 10

Groundwater change in the Murray basin from long-term in-situ monitoring and GRACE estimates

Marc Leblanc, Sarah Tweed, Guillaume Ramillien, Paul Tregoning,
Frédéric Frappart, Adam Fakes & Ian Cartwright

ABSTRACT

We present observations of groundwater change in a key, semi-arid, agricultural region of Australia, the Murray Groundwater Basin. Time series of in-situ groundwater levels archived in Government databases were compiled for all the States sharing this groundwater basin. A high quality subset of this dataset in a region affected by dryland salinity provides long-term in-situ observations on the respective impact of (1) increased recharge after deforestation and (2) the recent multi-year drought on groundwater levels. A change in the long-term dynamic of the water table is observed since the beginning of the drought in 1997 (1994 in some regions). The analysis of the bore data first showed a regional rise of the water table by ~5 cm/a from 1980 to 1992 followed by a regional decline at a rate of 17 cm/a from 1997 to 2009. Time series of groundwater storage anomalies obtained from a combination of total water storage using space gravimetry (GRACE) and soil moisture estimates from hydrological models also indicate a strong decline of the water table from 2002 to 2010. From August 2002 to December 2010, GRACE-based estimates indicate a groundwater loss of $\sim18 \pm 1.3$ mm/a which equates to a total loss of groundwater of $\sim45 \pm 3$ km^3 over the $\sim300{,}000$ km^2 hydrogeological basin. These observations suggest that the impact of the drought on groundwater recharge counterbalanced and surpassed the impacts of past land-clearing and has brought a temporary halt to dryland salinity.

10.1 INTRODUCTION

10.1.1 Purpose and scope

The Murray Groundwater Basin is a very valuable but equally sensitive water resource that covers about a third of the Murray-Darling drainage basin (Fig. 10.1). There are two main pressing issues relating to groundwater management in this region. First, increased groundwater recharge inherited from the historical clearing of the region continues to influence groundwater levels leading to dryland salinity in many areas. Second, the demand for groundwater is currently increasing, mostly to paliate for the low reliability of surface water resources. However, sustainable yields are already exceeded in some aquifers and it is not well known how resilient the aquifers of the region are to prolonged droughts. The main aim of this study is to provide long-term observations of ground-water dynamics in the Murray Groundwater Basin. These observations are then used to estimate the respective impact from past land clearing and the recent drought on ground-water resources in the basin. For this we used two main sources of information: (1)

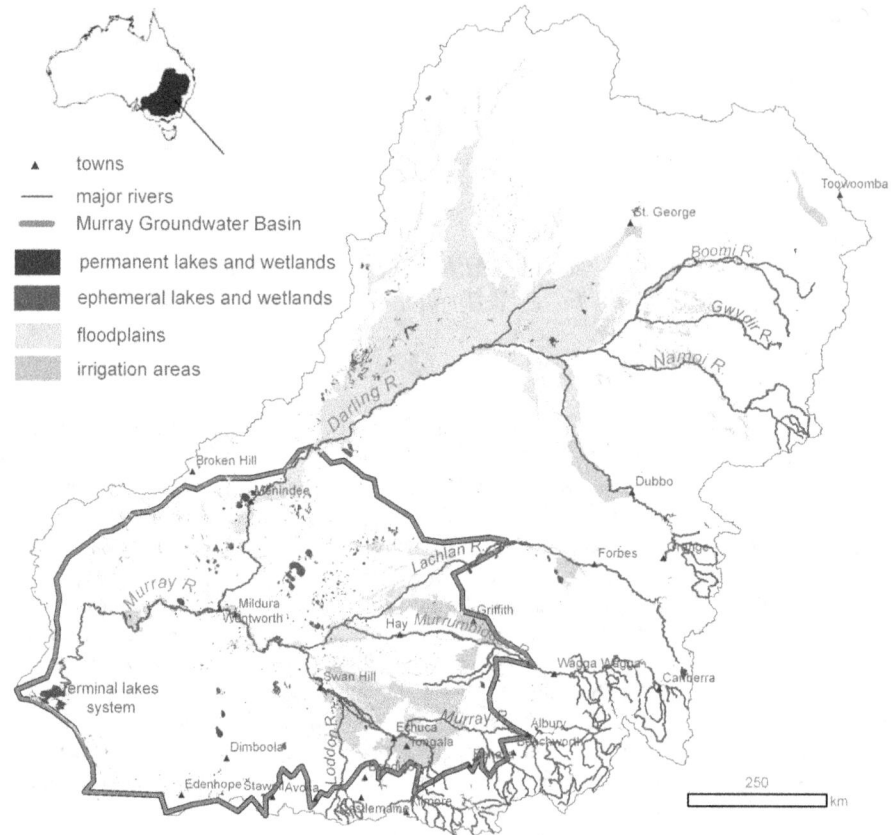

Figure 10.1.   Location of the Murray Groundwater Basin in the Murray-Darling drainage Basin, Australia.

in-situ point observations of groundwater levels at monitoring bores and (2) regional estimates of variations in water storage from space gravimetry GRACE (GRGS) and a hydrological model (GLDAS).

10.1.2   Study area description

Water resources in the Murray-Darling Basin

The Murray-Darling Basin (MDB) covers 1.06 M km^2 or approximately one-seventh (14%) of Australia (Fig. 10.1). The topography of the MDB is dominated by vast plains, bounded to the east and south by the Great Dividing Range that has undulating hills reaching an elevation of up to ~2200 m above sea level. The MDB is the nation's food bowl; agriculture is the dominant economic activity covering ~80% of the basin and accounting for ~40% of Australia's total agricultural production (Pink 2008). Land use across the basin is dominated by farming land (67%), native and plantation forests. The main agricultural activity is livestock production (cattle and sheep) while other key land uses include irrigated dairy, cotton, rice, wheat, corn, and horticulture (grapes, citrus and other fruit trees). There is a wide range of climatic conditions in the MDB that follow north-south and east-west gradients. In the eastern uplands (Great Dividing Range), the climate varies from temperate in the south to mountainous and subtropical in the north. The vast plains of the west, which occupy most of the MDB, have a hot semi-arid to arid climate. Rainfall is summer dominated in the north and winter dominated in the south. Fig. 10.2 shows the spatial distribution of mean annual rainfall across the MDB and highlights the strong influence of the topography (high rainfall along Great Dividing Range to the east). The inter-annual variability in rainfall is noticeably large (Fig. 10.2).

After European settlement in the early 1800s, much of the land in the MDB was cleared for new farming areas. Land clearing was associated with an increase in drainage and groundwater recharge, resulting in a subsequent rise of the water table. For example, in the south-eastern Murray Groundwater Basin, where annual rainfall is between 600–800 mm, groundwater recharge increased from <1 to 4–90 mm/a after land clearance (Cartwright et al. 2007). In areas of secondary salinity, saline water is remobilised and brought to the surface by the rising water table or by lateral flow in the unsaturated zone. This can lead to the degradation of soil and surface water resources through the formation of salt scalds at the surface and increased river salinity (Fig. 10.3). Land clearance in much of the MDB happened more than 50–100 years ago but, due to the slow nature of unsaturated zone and groundwater flow, human induced salinization has continued to expand. Affected land in the MDB (not necessarily all in the MDB), was estimated at 640,000 ha in 1997 (NLWRA 1997). Such widespread impacts mean that in addition to the degradation of water and land resources, the consequent impacts from dryland salinity, including economic (decline in capital value of land and infrastructure damage) and biodiversity (loss of flora and habitats for fauna) costs, are felt at both local and regional levels (ANRA 2010). Some catchment mitigation strategies for salinity have involved revegetation of agricultural areas and saline groundwater interception schemes.

The Murray Groundwater Basin (MGB)

The MGB is a closed groundwater basin covering an area of ~300,000 km^2 (Brown 1989). The groundwater flows towards the Murray River, which acts as the only outlet for groundwater transport from the basin (Evans and Kellett 1989). The saucer-shaped basin comprises 200–600 m of Cainozoic unconsolidated sediments and sedimentary rocks forming the major aquifers; Renmark Group, Murray Group, Pliocene Sands and Shepparton Formation (Evans and Kellett 1989). In this study we examine three of these aquifers (Murray Group, Pliocene Sands and Shepparton Formation), in regions where they are the water table aquifers. The extent of these water table aquifers are presented in Fig. 10.4a, and a cross-section illustrating the relationship between the different aquifers of the MGB is shown in Fig. 10.4b.

Pliocene Sands Aquifers

In the western region of the MGB the Pliocene Sands aquifer is the Loxton-Parilla Sands, and in the east it is the Calivil Formation. The generally unconfined aquifer of the Loxton-Parilla Sands comprises fine to medium sand with minor clay and silt, and averages 60 m in thickness (Evans and Kellett 1989). In the north-western MB this aquifer is confined in areas by the overlying Quaternary Blanchetown clay (Evans and Kellett 1989). The upper parts of the Loxton-Parilla Sands laterally grade into the Shepparton Formation in the eastern MB. The Loxton-Parilla Sands also overlie and laterally grade

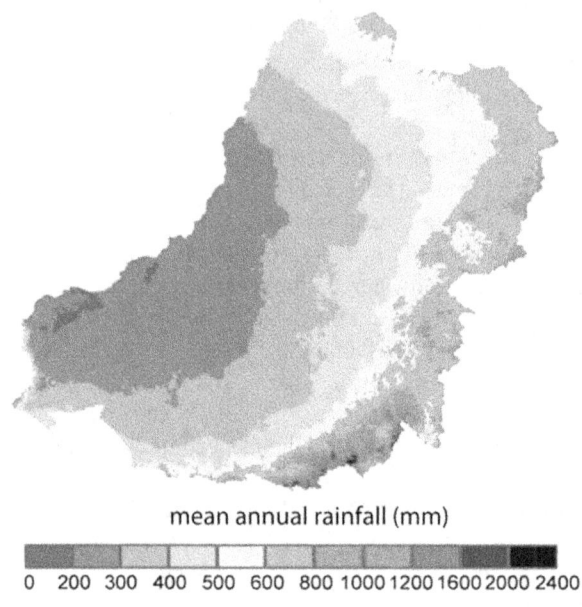

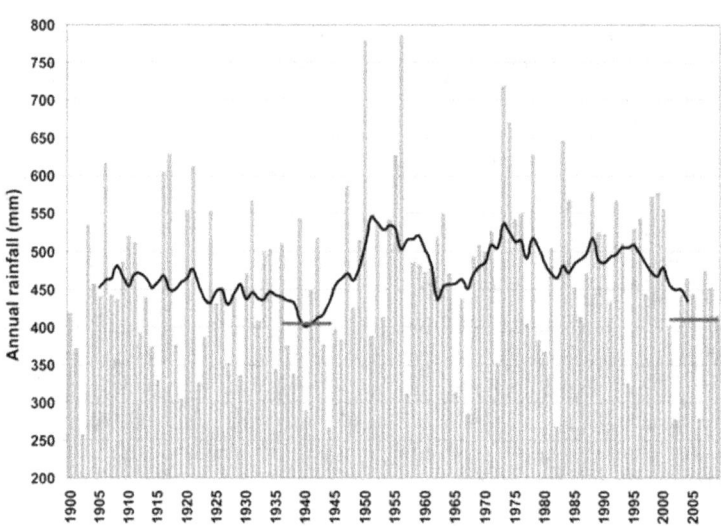

Figure 10.2.   Top: spatial distribution of average annual rainfall for the 1900–2009 period. Bottom: mean annual rainfall over the Murray-Darling Basin for individual years from 1900 to 2009. Also shown are the 11-year running mean (solid black) and the two lowest 9-year means on record: 1937–1945 (395 mm) and 2001–2009 (406 mm). Data courtesy of the National Climate Centre, Australian Bureau of Meteorology, Melbourne, Australia.

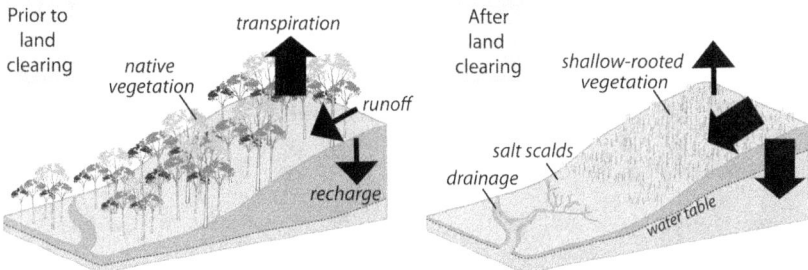

Figure 10.3. Schematic diagram (after DARA 1989) of processes leading to dryland salinity after the clearance of native vegetation for agricultural practices.

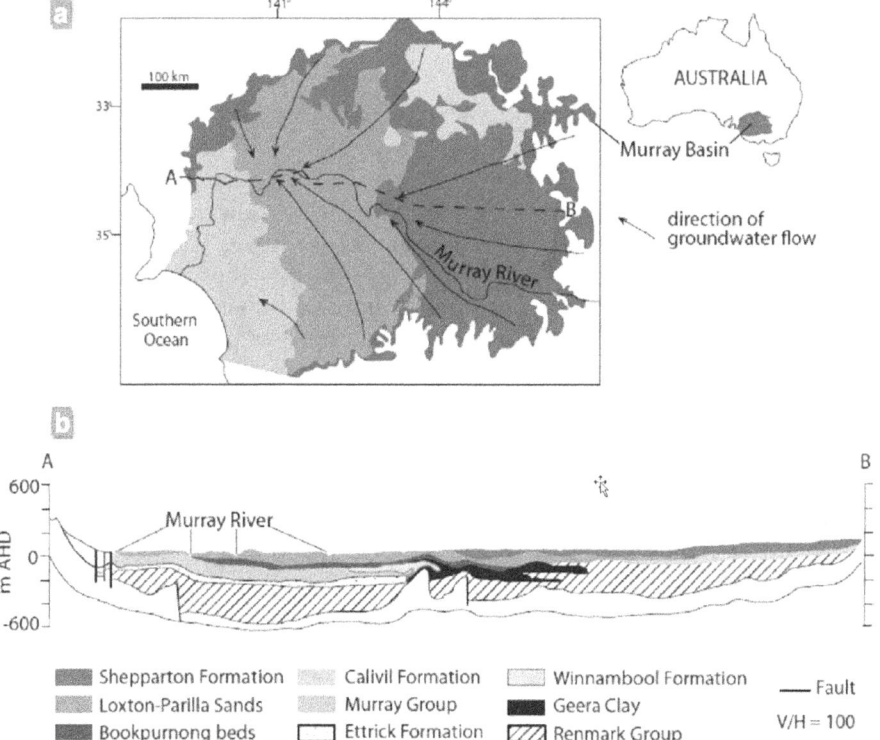

Figure 10.4. a) Location of the major water table aquifers within the Murray Groundwater Basin and groundwater flow directions. b) cross section showing the relationship between the different aquifers and confining units from west to east in the Murray Groundwater Basin (after Evans and Kellet 1989).

into the Calivil Formation in the eastern MGB. The water table is present in the Loxton Parilla Sands except in the western region where the sands overlie the unconfined Murray Group aquifer and the water table lowers below the base of the Loxton Parilla Sands (Evans and Kellett 1989). The Calivil Formation consists predominantly of coarse to granular quartz sands with lenses of kaolin and carbonaceous clay (Hennessy et al. 1994), and has an average thickness of 60 m, but can range up to 100 m (Brown 1989).

This formation is hydraulically connected to the underlying Renmark Group aquifer and the lateral deposits of the Loxton-Parilla Sands. In areas where the Calivil has coarser grained deposits, the aquifer is hydraulically connected to the overlying Shepparton Formation, and these regions generally coincide with the basin margins (Evans and Kellett 1989). In the northeastern margin of the MGB, the Shepparton is unsaturated and the water table is within the Calivil aquifer.

Shepparton Formation

Conformably overlying the Calivil Formation in the eastern MGB are the Pliocene to Quaternary fluvio-lacustrine deposits of the Shepparton Formation. In order of abundance, this aquifer comprises clay, slit, and sand (Evans and Kellett 1989). Hydraulic connectivity between the sand lenses is non-continuous resulting in vertical and horizontal heterogeneity of flow. This regional aquifer increases in thickness towards the basin's centre, from tens of meters up to 70–80 m (Brown and Stephensen 1991).

Murray Group Aquifer

The Murray Group aquifer is a middle Tertiary marine limestone and calcarenite aquifer located in the western MGB (Evans and Kellett 1989). The limestone is over 100 m thick in most areas (Brown 1989), and previous work shows data from bores screened in the aquifer up to 150 m depth (Leaney et al. 2003). This aquifer overlies a confining layer formed by the clay and marl of the Winnambool and Ettrick Formations (Evans and Kellett 1989), and is laterally separated from the Renmark Group aquifer by the Geera Clay (Brown 1989). The eastern section of this aquifer is confined to semi-confined predominantly by the clay and silt deposits of the Pliocene Bookpurnong beds (Evans and Kellett 1989). The western section of this aquifer is unconfined beneath the Pliocene Sands aquifer; which is largely unsaturated (Leaney et al. 2003).

10.1.3   Methodology

In-situ groundwater levels

Regional decadal trends

Archive groundwater level data were sourced from the Government departments of the States in the Murray Groundwater Basin (NSW, Department of Water and Energy; VIC, Department of Sustainability and Environment; and SA, Department of Water Land and Biodiversity Conservation) and collated in a common database. We used this database to generate decadal trends of groundwater levels across the entire Murray Groundwater Basin. Using a least square fit, linear trends were calculated at each monitoring bore for two periods: 1980–1992 and 1997–2009. The time between the two periods (1993–1996) was not analysed because of the spatial inconsistency in the onset of the drought across the basin (Leblanc et al. in press). The trend analysis was performed on a subset of bores matching the follow criteria: 1) each selected bore must have, at least, 75% of the years represented in each 13 year epoch; and 2) only Government observation bores (production bores excluded) with an average saturated zone ≤30 m from the bottom of the screened interval were analysed. Deeper bores were excluded as they can reflect processes occurring on longer time scales (Fetter 2001). In total, 354 representative bores for the unconfined aquifers across the Murray Groundwater Basin were used for the decadal trend analysis.

Detailed observations at selected sites

We also used this database to find high quality point observations that showed the respective long-term impacts of past land clearance and of the recent drought on groundwater levels. Accordingly, we searched the database for a selection of bore hydrographs that met the following criteria: 1) be a long time series; 2) be located at the edge of the Murray Groundwater Basin where the problems of dryland salinity were clearly identified. We also searched for an updated time series of groundwater levels of the bores used by Allison et al. (1990) to demonstrate increase in groundwater recharge due to land clearing.

GRACE estimates of changes in groundwater storage

Since its launch in March 2002, the Gravity Recovery and Climate Experiment (GRACE) mission has been acquiring data on continental water storage anomalies which have been increasingly used for large-scale hydrological and hydrogeological applications. GRACE provides global mapping of the time-variations of the gravity field at an unprecedented resolution of ~300 km (for harmonics coefficients of the geopotential developed up to degree 60) and a precision of 1.5 cm of water equivalent thickness when averaged over regions of a few hundred square-kilometres (Ramillien et al. 2008; Schmitt et al. 2008). These small temporal changes in the Earth's gravity field are mainly due to the redistribution of mass inside the fluid envelope of the Earth caused, at daily to decadal timescales, by tectonic or glacial isostatic adjustment signals but also changes in surface loads such as atmosphere, oceans and continental water storage (Tapley et al. 2004). The effects of atmospheric mass, ocean tides and barotropic signals are accounted for using oceanic and atmospheric models in the reduction of the raw observations (Bettadpur 2007). In large regions like Australia where there are no significant tectonic or glacial isostatic adjustment signals, the remaining GRACE data should mainly correspond to changes in total continental water storage (TWS) which is an integrated measure of all the major water stores: reservoirs and lakes, soil moisture, groundwater, snow, ice and biomass. In the semi-arid Murray Groundwater Basin variations in biomass, snow and ice are negligible compared to other stores and are well below the detection threshold of the GRACE satellites (Rodell et al. 2005). Similarly there are no major surface water reservoirs in the Murray Groundwater Basin; hence the GRACE TWS signal over this region corresponds mostly to change in soil moisture and groundwater storage.

We obtained a GRACE TWS time series over the Murray Groundwater Basin from solutions produced by the Groupe de Recherche en Géodesie Spatiale (GRGS). These solutions are an average of 10 days of GRACE observations (Bruinsma et al. 2010). Atmospheric mass, ocean tides and barotropic signals are accounted for in the GRGS solutions using the European Centre for Meteorological Weather Forecasting reanalysis, Finite Element Solution 2004 (FES2004) (Le Provost et al. 1998) and the MOG2D-G barotropic models (Carrère and Lyard 2003). GRACE estimates of changes in groundwater storage were obtained by subtracting soil moisture storage from the GRACE TWS time series. Variations in soil moisture storage were estimated for the groundwater basin using the NOAH land surface model (Ek et al. 2003), with the simulations being driven (parameterization and forcing) by the Global Land Data Assimilation System. The NOAH model simulates surface energy and water fluxes/budgets (including soil moisture) in response to near-surface atmospheric forcing and depending on surface conditions (e.g. vegetation state, soil texture and slope) (Ek et al. 2003). The NOAH model outputs of soil moisture estimates have a 1° spatial resolution and, using four soil layers, are representative of the top 2 m of the soil.

10.1.4   Relevance to GRAPHIC

This chapter contributes to the GRAPHIC project by providing an Australian case study which gives insights into major water resources issues in the Australian continent and materials for comparison with other countries or regions.

First, the Murray Groundwater Basin has experienced a general rise of the water table after land clearing. This rise of the water table induced widespread dryland salinity problems degrading vast areas of farming land and surface water bodies. Second, this key agricultural region has experienced a continued increase in water demand which is increasingly difficult to supply given the very high hydroclimatic variability. Third, it is often considered that groundwater has the capacity to help meet water needs during periods of drought when surface water resources are dwindling. This region has recently been affected by one of the most severe and prolonged multi-year droughts on record for the 20th and 21st centuries. It offers a unique opportunity to study the effects of a prolonged rainfall deficit, and the coupled pressures of climate variability (drought) and human activities (land clearing) on groundwater resources. Finally, the Murray Groundwater Basin is also one of the few semi-arid regions equipped with a relatively good monitoring network for groundwater at a regional scale. We therefore used this opportunity to test the capability of the GRACE mission to provide regional estimate of change in groundwater storage so that it can be applied for the monitoring of insufficiently instrumented regions.

10.2   RESULTS AND DISCUSSION

10.2.1   Long-term observations from in situ hydrographs

Regional trends across the basin

Long-term trends in groundwater levels across the Murray Groundwater Basin highlight a regional rise of the water table during the period 1980–1992; linear trends for this period range from −0.16 to 0.63 m/a with a median value of 0.05 m/a (Fig. 10.5). This regional rise is attributed to the increase of groundwater recharge after land clearance. Following the onset of the drought in 1994–1997, trends in groundwater elevations clearly reversed across the basin. Linear trends for the 1997–2009 period range from −0.88 to 0.50 m/a with a median value of −0.17 m/a (Fig. 10.5). Local examples from two regions in the Murray Groundwater Basin are presented below.

South eastern Murray Groundwater Basin

We found in the Government databases a cluster of monitoring bores in an area affected by dryland salinity that provides long-term continuous groundwater hydrographs showing the respective impacts from past land clearance and the recent drought on groundwater dynamics. This cluster of monitoring bores is located in the Benalla region close to the south-eastern border of the Murray Groundwater Basin (Fig. 10.6). Groundwater hydrograph data during ~1974–2010 (from VWRDW 2010), indicate most groundwater levels in the Benalla region show a steady increase of 0.05–0.42 m/a

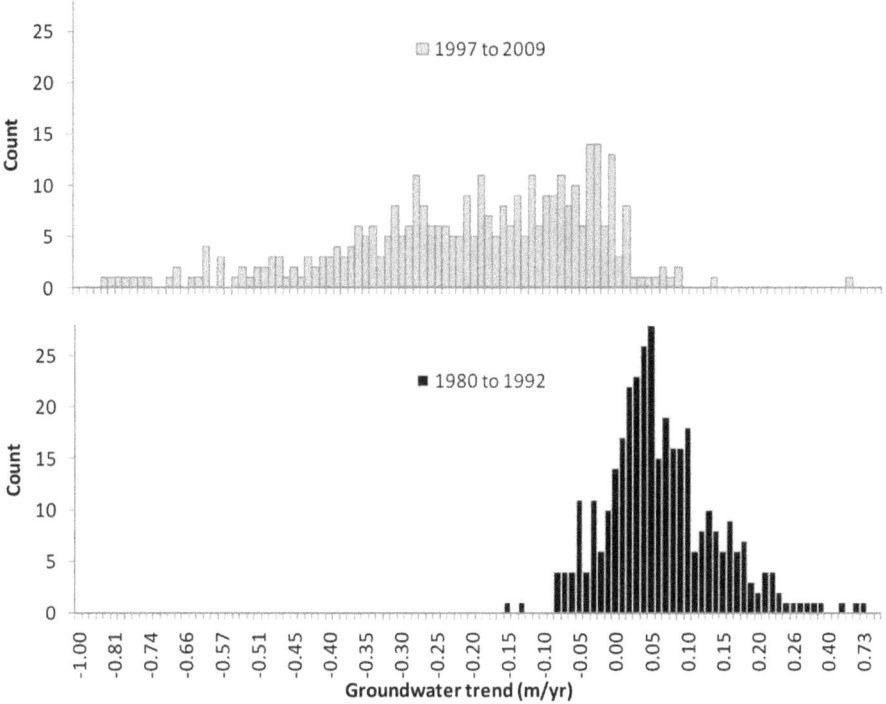

Figure 10.5.   Histogram of groundwater trends between the years 1980-1992 and 1997-2009 for bores located in the Murray Groundwater Basin.

up until ~1994/7, when the first impacts of drought conditions are reflected in the groundwater system (Fig. 10.7). Typical of many regions in the Murray Groundwater Basin, the increases in groundwater elevations, observed in both shallow and deeper aquifer systems prior to 1994/7, reflect the long term impacts of land clearance in the region throughout the last century (e.g. Allison et al. 1990). The rising water tables resulted in salt scalds often observed along the foothills (Fig. 10.6). From 1997–present (2010), the groundwater elevations of shallow and hydraulically well-connected deeper groundwater systems declined by 0.2–0.38 m/a due to drought conditions. Local variations in the stratigraphy (highlighted in Fig. 10.6b) cause heterogeneous connectivity between shallow and deep aquifers in the region. Some deeper bores presented in Figure 10.7 show a continued increase in groundwater elevations during the drought (e.g. bores 9–12). This continued rise in deeper bores could be due to a much slower response time of the groundwater system as the depth increases.

South western Murray Groundwater Basin

A study by Allison et al. (1990) in the south western Murray Groundwater Basin observed that the land clearing from 50 to 100 years ago resulted in increased recharge rates, which is reflected in the rising groundwater elevations. The original bores used in the study by Allison et al. (1990) are presented here (Fig. 10.8) with the time series of groundwater elevations extended to include the more recent drought impacts. Four groundwater bores are

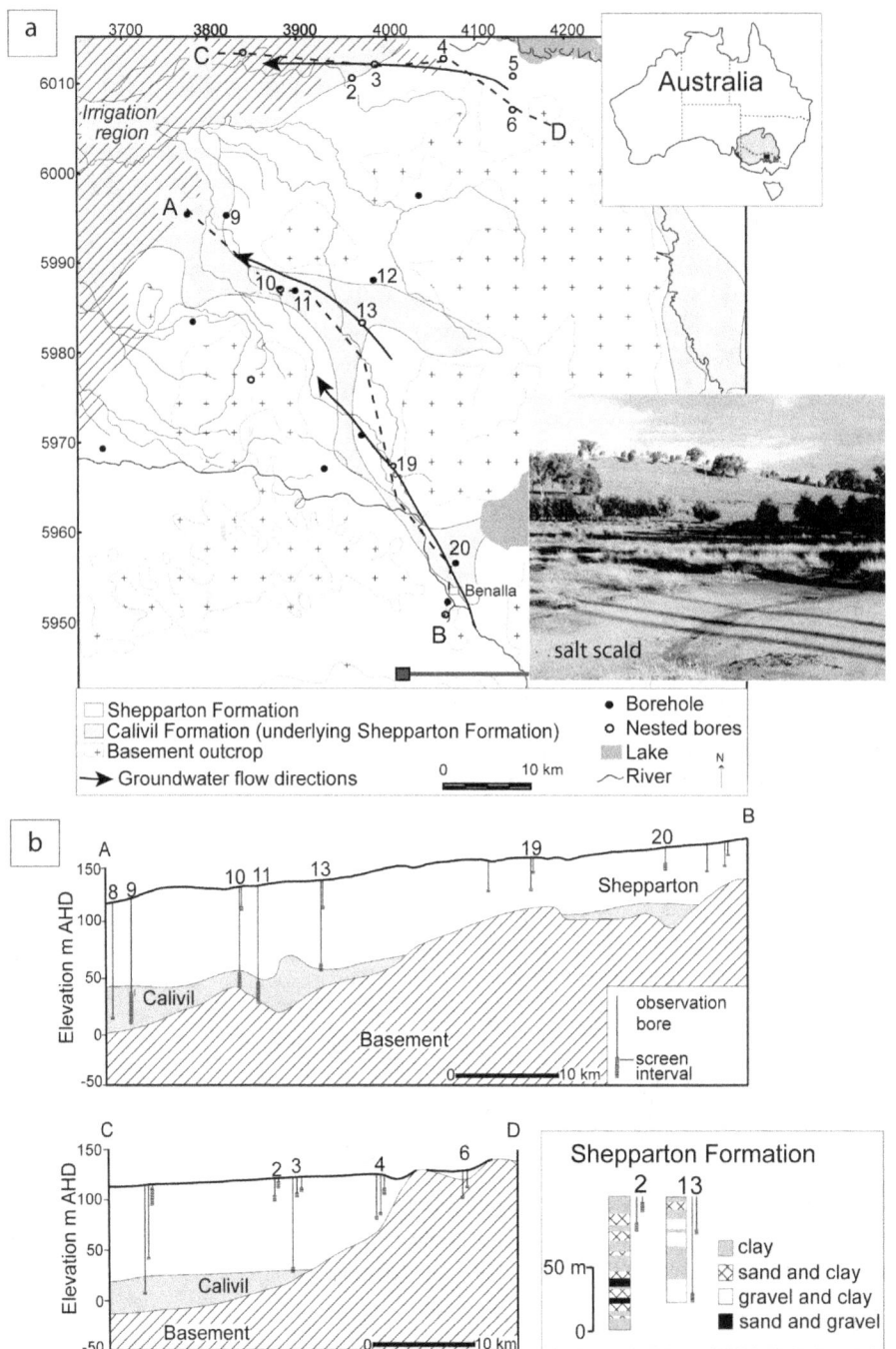

Figure 10.6. (a) Geology (after Hennessy et al. 1994), general groundwater flow directions for June-August 2003, and locations of bores (inserts: location of the Murray Groundwater Basin and Benalla region in south-eastern Australia; and local example of a salt scald due to dryland salinity (photo from English et al. 2004)). (b) Cross sections of the region and stratigraphic logs for two bores in the Shepparton aquifer, shows examples of the heterogeneous stratigraphy resulting in locally variable connection between shallow and deeper aquifer systems.

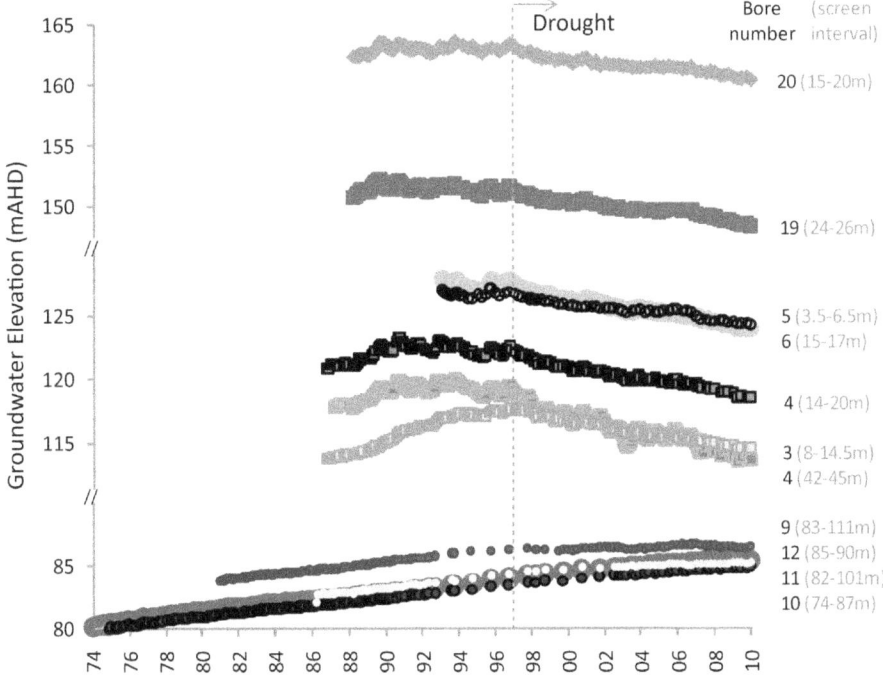

Figure 10.7. Groundwater hydrographs for selected bores in the south-west of the Murray Groundwater Basin (Benalla region).

presented, each highlighting variations in the response to environmental change. PAG 6 bore is located within a unit of low recharge rates (Renmark Group; Allison et al. 1990) and therefore showed relatively small increases due to land clearing between 1973–1994 (<0.03 m/a), and shows a continued rising trend during drought conditions. All other bores are screened in limestone aquifers. The shallow bore ARC 5, shows rising groundwater elevations to 1994, and a large decline during the drought. In comparison CTN 5 and SHG 7, which have deeper screen intervals, show little change and an increasing trend, respectively. These bores highlight local variations in the trend and magnitude of changes in groundwater levels in response to regional stresse (ie. land clearing and drought).

10.2.2   GRACE observations

Spatial average of Total Water Storage from GRACE

We used GRACE level-2 solutions from GRGS that are lists of Stokes coefficients (i.e., dimensionless spherical harmonics of the geopotential) from August 2002 to December 2010. They are estimated over 10-day intervals from Level-1B GRACE measurements, including the accurate inter-satellite distance and velocity variations. After removing the temporal mean over 2003–2010, the corresponding residuals should represent the vertically-integrated variations of the Total Water Storage (TWS) over continents, that include changes in surface waters, snow coverage, soil wetness and

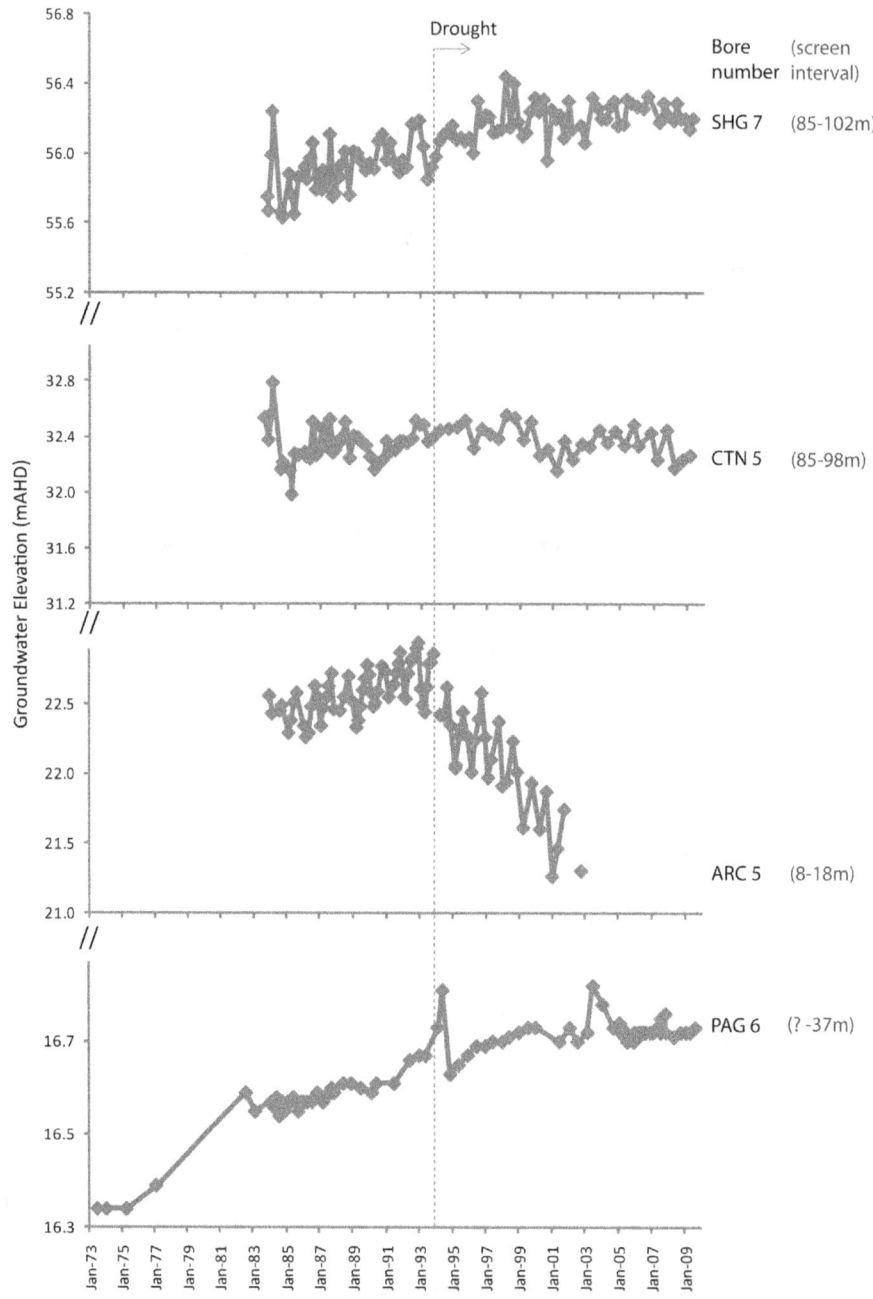

Figure 10.8. Groundwater hydrographs for selected bores in the south-west of the Murray Groundwater Basin – update on reference bores used in Allison et al. (1990).

groundwater (Figure 10.9). Stokes coefficients are converted into water mass coefficients in unit of equivalent-water height by simple linear filtering. Spatial averages in a given basin are computed as the scalar product of the water mass coefficients for each period and the coefficients of the geographical mask of the basin. GRGS solutions are provided up to degree $N = 50$, so their spatial resolution remains about 400 km.

Error budget

In order to establish the error balance of estimated regional averages over the Murray Groundwater Basin, different types of error are computed from the GRACE data. The formal errors are determined from the formal uncertainties on the Stokes coefficients that are adjusted using the Level-1 GRACE observations. Averaged over the Murray Groundwater Basin, these errors are typically less than ±25 mm of equivalent-water height (see also Llubes et al. 2007; Brown & Tregoning 2010). We computed the omission errors induced by the truncation of the gravity coefficients at degree and order 50–60. These errors represent the missing high-frequency signals that GRACE solutions cannot describe. Once they are averaged over basins, errors of truncation do not exceed ±2 mm of equivalent-water height. Leakage (contamination) errors are mostly caused by hydrological signals mostly in tropical regions and other strong geophysical signals (e.g. melting of Greenland icesheet) outside the considered basin. By using monthly outputs from the WaterGAP Global Hydrological Model (WGHM) over the period 2002–2006 (Döll et al. 2003; Ramillien et al. 2006), we found that leakage of continental water storage into the Murray-Darling basin create seasonal variations with amplitudes reaching a maximum of ±25 mm. To summarize, the balance of the error for the Murray-Darling drainage basin is around ±52 mm of equivalent-water height for each period of time.

Groundwater estimates

Regional estimates of changes in groundwater storage were obtained by subtracting soil moisture storage from the GRACE TWS time series (Fig. 10.10). These satellite-based

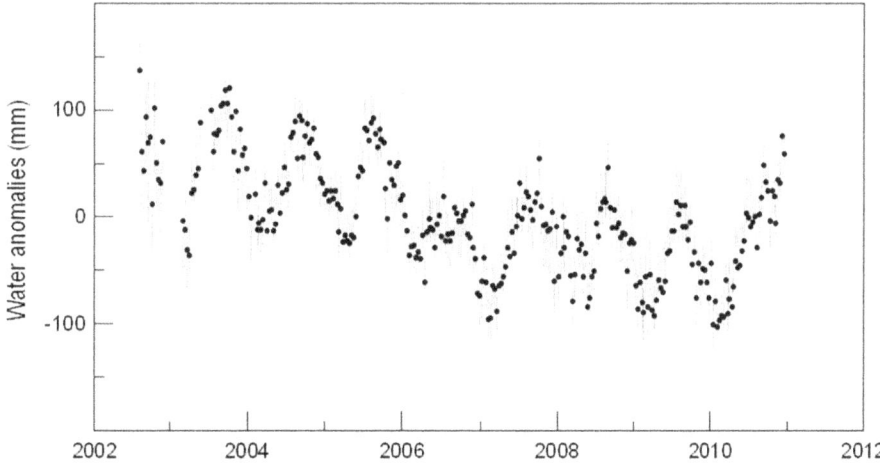

Figure 10.9.   Terrestrial water storage anomalies in the Murray Groundwater Basin from the GRGS RL02 solutions in mm from August 2002 to December 2010.

estimates indicate a seasonal amplitude of the groundwater storage ranging from 60 to 90 mm with an average of 80 mm. These regional space observations also indicate a strong and continuous decline in groundwater storage across the basin since the beginning of record in 2002. The linear trend using least squares on the GRACE-based estimates indicates a groundwater loss of 17.8 ± 1.3 mm/a or 5.3 ± 0.4 km^3/a from 2002 to 2010 (Fig. 10.10).

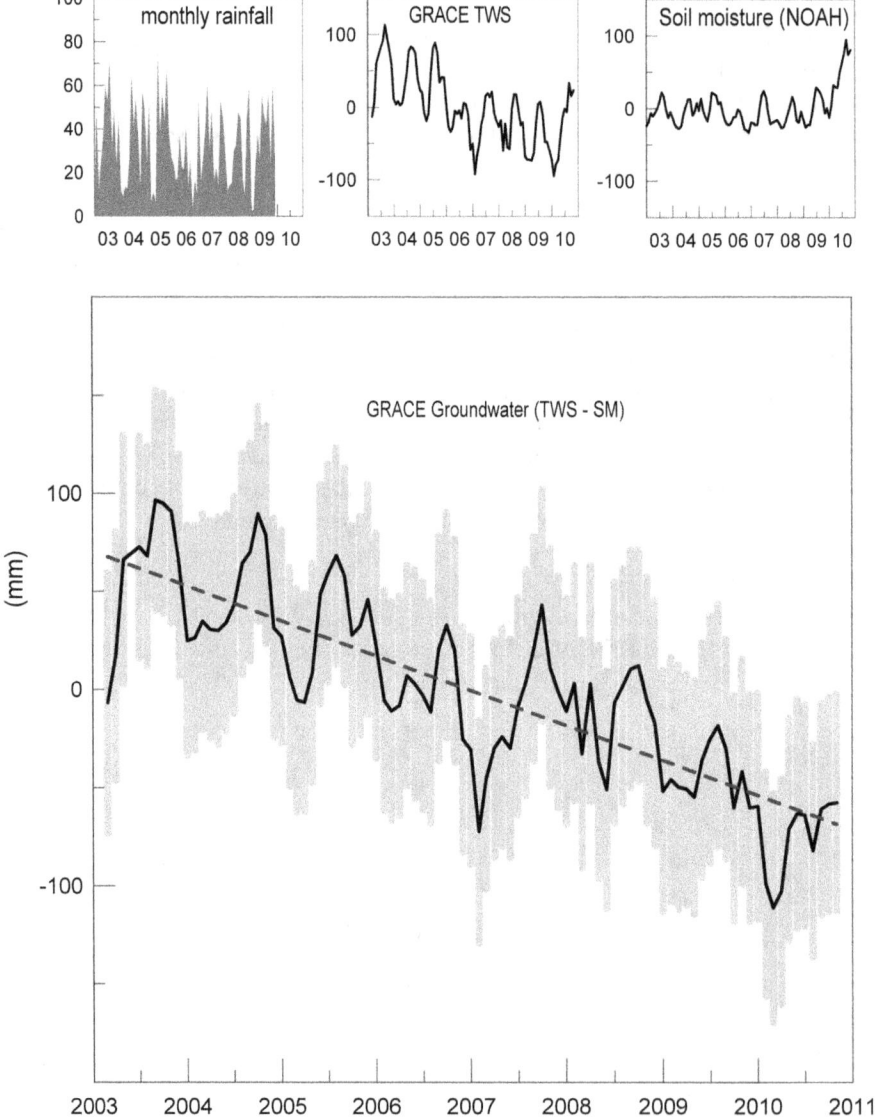

Figure 10.10.   (a) Monthly rainfall in mm from the Australian Bureau of Meteorology. Monthly anomalies of (b) total water storage derived from GRACE data (TWS; GRGS solutions); (c) soil moisture (SM; GLDAS ensemble mean); and (d) GRACE groundwater (TWS – SM) in mm from March 2003 to GRACE groundwater data (black line), uncertainty (grey bars), trend (dotted line) November 2010.

10.2.3   Discussion

Attribution of the groundwater loss during the recent drought

Total groundwater pumping in the MGB was estimated to be below $0.4\,km^3/a$ in 2004–2005 (CSIRO 2008). This pumping rate represents less than 10% of the decline rate in groundwater storage observed from 2003 to 2008 in the GRACE-based observations. In the absence of an important increase in pumping during the drought, the observed decline can only be explained by a reduction of the groundwater recharge while a strong natural discharge is maintained to create a deficit in the groundwater storage. Capillary rise from the saturated to the unsaturated zone and subsequent evapotranspiration may go a long way in explaining strong natural discharge. Deep rooted trees are widespread across the MGB and they could explain why a vast amount of groundwater may still be transpired during the drought.

The average annual rainfall deficit of the recent drought is similar to the one observed during the 1935–1945 drought. However the recent drought has, in comparison, led to a much stronger decrease in runoff values. This additional decrease in runoff during the recent drought can be explained by a change in rainfall patterns at various time scales: lower inter-annual variability, lower autumn and winter rainfall (Potter et al. 2010). This change in the rainfall pattern has most probably induced a strong reduction in the groundwater recharge, and explains why the loss of groundwater during the drought is so high.

Since our study, above average rainfall in 2010 and 2011 brings the end to this prolonged drought. Large flooding events occurred in south west Queensland and Victoria in early 2011. As of mid-2011, public water storage across the MDB is around full capacity.

Land surface versus climatic controls on groundwater

The respective impact of land use change and hydroclimatic variability on water resources can be complex. In a large region of West Africa in the Sahel it was found that, despite the long lasting droughts of the 1970s and 1980s, surface and groundwater resources increased following intensive land clearing (*Leduc* et al. 2001; *Leblanc* et al. 2008; Favreau et al. 2009; Favreau et al. 2011). Land clearing in the Murray-Darling Basin, which started with European settlement in the early 1800s caused a general rise of the water table that is still being observed during the second half of the 20th century, and subsequently led to the appearance of dryland salinity (Allison et al. 1990). The observed decline of groundwater levels and TWS show that the ongoing multi-year drought in the Murray-Darling Basin has, at least temporarily, reversed and surpasses the long-term groundwater trend inherited from land clearance and may induce a temporary halt of secondary salinity processes.

10.3   POLICY-RELEVANT RECOMMENDATIONS

Following from the high rainfall variability and largely arid conditions described above, water resources in the Murray-Darling are scarce and highly unreliable. The five Australian States and Territory that share this vast drainage system (Queensland, New South Wales, Victoria, South Australia and the Australian Capital Territory) have extensively exploited the basin's water resources since the second half of the 20th Century. To support agricultural activities, the MDB uses ~60% of Australia's total agricultural water consumption. The basin also meets the water needs of ~2 million people in addition to contributing to the water supply of Adelaide (population 1.4 million). Most water used in the MDB is sourced from surface water (84%). To satisfy high water demands a vast network of public and private infrastructure, including large reservoirs, weirs, diversion channels as well as farm dams has been progressively established across the entire basin. Today, the total surface water storage capacity (including farm storages) has reached ~35 km^3; that is 150% of the average annual water availability in the basin (GLOPACHA; WWDR). Total storage capacity was only ~2 km^3 in 1930 and most of the infrastructure development took place in the second half of the 20th Century. Groundwater currently represents the remaining 16% of total water use in the MDB but, under current water sharing arrangements, groundwater use could increase to be over 25% of total water use by 2030.

Maintaining a healthy basin and environmental flows has long been a challenge in the Murray-Darling Basin with intense water diversions (large reservoirs, dams, weirs), increasing demands from irrigation and urban areas, and the difficulties in sharing water allocations within the five States governed by separate legislation and policies. In addition, the current status of water resources is dire due to the impacts of probably the most severe hydrological drought since records began in the late 19th Century. Experts and Governments agree that urgent actions are required to save this important water system; for its cultural, ecological, and economic value. In an unprecedented effort in the history of Australia, the Australian Government is attempting to save the basin's health and to prepare the region for climate change by introducing a new integrated and sustainable approach to the management of the basin's water resources (http://www.thebasinplan. mdba.gov.au/). As a first step towards that plan, a discussion document – the "Guide to the proposed Basin Plan" was released in October 2010 in combination with a series of community meetings. The document outlined plans to return a long-term average of 3,000–4,000 GL/a back to the environment through 22–37% reductions in water use. The proposed range was at the lower end of the estimated environmental needs (3,000–7,600 GL/a) and represented an explicit compromise designed to avoid unacceptable social and economical consequences. In 'The Basin Plan', a cut in groundwater extraction of 186 GL/a was proposed to bring all extraction from all managed groundwater systems back to acceptable extraction rates. This total number includes wide ranging cuts, from possible increases in groundwater use to 40% reductions, reflecting the variable degree of overexploitation in the different aquifers. The proposed targets generally appeared to be accepted as more or less appropriate by environmentalists and ecologists, but were poorly received by irrigation interest groups and farming communities. While the hydrological modelling underpinning the proposed plan generally appears to be accepted, the estimation of the ecological benefits and socio-economic implications appear to have been met with much scepticism. In view of recent protests, it is clear that the 'Basin Plan' has become a contentious political issue for the Governments. Significant delays and most probably a 'watering down' of the proposed measures can be expected for the 'Basin Plan'. This is not so surprising, and in fact there is a long history of complex interactions between politics and water management in the Murray-Darling Basin (Connell 2007).

Achieving the right balance between environmental flows and socio-economic needs is proving difficult. Successful water planning will need to balance cultural and ecological values with food production, account for high natural variability and uncertainty in climate change projections, as well as address past mistakes and be cognisant of hydrological change.

10.4   FUTURE WORK

A key scientific challenge in the Murray Groundwater Basin is to assess the future impact of climate change on groundwater. If global climate models (GCM) remain the best tool to assist us in assessing future climate change, our case study is a good example of the difficulties in obtaining reliable estimates from these models. For the region, climate models are fairly consistent in predicting the change in annual average temperature. However, after the first half of the 21st century this estimate depends strongly on the emission scenarios for greenhouse gases (CSIRO and BoM 2007). Failure to establish policies and reduce global greenhouse gases emissions is one possible scenario. In fact, since 1990, observed fossil fuel emissions are at the upper limit of all emission scenarios and climate change and associated hydroclimatic impacts under such scenario will be stronger than most projections currently available (Meinshausen et al. 2009). Modelling rainfall is harder than temperature so the range of uncertainties attached to future rainfall projections is even larger. The pattern of floods and droughts in the MDB and other parts of Australia may often be linked to sea surface temperature anomalies in the Pacific (oscillation between El Niño and La Niña conditions or ENSO) and Indian Oceans. Climate models do not currently have the capability to predict these interannual oceanic anomalies (Meyers et al. 2007).

The effects of climate change on groundwater resources do not necessarily mirror those of surface water resources. In particular in drier environments, groundwater recharge can be a highly episodic event associated with a small number of large storms. Therefore, even when a GCM predicts a drier average climate, increased rainfall intensity could potentially lead to an increase in groundwater recharge rates (McCallum et al. 2010). Crosbie et al. (2010) estimated that, under emission scenarios adopted for the IPCC 4th assessment, average recharge across the Murray-Darling Basin would increase by 5%; with results ranging from −12% to +32% depending on the GCM used. Only the very southern margin of the basin was predicted to experience an actual decrease in recharge. This was despite a generally predicted reduced mean annual rainfall, and reflects the fact that GCMs generally predict greater rainfall intensity associated with intensification of the hydrologic cycle. Increases in rainfall intensity have indeed been observed in other regions (e.g. Huntington 2006). Trend analyses for Australia are more ambiguous, partly because of the additional influence of ocean circulation patterns on rainfall intensity. Dryland and river salinity are primarily attributed to increased groundwater recharge, and therefore might be expected to show similar changes. This stands in contrast to the analysis of Austin et al. (2010), which did not account for any effect of changed rainfall intensity and predicted a salt load reduction of 34–49%. It is noted however that in either case the expected recharge change is smaller than that which would accompany the removal or introduction of deep-rooted vegetation (Crosbie et al. 2010).

ACKNOWLEDGEMENTS

This project was supported by a research grant from the Australian Research Council (DP0985080). We also thank an anonymous reviewer for her/his constructive comments that helped improve the manuscript.

REFERENCES

Alcolea, A., Carrera, J., & Medina, A. (2006) Pilot points method incorporating prior information for solving the groundwater flow inverse problem. *Advances in Water Resources*, 29, 1678–1689.

Brown, C.M. (1989). Structural and stratigraphic framework of groundwater occurrence and surface discharge in the Murray Basin, southeastern Australia. *BMR Journal of Australian Geology and Geophysics*, 11, 127–146.

Brown, N. & Tregoning, P. (2010) Quantifying GRACE data contamination effects on hydrological analysis in the Murray-Darling Basin, southeast Australia. *Australian Journal of Earth Sciences*, 57, 329–335.

Bruinsma, S., Lemoine, J., Biancale, R. & Valès, N., 2010. CNES/GRGS 10-day gravity field models (release 2) and their evaluation. *Advances in Space Research*, 45(4), 587–601.

Carrera, J., Alcolea, A., Medina, A., Hidalgo, J. & Slooten, L. 2005. Inverse problem in hydrogeology. *Hydrogeology Journal*, 13, 206–222.

Connell, D. (2007), Water Politics in the Murray-Darling Basin, The Federation Press, Sydney.

Cresswell R.G., Dawes W.R., Summerell G.K., and Walker G.R. 2003. Assessment of salinity management options for Kyeamba Creek, New South Wales: Data analysis and groundwater modelling. CSIRO Land and Water Technical Report 26/03.

Crosbie R, McCallum J, Walker G, Chiew F. 2010. Modelling climate-change impacts on groundwater recharge in the Murray-Darling Basin, Australia. Hydrogeology Journal 18: 1639–1656.

CSIRO. 2008. Water availability in the Loddon-Avoca. A report to the Australian Government from the CSIRO Murray-Darling Basin Sustainable Yields Project. CSIRO, Australia. 123pp.

CSIRO and Australian Bureau of Meteorology. (2007) *Climate Change in Australia*. Techical report. CSIRO. [accessed January 2011] Available from: www.climatechangeinaustralia.gov.au/

Döll, P.F., Kaspar, F. & Kaspar, B. (2003) A global hydrological model for deriving water available indicators: model tuning and validation. *Journal of Hydrology*, 270, 105–134.

Ek, M.B., Mitchel, K.E., Lin, Y., Rogers, E., Grunmann, P., Koren, V., Gayno, G. & Tarpley, J.D. (2003) Implementation of Noah land surface model advances in the National Centers for Environmental Prediction operational mesoscale Eta model. *Journal of Geophysical Research Atmospheres*, 108(D22), 8851.

Evans, W.R. & Kellet, J.R. (1989) The hydrogeology of the Murray Basin, southeast Australia. *BMR Journal of Australian Geology and Geophysics*, 11, 147–165.

Fetter, C.W. (2001) *Applied Hydrogeology*. 4th edition. Upper Saddle River, Prentice-Hall. 598 pp.

Frappart, F., Ramillien, G., Leblanc, M., Tweed, S.O., Bonnet, M.-P. & Maisongrande, P. (2011) An independent Component Analysis approach for filtering continental hydrology in the GRACE gravity data. *Remote Sensing of Environment*, 115(1), 187–204.

Gupta, H.V., Sorooshian, S. & Yapo, P.O. (1998) Toward improved calibration of hydrologic models: Multiple and noncommensurable measures of information. *Water Resources Research*, 34(4), 751–763.

Hekmeijer, P. & Dawes, W. (2003a) Assessment of salinity management options for South Loddon Plains, Victoria: Data analysis and groundwater modeling. CSIRO Land and Water Technical Report 24/03.

Hekmeijer, P. & Dawes, W. (2003b) Assessment of salinity management options for Axe Creek, Victoria: Data analysis and groundwater modelling. CSIRO Land and Water Technical Report 22/03, MDBC Publication 08/03, 40 pp.

Hennessy, J., et al. (1994) Wangaratta hydrogeological map (1:250,000 scale) Murray Basin Hydrogeological map series. Canberra, ACT, Australian Geological Survey Organisation.

Herczeg, A.L., Dogramaci, S.S. & Leaney, F.W. (2001) Origin of dissolved salts in a large, semi-arid groundwater system: Murray Basin, Australia. *Marine Freshwater Research*, 52, 41–52.

Ife, D. & Skelt, K. (2004) *Murray Darling Basin Groundwater Status 1990–2000. Murray Darling Basin Commission, publication 32/04, Canberra.* ISBN 1876830948. [accessed November 2010] Available from: http://www.mdbc.gov.au

Kirby, M., Evans, R., Walker, G., Cresswell, R., Coram, J., Khan, S., Paydar, Z., Mainuddin, M., McKenzie, N. & Ryan, S. (2006) *The shared water resources of the Murray-Darling Basin. Murray-Darling Basin Commission, Publication 21/06, Canberra.* ISBN 192103887X. [accessed January 2011] Available from: http://www.mdbc.gov.au

Leaney, F.W., Herczeg, A.L. & Walker, G.R. (2003) Salinization of a fresh paleao-ground water resource by enhanced recharge. *Ground Water*, 41(1), 84–92.

Leblanc, M.J., Tregoning, P., Ramillien, G., Tweed, S.O. & Fakes, A. (2009) Basin-scale, integrated observations of the early 21st century multiyear drought in southeast Australia. *Water Resources Research*, 45, W04408. doi:10.1029/2008WR007333

Leblanc, M., Tweed, S., Van Dijk, A. & Timbal, B. (2011, January) Historical and future hydrological changes in the Murray-Darling Basin: A review. *Global and Planetary Change* (in press).

Leblanc, M., Tweed, S., Van Dijk, A. & Timbal, B. (2012). *The Murray-Darling Basin: A food bowl in crisis.* The United Nations World Water Development Report 4. Case Study Volume. UNESCO, Earthscan, Paris.

Macumber, P.G. (1999) *Groundwater flow and resource potential in the Bridgewater and Salisbury West GMAs.* Melbourne, Phillip Macumber Consulting Services. 88 pp.

McCallum, J., Crosbie, R., Walker, G. & Dawes, W. (2010) Impacts of climate change on groundwater in Australia: a sensitivity analysis of recharge. *Hydrogeology Journal*, 18, 1625–1638.

Meinshausen, M., Meinshausen, N., Hare, W., Raper, S.C.B., Frieler, K., Knutti, R., Frame, D.J., Allen, M.R. (2009) Greenhouse-gas emission targets for limiting global warming to 2°C. *Nature*, 458, 1158–1162.

Meyers, G., McIntosh, P., Pigot, L. & Pook, M. (2007) The Years of El Niño, La Niña, and Interactions with the Tropical Indian Ocean. *Journal of Climate*, 20, 2872–2880.

Petheram C., W. Dawes, G. Walker and R. B. Grayson. 2003. Testing in class variability of groundwater systems: local upland systems. *Hydrological Processes*, 17, 2297–2313.

Ramillien, G., Famiglietti, J.S. & Wahr, J. (2008) Detection of continental hydrology and glaciology signals from GRACE: A review. *Surveys in Geophysics*, 29(4–5), 361–374.

Ramillien, G., Lombard, A., Cazenave, A., Ivins, E.R., Llubes, M., Remy, F., & Biancale, R. (2006) Interannual variations of the mass balance of the Antarctica and Greenland ice sheets from GRACE. *Global and Planetary Change*, 53(3), 198–208.

Rodell, M., Chen, J., Kato, H., Famiglietti, J.N., & Wilson, C. (2007) Estimating groundwater storage changes in the Mississippi River basin (USA) using GRACE. *Hydrogeology Journal*, 15, 159–166.

Schmidt, R., Flechtner, F., Meyer, U., Neumayer, K.-H., Dahle, Ch., Koenig, R. & Kusche, J. (2008) Hydrological Signals Observed by the GRACE Satellites. *Surveys in Geophysics*, 29, 319–334.

Singh, A., Minsker, B.S. & Valocchi, A.J. (2008) An interactive multi-objective optimization framework for groundwater inverse modeling. *Advances in Water Resources*, 31, 1269–1283.

Smitt, C., Doherty, J., Dawes, W. & Walker, G. (2003) *Assessment of salinity management options for the Brymaroo catchment, South-eastern Queensland.* CSIRO Land and Water Technical Report 23/03. Murray Darling Basin Commission, Canberra. ISBN: 1 876 830 50 6. [accessed January 2011]. Available from http://www.clw.csiro.au/publications/technical2003/tr23-03.pdf

Strassberg, G., Scanlon, B.R. & Rodell, M. (2007) Comparison of seasonal terrestrial water storage variations from GRACE with groundwater-level measurements from the High Plains Aquifer (USA). *Geophysical Research Letters*, 34, L14402. [doi:10.1029/2007GL030139

Tregoning, P., Watson, C., Ramillien, G., McQueen, H. & Zhang, J. (2009) Detecting hydrologic deformation using GRACE and GPS. *Geophysical Research Letters*, 36, L15401. doi:10.1029/2009GL038718

Urbano, L.D., Person, M., Kelts, K. & Hanor, J.S. (2004) Transient groundwater impacts on the development of paleoclimatic lake records in semi-arid environments. *Geofluids*, 4, 187–196.

Yadav, M., Wagener, T. & Gupta, H. (2007) Regionalization of constraints on expected watershed response behavior for improved predictions in ungauged basins. *Advances in Water Resources*, 30, 1756–1774.

Temperate Climates

CHAPTER 11

Impact assessment of combined climate and management scenarios on groundwater resources. The Inca-Sa Pobla hydrogeological unit (Majorca, Spain)

Lucila Candela, Wolf von Igel, F. Javier Elorza &
Joaquín Jiménez-Martínez

ABSTRACT

The impacts of climate change and management on groundwater resources of the Inca-Sa Pobla hydrostratigraphic unit, Majorca, Spain have been investigated for the year 2025 under the greenhouse gas scenario A2 developed by the ICCP. Future climate change impact on water resources at the study scale was based on the downscaled HadCM3 (General Circulation Model) coupled to MODFLOW and Visual BALAN numerical models. Results of simulation show a 17% reduction on natural recharge, implying a decrease of the springs flow rate discharging into the aquifer associated wetland. Even under the climate change scenarios uncertainty, to preserve the spring discharge at the current level and avoid the wetland from drying up, a management options based on the decrease of groundwater abstraction and wells re-allocation for irrigation is needed, as assessed by sensitivity analysis.

11.1 INTRODUCTION

Current sustainable water management in Mediterranean countries, especially on islands, is of great concern because of the increasing competition for water resources between agriculture and tourism, and the Water Framework Directive (WFD) requires good ecological and chemical status of water bodies and associated aquatic systems. Additionally, climate change projections from atmosphere-ocean general circulation models (GCMs) forecast an increase of temperature, decrease of precipitation, and variability of extreme events under the A2 green-house gas scenario (IPCC 2001). The A2 scenario focuses on economic growth, increasing population, and CO_2 emissions are the highest of all scenario families. The combined pressures of future climate change and increasing water demand requires planning strategies to preserve water resources.

There are a growing number of publications on studies evaluating the impact of climate change on groundwater (Green et al. 2011), recharge and water resources (Kundzewick and Somlyódy 1997; Arnell 1998; Allen et al. 2004; Wilby et al. 2006); assessing the interrelationship between climate change and land-use change (Kracauer et al. 1997; Serra-Capdevila et al. 2007; Woldeamlak et al. 2007); on setting the framework for integrated assessment of physical (climate, vegetation and soil); and societal change (Green et al. 2007a; Risbey et al. 2007; Ruth et al. 2007). Yet, only a small number of publications analyze combinations of climate and management scenarios (Candela et al. 2009).

In the Inca Sa Pobla area of Majorca island, (Balearic Islands, western Mediterranean) there is a growing concern about the possible impacts of climate change on water-resource availability. The concern stems from a growing competition for scarce water resources among users (agriculture, tourisms domestic supply and ecosystems) and a possible deterioration of the S'Albufera coastal wetlands that are associated with the coastal aquifer (PHIB 1999).

The principal objective of the study was to quantify the response in the hydrologic system to climate scenario A2, specifically to changes in the springs discharge to S'Albufera wetland. The springs constitute a natural drain of the aquifer outflowing into the wetland, and the basic assumption is that the 1986–2005 period represents the minimum mean flow to preserve the ecological status of the wetland. The future scenario for 2025 developed by the IPCC was selected for the climatic series generation, as it represents the possible extreme situation to assess changes in groundwater recharge.

11.1.1   Description of the study area: the Inca-Sa Pobla hydrogeological unit

The Inca-Sa Pobla Plain (Figure 11.1) is located in the northern part of the Majorca Island (Balearic Islands, Spain). Although, traditionally characterized by an important irrigated agricultural activity, a growing tourism industry since the 1980s has helped double the resident population during summer periods and has changed the predominant economic activity. Water resources mainly rely on groundwater, and increasing pressures from growing extractions for municipal water supply and prior intensive agricultural practices have deteriorated the quality and quantity of groundwater resources.

Mean annual temperature and precipitation are 17°C and 600 mm/a respectively (1986–2005); potential evapotranspiration (Penman-Monteith) accounts for 1,000 mm/a. It is important to notice that 80% of precipitation occurs in spring and autumn.

From the geologic point of view (Figure 11.1), the Inca-Sa Pobla constitutes a northeast/south-west tectonic subsiding sedimentary basin, filled with post-tectonic Miocene, Pliocene, and Quaternary materials (sand, gravel, conglomerates, limestones, and marls) of continental origin. Marine influence on sediments is observed near the Bay of Alcudia. Due to the numerous lateral facies changes, the stratigraphy of the basins is rather complex.

The Inca-Sa Pobla hydrogeological unit is 360 km^2 and composed of three tectonic sub-basins that are hydraulically connected in a greater or lesser extension: S'Albufera (Mediterraneum coast), Sa Pobla (northeast), and Inca (southwest). The unit is characterised by four different hydrostratigraphic units and three aquitard units, grouped into an upper and lower aquifer system and separated by the lower Pliocene grey marls (Figure 11.1). The upper water-table aquifer consists of Quaternary (Q) silts and gravels and Pliocene (PL) calcareous sandstones; the lower aquifer of Messinian and Tortonian limestones is only confined in the central part of the sub-basins by Pliocene marls that pinch out towards the borders. The northern part of the basin is limited by Lias limestones and dolostones which always constitute an unconfined aquifer that is laterally connected with the Plio-Quaternary and the Miocene formations of the Sa Pobla plain.

The Quaternary sand and gravel deposits and the Pliocene calcareous sandstones constitute an important unconfined aquifer hydraulically connected with all aquifer formations in the area. According to PHIB (1999) Quaternary thickness varies between several metres up to 50 m, with the greatest thickness in the centre of the basin. Hydraulic conductivity ranges between 8 and 15 m/d, and up to 300 m/d near the S'Albufera area.

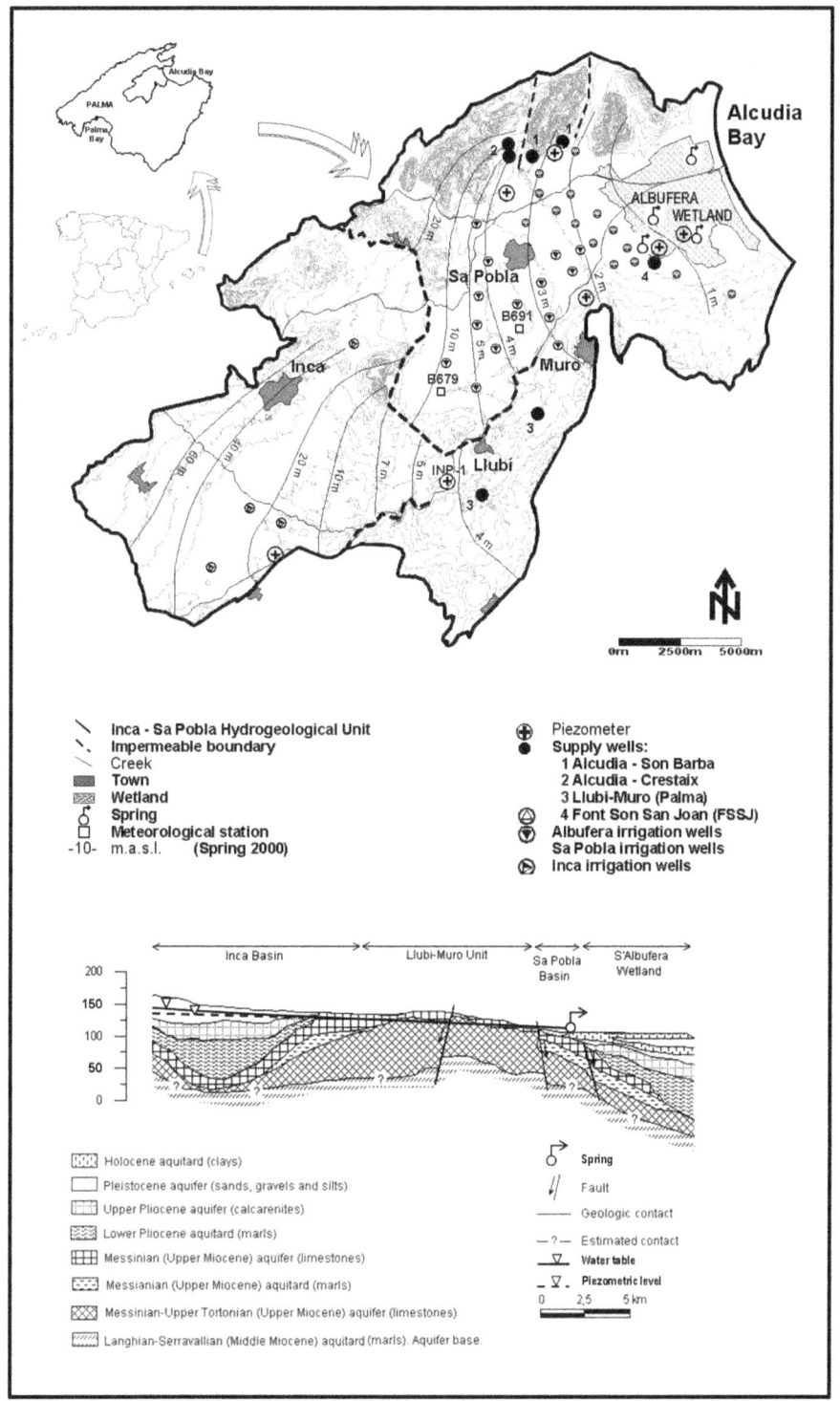

Figure 11.1. Map of the study area illustrating wells, piezometers and meteorological stations location and piezometric level of the Plio-Quaternary aquifer. The SW-NE geologic cross-section shows the different hydrostratigraphic units (modified from Candela et al. 2009).

Pliocene calcareous sandstones are covered by Quaternary deposits and only outcrop on the southern fringe. The unconfined upper Pliocene aquifer is directly connected to the Quaternary and Messinian when the lower Pliocene grey marls disappear (Sa Pobla area). Its thickness shows a maximum of 75 m at the southern edge of the basin. The transmissivity range is from 1,000 to 5,000 m²/d.

The lower aquifer of Messinian and Tortonian age is composed by limestone. The Messinian can also constitute an unconfined or semiconfined aquifer, with thickness never exceeding 100 m. The Tortonian outcrops at the southern border of the basin; its thickness varies from a few metres to more than 200 m. Their transmissivity is about 500–5,000 m²/d and sometimes even higher than 10,000–15,000 m²/d.

The upper aquifer is the main focus of the study. Groundwater from this unit generally flows from the SW to NE and discharges into the Alcudia Bay through the S'albufera wetland. The average groundwater level is above sea level but seasonal variations of up to 2 m exist. Recharge to the hydrogeological unit is mainly by rainfall infiltration, irrigation return flow, and lateral inflow from the Tramuntana range limestones. Most of the recharge occurs during spring and autumn, and is on average 54 Mm³/a for the last 25 years (PHIB 1999). Discharge is produced by springs draining into the S'Albufera wetland (17 Mm³/a) and lateral flow to the sea. Groundwater abstraction for agricultural activities and municipal supply also constitute important outflows (36.5 Mm³/a) as the aquifer is intensively exploited.

11.2   METHODOLOGY

The methodology includes the following steps: develop a conceptual and numerical hydrogeological model; obtain climate change forecasts (2025) from downscaled GCM models; use the model to assess future natural recharge scenarios at local scale; and evaluate climate change impacts on springs discharge through groundwater modelling and management scenarios.

11.2.1   Recharge estimation

Distributed groundwater recharge from precipitation and irrigation return was obtained through a daily soil-water balance with Visual BALAN 1.0 (Samper et al. 1999), which is a physically based code that has been successfully applied to many case studies (García-Santos et al. 2005; Castañeda and Garcia-Vera 2008; Sena and Molinero 2009). Visual BALAN 1.0 computes water mass balance taking into account water infiltration through soil and the properties of the unsaturated zone and aquifer at daily time step. According to soil characteristics, soil use (forest, urban, irrigated and non-irrigated farmland) and meteorological data from seven meteorological stations, thirteen different recharge zones were initially defined. The background recharge representative of the entire study area was defined as estimated by the numerical model using the precipitation records (1986–2005) from the B679 meteorological station (Figure 11.1), soil parameters, and historic irrigation patterns. This simplification reduced the definition of recharge areas to three different categories (dryland, irrigated land, and lias dryland recharge). Soil parameters were obtained from existing studies (PHIB 1999) while potential evapotranspiration was estimated according to Penmann equation.

Impact of climate change scenario on groundwater recharge was assessed by comparing future climate changes versus the historical input (1986–2005). Recharge for the three defined areas and climate scenario (precipitation and temperature for 2025 time horizon) was estimated through Visual BALAN.

11.2.2   Groundwater flow simulation model

The flow numerical model of the upper aquifer (Plio-Quaternary) was developed and implemented in MODFLOW-2000 (Harbaugh et al. 2000). The model was defined by 88 rows and 136 columns in a 250×250 m four-layered grid system. The top layer (layer 1) is a confining 3 m thick clay layer, only outcropping at the S'Albufera area. The Plio-Quaternary unit constitutes Layer 2, the outcropping upper aquifer, where springs–groundwater interactions occur. Layer 3 is the aquitard (grey marls) and Layer 4 is the lower aquifer. All model boundaries, other than a small section in the northern part (Tramuntana range) and the sea shore were considered impervious. The sea boundary was simulated using a constant head condition based on a 25 cm average tidal fluctuation. Lateral input was modeled by seven injection wells and discharge through springs was simulated by a drain condition in layer 2. For a detailed model description the reader is submitted to Candela et al. (2009).

Daily groundwater recharge was independently estimated with the Visual BALAN 1.0 linked to MODFLOW as input. Piezometric level and spring discharge into the wetland were the simulated output. The model was performed in transient state at monthly steps for 1986–2002; validation period was 2003–2005.

11.2.3   Climate change scenarios. Statistical downscaling

To simulate future climate change scenarios the HadCM3 GCM (http://www.meto.gov .uk/research/hadleycentre/; Pope 2000; Gordon 2000) widely used during the IPCC-3 was applied. Although several time-slices and emission scenarios were assessed, only results from the future Emission Scenario–SRES A2 (Medium-High) and the time slice 2012–2037, here after referred to as "2025" are here presented. Selection of the time slice was based on the consideration that results for this horizon would be relevant to water resource planners, and more reliable than longer term climate predictions due to uncertainties of specifying future emissions or modeling impacts. HadCM3 simulations provide 50 daily values for climatic variables, including precipitation, temperature and global solar radiation.

To generate future precipitation and temperature series relevant to the scale of study, results of the HadCM3 were statistically downscaled to generate local precipitation and temperature climate series (Hewiston and Crane 1996). The downscaling procedure applied first develops a stochastic model that reproduces several statistics of the observed historic data. Next the mean and variance of the calibrated stochastic model are perturbed according to the relative changes in these statistics predicted by the GCM. The method applied acknowledges that GCM only accurately predict the changes in monthly mean and variance of temperature and precipitation along time and is capable to incorporate trend effects in rainfall and temperature and consequently to generate series with statistical properties influenced by climate change. This downscaling technique has been also applied by researchers in similar case studies with good results (Bouraoui et al. 1999; Aronica et al. 2006).

The model applied for the stochastic rainfall generator is the well-known chain-dependent-process stochastic model for daily precipitation (Champan 1997; Wilks 1992; Kottegoda et al. 2003) structured in a two-state architecture: a first-order non-stationary Markov chain for modelling the occurrence of rainfall events (presence of wet/dry days) and a probabilistic Weibull distribution model for modelling the rainfall amount. For the generation of synthetic temperature series a classical Auto Regressive Moving Average model (ARMA) has been applied (Maidment 1993). For the Inca Sa-Pobla area, the rainfall and temperature generator was calibrated against a 25-year period (1980–2005) daily rainfall and temperature series from the B-679 and B-691 meteorological station respectively (Fig. 11.1).

Meteorological series of climate change scenarios

The stochastic precipitation model based on observation data (1980–2005) at station B-679 was used to generate 200 series of future daily precipitation and temperature of 50 years long. From the computational point of view, it is too time-consuming to estimate aquifer recharge and to run the groundwater flow simulation model from 200 synthetic series for each scenario. In order to overcome this problem only one series (one P and T realization) was selected as representative of the Inca-Sa Pobla unit. Series with mean precipitation and a standard deviation with values around ±3% of the downscaled historical series, and with the lowest normalized mean and variance objective function (Kornecki et al. 1999) were selected for aquifer recharge estimation.

For temperature generation, data from the B691 meteorological station were used, as being the only available time series. Daily temperature were used for the ARMA (1,1) parameters definition. For the 2025 horizon scenarios simulations, temperature series for each emission scenario was defined averaging the daily temperature of the 200 realizations for scenario A2.

11.2.4   Groundwater abstraction scenarios

Scenarios for future groundwater abstraction and demand projections were stated according to the Hydrological Plan of the Balearic Islands (PHIB 1999), key stakeholders and expert's analysis. For the purpose of simulation scenarios abstraction options based on sustainable development (decrease of exploitation for water supply or agriculture) or changes only conditioned by climate change were defined.

For municipal water supply, water abstraction scenarios were:

- 1: Future demand and total extraction would account for $14.8 \, Mm^3/a$, the historical abstracted volume.
- 2: Sustainable scenario. Compliance with the WFD measures and reduction to $8.1 \, Mm^3/a$. The following measures are foreseen: restriction to $1.5 \, Mm^3/a$ of abstraction at the Font the San Joan (FSSJ); spatial re-allocation of pumping wells or interdiction of pumping from wells.

Abstraction scenarios for irrigation purposes defined according the CAP (Common Agricultural Policy) and local trends. Scenarios assume constant crop-type distribution and proportional reduction of cultivated surface.

- A: Maintenance of the existing agricultural pattern (4,300 ha – 21.7 Mm^3/a).
- B: Decrease of irrigated surface by 20% (3,400 ha) and resulting abstraction of 17.0 Mm^3/a.
- C: Agricultural land reduction by 40% (2,600 ha) and abstracted volume 11.9 Mm^3/a.

Only two climate scenarios were considered for recharge: actual climate (γ), and under SRES A2 scenario (α).

11.2.5   Sensitivity and uncertainty analysis

The objective of sensitivity analysis (SA) was to assess the impacts on spring discharge of pumped water reduction from well groups located in different areas of the study zone. The objective functions for SA were the mean and standard deviation of the monthly spring flow rate between October 2000 and September 2005; a variation of 1.5 Mm^3/a of water extraction by a group of wells was the value considered in the analysis. The method being applied is a modified variance-based method known as Fourier Amplitude Sensitivity Test (FAST) (Mira et al. 2004), a global sensitivity analysis procedure that gives an estimation of both individual effects (main effects) and joint effects (interactions between the input variables).

The purpose of uncertainty analysis was to assess the uncertainty of a subset of groundwater model input parameters and its propagation in the simulated outcomes. The analysis performed for dry-land and groundwater extraction for historical data are presented here as an example of the methodological approach. The uncertainty analysis was addressed by applying a Latin Hypercube sampling methodology (McCay et al. 1979), using the Sandia's Latin Hypercube Sampling Software (Wyss and Jorgensen 1998).

11.2.6   Impact assessment by coupling climate and abstraction scenarios

For future impacts assessment at the Inca-Sa Pobla unit, a combination of previously defined scenarios for climate change (no change, A2), municipal water supply (sustainable, growth) and agriculture (light and strong reduction) were simulated with MODFLOW. Each simulation was executed in transient mode for a monthly time period over 25 years, and was re-run with the initial piezometric condition of the previous run final conditions in order to achieve long-term equilibrium.

11.3   RESULTS AND DISCUSSION

11.3.1   GCM and local predictions

For the time horizon 2025 HadCM3 (A2) predicts a slight decrease of 2% in the mean annual precipitation compared to the historical control period (B-679). On a monthly basis, the model shows a slight increase in the precipitation during spring and autumn and a decrease in winter and summer. Mean annual temperature will increase by 0.7°C with regard to the control period; temperature shows an increase trend all along the year, especially during winter time.

The mean precipitation of the selected downscales stochastic series (578 mm) do not vary significantly from the historical data (584 mm, B-679). On a monthly basis nevertheless, as predicted by the GCM, precipitation patterns vary. Winter precipitation diminishes while autumn precipitation augments compared to the historic observations.

Temperature series indicate an overall increase of annual temperature of 0.7°C above the historical mean. On a monthly basis temperature increases in autumn and winter and decreases in summer.

11.3.2   Climate change impact on groundwater resources and natural recharge

Comparison of historical results (1986–2005) and A2-2025 scenario shows a decrease of spring flow discharge from 17 to 12 Mm³/a, increase of seawater intrusion (Figure 11.2) and a slightly negative aquifer water balance, which may be related to the associated uncertainties of simulation. Additionally, a generalized groundwater level drawdown can be expected, reaching around 3 m at piezometer INP-1 (Figure 11.3).

The most important parameter conditioning aquifer water resources are changes of natural recharge. Under future scenario, recharge amount decreases in the three previously defined recharge areas although to a different extent. For the quaternary dry-land, recharge decreases by 21%; for irrigated quaternary recharge decreases by 13%; and for lias dry-land recharge decreases by 7%. Total recharge to the Inca-Sa Pobla hydrological unit presents an important reduction from the historical mean (55.1 Mm³/a) decreasing up to 45.9 Mm³/a (−17%) for SRES A2.

Monthly recharge decrease for the quaternary dryland area is most significant during winter and autumn when recharge rates are highest. This is in agreement with the

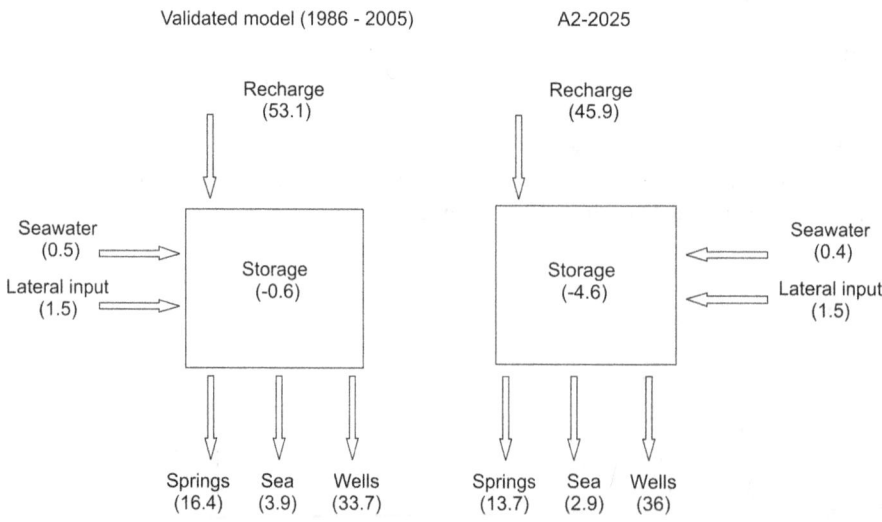

Aquifer balance (Mm³/yr)

Validated model (1986 - 2005)   A2-2025

Recharge (53.1)   Recharge (45.9)

Seawater (0.5)   Storage (-0.6)   Storage (-4.6)   Seawater (0.4)

Lateral input (1.5)   Lateral input (1.5)

Springs (16.4)   Sea (3.9)   Wells (33.7)   Springs (13.7)   Sea (2.9)   Wells (36)

Figure 11.2.   Comparison of aquifer balance results for the historical period and A2 (2025) scenario.

A2-2025

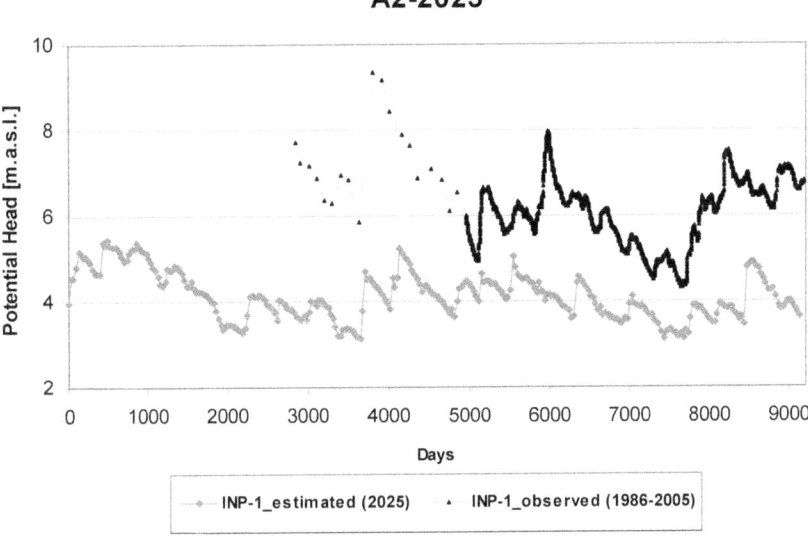

Figure 11.3. Simulation of piezometric level results for piezometer INP-1 according to future changes of water recharge (A2, 2025).

predicted increase in temperatures during these seasons. Since the soil in quaternary dryland areas has a higher retention capacity the effect of increased evapotranspiration is more acute than on the lias dryland area where soil is less developed and water infiltrates and deeply percolates faster.

11.3.3 Sensitivity analysis of water abstraction spatial location

Sensitivity analysis indicates that well-groups with the highest influence on mean spring discharge are agricultural wells near the Albufera wetland (27%) and in Sa Pobla (20%) (Figure 11.4). For water supply, most sensitive well-groups are Font de Son San Joan (15%, FSSJ) and Son Barba (13%). Results clearly indicate that the efficiency of adaptive measures to be taken is dependent on the spatial situation of pumping wells.

Mean spring discharge rate and can be approximated by a normal distribution (13.9±2.0 Mm³/a). Mean and minimum spring discharge rates from the validated model (historical period) account for 18.4 and 14.1 Mm³/a, respectively, however, monthly flow rate may range between 6 and 15 Mm³/a. As assessed by the uncertainty analysis, a reduction in the volume of water abstracted only based on minimum or mean monthly spring historical discharge could have a significant impact on future measures to preserve the wetland.

11.3.4 Impact of combined climate change and management scenarios on spring flow rate

Possible combination of climate change scenarios and management actions leads to 6 options (Table 11.1). Future extraction scenarios for the different management actions vary between a minimum of 20.7 Mm³/a under a sustainable supply scenario and a scarce

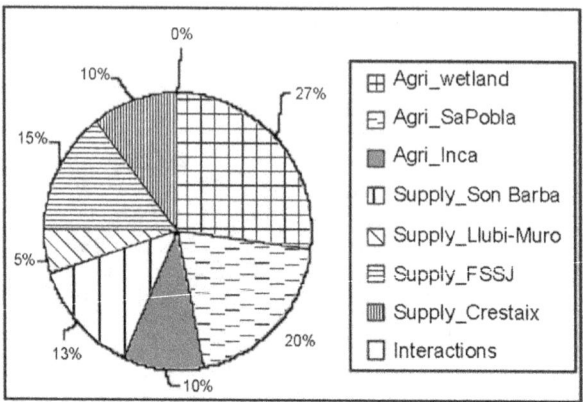

Figure 11.4.   Results from sensitivity analysis showing the relative contribution of abstraction wells to spring discharge.

Table 11.1.   Annual spring discharge flow obtained from coupled climate and management scenario.

Baseline Scenario $1A\gamma$ (17.6 Mm^3/a)	Spring flow (Mm^3/a)	Spring flow reduction (%)	Irrigation abstraction (Mm^3/a)	Supply abstraction (Mm^3/a)
$1A\alpha$	12.3	30.1	21.7	14.8
$1B\alpha$	14.0	20.5	17	14.8
$1C\alpha$	16.2	8	11.9	14.8
$2A\alpha$	16.5	6.2	21.7	8.1
$2B\alpha$	18.3	*	17	8.1
$2C\alpha$	20.7	*	11.9	8.1

α: A2 climate scenario for 2025; 1, 2: supply scenario; A, B, C: irrigated agriculture scenario; *: spring flow rate show no reduction

agriculture scenario ($A\gamma$), and 36.5 Mm^3/a under existing supply and agriculture scenario. When combining the proposed management actions with climate change scenarios, the estimated mean springs flow varies between a minimum of 12.3 Mm^3/a and a maximum of 20.7 Mm^3/a.

Under A2 scenario, spring discharge levels are maintained only if water supply extractions are reduced and agricultural activity decreases (Figure 11.5). To preserve historical spring flow rate, and consequently the wetland sustainability, total extraction must be reduced between 25 and 29 Mm^3/a. Since under greater extraction rates spring discharge may still be higher, spatial location of the abstraction wells is also a relevant parameter controlling spring discharge.

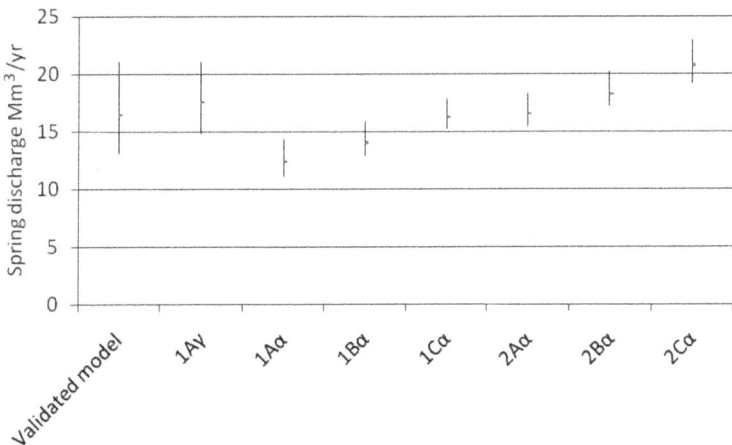

Figure 11.5.   Estimated spring discharge (maximum, minimum and average) for A2 and time slice 2025. 1A: baseline; 1,2: supply scenarios; A, B, C: irrigation scenarios.

11.4   CONCLUSIONS AND RELEVANCE FOR GRAPHIC

The work here presented constitutes an attempt to develop an integrated approach of climate change impact and management scenarios on groundwater at local scale by linking a GCM with a groundwater numerical model. The GCM scale is obviously improper to be directly used at basin scale due to the coarse resolution. As more detailed and quantitative information is needed at appropriate spatial resolution for local climate impact assessment, future climate series were derived downscaling the outputs for temperature and precipitation from the HadCM3 model using a stochastic weather generator. However, the intrinsic uncertainty existing in projected climate change is still high (Chen et al. 2011), as demonstrated by the fact that current actual emissions trajectories exceed the reference case assumptions of scenarios produced even just a few years ago. As a consequence, results must be taken as indicative of the system potential response.

Also, the stochastic rainfall generator developed for this study is not capable of generating series correlating the precipitation events between the meteorological stations, and given that the natural spatial and temporal variability of temperature and precipitation, added uncertainty must be included in final results by simplifying stochastic precipitation series generation and by selecting only one meteorological station and only one time series. Appropriate methods considering occurrence of joint variables appears to be the best approach.

Sensitivity analysis allowed identifying group of wells where management options would be most efficient to preserve the wetland, either by wells re-allocation or abstraction reduction. According to simulation results, for the range of abstraction rates analysed here, the spring discharge is approximately proportional to the extraction rate, with a mean flow increase of $0.4\,Mm^3/a$ for each Mm^3/a decrease of extractions. Therefore, as changes in recharge are foreseen in the future, which may impact available water resources; changes in groundwater abstraction related to land use (reduction between 20 and 40% of irrigated area) represent the only options for planned adaptive measures. Otherwise, external water sources will be needed to meet the imbalance from future water demand.

This case study highlights the strong relationship that exists between groundwater recharge and management strategies to preserve groundwater resources under predicted climate change in a Mediterranean island. The geographic location and the importance of groundwater resources as well as the applied methodological approach make this application a good example of the philosophy and approach of the GRAPHIC programme.

REFERENCES

Allen, D.M., Mackie, D.C. & Wei, M. (2004) Groundwater and climate change: A sensitivity analysis for the Grand Forks aquifer, southern British Columbia, Canada. *Hydrogeology Journal,* 12, 270–290.

Arnell, N.W. (1998) Climate change and water resources in Britain. *Climatic Change,* 39, 83–110.

Aronica, G., Corrao, C., Amengual, A., Alonso, S. & Romero, R. (2005) Water resources evaluation under climatic trend effects in mediterranean catchments. *Geophysical Research Abstracts,* 7, 04091.

Bates, B.C., Charles, S.P., Sumner, N.R. & Fleming, P.M. (1994) Climate change and its hydrological implications for South Australia. *Transactions of the Royal Society of South Australia,* 118, 35–43.

Bouraoui, F., Vachaud, G., Li, L.Z.X., Le Treut, H. & Chen, T. (1999) Evaluation of the impact of climate changes on water storage and groundwater recharge at the watershed scale. *Climate Dynamics,* 15, 153–161.

Brouyère, S., Carabin, G. & Dassargues, A. (2004) Climate change impacts on groundwater resources: Modelled deficits in a chalky aquifer, Geer basin, Belgium. *Hydrogeology Journal,* 12, 123–134.

Burns, D.A., Klaus, J. & McHale, M.R. (2007) Recent climate trends and implications for water resources in the Catskill Mountain region, New York, USA. *Journal of Hydrology,* 336, 155–170.

Candela, L., von Igel, W., Elorza, F.J. & Aronica, G. (2009) Impact assessment of combined climate and management scenarios on groundwater resources and associated wetland (Majorca, Spain). *Journal of Hydrology,* 376, 510–527.

Castañeda, C. & García-Vera, M.A. (2008) Water balance in the playalakes of an arid environment, Monegros, NE Spain. *Hydrogeology Journal,* 16, 87–102.

Chapman, T. (1997) Stochastic models for daily rainfall in Western Pacific. *Mathematics and Computers in Simulation,* 43, 351–358.

Chen, J., Brissette, F.P. & Leconte, R. (2011) Uncertainty of downscaling method in quantifying the impact of climate change on hydrology. *Journal of Hydrology,* 401, 190–202.

Cuculeanu, V., Tuinea, P. & Balteanu, D. (2004) Climate change impacts in romania: Vulnerability and adaptation options. *Geological Journal,* 57(3), 203–209.

García-Santos, G., Mazol, V., Morales, D., Gómez, L.A., Pisani, B. & Samper, J. (2005) Groundwater recharge in a mountain cloud laurel forest at Garajonay National Park (Spain). *Geophysical Research Abstracts,* 7, 00942.

Green, R., Bates, C., Charles, S.P. & Fleming, P.M. (2007b) Physically based simulation of potential effects of carbon dioxide altered climates on groundwater recharge. *Vadose Zone Journal,* 6, 597–609.

Green, T.R., Taniguchi, M. & Kooi, H. (2007a) Potential impacts of climate change and human activity on subsurface water resources. *Vadose Zone Journal,* 6, 531–532.

Green, T.R., Taniguchi, M., Kooi, H., Gurdak, J.J., Allen, D.M., Hiscok, K.M., Treidel, H. & Aureli, A. (2011) Beneath the surface of global change: Impacts of climate change on groundwater. *Journal of Hydrology,* 405, 532–560.

Hagg, W., Braun, L.N., Huhn, M. & Nesgaard, T.I. (2007) Modelling of hydrological response to climate change in glacierized Central Asia Catchments. *Journal of Hydrology*, 332, 40–53.

Harbaugh, A.W., Banta, E.R., Hill, M. & McDonald, M.G. (2000) *MODFLOW-2000, the U.S. Geological Survey modular ground-water model – user guide to modularization concepts and the ground-water flow process*. 0-121. 2000. U.S. Geological Survey. Open-File Report 00-92.

Hewiston, B.C. & Crane, R.G. (1996) Climate downscaling: Techniques and application. *Climate Research*, 7, 85–95.

Hsu, K.C., Wang, C.H., Chen, K.C., Chen, C.T. & Ma, K.W. (2007) Climate-induced hydrological impacts on the groundwater system of the Pingtung Plain, Taiwan. *Hydrogeology Journal*, 15, 903–913.

IPCC. (2001) Projections of Future Climate Change, Chapter 9. In: Houghton, J.T., Ding, Y., Griggs, D.J., Noguer, M., Van Der Linden, P.J., Dai, X., Maskell, K. & Johnson, C.A. (eds.) *Climate Change 2001: The Scientific Basis. Contribution of Working Group 1 to the Third Assessment Reports of the Intergovernmental Panel on Climate Change*. Cambridge University Press, Cambridge, UK and New York, NY. 526–582.

Kornecki, T.S., Sabbagh, G.J. & Storm, D.E. (1999) Evaluation of runoff, erosion, and phosphorus modeling system – SIMPLE. *Journal of the American Water Resources Association*, 35, 807–820.

Kottegoda, N.T., Natale, L. & Raiteri, E. (2003) A parsimonious approach to stochastic multisite modelling and disaggregation of daily rainfall. *Journal of Hydrology*, 274, 47–61.

Kovalevskii, V.S. (2007) Effect of climate changes on groundwater. *Water Resources*, 134(2), 140–152.

Kracauer Hartig, E., Grozev, O. & Rosenzweig, C. (1997) Climate change, agriculture and wetlands in Eastern Europe: Vulnerability, adaptation and policy. *Climatic Change*, 36, 107–121.

Kundzewick, Z.W. & Somlyódy, L. (1997) Climatic change impact on water resources in a system perspective. *Water Resources Management*, 11, 407–435.

Maidment, D.R. (1993) *Handbook of Hydrology*. McGraw-Hill. New York, NY, USA. 1424 pp.

Mckay, M.D., Beckman, R.J. & Conover, W.J. (1979) Comparison of 3 methods for selecting values of input variables in the analysis of output from a computer code. *Technometrics*, 21, 239–245.

Mira, J., Elorza, F.J., Bolado, R. & García-Martos, C. (2004) *The analytical identification of main effects and interactions when using the FAST sensitivity analysis technique*. Technical Report 04/1, Lab. de Estadística – DMAMI, UPM, Madrid, 2004. 60 pp.

PHIB. (1999) Propuesta del Plan Hidrológico de las Islas Baleares (National Water Plan. Balearic Islands). Govern Balear, Febrero 1999, 236 pp.

Pope, V.D., Gallani, M.L., Rowntree, P.R. & Stratton, R.A. (2000) The impact of new physical parametrizations in the Hadley Centre climate model – HadCM3. *Climate Dynamics*, 16, 123–146.

Risbey, J.S., Hamza, K. & Marsden, J. (2007) Use of climate scenarios to aid in decision analysis for interannual water supply planning. *Water Resources Management*, 21, 919–932.

Rosenberg, N.J., Epstein, D.J., Wang, D., Vail, L., Srinivasan, R. & Arnold, J.G. (1999) Possible impacts of global warming on the hydrology of the Ogallala Aquifer region. *Climatic Change*, 42, 677–692.

Ruth, M., Bernier, C., Jollands, N. & Golubieswski, N. (2007) Adaptation of urban water supply infrastructure to impacts from climate and socioeconomic changes: The case of Hamilton, New Zealand. *Water Resources Management*, 21, 1013–1045.

Samper, J., Llorenç, H., Arés, J. & García, M.A. (1999) *Manual del usuario del programa Visual BALAN V. 1.0. Código interactivo para la realización de Balances Hidrológicos y la estimación de la recarga* [Visual Balan 1.0 manual. Interactive code for the hydrologic balance and recharge estimation]. La Coruña, Spain, Enresa. 132 pp.

Scibeck, J., Allen, D., Cannon, A. & Whitfield, P.H. (2007) Groundwater-surface water interaction under scenarios of climate change using a high-resolution transient groundwater model. *Journal of Hydrology*, 333, 165–181.

Sena, C. & Molinero, J. (2009) Water resources assessment and hydrogeological modelling as tool for the feasibility study of a closure plan for an open pit mine (La Respina Mine, Spain). *Mine Water Environment*, 28, 94–101.

Serrat-Capdevila, A., Valdés, J.B., González Pérez, J., Baird, K., Mata, L.J., Maddock, T. III (2007) Modeling climate change impacts – and uncertainty– on the hydrology of a riparian system: The San Pedro Basin (Arizona/Sonora). *Journal of Hydrology*, 347, 48–66.

Sumner, G.N., Romero, R., Homar, V., Ramis, C., Alonso, S. & Zorita, E. (2003) An estimate of the effects of climate change on the rainfall of Mediterranean Spain by the late twenty first century. *Climate Dynamics*, 20, 789–805.

Wilby, R.L., Whitehead, P.G., Wade, A.J., Butterfield, D., Davis, R.J., Watts, G. (2006) Integrated modelling of climate change impacts on water resources and quality in a lowland catchment: River Kennet, UK. *Journal of Hydrology*, 330, 204–220.

Wilks, D.S. (1992) Adapting stochastic weather generation algorithms for climate change studies. *Climatic Change*, 22, 67–84.

Wilks, D.S. (1998) Multisite generation of daily stochastic precipitation generation model. *Journal of Hydrology*, 210, 178–191.

Woldeamlak, S.T., Batelaan, O. & De Smedt, F. (2007) Effects of climate change on the ground system in the Grote-Nete catchment, Belgium. *Hydrogeological Journal*, 15, 891–901.

Wyss, G.D. & Jorgensen, K.H. (1998) *A user's guide to LHS: Sandia's Latin Hypercube Sampling Software*. Technical Report SAND98-0210. Albuquerque, NM, Sandia National Laboratories.

CHAPTER 12

The effect of climate and anthropogenic sea level changes on Israeli coastal aquifers

Yoseph Yechieli, Uri Kafri & Eyal Shalev

ABSTRACT

This review summarizes several aspects related to the effect of climate and anthropogenic sea level changes on the Israeli coastal aquifers of the Mediterranean Sea (rise of ~1 cm/a) and the Dead Sea (drop of ~1 m/a). Numerical simulations show that the effect of global future sea-level rise will depend on the specific configuration of the aquifer and its connection to the sea. An important factor is the coastal topography next to the shoreline, whereby in the case of a steep coastal topography, no significant salinization is expected due to sea-level rise. Reduced recharge due to climate change or over-exploitation of groundwater is also expected to enhance the inland shift of the fresh-saline water interface.

A significant fast response of groundwater system to changes in the base levels is observed in the Dead Sea coastal aquifer, both in simulations and in the field. This is exhibited by the drop of the adjacent groundwater levels as well as by the shift of the fresh-saline water interface. Also, in most parts of the Dead Sea aquifer, a fast freshening process is taking place due to the drop of groundwater level and seaward retreat of the interface.

12.1 INTRODUCTION

The purpose and scope of the present paper is to provide tools to forecast future reaction and behaviour of hydrological systems in response to climate changes, base level changes and anthropogenic activity. Coastal aquifers, connected to the seas or to endorheic lakes and inland seas were in the past, as well as presently, affected by their level variations as a result of climate changes. Examples of such variations in the past from all over the world are summarized by Kafri and Yechieli (2010). The present review summarizes our recent studies that were conducted in the last five years, such as by Kiro et al. (2008), Yechieli et al. (2009a,b) and Yechieli et al. (2010).

Sea levels have changed as a result of alterations in climatic conditions during the geological history. Currently, there is a common understanding which claims that the climate is largely influenced by human activity leading to global warming. As a result, polar caps and on-land glaciers are melting and raising the sea level. A sea-level rise of ~0.5 m by the year 2100 was suggested by Warrick et al. (1996) and of 0.5 to 1.4 m by Rahmstorf (2007).

The main effects of sea level changes on the adjoined groundwater systems are their salination due to a sea level or a saline lake level rise or the opposite freshening process in the case of level drop. Subsurface seawater intrusion is a concern and is thus an extremely important factor in management of coastal aquifers (i.e. van Dam 1999).

Its rate is usually determined by the increase of groundwater salinity (e.g. Melloul and Zeitoun 1999). An estimation of the penetration rate can be assessed by the analysis of radiocarbon and tritium isotopes from groundwater in monitoring wells (Yechieli et al. 2001; Sivan et al. 2005).

The occurrence of a fresh-saline water interface in coastal aquifers exists due to density differences between the fresh and saline water bodies. A mixing or transition zone occurs between the fresh and saline water due to dispersion (Henry 1964; Lee and Cheng 1974; Pinder and Cooper 1970; Voss 1984; Sanford and Konikow 1985; Galeati et al. 1992), and a saline water circulation beneath the transition zone (Cooper 1959). The above studies have also modelled the transition zone in steady-state and transient conditions, and others studied the effect of sea level changes on the fresh-saline interface in a confined aquifer using sharp interface models (e.g., Essaid 1990; Harrar et al. 2001) and density-dependent flow models (e.g., Kooi et al. 2000; Chen and Hsu 2004).

Additionally, several studies have discussed the effect of the sea-level rise on ground-water flow and sea water intrusions (i.e. Sherif and Singh 1999; Oude Essink 1996). The expected 50 cm sea-level rise, during the coming century, will cause an additional inland seawater intrusion of 9 km in the Nile Delta and 0.4 km in the Bengal Bay (Sherif and Singh 1999) and an increase in groundwater salinity in the Netherlands close to the shore (Oude Essink 1996).

The objective of this summary is to describe the possible effects of variations of climate and precipitation as well as of anthropogenic activity on sea level variations and in turn on variations of the hydrogeological configuration of coastal aquifers. The Mediterranean coastal aquifer can serve as a good example of many coastal aquifers around the world whereas the Dead Sea provides a unique opportunity to study in real time the response of the adjoined groundwater systems to sea level changes in the field.

12.1.1   Description of the area: the Israeli Mediterranean and the Dead Sea coastal aquifer systems

The study was conducted in two different coastal aquifer systems, one adjoined to the Mediterranean Sea and the other to the Dead Sea. Both aquifer systems consist of several sub-aquifers, separated by less permeable layers. The description of the two systems is as follows:

The Israeli Mediterranean coastal aquifer system

The Israeli Mediterranean coastal aquifer system extends over more than 120 km along the Mediterranean coast (Figure 12.1). Its thickness decreases eastwards from ~200 m near the coastline to a few tens of meters and less near its eastern boundary (Figure 12.2). The aquifer, belonging to the Kurkar Group, consists of inter-layered sandstone, calcare-ous sandstone, siltstone, and red loam which alternate with continental and marine clays of Pleistocene age, overlying impervious marine clays of the Saqiye Group of Pliocene age (Issar1968). East of the shoreline, to a distance of 5 to 8 km, clay inter layers subdi-vide the aquifer into 4 sub-aquifers. Some of the lower sub-aquifers are confined while the upper ones are phreatic (Nativ and Weisbrod 1994). The upper sub-aquifers are con-nected to the sea whereas the connection of the lower sub-aquifers to the sea is debatable. Some studies claim that they are connected to the sea (e.g. Kapuler and Bear1970), while

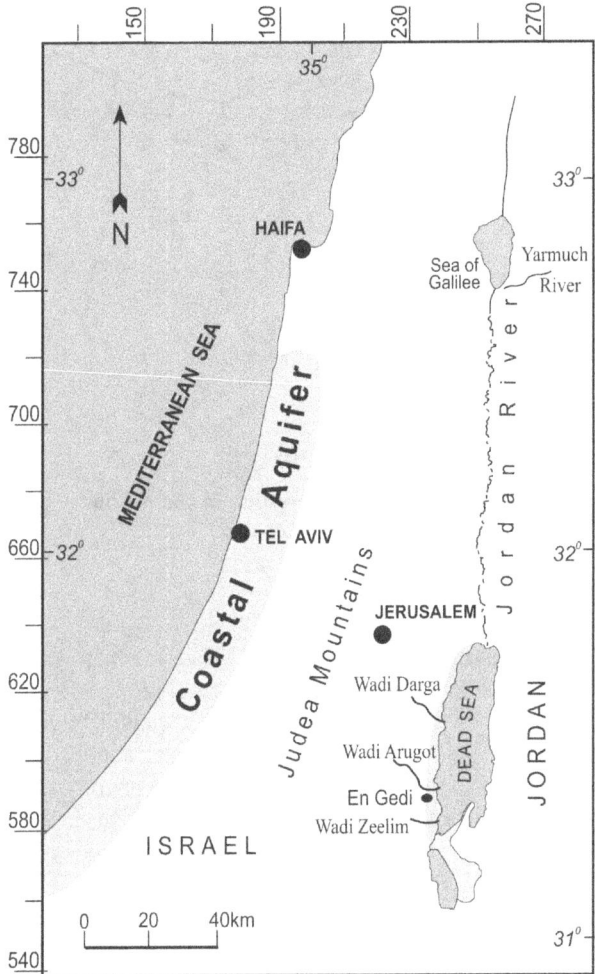

Figure 12.1.   Location map of the Mediterranean and the Dead Sea coastal aquifers. The relevant study areas are signed as narrow strips along the coastal area of the Dead Sea and the Mediterranean Sea.

others suggest that they are in places blocked to the sea (Kolton1988; Kafri and Goldman 2006). The precipitation over most of the coastal aquifer is around 600 mm/a and the recharge coefficient was taken to be about 0.3 (Gvirtzman 2002; Bruce et al. 2007). Thus, the recharge in most parts of the aquifer is around 200 mm/a.

Basic information on the fresh-saline water interface is obtained from hundreds of wells, from Time Domain ElectroMagnetic studies (TDEM) (Goldman et al. 1989) and from hydrochemical and hydrological monitoring. In several areas, the fresh-saline water interface has moved inland over a distance of more than 1 km, as a result of over pumping during the last decades (Melloul and Zeitoun 1999). However, since the coverage of the monitoring well network is not sufficient, the actual extent of the inland interface penetration is not known for the entire aquifer system.

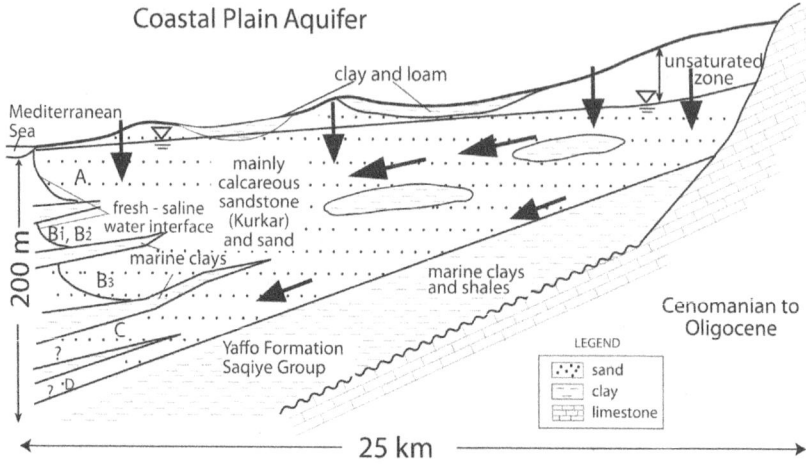

Figure 12.2.   Hydrogeological schematic cross-section of the Mediterranean coastal aquifer of Israel (modified after Ecker 1999).

The Dead Sea coastal aquifer system

The Dead Sea is situated at the deepest part of the Dead Sea Rift System. It is a terminal endorheic lake with no surface or subsurface water outflow. The Dead Sea groundwater system consists of two main aquifer systems (Figure 12.3), namely the Upper Cretaceous Judea Group Aquifer and the Quaternary alluvial coastal aquifer (Arad and Michaeli 1967; Yechieli et al. 1995). The coastal aquifer consists mainly of clastic sediments (gravel, sand and clays) deposited in fan delta environments and chemical deposits (aragonite, gypsum and salts with clay alternations) formed in a hyper-saline lacustrine environment of deposition. The alternations between the gravel and clay horizons subdivide the aquifer into several sub-aquifers that differ in their groundwater level and chemical composition. This aquifer system is bounded by normal faults, which set Cretaceous carbonate rocks of the Judea Group against Quaternary alluvial and lacustrine sediments. The recharge of the aquifer is mainly through lateral flow from the Judea Group aquifer, which is replenished in the highlands some 10 to 30 km to the west and by flash floods. Direct recharge is negligible because of the arid climate and high evaporation in the Dead Sea region.

The Dead Sea salinity and density are 340 g/l and 1.24 kg/l, respectively (Lensky et al. 2005). The extremely high density of the Dead Sea induces a very shallow inter-face between the fresh water and the brine bodies. According to the Ghyben-Herzberg approximation, the depth of the interface near the Dead Sea is 4.35 times that of the groundwater head above the Dead Sea level as compared to 40 times of the head above sea level near any normal ocean water (Yechieli 2000).

The situation in the Dead Sea is different from that of the Mediterranean Sea following the current rapid drop of its level by 1m/a due to its negative water balance. This drop is a result of the considerable exploitation and diversion of water upstream from the Sea of Galilee, the Jordan and the Jarmuch rivers since the 1960's, and due to the Dead Sea brine pumping of the Israeli and Jordanian Dead Sea Industries (Lensky et al. 2005). Fluctuation of the Dead Sea occurred also during the Pleistocene and

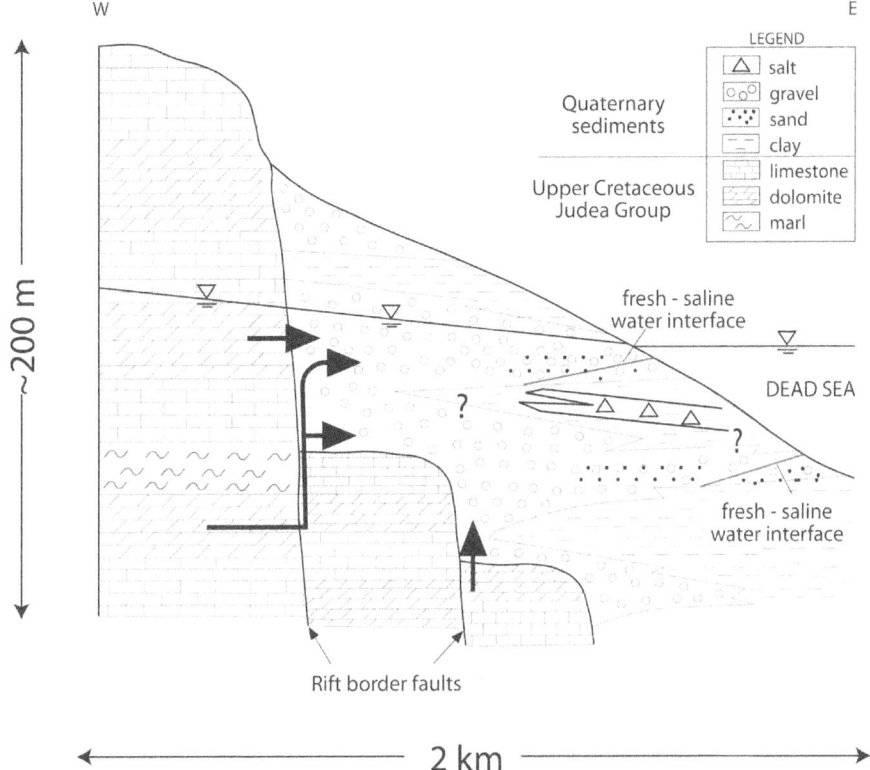

W   E

LEGEND

salt	
gravel	
sand	
clay	
limestone	
dolomite	
marl	

Quaternary sediments

Upper Cretaceous Judea Group

fresh - saline water interface

DEAD SEA

fresh - saline water interface

~200 m

Rift border faults

2 km

Figure 12.3. Hydrogeological schematic cross-section of the Dead Sea coastal aquifer (modified after Yechieli 2006). The arrows denote the possible flow into the coastal aquifer from the Judea Group aquifer. After Yechieli et al., Response of the Mediterranean and Dead Sea coastal aquifers to sea level variations. Water Resour Res, 46: W12550, 11 PP, doi:10.1029/2009WR008708; Copyright [2010] American Geophysical Union.

Holocene times (Bartov et al. 2002; Bookman et al. 2004), assumingly affecting the groundwater system (Yechieli et al. 2009a). The hydraulic connection between the Dead Sea and the adjoining groundwater systems is expressed by the relatively fast (a few days) groundwater level response to the Dead Sea level changes (Yechieli et al. 1995).

12.1.2 Relevance for GRAPHIC

The present study deals with the response of groundwater systems to climate and anthropogenic sea level change which is the main topic of GRAPHIC.

The combination of both studies allows a unique opportunity to understand the process in sea-aquifer system. Such changes are very difficult to be traced in the ocean system because of the slow rate of processes that will require many years to be noticed. On the other hand, the Dead Sea system, although being a very unusual system, provides an excellent field laboratory where the processes are rapid enough to be monitored and examined within a relative short period. The extrapolation from the Dead Sea system case to other oceanic systems is not trivial and should be done with caution.

The present study, thus, attempts to provide tools which might enable the study and monitoring of hydrological systems which are adjoined to the sea and track their changes with time.

12.2   METHODOLOGY

The response of groundwater systems to changes in sea levels were studied using both field data and hydrological simulations.

12.2.1   Field studies

The field studies were conducted mainly in the Dead Sea coastal aquifer, where significant changes currently occur and are expected to continue within a short time due to the rapid drop of the lake's level. Water levels were measured in several exploration boreholes, located at different distances from the shoreline, using a manual water level meter.

The location of the fresh-saline water interface was measured with an *in situ* electrical conductivity (EC) device in both the Dead Sea and the Mediterranean coastal aquifers. EC profiles were conducted at 1–2 m depth intervals, or less in specific depths of interest, between the water table and the bottom of the boreholes. Usually, the EC values correlate well with salinity values. The situation is, however, more complicated in the Dead Sea area. At very high salinities, exceeding half of that of the Dead Sea brines, there is no correlation between EC and salinity (Yechieli 2000). Therefore, the location of the fresh-saline water interface can only be inferred from measurement of EC values below 160 mmho/cm, according to calibration curves given by Yechieli (2000) which provide the relationship between EC and salinity. The EC profiles were conducted for six years in order to examine the response of the interface to change in the declining Dead Sea levels.

12.2.2   Numerical simulation of the Mediterranean coastal aquifer system

The response of groundwater to a consistent rise of Mediterranean Sea level was modelled by the FeFlow, a finite element simulator (Diersch and Kolditz 2002), that solves the coupled variable density ground water flow and solute transport. The specifics of the boundary conditions and hydraulic parameters in these simulations are given in Yechieli et al. (2010). The sea water level rise was set at a rate of 1 cm/a for 100 years. Steady state simulations were run for two possible coastal topographic configurations, which probably represent the extreme situations in many coastal aquifers: a) low angle coastal slope of about 2.5‰, and b) a vertical coastal cliff of 30 meters.

Additional simulations included: 1) Two different recharge values were applied: 0.2 m/a, representing the average present day conditions, and a lower rate of 0.1 m/a representing drier climatic conditions. 2) Imposing a pumping value of 5 MCM/a (million cubic meters per annum) per km width, conducted from a single point, located about 3 km from the shore. 3) Subdivision to a lower confined and an upper phreatic sub-aquifers by a confining clay layer subdividing the sea side portion of the aquifer into two sub-aquifers.

12.2.3   Numerical simulation of the Dead Sea aquifer system

The simulations of the Dead Sea were also run with FeFlow. The specific conditions of the simulation are given in Yechieli et al. (2010). A steady-state simulation was performed in order to generate the initial flow and concentration distribution at time zero. A transient simulation was then performed whereby the Dead Sea level declines at a rate of 1m/a for 20 years, reaching a total decline of 20 m.

12.3   RESULTS AND DISCUSSION

This review summarizes the results of several simulations and data sets studied by Kiro et al. (2008), Yechieli et al. (2009a), Yechieli et al. (2009b) and Yechieli et al. (2010).

12.3.1   The Mediterranean coastal aquifer system

While the levels of the Mediterranean Sea show a minute rise of some 10 to 20 cm in the last 20 years (Rosen 2004), groundwater monitoring adjacent to the shoreline does not show a measurable consistent change in groundwater levels during this time span. Moreover, no significant change was observed in the location of the fresh-saline water interface that can be related to sea-level rise (pers. com. V. Friedman, Israeli Hydrological Service). Indeed, such effects were not expected due to seasonal variations, coupled with considerable anthropogenic effect (over-pumping and artificial recharge), which mask the expected little effect of sea-level rise. Moreover, the effects of tidal fluctuations on the fresh-saline water interface, which are intensified artificially in monitoring boreholes (Shalev et al. 2009), is an additional process that masks the possible detection of small changes.

Numerical modelling was conducted in order to forecast the expected hydrological response in the next 100 years due to an expected extreme case of ~1m rise in sea level (Figures 12.4 and 12.5). In general, the sea-level rise is assumed to result in a similar change in the nearby groundwater water levels. Since the specific conditions along the Mediterranean coast vary, the simulations were conducted for two different coastal topographic configurations, and two different climatic conditions. Simulation show that for the case of a steep coastal topography, simulated as a vertical cliff (Figure 12.4b), the fresh-saline water interface is expected not to move significantly inland, namely by only 1 meter. For a milder slope of 2.5‰, the interface is expected to move inland to a distance of some 400 m, which coincides with the expected shift of the shoreline (Figure 12.4c). Therefore, the actual distance depends on the slope of the topography, whereby a milder slope would yield a farther inland shift of both the shoreline and the location of the interface. In the Israeli Coastal Aquifer, where the coastal topography slope is usually about 1–2%, the interface is expected to move some 50 to 100 m inland.

The possible climate change, namely a decrease in precipitation and thus of aquifer recharge, was also simulated. Since the expected change in recharge is not known, the present study examined an extreme reduction of 50% of aquifer recharge. This, by itself, is expected according to the simulations to lower the water table and to significantly move the interface inland to a distance of about 1,200 meters (Figure 12.4c). A sea-level rise coupled with such conditions will assumingly result in a farther inland interface shift.

Additional simulations were conducted in order to examine the prevailing case in many coastal aquifers, where over pumping exists, causing upconing of the encroaching seawater. These simulations, thus, include a pumping well of 5 MCM/a, for a strip of 1 km width at a distance of 3 km from the shoreline. This value, which is quite extreme for a normal pumping regime in the study area, was taken in order to test the situation

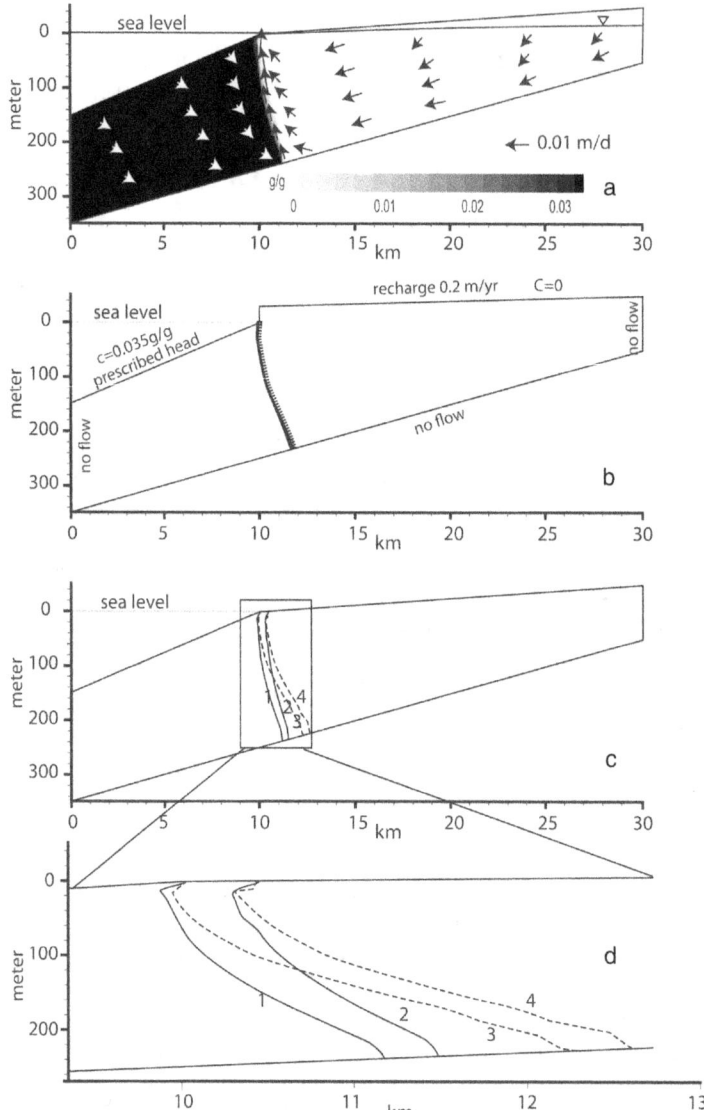

Figure 12.4. Hydrological steady state simulations of the expected changes in the case of Mediterranean sea-level rise. The value that represents the fresh-saline water interface is 50% Mediterranean seawater, namely, ca 11 g/l Cl.

After Yechieli et al., Response of the Mediterranean and Dead Sea coastal aquifers to sea level variations. Water Resour Res 46; W12550, 11 PP, doi:10.1029/2009WR008708; Copyright [2010] American Geophysical Union.

a. Basic simulation of the Mediterranean coastal aquifer showing the fresh and saline water bodies and the interface in between. Also shown are the flow velocity arrows.

b. Simulation of the case of a steep (cliff) topography. The effect of sea-level rise on interface location is not observed.

c. Simulation of the case of a mild topography with a slope of 2.5‰. Sea-level rise of 1 m exhibits the shift of the fresh-saline water interface inland after 100 yr by 400 m (line 1 is at T = 0 yrs and line 2 is at T = 100 yrs). Also exhibited is the shift of the interface in the case of decrease in recharge by 50%, before and after sea-level rise (lines 3 to 4, respectively).

d. Blow-up of figure 4c, showing the changes in the location of the fresh-saline water interface.

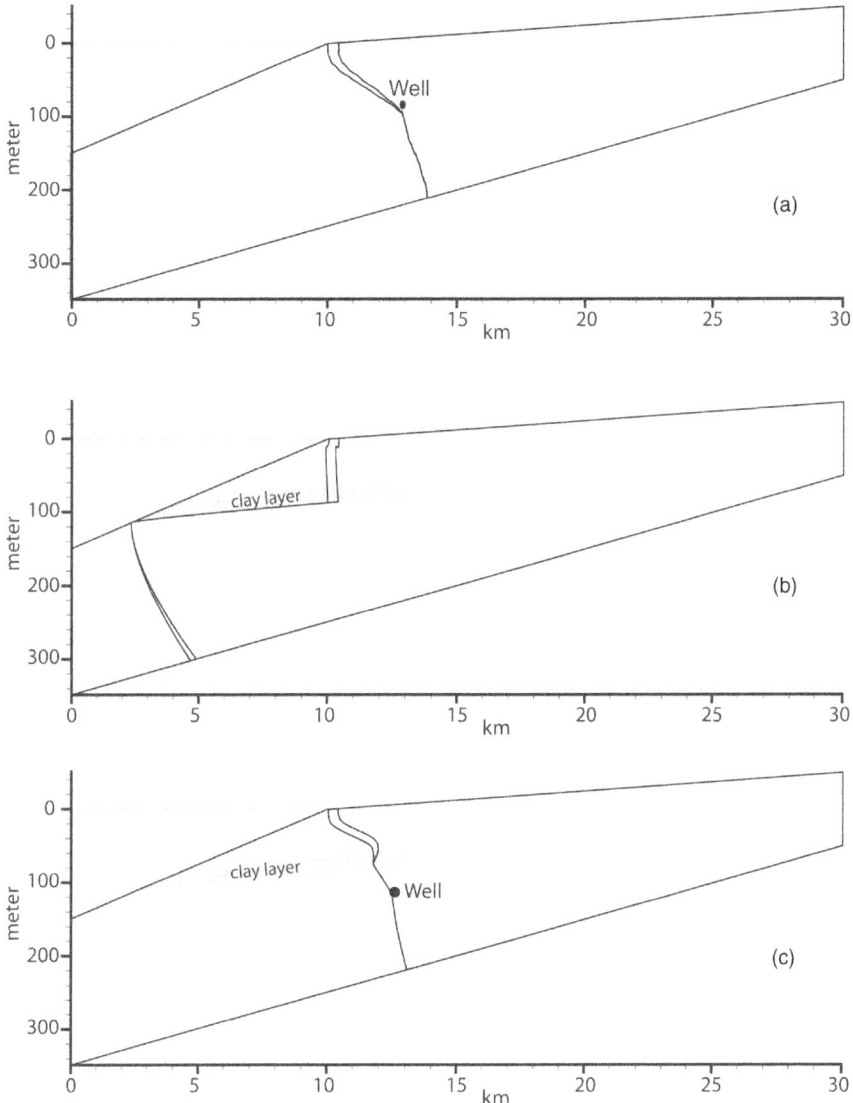

Figure 12.5. Simulations of the expected interface location in the Mediterranean coastal aquifer following a sea-level rise (after 100 yrs) under the following conditions:

a. Coupled with over pumping.
b. The case of two sub-aquifers, separated by a clay aquiclude, simulating the condition of the upper two sub-aquifers of the coastal aquifer (see figure 2).
c. The case of two sub-aquifers coupled with pumping from the lower sub-aquifer.

After Yechieli et al., Response of the Mediterranean and Dead Sea coastal aquifers to sea level variations. WATER RESOURCES RESEARCH, V. 46, W12550, 11 PP, doi:10.1029/ 2009WR008708; Copyright [2010] American Geophysical Union.

of extreme cases. It was done despite the fact that the average pumping in the Israel coastal aquifer is some 2 MCM/a per km wide strip, perpendicular to the flow direction. However, an increase of pumping may occur in the future due the water shortage in this region. In the case of a sea-level rise, coupled with pumping, some change in the geometry of the interface is noticed, but the location of the interface toe remained at the same place (Figure 12.5a). Lower values of pumping (e.g. 2–3 MCM/a) yield a smaller shift of the interface.

Simulations were also conducted in order to test the actual configuration of most coastal aquifers which are subdivided to sub-aquifers (Figure 12.5b). The simulations in the present study were conducted for the relatively simple case of only two sub-aquifers, where the upper one is phreatic and the lower one is confined.

The above simulation resembles the situation of two upper sub-aquifers in the Israeli coastal aquifer (Figure 12.2), whereby the lower sub-aquifers, which assumingly are not connected to the sea were not considered in these simulations. In these simulations, the phreatic aquifer behaves quite similar to the previous simulation, as expected. In the confined aquifer, only a very small change in the location of the interface is noticed.

The effect of pumping was also examined for a two sub-aquifers configuration where pumping is executed from the lower sub-aquifer. The simulations exhibit a significant change, whereby the interface in the lower sub-aquifer moves inland (Figure 12.5c). Similar to the case of one undivided aquifer, the sea-level rise is expected to shift the interface farther inland but the interface toe is expected to remain at the same place. The suggested interpretation is that in this case the pumping well is the sole fresh water output point of the discussed sub-aquifer. Thus, the interface change is confined to the pumping well proper area.

All the above results suggest that the expected sea-level rise by itself will not change considerably the location of the fresh-saline water interface in the Israeli coastal aquifer due to its relative steep coastal topography. On the other hand, a significant reduction of recharge due to climate change might cause an inland shift of the fresh-saline water interface. These results agree with those of Sherif and Singh (1999) which show that huge flat deltas close to the oceans are expected to be more vulnerable to significant sea water intrusion due to sea-level rise. Similarly, low lying deltaic areas, such as in The Netherland, are also expected to be effected by sea-level rise (Oude Essink 2001).

The rate of seawater intrusion into the Mediterranean coastal aquifer was also examined by analysis of radioactive isotopes (Yechieli et al. 2009b). This rate depends on the connection between the different sub-aquifers and the sea and the rate of sea-level rise since the last glacial maximum. In the upper sub-aquifers, which are connected to the sea, young ages were found for both fresh and saline water. On the other hand, older ages were found in groundwater hosted in the lower sub-aquifers implying a poor connection or blockage to the sea.

The first given example is of boreholes at a distance of up to 700 m from the shoreline to different sub-aquifers (Figure 12.6a), separated by clayey layers. Saline groundwater was found in each of the sub-aquifers. In the upper sub-aquifer, salinity was similar to that of seawater while in the lower sub-aquifers the salinity was about half that of the present Mediterranean Sea. These high salinities imply on a connection with the sea in all sub-aquifers. The interesting fact here is the variation in the apparent ^{14}C ages of the various saline groundwaters, which decrease from the lower sub-aquifer to the upper one (Yechieli et al. 2009b). Despite the fact that exact dating of this

a. Nitzanim

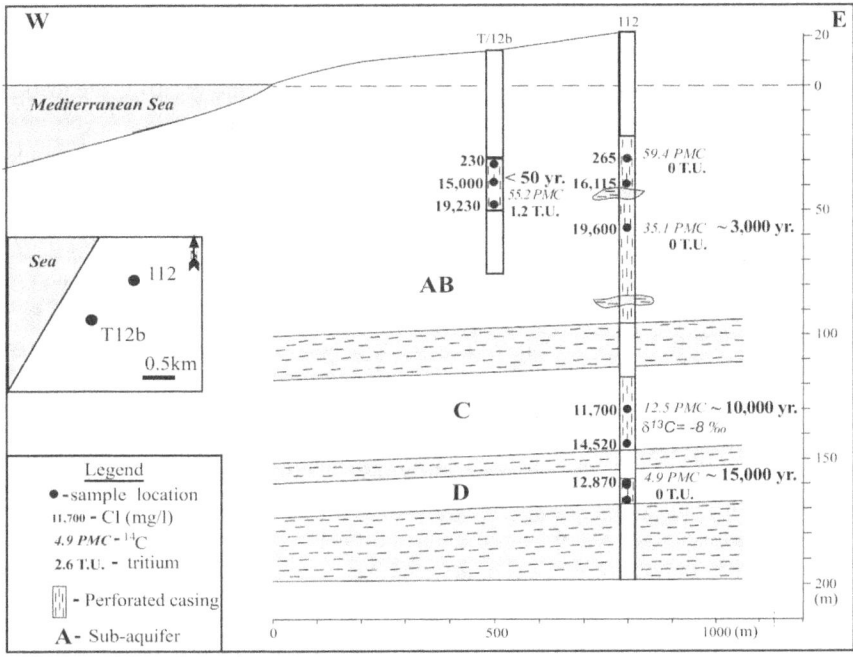

b. Rishon Letzion

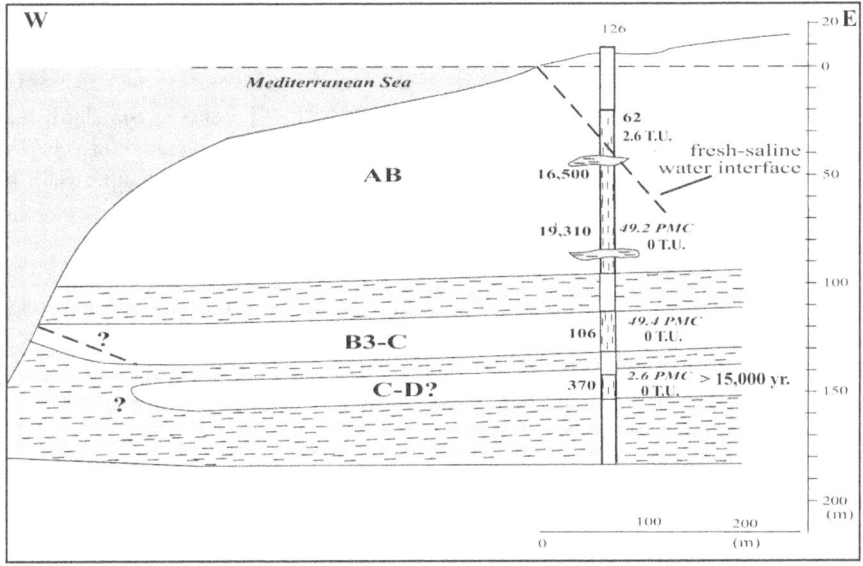

Figure 12.6. Cross sections of the Mediterranean coastal aquifer, exhibiting the cases of connection (a) or disconnection (b) to the sea. (a) Nitzanim area: Evidence of seawater intrusion also into the lower sub-aquifers (modified after Yechieli et al. 2009b). (b) Rishon Letzion area: No seawater intrusion is evident in the lower sub-aquifers (modified after Yechieli and Sivan 2011).

water is difficult, due to mixing of fresh and saline water of different ages and water-rock interaction, a rough estimation yields an age of >10,000 years (~5–10 PMC) in the lower sub-aquifers and probably several tens of years in the upper sub-aquifer.

The second example is of a borehole close to the shoreline (~50 m, Figure 12.6b). Here, the hydrogeological sequence from bottom to top is as follows: The lowermost sub-aquifer contains fresh groundwater, although it is very close to the sea. This fact implies that there is no simple connection to the sea. The blockage from the sea is also evidenced by the low ^{14}C values and the absence of tritium, indicating slow groundwater flow and no direct flow to the sea in this sub-aquifer. The overlying sub-aquifer, contains also fresh groundwater of a considerably younger age, implying a faster fresh water flow and thus probably a better connection to the sea. The overlying sub-aquifers which are separated from the ones below by a distinctive clay layer, contain saline water close to seawater salinity. This saline sequence is younger (49 PMC), implying quite recent seawater intrusion to this location. The difference between the ^{14}C value of the saline groundwater in this sub-aquifer and recent seawater (~50 versus ~100 PMC) is mostly due to oxidation of old organic matter and not to long periods of decay (Sivan et al. 2005). The ages of the fresh groundwater in this aquifer are mostly younger than 50 years according to the high ^{14}C values and the abundance of tritium in most samples (Bruce et al. 2007).

As shown in the above two examples, the connection of the different sub-aquifers to the sea depends on the specific hydraulic properties of the aquifer. A seaward gradual grain size reduction, as often occurs, reduces in turn the hydraulic properties of the aquifer up to a complete lateral blockage of the aquifer to the sea, as suggested by Kolton (1988), and as proven in some areas by the time domain electromagnetic (TDEM) method and geochemical data (Kafri and Goldman 2006; Yechieli et al. 2009b).

12.3.2   The Dead Sea coastal aquifer

The response of the groundwater system to the Dead Sea level drop was examined by both numerical simulations and field monitoring (Figures 12.7 and 12.8). The Dead Sea level drop significantly effects the position of the fresh-saline water interface (Figure 12.7b), which is consistent with the field observation. The change in groundwater heads (Figure 12.8) is a function of the permeability of the aquifer and distance from the sea shore. Boreholes adjacent to the shoreline exhibited water levels drop close to that of the Dead Sea levels (Figure 12.8a, Kiro et al. 2008). Along the Arugot and Darga alluvial fans, which consist mostly of high permeability gravel, the groundwater level drop was similar to that of the Dead Sea level up to a distance of 700 m inland (Figure 12.8a,c). In the Zeelim boreholes, which are located in the alluvial fan of Wadi Zeelim, more distant from the shoreline (~4 km), the drop of the groundwater level is significantly lower than that of the Dead Sea level (Figure 12.8d).

Boreholes outside the main stream courses have shown, in general, smaller water level drops as compared to that of the Dead Sea (i.e. EG-3a and the Tureibe boreholes; Figure 12.8b,e). This is explained by their relatively lower permeability being located outside the main alluvial fans. It should be noted that the distance from the monitoring boreholes to the shoreline increases with time due to the retreat of the shoreline, and in turn also affects the hydraulic gradient.

The drop of the Dead Sea level, and the eastward movement of the shoreline, also effects the location of the fresh-saline water interface (Figure 12.9). The decline of the

Dead Sea is expected to result in a drop in the location of the interface which is to be exhibited in monitoring boreholes by a downward shift of the EC profiles with time (Figure 12.9). The EC profiles in boreholes near the Dead Sea, at a distance of some 70 m, indeed show a significant change in the last 3–5 years (Figure 12.9a). On the other hand, no significant change in the interface location was observed in Darga borehole, which is more distant (about 700 m) from the shoreline in the alluvial fan of Wadi Darga (Figure 12.9b).

The fast decline of the fresh-saline water interface is resulting in a fast rate of flushing and freshening in some parts of the Dead Sea coastal area (Kafri et al. 1997; Kiro et al. 2008). The rate of this process is probably controlled by the hydraulic properties of the sediments near the Dead Sea, whereby the gravel in the Arugot alluvial fan allows a more rapid flushing, while in other locations a slower flushing process is expected.

It is interesting to note that the penetration of the Dead Sea brine into the aquifer continues even in the present condition of rapid lake level drop (Kiro et al. 2008). This means that the circulation of the saline water body occurs below the fresh-saline water

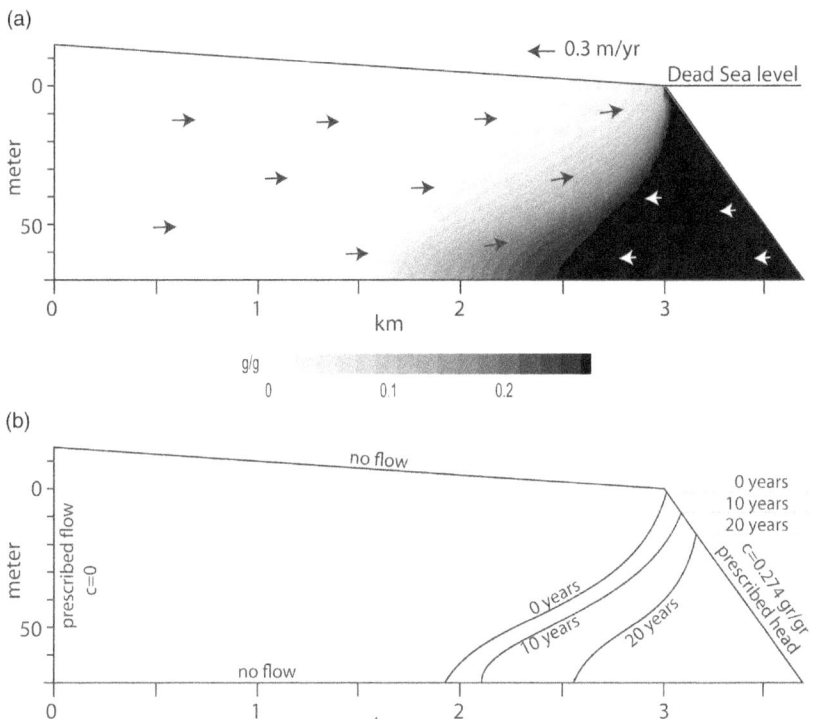

Figure 12.7. Numerical simulations of the expected changes in the case of Dead Sea level drop. The hydraulic parameters were taken to be similar to those at the Wadi Arugot. The value that represents the fresh-saline water interface is 50% Dead Sea water, namely 110 g/l Cl.

a. Basic simulation of the Dead Sea coastal aquifer showing the fresh and saline water bodies and the interface in between.
b. Simulation of the case of a drop in the Dead Sea level by 20 m.

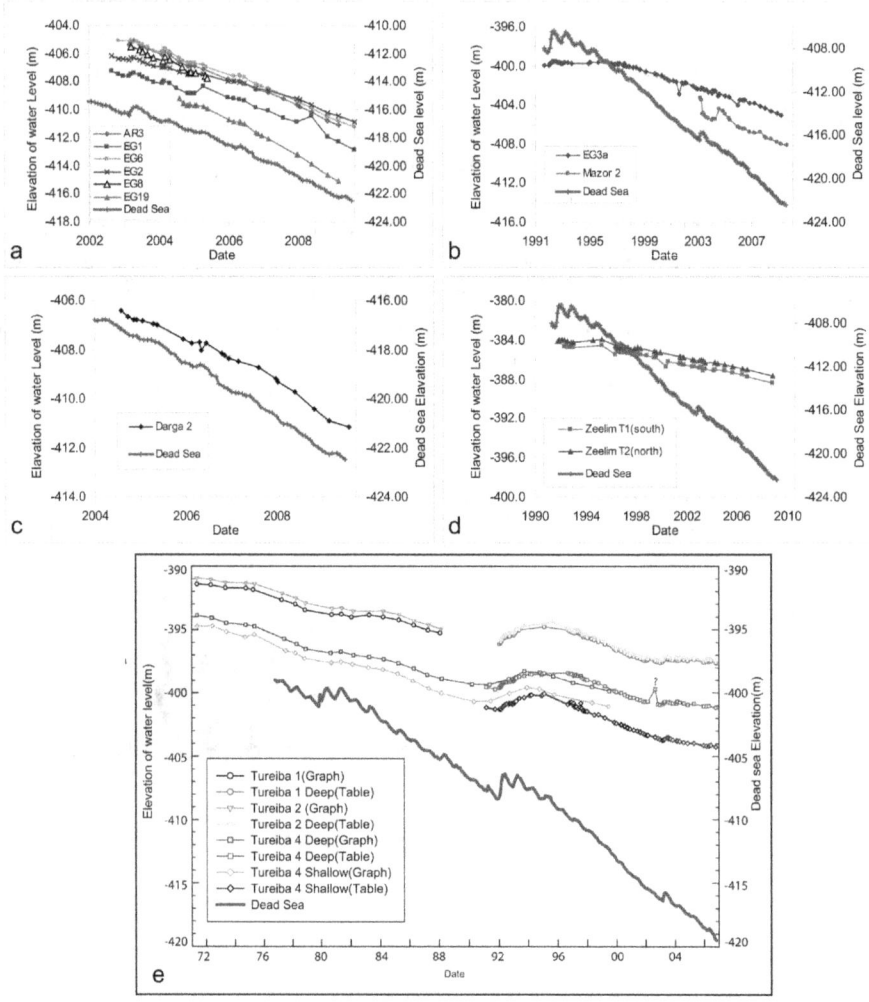

Figure 12.8. Groundwater levels in monitoring boreholes in the Dead Sea coastal aquifer at different locations and distances from the shoreline along with the Dead Sea level.

a. Boreholes located at Wadi Arugot, ~500 m from the Dead Sea shoreline.
b. EG-3a , ~900 m from the shoreline, located 3 km south from Wadi Arugot.
c. Darga 2, ~700 m from the shoreline, located in Wadi Darga.
d. Boreholes located ~4 km from the shoreline, in Wadi Zeelim.
e. Boreholes located ~500 m from the shoreline, 10 km north of Wadi Darga.

After Yechieli et al., Response of the Mediterranean and Dead Sea coastal aquifers to sea level variations. Water Resour Res 46; W12550, 11 PP, doi:10.1029/2009WR008708; Copyright [2010] American Geophysical Union.

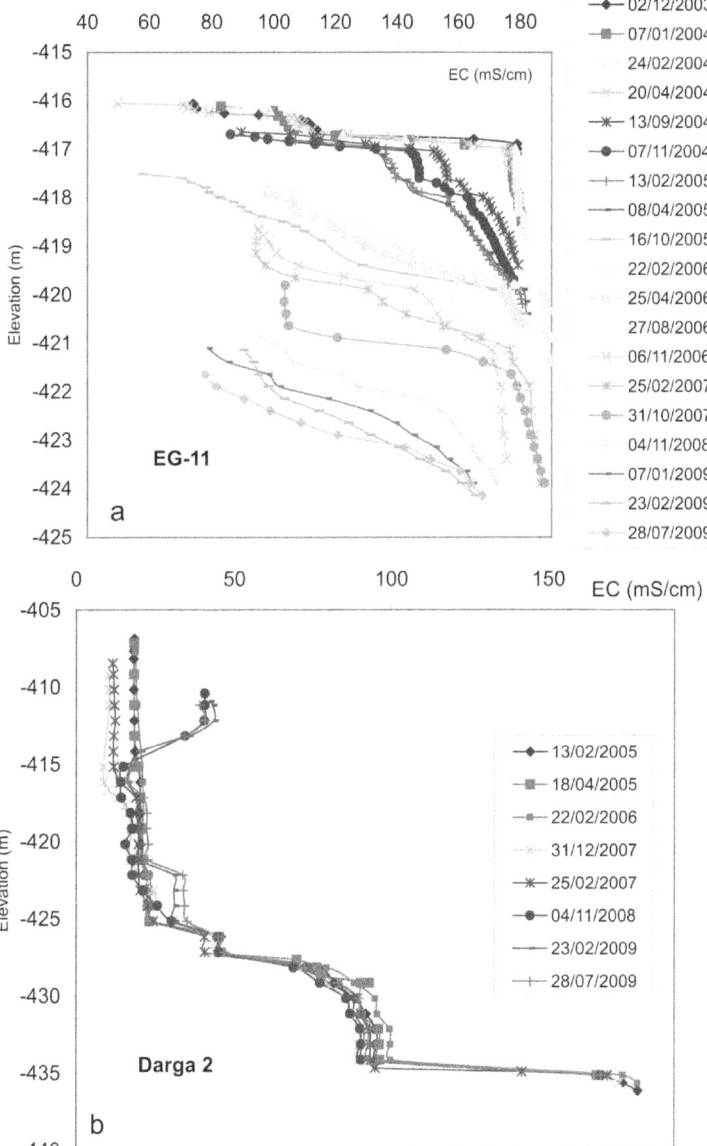

Figure 12.9. EC profiles of in two boreholes in the Dead Sea coastal aquifer.

a. EG 11 borehole at a distance of 70 m from the shoreline (modified after Kiro et al. 2008). Significant changes in the location of the fresh-saline water interface are observed.

b. Darga 2 borehole, at a distance of 700 m from the shoreline, exhibiting only very small changes in the depth of the interface.

After Yechieli et al., Response of the Mediterranean and Dead Sea coastal aquifers to sea level variations. Water Resour Res. 46; W12550, 11 PP, doi:10.1029/2009WR008708; Copyright [2010] American Geophysical Union.

interface, similar to the situation next to the ocean and other saline water bodies. This circulation is implied by both hydrological simulation and chemical properties of the saline water (Kiro et al. 2008).

Similar to some extent to the situation in the Mediterranean coastal aquifer, the ages of groundwater in the Dead Sea coastal aquifer also vary with depth and distance to the shoreline. Although less data is available in the Dead Sea area, the general trend here is that older saline water (lower concentration of ^{14}C and no tritium) is found deeper and farther from the shoreline (Avrahamov et al. 2010; Yechieli and Sivan 2011).

The expected response of the groundwater to variation of the Dead Sea level was also examined (Figure 12.10) for the predicted additional decline of ca 100 m in the coming 200 years from now, and past configuration of paleo Lisan lake at 180 m below sea level some 20,000 year ago (Yechieli et al. 2009a). In addition, the effects of base (sea or lake) level changes on the location and elevation of the groundwater divide between the Mediterranean Sea and the Dead Sea were also studied. The simulations of the hydrological situation were performed using the FEFLOW groundwater modelling code. In the first case, a resultant decline of groundwater level is encountered, progressively up to several km from the present shore line inland. The hydraulic gradient will increase and thus the discharge to the lake is also expected to slightly increase, on the expense of the storage and due to a small enlargement of the intake area (Yechieli et al. 2009a). The groundwater divide location will somewhat shift by ca 600 m to the west. In the second case, during Lisan Lake time, the flow to the lake was found to be similar to the present one, assuming a similar amount of recharge. It could have been higher in case of higher recharge. The fresh-saline water interface of the Lisan Lake system was found to be some 2 to 3 km west of the present day Dead Sea interface location.

The geological structure was also found to be a significant factor controlling the location of the groundwater divide. This structural effect is manifested in the permeability distribution, the location of the recharge zone and the geometry of the base of the aquifer. Numerical and analytical simulations showed that a drop of the base level has no considerable effect on the location of the divide in the case where there is a distinct anticline between both base levels, whereas without this structure the water divide could have been shifted several kilometres westward.

12.4   SUMMARY AND CONCLUSION

The response of two coastal aquifers, namely the Mediterranean and the Dead Sea coastal aquifers, to climatically and anthropogenic driven fluctuations of sea (base) level is preliminary analyzed herein. Both are typical coastal aquifers but they differ in being subjected to a future slow Mediterranean sea-level rise (1 m/a) in the case of the first aquifer and to a rapid (1 m/a), already occurring, decline of the Dead Sea in the case of the latter. The Mediterranean aquifer is a good representative of many coastal aquifers near the ocean around the world while the Dead Sea provides an almost unique opportunity to study the response of groundwater to endorheic lake or sea level changes in the field. The applicability of the unique Dead Sea system to other coastal aquifers is more complicated but it is worth the effort since it is probably one of the only places in the world where actual changes in groundwater system due to changes in base level can be monitored in the field.

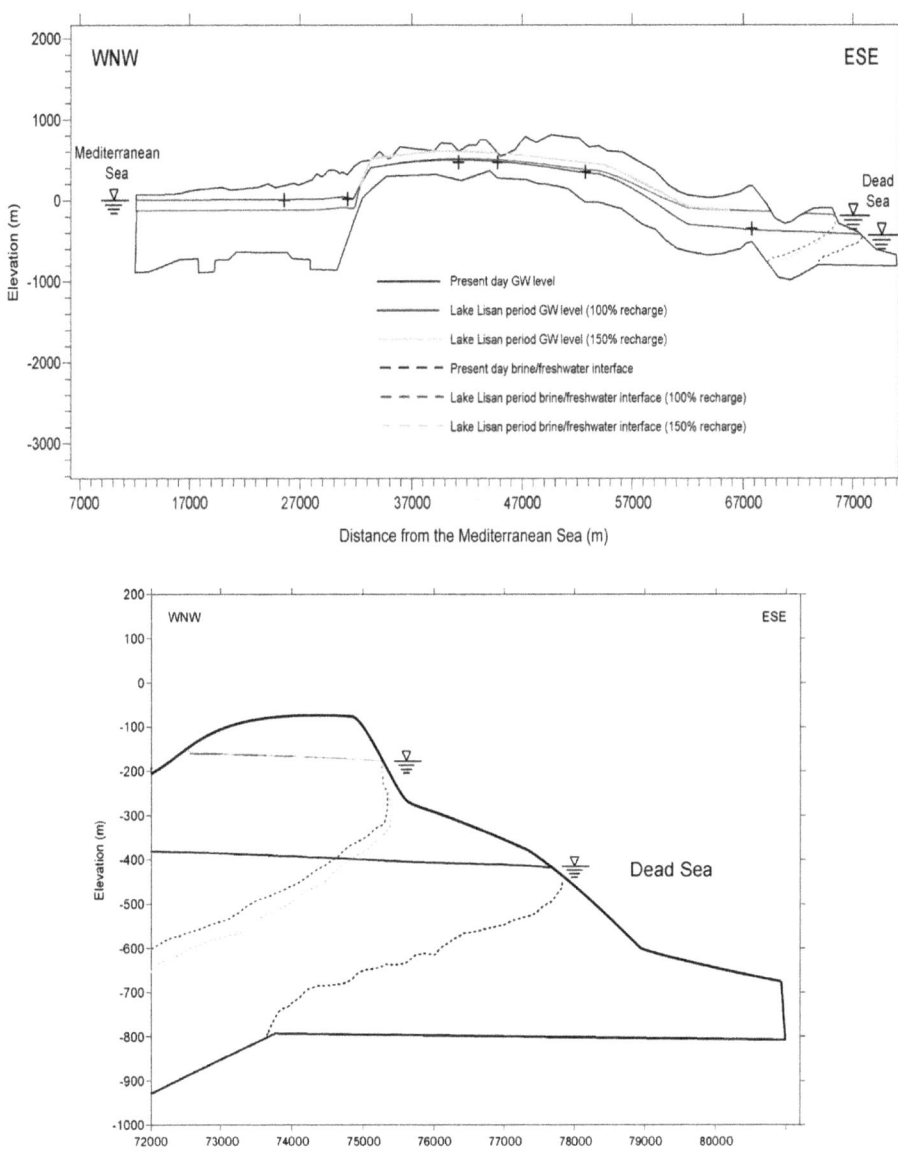

Figure 12.10.   a. Simulation of Lake Lisan period (lake level at 250 m above the present DS elevation, some 20,000 years ago) as compared with the present situation. Two recharge scenarios of 100% and 150% are simulated. b. Close-up of the simulated brine/freshwater interface for the Lake Lisan period (100% and 150% recharge scenarios) as compared to the present day simulated interface)Modified after Yechieli et al. 2009a).

The fast response of the Dead Sea coastal aquifer to change in the base level could imply that the response will also be fast in the Mediterranean coastal aquifer since the hydraulic conductivity in both aquifers is quite similar. However, due to the slow rate of change of the base level, the change in the groundwater levels and interface locations is not expected to be noticed here within a short time span.

The simulations of the Mediterranean coastal aquifer system yield the following results: In the case of a steep coastal topography, simulated as a cliff, the shoreline and thus the fresh-saline water interface are not expected to shift significantly inland. A considerable inland shift of the shoreline, accompanied by an inland shift of the interface, is expected in the case of a low angle almost flat coastal topography. Reduced recharge due to climate change or over exploitation of groundwater is also expected to enhance the inland shift of the interface.

In the case of the Dead Sea coastal aquifer system, the response to the current rapid Dead Sea level decline is already observed and monitored. The magnitude of groundwater level drop relatively to that of the Dead Sea is controlled by the aquifers permeability and the distance from the shoreline. Groundwater level drop is greater and detected more inland in high permeability alluvial fan areas as compared to low permeability areas. Regarding the fresh-saline water interface, its drop is observed by using repeated EC profiling in boreholes. In addition, a continuous freshening of the overlying portion of the aquifer as well as the vadose zone takes place.

These monitored processes can, in turn, be cautiously extrapolated or projected to long-term variations created by climate changes. Moreover, these can also shed light and assist in reconstructing paleo groundwater systems regarding their flow gradients, groundwater divide location, interface location, refreshening or flushing processes.

12.5   POLICY RECOMMENDATIONS

Based on the forecast that predicts a rather slow sea-level rise in the coming century, it is assumed that flat lying coastal aquifers such as deltas or atolls and most of the "normal" coastal aquifers in moderate to steep slope coastal plains will be subjected to slow rate effects of the above. Thus, the name of the game is going to be to use high resolution methods and long term time series in order to be able to notice even small changes. This is mostly essential for the above "normal" coastal aquifers where the expected effect is small and difficult to be recognized.

In normal coastal aquifers, the effect of pumping is expected to be much more severe than that of sea-level rise and climate change. Pumping should be regulated and its effect should be monitored. Thus, an effort should be made to differentiate between the anthropogenic and the natural effects in order to obtain the net effect of the latter.

Special attention should be given to the differentiation of different sub-aquifers within the coastal system, to their interconnection and to their connectivity or blockage to the sea. The above factors are expected to be responsible to changes in the effects on different sub-aquifers.

It is recommended to combine field work, geophysical measurements, hydrological simulation and chemical and isotopic analyses all of which will yield better interdisciplinary understanding of the hydrological systems, especially in cases of dynamic conditions.

ACKNOWLEDGEMENTS

We thank Haim Hemo and Vladimir Friedman for their technical help. Batsheva Cohen and Nili Almog are thanked for their help with the figures. We thank the Hydrological Service of Israel for providing the Dead Sea level.

REFERENCES

Arad, A. & Michaeli, A. (1967) Hydrogeological investigations in the western catchment of the Dead Sea. *Israel Journal of Earth Sciences*, 16, 181–196.

Avrahamov, N., Yechieli, Y., Lazar, B., Lewenberg, O., Boaretto, E. & Sivan, O. (2010) Characterization and dating of saline groundwater in the Dead Sea area. *Radiocarbon*, 52(3), 1123–1140.

Bartov, Y., Stein, M., Enzel, Y., Agnon, A. & Reches, Z. (2002) Lake levels and sequence stratigraphy of Lake Lisan, the late Pleistocene precursor of the Dead Sea. *Quaternary Research*, 57, 9–21.

Bookman (Ken-Tor), R., Enzel, Y., Agnon, A. & Stein, M. (2004) Late Holocene levels of the Dead Sea. *Geological Society of America Bulletin*, 116, 555–571.
This is the correct journal name

Bruce, D.L., Yechieli, Y., Zilberbrand, M., Kaufman, A. & Friedman, G.M. (2007) Delineation of the coastal aquifer of Israel based on repetitive analysis of ^{14}C and tritium. *Journal of Hydrology*, 343, 56–70.

Chen, B.F. & Hsu, S.M. (2004) Numerical study of tidal effects on seawater intrusion in confined and unconfined aquifers by time-independent finite-difference method. *Journal of Waterway Port, Coastal, and Ocean Engineering*, 130, 191–206.

Cooper, H.H. (1959) A hypothesis concerning the dynamic balance of fresh water and salt water in a coastal aquifer. *Journal of Geophysical Research*, 64, 461–467.

Diersch, H.-J.G. & Kolditz, O. (2002) Variable-density flow and transport in porous media: Approaches and challenges. *Advances in Water Resources*, 25, 899–944.

Essaid, H.I. (1990) A multilayered sharp interface model of coupled freshwater and saltwater flow in coastal systems: Model development and application. *Water Resources Research*, 26, 1431–1454.

Galeati, G., Gambolati, G. & Neuman, S.P. (1992) Coupled and partially coupled eulerian-lagrangian model of freshwater–seawater mixing. *Water Resources Research*, 28(1), 149–165.

Goldman, M., Arad, A., Kafri, U., Gilad, D. & Melloul, A. (1989) Detection of freshwater/seawater by the time domain electromagnetic (TDEM) method in Israel. *Naturvet Tijdschr*, 70, 339–344.

Gvirtzman, H. (2002) *Israel Water Resources*. Jerusalem, Yad Ben-Zvi Press. pp. 64–65 (in Hebrew).

Harrar, W.G., Williams, A.T., Barker, J.A. & Camp, M.V. (2001) Modelling scenarios for the emplacement of palaeowaters on aquifer systems. In: Edmunds, W.M. & Milne, C.J. (eds.), *Palaeowaters in Coastal Europe: Evolution of Groundwater Since the Late Pleistocene. Geological Society, London, Special Publications*, 189, 213–229.

Henry, H.R. (1964) Effects of dispersion on salt encroachment in coastal aquifers. In: Cooper, H.H. (ed.) *Sea Water in Coastal Aquifers: US Geological Survey Water Supply Paper*, 1613-C,. C71–C84.

Issar, A. (1968) Geology of the central coastal plain of Israel. *Israel Journal of Earth Sciences*, 17, 16–29.

Kafri, U. & Goldman, M. (2006) Are the lower subaquifers of the Mediterranean coastal aquifer of Israel blocked to seawater intrusion? Results of a TDEM (time domain electromagnetic) study. *Israel Journal of Earth Sciences*, 55, 55–68.

Kafri, U., Goldman, M. & Lang, B. (1997) Detection of subsurface brine, fresh water bodies and the interface configuration in between by the TDEM method in the Dead Sea Rift, Israel. *Environmental Geology*, 31, 42–49.

Kafri, U. & Yechieli, Y. (2010) *Groundwater Base Level Changes and Adjoining Hydrological Systems*. Berlin, Heidelberg, Springer-Verlag. 171 pp.

Kapuler, Y. & Bear, J. (1970) *Numerical solution to the movement of the interface in a multi-layered coastal aquifer*. Tahal, Water Planning for Israel , Tel Aviv. Rep. 01/74/70 (in Hebrew).

Kiro, Y., Yechieli, Y., Shalev, E., Lyakhovsky, V. & Starinsky, A. (2008) Time response of the water table and saltwater transition zone to a base level drop. *Water Resources Research*. 44, w12442. : doi:10.1029/2007WR006752

Kolton, Y. (1988) Examination of the connection between seawater in the Pleistocene aquifer in the Mediterranean shelf of central Israel. Tahal, Water Planning for Israel, Tel Aviv. Rep. 01/88/31 (in Hebrew).

Kooi, H., Groen, J. & Leijnse, A. (2000) Modes of seawater intrusion during transgressions. *Water Resources Research*, 36(12), 3581–3589.

Lee, C.H. & Cheng, R.T.S. (1974) On Seawater encroachment in Coastal Aquifers. *Water Resources Research*, 10(5), 1039–1043.

Lensky, N., Dvorkin, Y., Lyakhovsky, V., Gertman, I. & Gavrieli, I. (2005) Water, salt, and energy balance of the Dead Sea. *Water Resources Research*. [Online] 41, W12418. Available from: doi:10.1029/2005WR004084.

Melloul, A.J. & Zeitoun, D.G. (1999) A semi-empirical approach to intrusion monitoring in Israeli coastal aquifer. In: Bear et al. (eds.) *Seawater Intrusion in Coastal Aquifers – Concepts, Methods and Practices*. The Netherlands, Kluwer Academic Publishers. pp. 543–558.

Nativ, R. & Weisbrod, N. (1994) Hydraulic connections among sub-aquifers of the Coastal Plain Aquifer, Israel. *Groundwater*, 32, 997–1007.

Oude Essink, G.H.P. (1996) Impact of sea level rise on groundwater flow regimes: A sensitivity analysis for the Netherlands. *PhD Thesis*. Delft University of Technology, Delft, The Netherlands. 411 pp.

Oude Essink, G.H.P. (2001) Saltwater intrusion in a three dimensional groundwater system in the Netherland: A numerical study. *Transport in Porous Media*, 43, 137–158.

Pinder, G.F. & Cooper, J.R. (1970) A Numerical technique for calculating the transient position of the salt water front. *Water Resources Research*, 6(3), 875–882.

Rahmstorf, S. (2007) A semi-empirical approach to projecting future sea-level rise. *Science*, 315, 368–370.

Rosen, D.S. (2004) Sea level rise and assessment of its impact on the state of the beaches at the Mediterranean coast of Israel. Beaches 2004, Report of the Society for protection of Nature pp 21-28 . Isr. (in Hebrew).

Sanford, W.E. & Konikow, L.F. (1985) *A two-constituent solute-transport model for ground water having variable density*. Water Resources Investigations, US Geological Survey, 88 pp.

Shalev, E., Lazar, A., Wollman, S., Kington, S., Yechieli, Y. & Gvirtzman, H. (2009) Freshwater–saltwater mixing zone in coastal aquifers: Biased vs. reliable monitoring. *Groundwater*, 47(1), 49–56.

Sherif, M.M. & Singh, V.P. (1999) Effect of climate change on sea water intrusion in coastal aquifers. *Hydrological Processes*, 13, 1277–1287.

Sivan, O., Yechieli, Y., Herut, B. & Lazar, B. (2005) Geochemical evolution and timescale of seawater intrusion into the coastal aquifer of Israel. *Geochimica et Cosmochimica Acta*, 69, 579–592.

van Dam, J.C. (1999) Exploitation, restoration and management. In: Bear et al. (eds.) *Seawater Intrusion in Coastal Aquifers – Concepts, Methods and Practices*. The Netherlands, Kluwer Academic Publishers. pp. 73–126.

Voss, C.I. (1984) *SUTRA: Finite-element simulation model for saturated–unsaturated fluid-density-dependent groundwater flow with energy transport or chemically reactive single-species solute transport*. Water Resources Investigation Report (*United States Geological Survey*), 84–4369. 429 pp.

Warrick, R.A., Oerlemans, J., Woodworth, P.L., Meier, M.F., Le Provost, C. (1996) Changes in sea level. In: Houghton, J.T., Meira Filho, L.G. & Callander, B.A. (eds.) *Climate Changes 1995: The Science of Climate, Contribution of Working Group I to the Second Assessment Report of the Intergovernmental Panel of Climate Changes.* Cambridge University Press, U.K. pp. 359–405.

Yechieli, Y. (2000) Fresh-saline water interface in the western Dead Sea area. *Groundwater*, 38, 615–623.

Yechieli, Y., Kafri, U., Wollman, S., Shalev, E. & Lyakhovsky, V. (2009a) The effect of base level changes and geological structures on the location of the groundwater divide, as exhibited in the hydrological system between the Dead Sea and the Mediterranean Sea. *Journal of Hydrology*, 378, 218–229.

Yechieli, Y., Kafri, U. & Sivan, O. (2009b(The inter-relationship between coastal sub-aquifers and the Mediterranean Sea, deduced from radioactive isotopes analysis. *Hydrogeology Journal*, 17, 265–274.

Yechieli, Y., Ronen, D., Berkovitz, B., Dershovitz, W.S. & Hadad, A. (1995) Aquifer characteristics derived from the interaction between water levels of a terminal lake (Dead Sea) and an adjacent aquifer. *Water Resources Research*, 31(4), 893–902.

Yechieli, Y., Shalev, E., Wollman, S., Kiro, Y. & Kafri, U. (2010) Response of the Mediterranean and Dead Sea coastal aquifers to sea level variations, *Water Resources Research.* [Online] 46, W12550. Available from: doi:10.1029/2009WR008708

Yechieli, Y. & Sivan, O. (2011) The distribution of saline groundwater and its relation to the hydraulic conditions of aquifers and aquitards: Examples from Israel. *Hydrogeology Journal*, 19(1), 71–81.

Yechieli, Y., Sivan, O., Lazar, B., Vengosh, A., Ronen, D. & Herut, B. (2001) Radiocarbon in seawater intruding into the Israeli Mediterranean coastal aquifer. *Radiocarbon*, 43, 773–781.

CHAPTER 13

Land subsidence and sea-level rise threaten fresh water resources in the coastal groundwater system of the Rijnland water board, The Netherlands

Gualbert Oude Essink & Henk Kooi

ABSTRACT

Large parts of the Netherlands are situated several meters below mean sea level. Saline groundwater is attracted from the sea and from deep saline aquifers, and subsequently, intrudes near-surface coastal groundwater systems. The salinization of the subsoil is caused by human-driven processes of land subsidence that have been going on for nearly a millennium. A database from TNO of chloride concentration measurements throughout the Dutch coastal zone documents these salinizing processes in the subsoil, though locally also freshening processes are taking place. Here the results are reported of a variable-density groundwater flow and coupled solute transport model in three-dimensions of the coastal groundwater system of the Rijnland Water Board. Computations demonstrate that an anticipated relative sea-level rise accelerates the ongoing salt water intrusion to the aquifers. On average, the aquifer will contain much more saline groundwater. More areas will experience higher rates of upward flow of groundwater, because the head difference increases between rising mean sea level and surface water levels of subsiding land. In addition, the amount of salt encroaching the low-lying areas increases significantly. Water management sectors like drinking water supply and agriculture will be affected. It is necessary to anticipate on these threats, already in the near future. Various technical measures to compensate the salinization are considered; extraction of saline groundwater seems to be the most feasible one.

13.1 INTRODUCTION

The Netherlands is a densely populated country of which some eight million people are living in the coastal zone. Water management of the low-lying coastal areas already faces serious difficulties due to the salinization of the subsoil (Dufour, 2000; Huisman et al., 1998; Pulles, 1985). Due to several natural and anthropogenic causes, Dutch coastal groundwater systems are threatened by an intrusion of saline water (Wesseling, 1980; Leenen, 1980). The salinization affects water management sectors such as agriculture, and domestic and industrial water supply. Water courses in low-lying areas are flushed frequently to dispose the excess of salt. Pumping wells for drinking water supply have been abandoned since the early 1920s due to the upconing of saline groundwater (Stuyfzand, 1996). A shortage of fresh water to flush water channels occurs during dry summers like the year 2003. In addition, there is the danger of salt damage to crops due to a high salt load from the deeper subsoil (Huisman et al., 1998; Pulles, 1985).

The prime causes of the current salinization problem are two anthropogenic activities that have greatly altered both land surface conditions and the groundwater system over the past millennium (Van de Ven, 1993). The first – drainage of peaty and clayey soil by digging channels and building dikes – is a slow and continuous process, leading to land subsidence by peat oxidation as well as compaction and shrinkage of clay. The greatest land subsidence (of about 1m per century) of this kind occurred in the western and northern parts of the Netherlands. The second process – land reclamation, notably drainage of lakes – caused a relatively abrupt changes in the land surface level, creating the well-known Dutch polders (Schultz, 1992; Van de Ven, 1993). The artificially lowered water table relative to surrounding land invigorated and deepened groundwater flow in polder areas, progressively transporting saline groundwater to shallower depths and the land surface in discharge zones. In the coastal zone, over time, a significant head difference developed between the high mean sea level and the controlled surface water level (Figure 13.1). Many low-lying polders are currently experiencing upward groundwater flow (seepage) with a high salinity. The salt originates from marine deposits of Pliocene to Early-Pleistocene origin as well as from transgressions of the sea during the Holocene (Beets et al., 1992; Post, 2003; Zagwijn, 1989). The complicated distribution of fresh, brackish and saline groundwater is related to the diversity of local flow systems associated with man-made topography during the past millennium (Post, 2003; Van de Ven, 1993). Various theories have been developed during these last decades about the occurrence of saline and brackish groundwater in the Dutch coastal aquifers. The physical processes under consideration are advection, dispersion, molecular diffusion and possibly chemical osmosis (Meinardi, 1991; Post, 2003; Post et al., 2003; Van Rees Vellinga et al., 1981; Volker and Van der Molen, 1991).

Within the EU Water Framework Directive (the aim is that in 2015 the status of all groundwaters and surface waters must be good), quality changes in groundwater bodies have to be monitored and assessed, and trends in groundwater pollution have to be identified and reversed. For that purpose, the Rijnland District Water Control Board started up a study to quantify salinity changes in their regional groundwater system. The objective of this study is to evaluate the future regional distribution of fresh, brackish and saline groundwater in this aquifer system in the Dutch coastal zone. A three-dimensional model was constructed of the regional area of the Rijnland District Water Control Board (called the Rijnland model). The model quantifies the effect of future developments on this aquifer system under stress. The numerical model predicts on a regional scale the movement of variable-density groundwater and coupled solute transport, during the coming 200 years. The regional model cannot predict transport of fresh, brackish and saline groundwater on a local scale of meters to tens of meters. Detailed processes such as upconing of saline groundwater under groundwater extraction wells or weak salty spots in the Holocene layer are not considered here. Models with fine discretized grids should be applied to quantify these local processes.

13.1.1   Relevance for GRAPHIC

The work presented here represents a case study which addresses prime themes of the GRAPHIC Project (UNESCO, 2008) that seeks to understand how groundwater resources may be threatened by human activities and the uncertain consequences of climate change. Emphasis in the current work is on the impact of anticipated relative sea-level

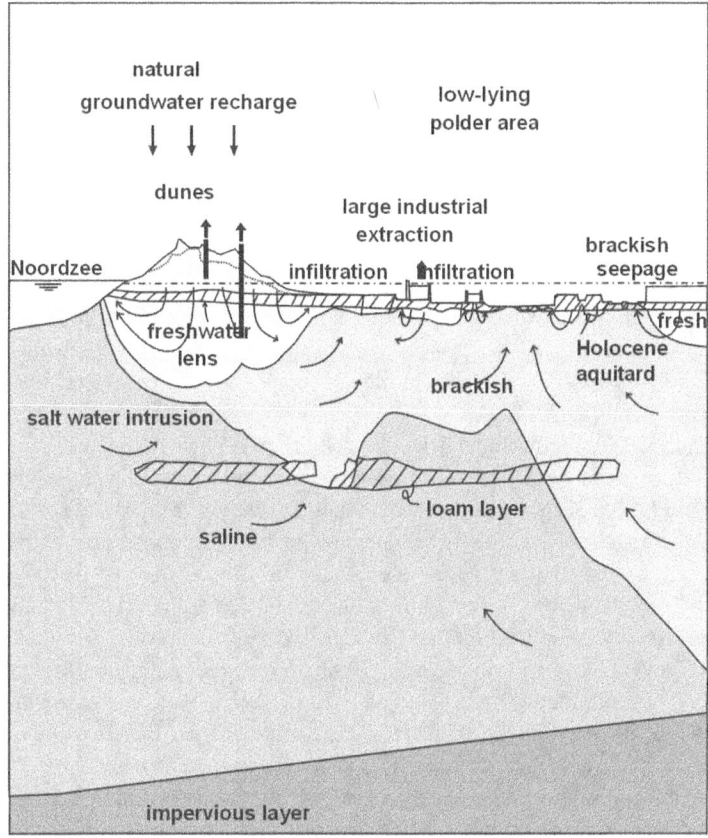

Figure 13.1.   Schematisation of the coastal hydrogeological situation in The Netherlands. A polder is an area which is protected from water outside the area and which has a controlled water level. In the dune area, surface water flows into the aquifer by natural recharge and infiltration through high lying pools. Groundwater from the fresh water lens is extracted for drinking water supply. Low-lying polders in the hinterland, that attract brackish seepage, have to be flushed to guarantee good water quality. Vertical axes: some 200 m; horizontal axes: some 50 kilometres.

rise – as one of the key expressions of climate change – on groundwater resources in the study area. The present work further illustrates and highlights how past anthropogenic stresses over long time scales play an essential part in current and projected changes in groundwater resources in The Netherlands.

In this article, salinizing and freshening processes in the Dutch coastal aquifers in general are first discussed to demonstrate the complexity of the Dutch situation. Then the area of the Rijnland Water Board is described and the development of the numerical model of the considered groundwater system is presented, followed by a discussion of some numerical results. Finally, conclusions are drawn on the impact of sea-level rise and land subsidence in the Rijnland Water Board groundwater system.

13.1.2   Salinizing and freshening processes in Dutch coastal aquifers

A large database from the TNO – Geological Survey of the Netherlands was analyzed to document salinising and freshening processes in the subsoil. More than 16,500 chloride concentration measurements in observation wells throughout the Dutch coastal zone have been considered. The deepest measurement in this so-called DINO database is just below –400 m N.A.P. (Normaal Amsterdams Peil, the reference level in The Netherlands, roughly equals Mean Sea Level). The first chloride concentration measurement started in 1853 AD. Time series of chloride measurements have been recorded in 2,126 wells. The longest record of chloride measurements lasted 72 years, whereas in one well the chloride concentration was measured seventeen times over a period of 13.5 years. The chloride concentration increases as a function of time in 1,010 wells, see Figure 13.2 (the location of the wells are marked with the symbol 'o'). In 979 wells, the chloride concentration decreases as a function of time (marked with '+' in Figure 13.2).

The groundwater regime in the Dutch deltaic area is diverse. Salinising processes in seepage areas exist next to freshening processes in infiltration areas (Figures 13.2 and 13.3). Evidence of upconing of brackish to saline groundwater due to heavy groundwater extraction rates has been detected in many groundwater extraction wells (Stuyfzand, 1996), especially in dune areas close to the sea as that of the Amsterdam Waterworks (Stuyfzand, 1993). In addition, the creation of the deep polders in the coastal provinces of Noord- and Zuid-Holland during the past centuries have caused a head difference of several meters (Van de Ven, 1993). On a regional scale, the chloride concentration in a vast number of wells situated in these provinces is increasing as a function of time, in a few cases even with thousands of mg chloride per liter in some tens of years (e.g. see the wells 14GP0028 and 19GP0170 in Figures 13.2 and 13.3). The two key processes of this salinizing trend are upconing of old saline groundwater at the interior of the coastal zone and an inflow of saline groundwater from the sea to the coastal aquifers. In addition, groundwater recovery in the dune areas that started in the 20th-century accelerated the upconing of sea water (Stuyfzand, 1993) (e.g. the wells 24FA3003 and 24HA3025). The chloride concentration in the Wieringermeerpolder (reclaimed in the period 1927–1930) is increasing significantly as a function of time (e.g. the wells 14FP0010 and 14GP0028), probably due to head differences created since the reclamation of the Wieringermeerpolder in the early 1930s. Locally, groundwater started to flow between various adjacent areas with different surface water levels. Meanwhile, a freshening process is going on in the upper part of the subsoil on the boundary between the Wieringermeerpolder and the former Wieringen island (e.g. wells 14EA3065, 14EP0061 and 14EP0063). At some places, fresh surface water from the IJssel Lake infiltrates to the shallow groundwater regime of the Province of Noord-Holland. In some hundreds of observation wells, the groundwater regime is freshening as surface water infiltrates the upper part of the subsoil (De Ruiter, 1991) (e.g. the well 31AP0033). In the already brackish shallow aquifers of the Province of Zeeland, groundwater flow is highly induced by the water levels in the surface water system.

13.1.3   Description of the area: the Rijnland Water Board

A characteristic feature at the western side of the Rijnland Water Board is the dune area (Figure 13.4a). Three major drinking water companies are active here: the Drinking

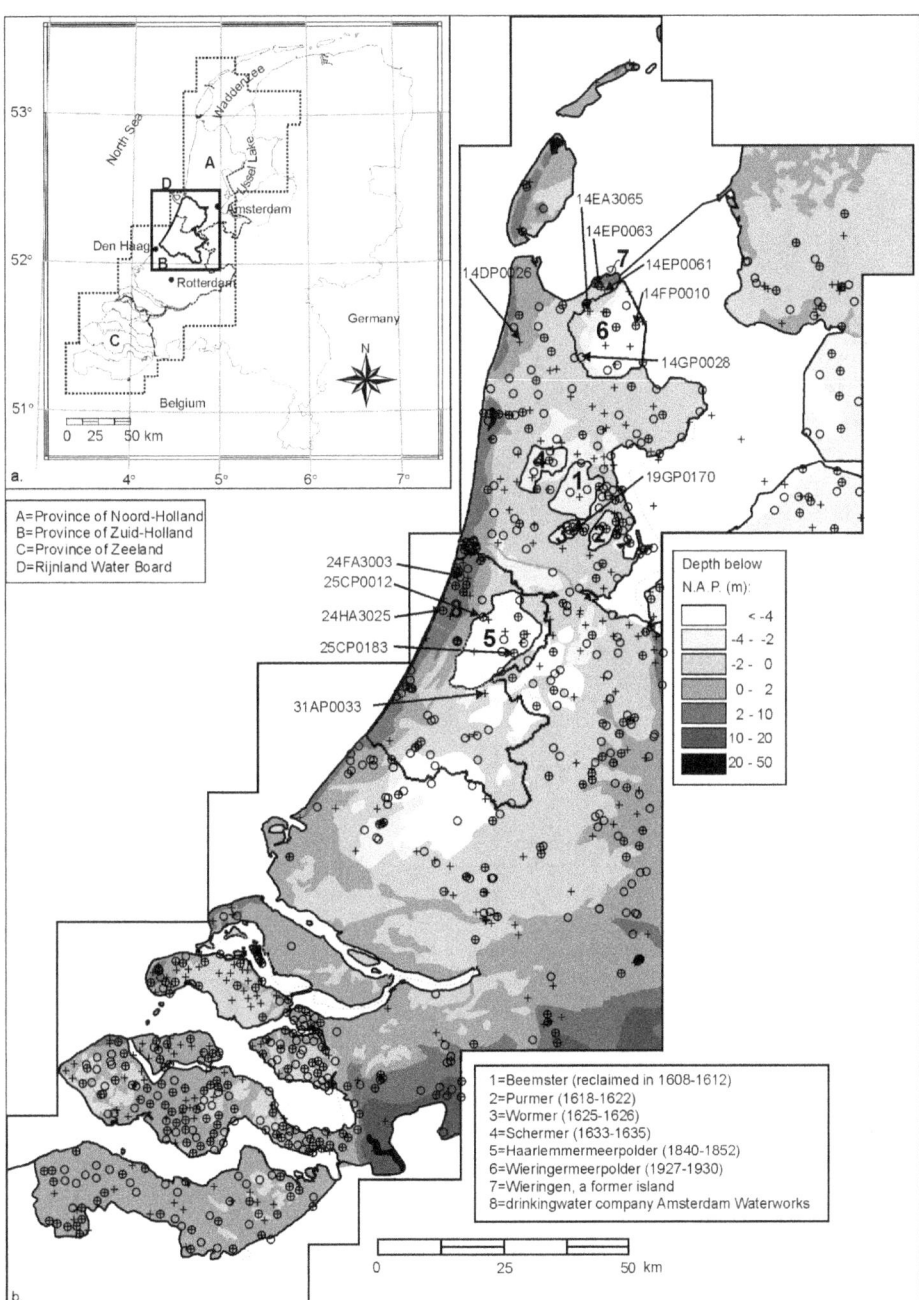

Figure 13.2.   a. Position of the Rijnland Water Board and the 3D variable-density groundwater flow model along the coastal zone of The Netherlands. b. Ground surface as well as the position of salinizing and freshening wells.

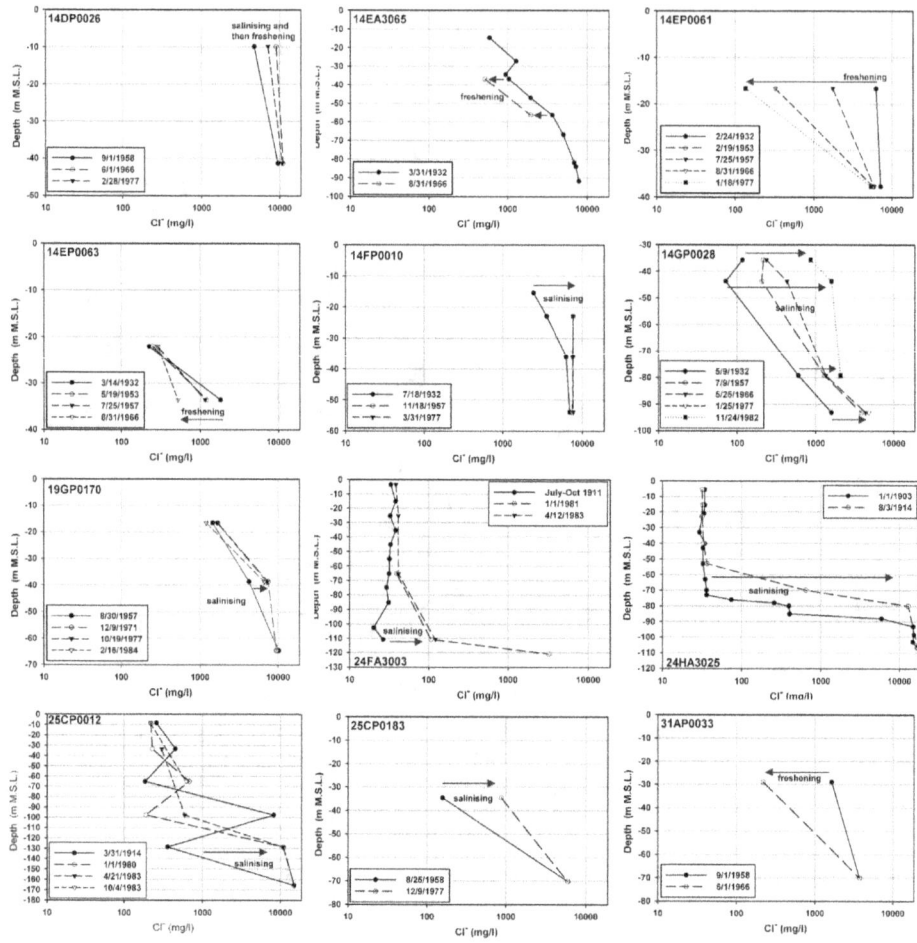

Figure 13.3.   The chloride concentration versus depth as a function of time in a selected number of observation wells in the Dutch coastal groundwater regime, indicating salinizing and/or freshening processes in the surrounding subsoil.

Water Company Zuid-Holland (DZH), the Amsterdam Waterworks (WLB) and the Water Company Noord-Holland (PWN). Phreatic water levels in the dune-areas can go up to more than 7 meters above mean sea level. At the inland side of the dune area, low-lying polder areas with controlled water levels occur. The lowest phreatic water levels in the water board itself can be found northwest of the city Gouda (down to nearly −7 m N.A.P.) and in the Haarlemmermeer polder, where the international airport Schiphol is located, with levels as low as −6.5 m N.A.P. Before the middle of the 19th century, a lake covered the Haarlemmermeer polder area. Due to flooding threats at the neighbouring cities, this lake was reclaimed during the period 1840–1852. Subsequently, heads changed relatively abrupt and a completely different groundwater flow regime was created regionally (Oude Essink, 1996). In addition, the polder Groot-Mijdrecht, which is situated outside the water board, is also mentioned here. Though the surface area of this polder is not large,

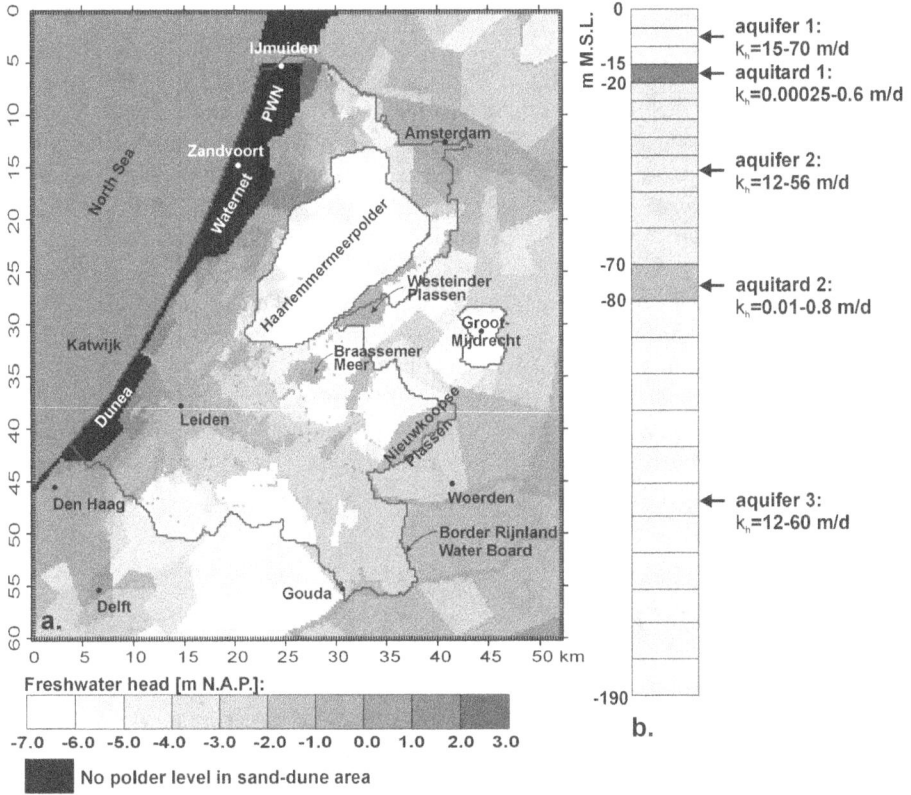

Figure 13.4.   Map of the Rijnland Water Board: position of some polder areas and the dune areas of the drinking water companies DZH, WLB and PWN. Large groundwater extractions from the lower aquifer system are taking place at: DSM/Gist-Brocades near Delft (11 million m³/a), Hoogovens near IJmuiden (20 million m³/a), drinking water company PWN (10 million m³/a) and southeast of Woerden (10 million m³/a): a. Phreatic water levels or polder levels in the area (note that in the sand-dune areas, no polder levels are given but a natural groundwater recharge is implemented). b. Simplified subsoil composition of the bottom of the Rijnland Water Board and the range of used hydraulic conductivites.

the phreatic water level is low (less than –6.5 m N.A.P.) and the Holocene aquitard on top of the groundwater system is very thin. As such, seepage in this area is very large (more than 5 mm per day) and groundwater from a large region around it is flowing towards the polder at a rapid pace (Van der Berg et al., 1976).

13.2   DESCRIPTION OF THE NUMERICAL METHOD

13.2.1   Numerical code

To quantify changes in this coastal groundwater flow regime, the processes transient variable-density groundwater flow and coupled solute transport in three dimensions are modelled. The modular variable-density groundwater flow and solute transport code MOCDENS3D has been used. It is based on MODFLOW (McDonald and Harbaugh, 1988) and MOC3D (Konikow et al., 1996), which has been adapted for density differences. MODFLOW (U.S. Geological Survey public domain) is the most widely used computer code for groundwater flow in porous media. Advective and hydrodynamic dispersive solute transport processes through porous media are modelled by a particle tracking technique (Peaceman, 1977) in combination with the finite difference method (Konikow et al., 1996), respectively. As chloride is the major conservative negative ion in Dutch coastal groundwater, the discussion about salinization is focused on that predominant solute. Under the given circumstances in the Dutch coastal aquifers, the Oberbeck-Boussinesq approximation is valid as it is suggested that the density variations (due to concentration changes) remain small to moderate in comparison with the reference density ρ throughout the considered hydrogeologic system (Holzbecher, 1998; Nield and Bejan, 1992). As such, a substantial simplification of the governing differential equations can be derived. For additional information on the mathematical formulation, see a.o. Bakker et al., 2004; Oude Essink, 2001 or Oude Essink et al., 2010.

13.2.2   Scenarios of sea-level rise and land subsidence

A future global mean sea-level rise of 0.47 m is expected between 1990–2100 (IPCC, 2001) with an uncertainty range from 0.09 to 0.88 m, depending on the different scenarios for the emission of greenhouse gases (IPCC, 2000). This rate is 2 to 5 times the rate experienced over the last century. A sea-level rise of 0.5 m per century is used in the modelling to match the most likely scenario. Note that other scenarios are also modelled, but not discussed here. Land subsidence is considered due to groundwater recovery, compaction and shrinkage of clay, and especially the oxidation of peat that differs from place to place: 1.0 m per century for the peat areas; no subsidence for the dune areas; and 0.3 m per century for the rest of the land surface (respectively 25%, 9% and 66% of the land surface of the Rijnland Water Board). The simulation time period is 200 years. During this time span, changes in groundwater recharge in dunes due to climate change or infiltration area reduction, changes in extraction rates, changes in land use and changes in shoreline position due to new coastal defence systems are not taken into account.

13.2.3   The 3D model

The Rijnland model represents a typical Dutch coastal region on the mainland. The groundwater regime consists of good permeable aquifers of Quaternary deposits, intersected by loamy aquitards and overlain by a Holocene aquitard of clay and peat (Dufour, 2000; Staalduinen et al., 1979). Hydrogeological data, such as location and hydraulic conductivity of formations, piezometric heads and chloride concentrations, are derived from the national geophysical databases REGIS, DINO and from the National Groundwater Model of The Netherlands NAGROM (Table 13.1). Figure 13.4b shows the simplified subsoil composition of coastal aquifer of the Rijnland Water Board including the applied hydraulic conductivity values.

An accurate initial density distribution is important in the model, as it has a major impact on the simulated variable-density groundwater flow. The present density distribution cannot be deduced by just simply simulate the saline groundwater system for many

Table 13.1. Geometry, subsoil, model and calibration parameters of the Dutch coastal aquifer. Sensitivity analyses have been executed on various subsoil and model parameters (Oude Essink, 1996).

Parameter (unit)	Value
total land surface of area of interest (km^2)	1,095
modeled area (km $\times$ km)	52 $\times$ 60
depth groundwater regime (m –N.A.P.)	190
range aquifers hydraulic conductivity (m/d)	
aquifer 1 [ground surface -15 m N.A.P.]	15–70
aquifer 2 [-20 m $- -70$ m N.A.P.]	12–56
aquifer 3 [-80 m $- -190$ m N.A.P.]	12–60
range aquitard hydraulic conductivity (m/d)	
aquitard 1 [-15 m $- -20$ m N.A.P.]	2.5 $\times$ 10^{-4}–0.6
aquitard 2 [-70 m $- -80$ m N.A.P.]	0.01 –0.8
effective porosity ($-$)	0.25
anisotropy (k_z / k_x)	0.1
longitudinal dispersivity α_L (m) [Gelhar et al. 1992]	1
cell size in horizontal plane (m)	250 $\times$ 250
cell size in vertical plane (m)	5 to 10
number of cells in x, y and z direction	209 $\times$ 241 $\times$ 24
total number of active cells	1,208,856
number of particles per cell [Konikow et al., 1996]	27
flow time step to recalculate flow equation (year)	1
number of head & concentration observations	1,632
characteristics head calibration (m)	
$|\Delta\phi|$ = absolute error	0.60
σ = standard deviation	0.77

hundreds of years with all changing actual load and concentration boundary conditions and wait till the composition of solutes does not change any more. The reason is that the present distribution of fresh, brackish and saline groundwater is still not in equilibrium, though the time scale of changes is often large, viz. tens of years. This time scale is often too large to detect possible trends in time, as most time series with measured chloride concentrations don't last so long. In this specific studied area only 411 time series are occurring, and the number of measurements in an observation filter is on the average small, most filters have only be monitored twice (see Figure 13.5a and 13.5b). In one observation filter there is measured 71 times during 459 days: the average chloride concentration varies between 15 and 23 mg Cl$^-$/l.

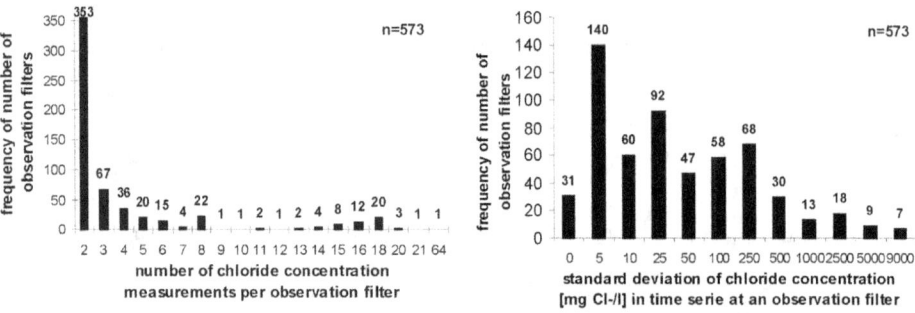

Figure 13.5.   a. Number of chloride concentration measurements per observation filter versus frequency of number of observation filters; b. Standard deviation of chloride concentration [mg Cl$^-$/l] in time serie at an observation filter.

In addition, temporal concentration fluctuations, e.g. due to seasonal and/or inter-annual hydrological variability are to be expected, but they don't seem to be spectacular. The interval between minimum and maximum measured chloride concentration in the observation filter (Figure 13.6a) is on the average small, whereas the deviation from the average chloride concentration in the observation filter is also small (Figure 13.6b). Figure 13.7 shows at two depths the computed chloride concentration as well as the observed chloride concentrations in the available observation wells.

Several processes initiated in the past can still be sensed and make the situation dynamic. For instance, a major change in boundary condition occurred in the early twelfth century, when the river Rhine was forced to change its course in southern direction due to damming up the old branch the 'Oude Rijn' (Van de Ven, 1993). Other human activities were the creation of polders, from the 17th century on, and groundwater extractions in the sand-dune areas. Therefore, from a practical point of view and based on the fact that the system is still dynamic, chloride (and thus density) measurements at the year 1990 AD are chosen as the initial situation. At places where only chloride concentrations were measured at earlier times than 1990 AD, and no 1990 AD data are available, the latest data are also taken into account. These data were inter- and extrapolated with the multi quadric biharmonic method of Hardy (Hardy, 1971). Though this initial chloride distribution is based on more than sixteen hundred measurements of chloride, errors occur, mainly because of a lack of enough data in deeper part of the aquifer system. Artificial inversions of fresh and saline groundwater occur in the numerical model, though they do not exist in reality. As a remedy, ten years are simulated under reference conditions (e.g. constant head at polders and the sea). This simulation time is necessary to smooth out unwanted, unrealistic density dependent groundwater flow that was caused by the numerical discretisation of the initial density distribution. The mean error between measured and computed chloride concentration is -102 mg Cl$^-$/l, the mean absolute error 559 mg Cl$^-$/l and the standard deviation 1,323 mg Cl$^-$/l.

Though molecular diffusion for porous media appears to be of negligible importance, D_m is taken equal to 10^{-9} m^2/s. The volumetric concentration expansion gradient β_C is 1.34×10^{-6} l/mg Cl$^-$. The concentration of saline groundwater in the aquifer does not exceed the value of sea water, 18,630 mg Cl$^-$/l. The corresponding density of that

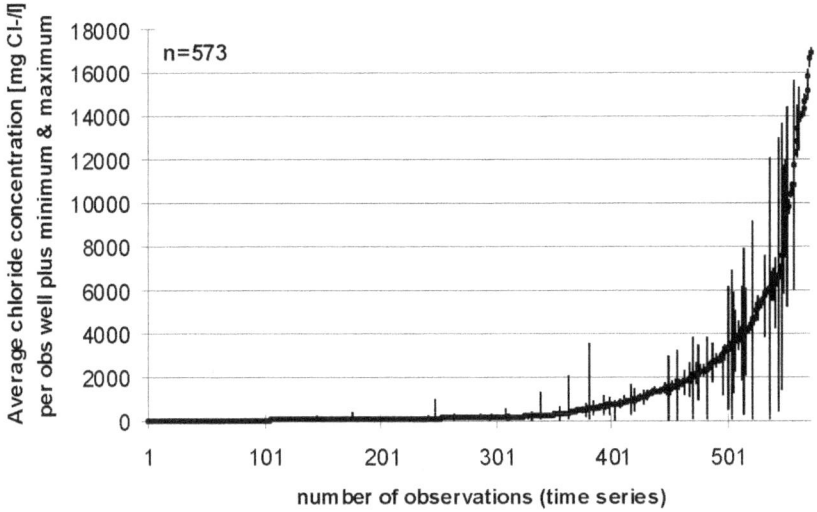

Figure 13.6. Number of time series observations versus the average chloride concentration [mg Cl⁻/l] per observation filter.

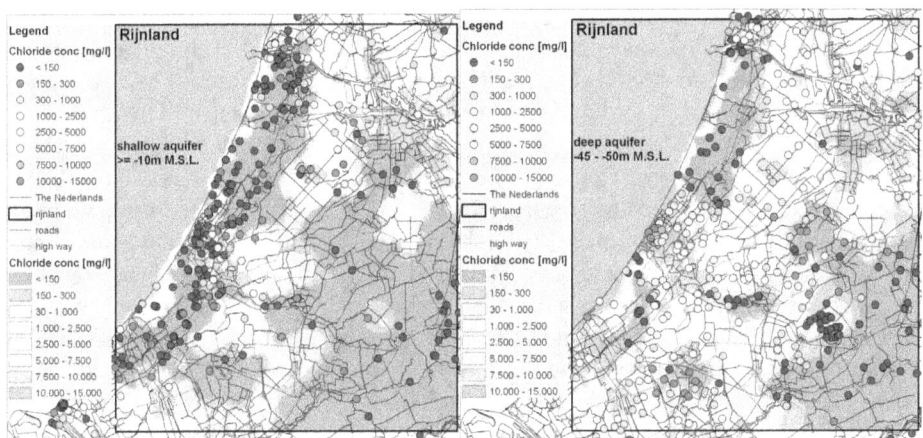

Figure 13.7. Initial chloride concentrations [mg Cl⁻/l] as well as chloride concentrations in observation wells in the studied area: a. shallow (higher than −10m M.S.L.) and b. deep (between −45m and −50m M.S.L.).

saline groundwater equals 1,025 kg/m³. On the applied time scale, the specific storativity S_s to model transient groundwater flow can be set to zero.

The groundwater system consists of a 3D grid of 52.25 km by 60.25 km (~3,150 km²) by 190 m depth and is divided into a large number of elements. Each element is 250 m by 250 m in horizontal plane. In vertical direction the thickness of the elements varies from 5 m for the ten upper layers to 10 m for the deepest fourteen layers. The grid contains

1,208,856 active elements: $n_x = 209$, $n_y = 241$, $n_z = 24$, where n_i denotes the number of elements in the i direction. Each element contains initially twenty-seven, which gives in total 32.6 million particles to solve the advection term of the solute transport equation. The flow time step Δt to recalculate the groundwater flow equation is one year. The convergence criterion for the groundwater flow equation (freshwater head) is equal to 10^{-8} m. Table 13.1 shows the used model parameters in the numerical model.

The bottom of the system is represented as a no-flow boundaries. Hydrostatic boundary conditions occur at the four sides of the model. At the top of the dune area, the natural groundwater recharge varies from 0.94 to 1.14 mm/day, which causes head elevations of several meters above N.A.P. The initial water level at the sea is set to 0.0 m N.A.P. for the year 2000 AD. The top system at the sea and the polder area are modelled in MODFLOW by the so-called 'general head boundary' package. The conductance of the general head boundary is related to the resistance of the drains. Used general head boundary conductances vary from 2.5 to 327 m²/d. The initial water levels in the polders units are equal to the phreatic water level in the considered polders units, which differs for each polder unit from +2.0 m near IJmuiden to −7.0 m N.A.P. northwest of Gouda (Figure 13.4). The relatively small seasonal fluctuations in the phreatic water level in the polders are neglected.

13.2.4 Calibration of the 3D model

Calibration is executed with equivalent freshwater heads in the hydrogeologic system, and to some extent with water fluxes and salt load fluxes in the Haarlemmermeer polder and the polder Groot-Mijdrecht. Calibration data has been derived from the water board Rijnland itself, ICW (1976) and the national geophysical database DINO and from NAGROM.

The model was mainly calibrated by comparing 1,632 measured and computed freshwater heads through trial-and-error (Figure 13.8). Note that the measured freshwater heads are corrected for density differences, so the chloride concentration at each observation location should be known (see Equation 9). The mean error between measured and computed freshwater heads is −0.16 m, the mean absolute error 0.61 m and the standard deviation 0.79 m. More than ten major iteration steps were executed, by one-by-one changing the hydraulic conductivities of the aquifers and the Holocene aquitard, the general head boundary conductances, the initial chloride concentration and the anisotropy (k_z / k_x). Though the mean absolute error is not really a small value based on the maximum measured values (of +6.22 m N.A.P. at the sand-dune area of WLB) and minimum measured values (of −8.51 m N.A.P. at the groundwater extraction of DSM/Gist-Brocades near Delft), it was accepted within the scope of this research.

As can be seen in Figure 13.9, the difference between measured and computed freshwater head are on the average small, though sometimes considerable in certain locations of the studied area. The differences are created due to a number of causes. Besides some initial data errors caused by the conversion from NAGROM to the MODFLOW format, the lack of detailed and exact information of the thickness and vertical hydraulic conductivity of the Holocene aquitard and the hydraulic conductances of the underlying aquifers is very likely causing a number of misfits. Subsequently, the resistance of the Holocene aquitard directly influences seepage and salt load values of the polders and surrounding lakes in the area. In addition, not taking into account the concept of small-scale water

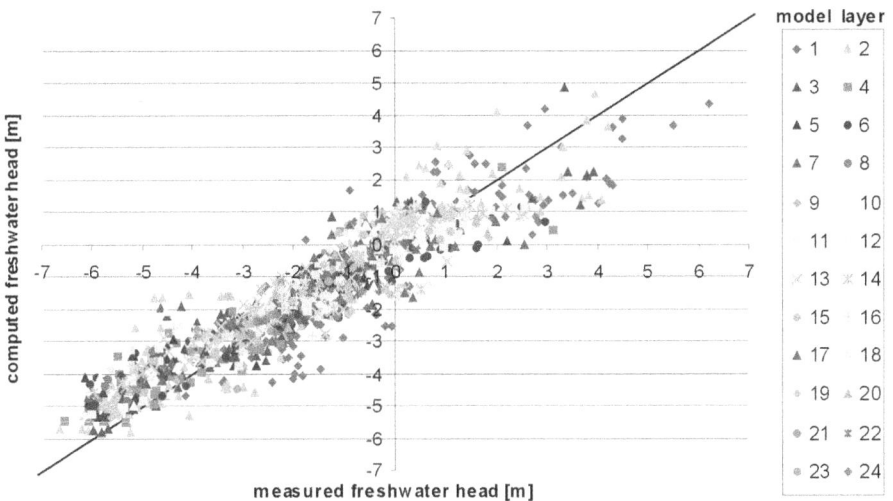

Figure 13.8.   Measured versus computed freshwater heads [m].

pools in sand-dune areas is also a source of errors. Furthermore, the size of the element is rather coarse for exactly modelling freshwater heads around large differences in ground surface and around extraction wells. For instance, the so-called embankment seepage between a high lying lake and the adjacent low-lying polder is rather difficult to model with elements of 250 by 250 m. In addition, modelling the drawdown near groundwater extraction wells with elements of this size is also not straightforward. Finally, an important cause of error is the initial density distribution. A significant number of density points with a saline groundwater density of 1,025 kg/m^3 is placed artificial in the North Sea, where measurements in the groundwater system are not available. These density points together have to fill a 3D density-matrix by mean of inter- and extrapolation, which results in a grid of more than 1.2 million elements. It is clear that errors in the generated density distribution are easily created. Though unwanted, unrealistic density dependent groundwater flow is smoothed out, it is likely that some artificial numerical inversions of fresh and saline groundwater are occasionally occurring in the numerical model. This can locally result in an incorrect density driven groundwater flow system.

The salt load fluxes to the main polder areas were used to calibrate the concentration data (Figure 13.9), especially in the top of the aquifer system. Data such as water fluxes (seepage) and salt load fluxes at polder level scale is collected from several reports and water board databases (ICW, 1976; Berg et al., 1976; Leenen, 1980; Wesseling, 1980). Some measured versus computed values are: 1) the Haarlemmermeer polder: total seepage 83,100 m^3/d versus 92,500 m^3/d (Figure 13.10) and total average salt load 61,000 ton/a (Figure 13.11) versus 62,900 ton/a; 2) the polder Groot-Mijdrecht: seepage 5.1 mm/d versus 2.8 mm/d (probably the Holocene aquitard was modelled too thick (thickness model layer is 5 m), and the resistance to flow through the aquitard was larger than in reality) and total average salt load 52,000 ton/a versus 47,000 ton/a. Knowing that the exact fluxes at polder level scale is rather difficult to determine, these values correspond quite well with each other.

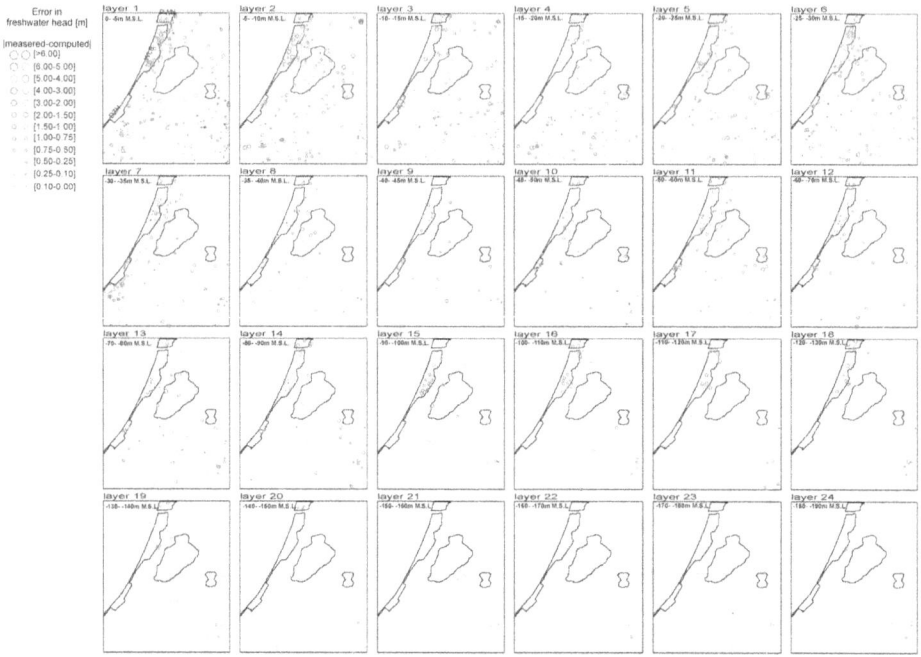

Figure 13.9.   Distribution of the absolute error in freshwater head (measured versus computed) in the studied area [m], for each model layer.

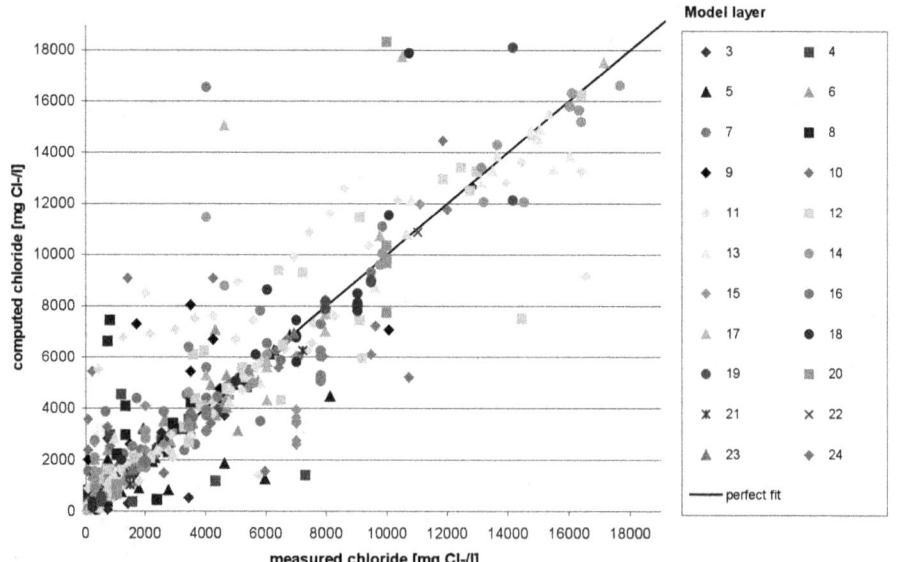

Figure 13.10.   Measured versus computed chloride concentration [mg Cl⁻/l].

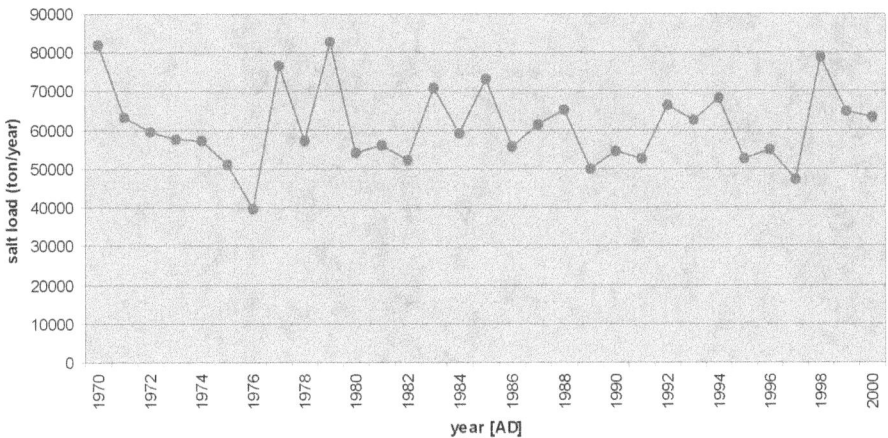

Figure 13.11.   Measured salt load in the Haarlemmermeer polder (ton/a) (source: Water board Rijnland, pers. comm.).

Quite a few numerical accuracy tests have been executed on various model parameters such as initial number of particles per element (viz. 1, 8 and 27), the convergence criterion for the groundwater flow equation (viz. 10^{-3}, 10^{-6} and 10^{-8}m) and the flow time step Δt to recalculate the variable-density groundwater flow equation (viz. 1 month, 0.5 year, 1 year and 2 years) (e.g. see also Oude Essink, 1996).

In spite of the uncertainties mentioned above, especially in geological subsoil and density parameters, it is from some point surprising that the measured and calculated values fit so well. Note that in addition this studied area is considered in the Netherlands as very densely measured.

To conclude: the computed heads, fluxes and chloride concentrations fit rather well the observed ones, and thus the 3D model can be used for comparing with each other forward simulations of different scenarios.

13.3   RESULTS AND DISCUSSION

13.3.1   Salinization of the groundwater system

Transient simulations of variable-density groundwater flow and solute transport during a simulation time of 200 years show that the solute distribution in the groundwater regime is at present not in a steady-state (Figure 13.12). The velocity field suggests that salt water is intruding the coastal aquifer up to rates of several tens of meters per year. The average salt content in the uppers parts of the aquifer of the low-lying polder areas increases. At the same time, in more detail, various areas are freshening. On the average, the seepage from the groundwater regime to the entire surface water system is predicted to increase only +7% over a period of 200 years. This overall increase in seepage quantity to the surface water system in the Water Board is not dramatic: The rise in freshwater head due to sea-level rise attenuates in landward direction. The effect of sea-level rise on the heads is hardly perceptible >15 km from the coastline. During

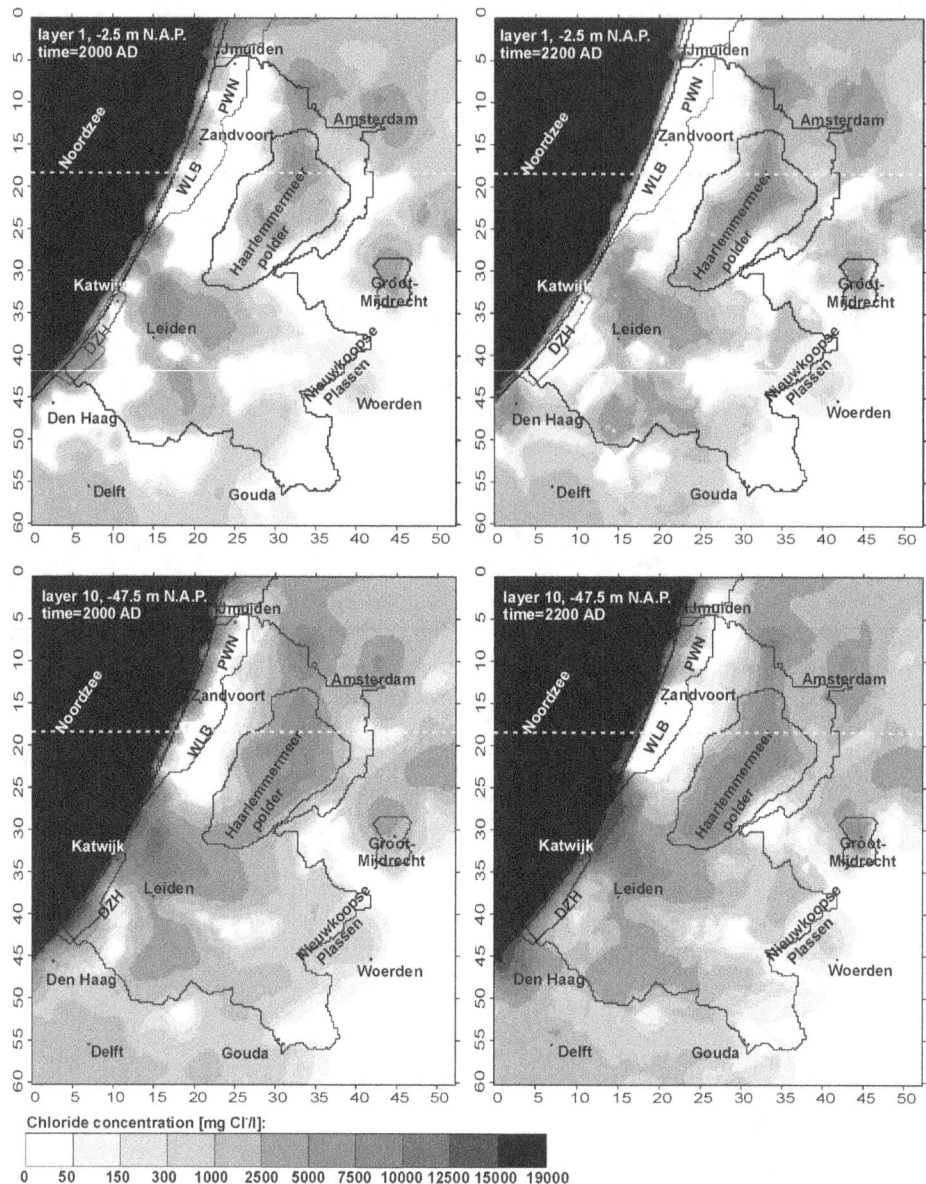

Figure 13.12.   Computed chloride concentration at the levels -2.5 and -47.5 m N.A.P. for the years 2000 and 2200 AD. A sea-level rise of 0.5 m per century and land subsidence are taken into account. The grey dashed line represents the position of the vertical 2D profiles.

the coming two centuries, about 10% (107 km^2) of the land surface of the Rijnland Water Board switches from infiltration to seepage areas (the so-called hinge areas). They are in most cases located in the coastal zone. Significant changes are predicted in the salt load to the surface water system: $+67\%$ in 2200, relative to the present and linearly increasing with time. Progressively higher salinities in the top aquifer mainly cause this phenomenon. The model shows that in polders within a few kilometres from the coastline and surface water level a few meters below mean sea level, the effect of a sea-level rise of 0.5 m per century should become significant within a few decades. Fresh water volumes in dune areas will reduce due to sea-level rise, though the effect of a possible increase in natural groundwater recharge due to climate change (Allen and Ingram, 2002) has to be superimposed, which may lead to larger fresh water volumes.

13.3.2   Compensating measures

Various measures have been considered to compensate the salinization of this groundwater system:

1.   Land reclamation offshore
2.   Inundation of the Haarlemmermeer polder
3.   Injection of fresh surface water
4.   Extraction of saline groundwater
5.   Creation of a physical barrier

It is obvious that some control measures are, for the time being, hypothetical. For instance, the international airport Schiphol Amsterdam cannot easily be removed from

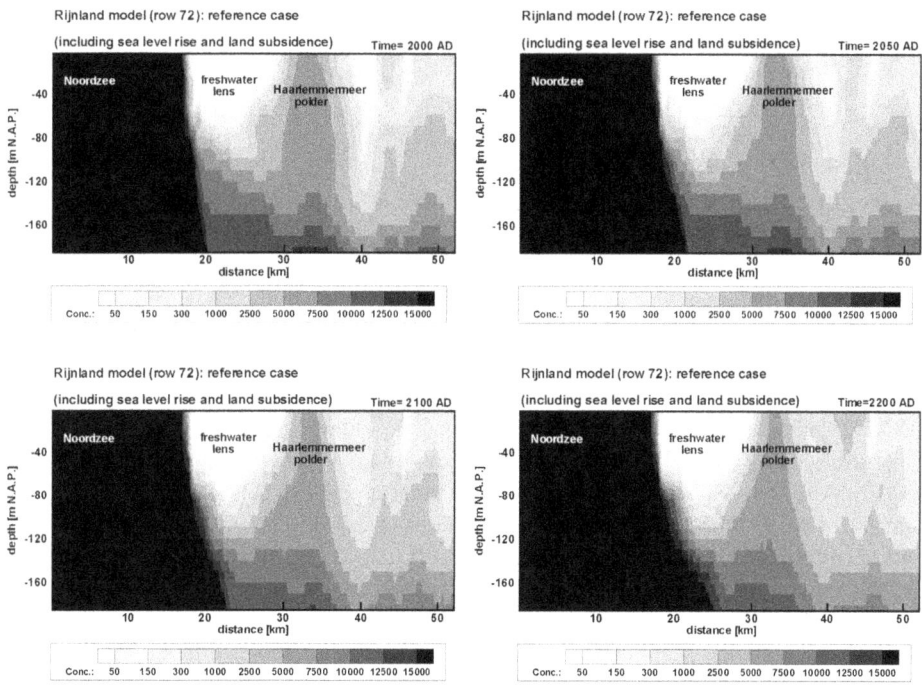

Figure 13.13.   Reference case: including the effect of sea-level rise and land subsidence on the fresh-salt distribution in the subsoil: 2D profile of the 3D model of the Rijnland Water Board. Moments in time are 2000 AD, 2050 AD, 2100 AD and 2200 AD.

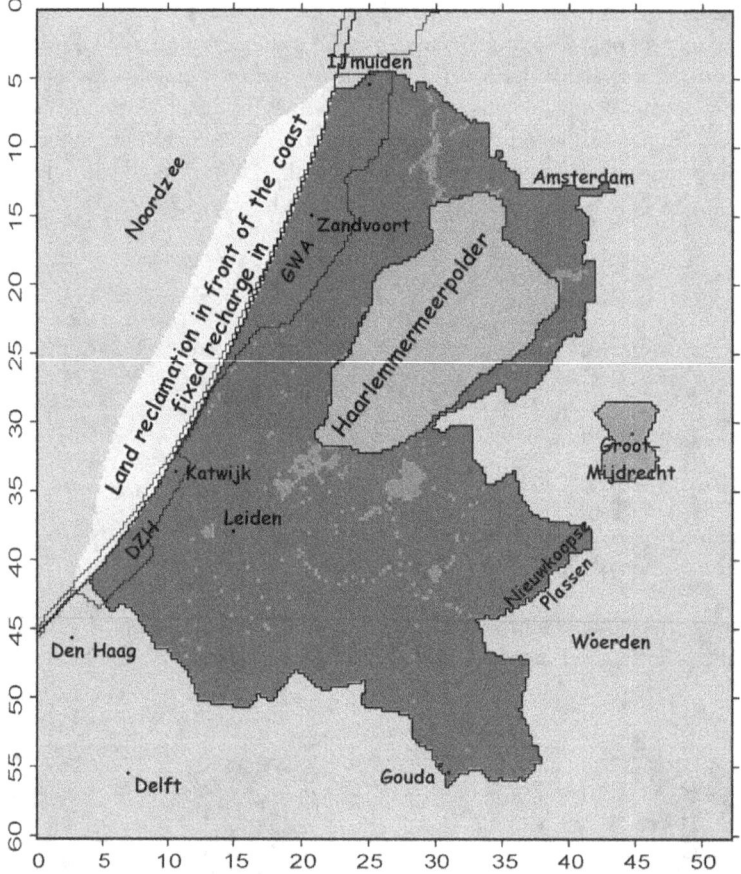

Figure 13.14.   Position of the reclaimed land in front of the coast: the fixed recharge into the groundwater system is equal to 1.06 mm/d.

the Haarlemmermeer polder and a physical barrier at depth along the entire Dutch coastal zone is also not easily constructed. However, land reclamation offshore is actually being implemented in the southern part of the study area, with the *Building with Nature* pro-gramme. In addition, the boundary conditions such as the rate of especially sea-level rise and land subsidence are still uncertain. For the time being, these suggested control measures have a more academic character, though Dutch water managers and scientists already pose that extraction of saline groundwater along the Dutch coast is possibly a feasible measure. The simulated control measures in this article will thus help managers and scientists to conceptualize their ideas.

Note that in all compensating measures, the hydrostatic boundary conditions remain the same. The attenuation of the head change due to measures is great in the studied area, and as such, the zone of influence of measures in the area is rather limited. For instance, the subsoil parameters are such that within 5.3–8 km from the coastline, less than 10% of the sea-level rise is noticed (Oude Essink, 1996, p. 308).

The situation with no measure is taken as the reference case. Figure 13.13 shows that the salt water tongue (concentration larger than 15,000 mg Cl⁻/l) moves various

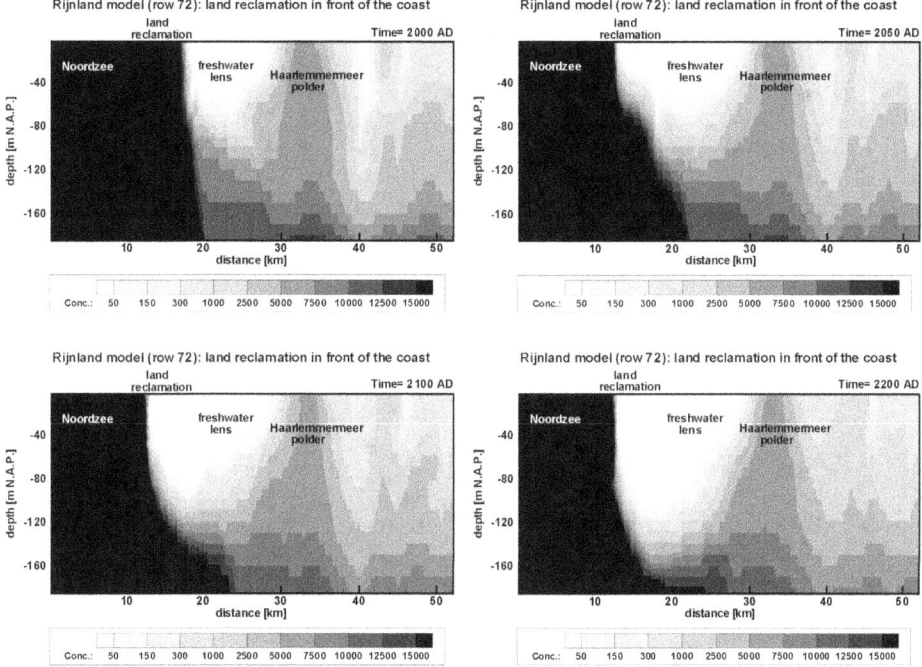

Figure 13.15.   The effect of land reclamation offshore on the fresh-salt distribution in the subsoil: 2D profile of the 3D model of the Rijnland Water Board. Moments in time are 2000 AD, 2050 AD, 2100 AD and 2200 AD.

kilometres eastwards during the coming 200 years. The top of the groundwater system will contain groundwater with a higher salt content, and consequently, the amount of saline seepage in the Haarlemmermeer polder increases. This area is particular interesting as the low-lying polder is situated close to the sea and salt groundwater can easily pass through the thin Holocene resistance layer.

The Figures 13.14 and 13.15 show what will happen when land would be reclaimed in front of the coast. The phreatic water level is supposed to be high in the new sandy area due to natural groundwater. As such, recharge increases the volume of the fresh water lens, saline groundwater is pushed with a greater groundwater velocity into the direction of the Haarlemmermeer polder. Consequently, the salt load will increase the coming centuries. Figure 13.16 shows what happens when the Haarlemmermeer polder would be inundated. Inundation obviously stops the salinization process of the Haarlemmermeer polder itself, but at other places in the Rijnland Water Board, the amount of saline seepage increases.

13.4   CONCLUSIONS

Based on these modelling results, it can be stated that the ongoing salinization of the Dutch subsurface is a slow, though continuous and practically irreversible process. The existing fresh groundwater resources are jeopardised by a combination of

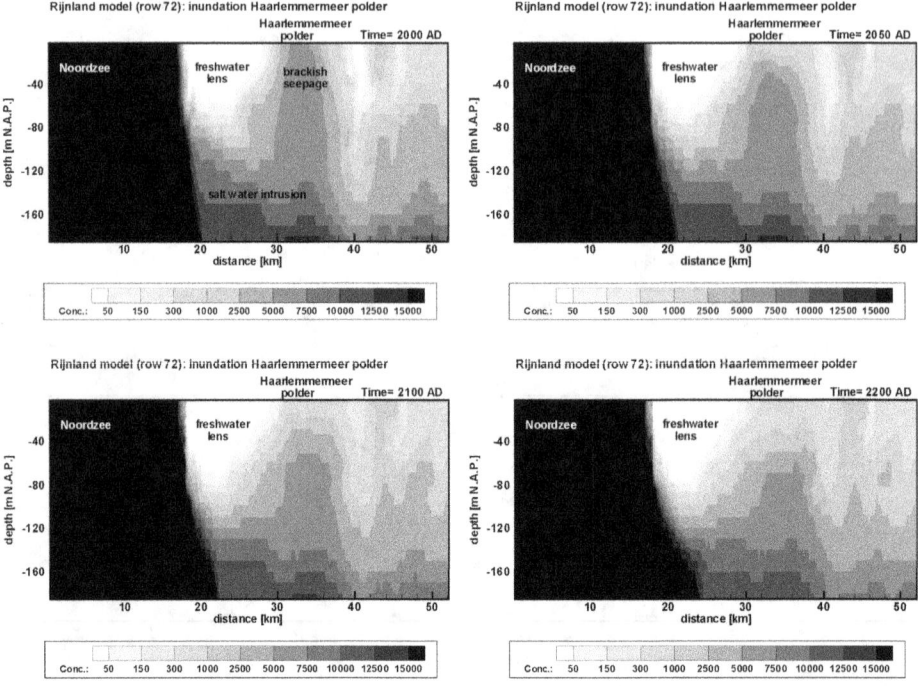

Figure 13.16.   The effect of inundation of the Haarlemmermeerpolder on the fresh-salt distribution in the subsoil: 2D profile of the 3D model of the Rijnland Water Board. Moments in time are 2000 AD, 2050 AD, 2100 AD and 2200 AD.

conditions and processes. Land surface settling and land reclamation in the past is still affecting the distribution of fresh, brackish and saline groundwater. Though these remain the dominant factors for the time being, the expected future sea-level rise of 0.5 m per century as well as the ongoing land subsidence of the low-lying coastal, peaty, areas definitely accelerate the rate of salinization of the Dutch subsoil. These developments will influence surface as well as coastal zone water management in the Netherlands on the medium long term. It may be necessary to enlarge the capacity of the fresh water supply and drainage systems in the coastal area.

Mitigating the salinization process on this regional scale appears to be difficult. The tested technical measurements, viz. land reclamation and inundation, have some positive effect on a local scale whereas at other places in the neighbourhood the situation gets worse.

This study shows the effects of sea-level rise and land subsidence on fresh groundwater resources in a large-scale Dutch aquifer system. The effects pattern for what might happen to fresh water resources in similar deltaic areas worldwide like the Mississippi, Nile, Mekong and Ganges deltas. Though characteristics of these deltaic areas are somewhat different, the general picture is that coastal zone groundwater management has to face serious impacts of future stresses.

The results in this study on the salinization of the groundwater system can be applied for water management issues on a regional scale. On a local scale, however, the flow processes in the top of the water system are so dynamic that the regional changes in seepage and salt load at the deeper groundwater system cannot be applied straight on. For instance, to estimate the salt load to the surface water system in a coastal area the complex behaviour of dynamic rainwater lenses throughout the season on agricultural plots should also be considered.

REFERENCES

Allen, M.R. & Ingram, W.J. (2002) Constraints on future changes in climate and the hydrologic cycle. *Nature*, 419, 224–232.

Bakker, M., Oude Essink, G.H.P. & Langevin, C.D. (2004) The rotating movement of three immiscible fluids. *Journal of Hydrology*, 287, 270–278.

Beets, D.J., Van der Valk, L. & Stive, M.J.F. (1992) Holocene evolution of the coast of Holland. *Marine Geology*, 103, 423–443.

Berg, van den C., Rijteman, P.E., Couwenhoven, T., Steenvoorden, J.H.A.M., Toussaint, C.G., Pomper, A.B., Van Rees Vellinga, E., Witt, H., Wesseling, J., Wit, K.E. & Wijnsma, M. (1976) *Hydrology and Water Quality of the Central Part of the Western Netherlands (in Dutch), ICW Regional Studies 9*. Wageningen, Institute for Land and Water Management Research (ICW).

De Ruiter, J.C. (1991) Some aspects of a hydrological systems analysis in the province of South Holland. In: Kozerski, B. & Sadurski, A. (eds.) *Proceedings of the 11th Salt Water Intrusion Meeting*. Technical University of Gdansk, Department of Hydrogeology and Water Supply, Gdansk, Poland. pp. 56–71.

Dufour, F.C. (2000) *Groundwater in the Netherlands – Facts and Figures*. Netherlands Institute of Applied Geosciences – TNO, Van der Rhee, Rotterdam, The Netherlands, 265 p.

Gelhar, L.W., Welty, C. & Rehfeldt, K.R. (1992) Critical review of data on field-scale dispersion in aquifers. *Water Resources Research*, 28(7), 1955–1974.

Hardy, R.L. (1971) Multiquadric equations of topography and other irregular surfaces. *Journal of Geophysical Research*, 76, 905–1915.

Holzbecher, E. (1998) *Modeling Density-driven Flow in Porous Media. Principles, Numerics, Software*. Berlin, Heidelberg, Springer Verlag.

Huisman, P., Cramer, W., Ee van, G., Hooghart, J.C., Salz, H. & Zuidema, F.C. (1998) *Water in The Netherlands*. Van de Rhee, Rotterdam, Netherlands Hydrological Society.

IPCC, (2000) Emissions Scenarios (Special Report), In: Nakicenovic, N. & Swart, R. (eds.). Cambridge, UK, Cambridge University Press, 570 pp.

IPCC, (2001) *Climate Change 2001: The Scientific Basis. Contribution of Working Group I to the Third Assessment Report of the Intergovernmental Panel on Climate Change* [Houghton, J.T.,Y. Ding, D.J. Griggs, M. Noguer, P.J. van der Linden, X. Dai, K. Maskell, & C.A. Johnson (eds.)]. Cambridge University Press, Cambridge, United Kingdom and New York, NY, USA, 881 pp.

Konikow, L.F., Goode, D.J. & Hornberger, G.Z. (1996) A three-dimensional method-of-characteristics solute-transport model (MOC3D): *United States Geological Survey Water-Resources Investigations Report 96-4267*, Reston, Virginia, 87 p.

Leenen, J.D. (1980) Results of a geo-electrical survey to the depth of the fresh water–salt water interface in the polder Groot Mijdrecht. In: Committee for hydrological research TNO (eds.) *Research on Possible Changes in the Distribution of Saline Seepage in the Netherlands*. The Hague, The Netherlands, TNO, Proc. and Inform. No. 26: 124–139.

McDonald, M.G. & Harbaugh, A.W. (1988) A modular three-dimensional finite-difference groundwater flow model. In: U.S.G.S. Techniques of Water-Resources Investigations, Book 6, Chapter A1, Reston, Virginia, 588 p.

Meinardi, C.R. (1991) The origin of brackish groundwater in the lower parts of The Netherlands. In: De Breuck, W. (ed.) *Hydrogeology of Salt Water Intrusion: A Selection of SWIM Papers*. Hannover, Germany, Verlag Heinz Heise.

Nield, D.A. & Bejan, A. (1992) *Convection in a Porous Media*. New York, Springer-Verlag.

Oude Essink, G.H.P. (1996) Impact of sea level rise on groundwater flow regimes: A sensitivity analysis for the Netherlands. *PhD Thesis*. Delft University of Technology, Delft, The Netherlands.

Oude Essink, G.H.P. (1998) MOC3D adapted to simulate 3D density-dependent groundwater flow. In: Poeter, E. (Ed), *Proceedings of the MODFLOW'98 Conference*. Golden, Colorado, USA, October 4–8, 1998, vol. I, Golden, Colorado, pp. 291–303.

Oude Essink, G.H.P. (2001) Salt water intrusion in a three-dimensional groundwater system in The Netherlands: A numerical study. *Transport in Porous Media*, 43(1), 137–158.

Oude Essink, G.H.P., van Baaren, E.S., de Louw, P.G.B. (2010) Effects of climate change on coastal groundwater systems: A modeling study in the Netherlands. *Water Resources Research*.: doi:10.1029/2009WR008719. http://www.agu.org/pubs/crossref/2010/2009WR008719.shtml

Peaceman, D.W. (1977) Fundamentals of numerical reservoir simulation (*Developments in* Petroleum Science volume 6), Amsterdam, The Netherlands, Elsevier Scientific Publishing Company, Amsterdam-Oxford-New York, ISBN 0-444-41578-5, 176 p.

Post, V.E.A. (2003) Groundwater salinization processes in the coastal area of the Netherlands due to transgressions during the Holocene. *PhD Thesis*. VU University, Amsterdam, The Netherlands.

Post, V.E.A., Van der Plicht, H. & Meijer, H.A.J. (2003) The origin of brackish and saline groundwater in the coastal area of the Netherlands. *Geologie en Mijnbouw*, 82(2), 133–147.

Pulles, J.W. (1985) Een beleidsanalyse van de waterhuishouding van Nederland [*Policy Analysis for the Water Banagement in the Netherlands (PAWN)*]. Den Haag, The Netherlands, Rijkswaterstaat, Hoofddirectie van de Waterstaat (in Dutch).

Schultz, E. (1992) Water management of the drained lakes in the Netherlands. . *PhD Thesis*. Delft University of Technology, Delft, The Netherlands (in Dutch).

Staalduinen van, C.J., Adrichem Boogaert van, H.A., Bless, M.J.M., Doppert, J.W.C., Harsveldt, H.M., Montfrans van, H.M., Oele, E., Wermuth & R.A., Zagwijn, W.H. (1979) The geology of The Netherlands. *Mededelingen Rijks Geologische Dienst*, 31, 9–49.

Stuyfzand, P.J. (1993) Hydrochemistry and hydrology of the coastal dune area of the Western Netherlands. *PhD Thesis*. VU University, Amsterdam, The Netherlands.

Stuyfzand, P.J. (1996) Salinization of drinking water in the Netherlands: Anamnesis, diagnosis and remediation. In: *Proceedings of the 14th Salt Water Intrusion Meeting (SWIM), June 17–21, 1996, Sveriges Geologiska Undersökning*, Malmö, Sweden, pp. 168–177.

UNESCO. (2008) *Groundwater Resources Assessment under the Pressures of Humanity and Climate Change (GRAPHIC): A Framework Document*. GRAPHIC Series Number: 2. Paris, France, United Nations Educational, Scientific, and Cultural Organization (UNESCO).

Van de Ven, G.P. (1993) *Man-Made Lowlands; History of Water Management and Land Reclamation in the Netherlands*. Utrecht, The Netherlands, Uitgeverij Matrijs.

Van Rees Vellinga, E., Toussaint, C.G. & Wit, K.E. (1981) Water quality and hydrology in a coastal region of the Netherlands. *Journal of Hydrology*, 50, 105–127.

Volker, A. & Van der Molen, W.H. (1991) The influence of groundwater currents on diffusion processes in a lake bottom: An old report reviewed. *Journal of Hydrology*, 126, 159–169.

Wesseling, J. (1980) Saline seepage in the Netherlands, occurrence and magnitude. In: Committee for hydrological research TNO (eds.). *Research on Possible Changes in the Distribution of Saline Seepage in the Netherlands*. The Hague, The Netherlands, TNO, Proc. and Inform. No. 26: pp. 17–33.

Zagwijn, W.H. (1989) The Netherlands during the tertiary and the quaternary: A case history of coastal lowland evolution. *Geologie en Mijnbouw*, 68, 107–120.

CHAPTER 14

Climate change impacts on valley-bottom aquifers in mountain regions: case studies from British Columbia, Canada

Diana M. Allen

ABSTRACT

Mountain watersheds exhibit geological, landscape, climate, and other characteristics that are distinctive from other types of watersheds/basins. As such, water management in regional mountain aquifer systems, particularly in light of future projected climate change, requires an understanding of hydrogeological processes, including the complex linkages between surface and subsurface hydrology. This paper provides a brief overview of hydrogeological processes in temperate mountain regions, focusing primarily on sources of recharge to valley-bottom (or basin-bottom) aquifers. Recharge modelling in two case study areas in British Columbia (BC), Canada indicates that diffuse recharge to valley-bottom aquifers in these settings is, in general, very low (<12% of annual precipitation in Oliver and <24% in Grand Forks). Under current climate conditions, all recharge models show peak diffuse recharge in March and April coincident with snowmelt. Recharge declines through the summer and fall months (July through October) and attains the lowest values in January and February (when the ground is frozen and covered by snow). Future climate change in southern BC, estimated from multiple downscaled global climate models (GCMs), suggests minor increases in annual recharge due the shift (earlier) of peak recharge from increased temperature affecting earlier valley-bottom snowmelt and ground thaw. Summer recharge, currently augmented by irrigation return flow, may become more so in future due to a longer growing season assuming irrigation efficiency does not improve and crop types remain the same. Valley-bottom aquifers can receive indirect recharge through interaction with rivers. In southern BC, climate change projections suggest that peak discharge will occur earlier, winter discharge may increase, and the summer low flow period will be longer and with lower baseflow. Groundwater levels in valley-bottom aquifers will respond accordingly.

14.1 INTRODUCTION

14.1.1 Purpose and scope

Mountain watersheds or basins are unique high-relief environments that are important sources of water for local ecosystems and human population, as well as for connected downstream basins (Viviroli et al. 2003). As suggested by Wilson and Guan (2004), mountain watersheds exhibit geological, landscape, climate, and other characteristics that are distinctive from other types of watersheds/basins.

The purpose of this paper is to first provide an overview of hydrogeological processes in temperate mountain regions as a basis for understanding how climate change may influence groundwater systems within them. Case study examples of two valley-bottom aquifer systems in southern British Columbia (BC), Canada highlight the complex inter-actions that need to be considered for climate change impacts and adaptation assessments.

14.1.2   Study area description: valley-bottom aquifers in mountain regions

Typically, mountain regions are characterized by moderate to high relief, steep slopes, deeply incised valleys, variable bedrock and alluvium geological conditions, extreme spatial and temporal climate variability, complex and dynamic surface water drainage, and variable population concentrated in the valley bottom (relying on both surface water and groundwater resources). Commonalities of many of these characteristics are evident within and between mountainous watersheds worldwide.

British Columbia (BC), Canada, the focus region for this paper, is characterized by several mountain ranges with elevations above 3000 metres, vast plateau areas, and deep valleys. Geology and geomorphology are highly variable; mountains expose volcanic, plu-tonic, and sedimentary bedrock (Holland 1976; Back et al. 1988). Evidence of Pleistocene glacial, fluvioglacial and glaciolacustrine activity can be found almost anywhere in the region (e.g., Clague 1977). Considerable Holocene alluvial deposits exist in the valleys of most large rivers (e.g., Clague 1977). The region is also one of the most hydroclimati-cally complex regions in North America owing to its complex relief and proximity to the Pacific Ocean, which give rise to strong precipitation and temperature gradients (Wade et al. 2001; Moore et al. 2008). Precipitation generally ranges from greater that 4000mm/ annum (a) in the west to less than 200mm/a, in a series of rain belts and rain shadows (Moore et al. 2008). During the winter months, frontal weather systems formed in the Pacific Ocean bring moisture into the region. These regional precipitation events may have a low precipitation rate and large spatial and temporal scales (10–300 km, 0.5–5 days). Except at low elevations along the coast, winter precipitation falls predominantly as snow. Local precipitation events result largely from convective systems, which may have high precipitation rates, but limited spatial and temporal scales (0.5–2 km, 5–100 minutes), and which derive much of their source water from local evapotranspiration. Convective events are common during summer; therefore, summers are relatively dry throughout the region and considerable climatic moisture deficits develop, particularly in parts of the southern interior of BC where summer temperatures are highest.

In mountain regions, two groundwater systems are often present: 1) a mountain bed-rock aquifer system and 2) a valley-bottom unconsolidated aquifer system. These two systems interact at the interface between the mountain bedrock and the valley-bottom sediments both at the surface and in the subsurface. Figure 14.1 shows a conceptual model of groundwater flow a mountain catchment – valley-bottom aquifer system. The case studies described in this paper are valley-bottom aquifers, typical of those found in temperate mountain regions. Thus, before giving details on the two case study areas, the various processes that control valley-bottom groundwater systems are discussed.

The valley-bottom (or basin-bottom) aquifers can be recharged by several mecha-nisms (Fig. 14.1):

  i) seepage from mountain streams and rivers (ephemeral and perennial);
 ii) groundwater flow from the adjacent mountain block;
iii) direct, diffuse recharge to the valley bottom;

iv) groundwater flow from up-gradient valley-bottom aquifers; and

v) interaction with rivers sourced from within (local) or outside the local watershed (allogenic).

i) Streams generated within the mountain block (either perennial or ephemeral) typically lose water as they traverse the mountain front zone. This process is inferred to be the result of the hydraulic conductivity contrast at the mountain front between low permeability bedrock and higher permeability alluvial sediments in the valley bottom (Wilson and Guan 2004). Stream loss may occur as "focused recharge" for major tributaries, as well as "diffuse recharge" for small (e.g., disappearing or ephemeral) streams or surface runoff across the mountain front (Wilson and Guan 2004).

ii) Groundwater flow across the mountain front (termed mountain block recharge; MBR) is thought to provide a fairly stable supply of groundwater to valley-bottom aquifers (Neilson-Welch and Allen submitted). Feth (1964) suggested that this mechanism of groundwater recharge may be important for valley-bottom aquifer replenishment, particularly in (semi) arid areas where there are annual moisture deficits in the valley bottom. It is important to note that this deep regional groundwater flow may not be uniform across the mountain front, but may be influenced by geological heterogeneity (e.g., fault and fracture zones). Thus, deep groundwater discharge to valley-bottom alluvial aquifers at the mountain front can produce diffuse (through the bedrock massive) or focused (through fault/fracture zones) mountain block recharge (Wilson and Guan 2004). MBR can also discharge directly to valley-bottom rivers and lakes, if the valley-bottom sediments are thin to absent.

iii) Diffuse recharge is thought to be generally quite low in valley-bottom aquifers in temperate mountain regions, particularly during the summer months (Scibek and Allen 2006; Liggett and Allen 2010; Smerdon et al. 2009; Toews and Allen 2009a).

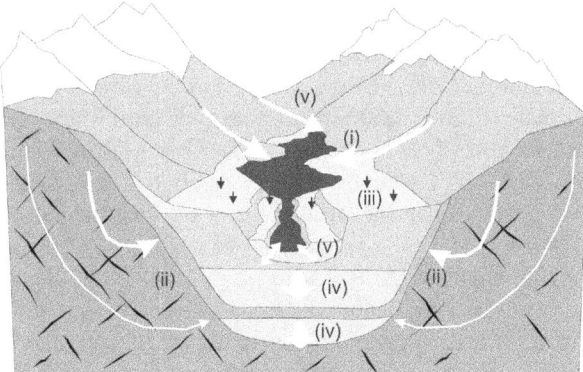

Figure 14.1. Conceptual model of groundwater flow systems in a mountain to valley-bottom system. Recharge to valley-bottom aquifers occurs as (i) seepage from mountain streams and rivers; (ii) from mountain block recharge (MBR); (iii) direct, diffuse recharge to the valley bottom; (iv) from groundwater flow from up-gradient valley-bottom aquifers; and (v) through interaction with rivers sourced from within or outside the local watershed (allogenic systems). Arrows scaled in width according to relative magnitude of contribution to valley-bottom recharge.

Assessments concerning the future impacts of climate change on valley bottom aquifers have largely been limited to modelling studies. The case studies described in this paper, stemming from the listed publications above, focus primarily on diffuse recharge, attempting to constrain through modelling the magnitude and timing of diffuse recharge, and also how recharge might be expected to change under future climate conditions. One additional consideration is that valley-bottom land use in many mountain areas is often agricultural. Therefore, while natural recharge may be very limited during the summer months when evapotranspiration (ET) is high, irrigation may augment the natural summer recharge through return flow.

iv) Within the valley bottom, particularly at the basin scale, there may be a series of interconnected aquifers of alluvial or glacial origin. These aquifers are fed not only locally through the various processes described above, but also by groundwater originating from up-gradient valley-bottom aquifers. Estimation of groundwater contributions from up-gradient aquifers requires detailed knowledge of the entire basin water balance, which can only reasonably be attained by modelling the entire basin and all of the individual aquifers.

v) Finally, many mountain valleys contain a main stem lake or river system. Recharge to local aquifers can derive from interaction with rivers or lakes in the valley bottom (Allen et al. 2010a). As well, these rivers often receive water from various catchments that may be situated at high elevation in remote areas of the watershed – allogenic source. Due to the generally high permeability of fluvial sediments in the valley bottom, the river is strongly hydraulically connected to the aquifer. This type of dynamic coupling between valley-bottom aquifers and the rivers that flow through them is pervasive in mountain regions. Understanding how the aquifer will respond to climate change must, therefore, consider changes to the hydrology at the larger watershed scale.

The two specific case study areas are situated in southern British Columbia, Canada (Fig. 14.2). These include Okanagan Basin (semi-arid to arid) and the Grand Forks aquifer (semi-arid). Okanagan Basin (8,000 km²) is characterised by a long, north-south trending valley (~200 km) that extends into Washington State, USA. Upland plateaus and mountains surround this long, narrow valley. Surface elevations in Okanagan Basin range from approximately 270 to 2,300 metres above sea level (masl). Much of the populated valley bottom area lies below 500 masl, which is less than 7 km wide. Okanagan Lake, also located in the valley bottom, stretches nearly half the length of the basin and averages about 3 km wide. Okanagan Basin has a dry continental climate with mild winters and hot summers. The valley bottom is semi-arid to arid, with a climate gradient trending along the length of the valley from north to south. Geologically, the bedrock forming the mountains surrounding the basin was carved into a deep valley during glaciation. The valley fill is comprised of unconsolidated Quaternary (including recent Holocene) sediments of variable composition, forming a network of interconnected aquifers. Increases in population, tourism, and agriculture (primarily orchards and vineyards) have led to an increased demand for water throughout the basin. Presently, much of the water for domestic, agricultural and industrial use is supplied from surface water, including streams and Okanagan Lake. Surface water sources in the basin are almost fully allocated, and many communities are turning to groundwater as a primary water supply source or to supplement existing supplies.

The second study area, the Grand Forks aquifer (34 km^2), is contained within the mountainous valley of the Kettle River, which drains an area of ~9,800 km^2. The Grand Forks valley is approximately 12 km long and about 4 km wide. This small valley is characteristic of many small valleys in BC. The Kettle River meanders through the valley bottom, which is comprised near surface of glacial outwash sediments. The climate is semi-arid and most rainfall occurs in summer months during convective activity. In the winter, much of the precipitation at high elevation is as snow, although the climate station at valley bottom records less snowfall. Groundwater is used extensively for irrigation and domestic use.

14.1.3   Methodology

The two case studies were undertaken to develop and implement a groundwater modelling approach for forecasting potential impacts of climate change on groundwater. The primary purpose of the modelling was to explore some of the recharge mechanisms

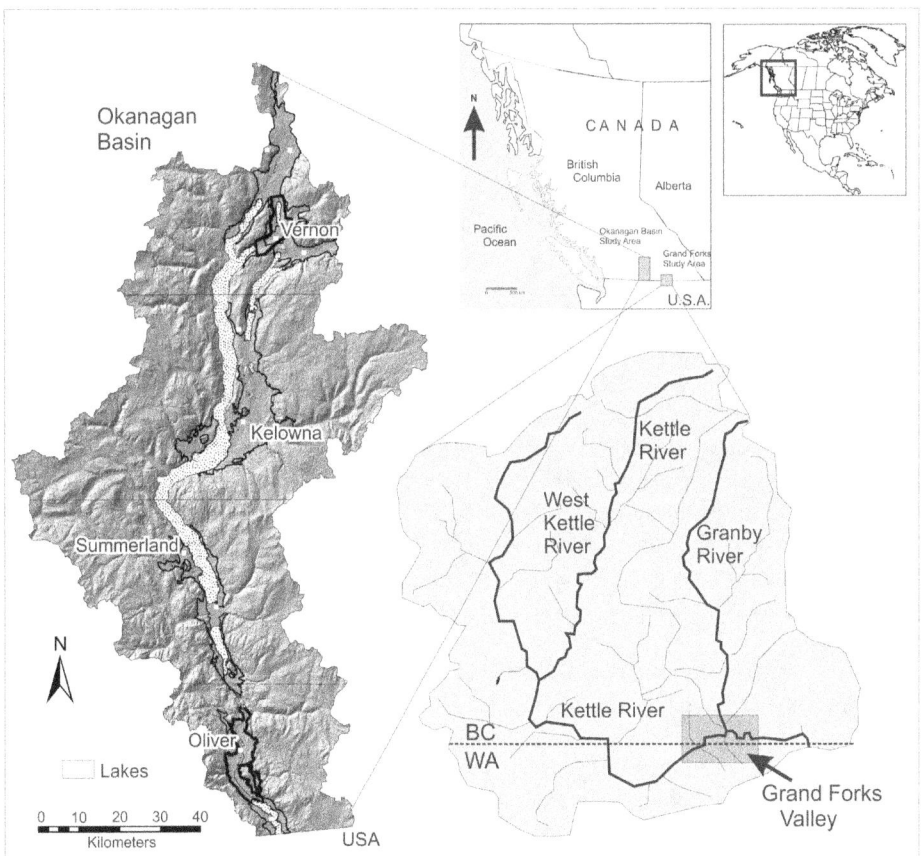

Figure 14.2.   Location maps for the two case study areas in south-central British Columbia, Canada: Okanagan Basin and the Grand Forks valley within the Kettle River watershed.

described above and how the magnitude and timing of recharge may change under future climate conditions.

The methodology used for each case study was generally consistent, although choice(s) of global climate model (GCM), downscaling method(s), and treatment of irrigation return flow varied between the study sites (as described in Results). To generate a weather time series for recharge simulations, GCM data were extracted from one or more models. The data were then downscaled to obtain shift factors that were then used to generate a weather series in a stochastic weather generator. These time series were then used to drive a one-dimensional recharge model. Recharge simulations were carried out for a series of 1-D soil columns, representative of soil and aquifer conditions in each aquifer. Spatially-distributed recharge was mapped using a geographic information system (GIS) from the results for each unique soil column, and then input to a groundwater flow model. The groundwater flow models for each area were developed using traditional numerical modelling techniques for aquifer systems, and include surface water boundary conditions, detailed subsurface lithology, and recharge boundary conditions for current and future climate periods. For the Grand Forks aquifer, which is in hydraulic connection with the Kettle River, shifts in stream discharge under future climate conditions were simulated. Recharge was also modified in the two case studies to account for irrigation.

14.1.4   Relevance for GRAPHIC

Mountains play a critical role in the hydrologic cycle, capturing precipitation by orographic effects, storing water in snowpack and mountain aquifers, and initiating transport of water from the surface to local and regional aquifers, and possibly even deeper to the upper crust of the Earth. Water management in such regional mountain aquifer systems, particularly in light of future projected climate change, requires an understanding of the hydrogeological processes in mountain regions, including the complex linkages between surface and subsurface hydrology.

The two case studies described in this paper attempt to illustrate the complexities of groundwater systems in mountain regions, specifically by exploring recharge mechanisms. As such, the case studies address two GRAPHIC themes: groundwater recharge, and modelling and simulation. While diffuse groundwater recharge is a focus, contribution to recharge from surface water is also addressed, as well as artificial recharge through irrigation. The sensitivity of recharge to changing climate conditions is examined by considering a range of GCMs. The use of numerical modelling in these studies permits assessment of climate change impacts on a range of aquifer processes in mountain regions.

The long term goal of the GRAPHIC Programme is to improve understanding of how groundwater interacts with the global water cycle, how it supports ecosystems and humankind and, in turn, responds to complex and coupled pressures of human activity and climate change. The case studies presented in this paper advance our understanding of groundwater systems in mountain regions and will allow for better planning through management of the resource. To date, these case studies have helped to inform water management at both a provincial and local level in British Columbia. It is hoped that this work will ultimately help groundwater management planning in other mountain regions through knowledge shared within the GRAPHIC community.

14.2. RESULTS AND DISCUSSION

The varied responses of aquifer systems in mountain settings, as described above, point to the complexity of the climate-groundwater-surface water interconnections. Results from the two case studies serve to illustrate these complex interactions, the magnitude of diffuse valley-bottom recharge historically and how it may change under changing climate conditions, as well as how changes in surface water hydrology may directly impact groundwater systems.

14.2.1 Okanagan Basin

Liggett and Allen (2010) simulated historical spatially- and temporally-varying recharge to the entire Okanagan valley bottom (regional scale; Fig. 14.2) using the software HELP (version 3.80D; Berger 2004). Simulations were carried out for one dimensional soil column models representative of different combinations of soil permeability, soil hydraulic conductivity, and depth to water table. Recharge results were then mapped throughout the valley bottom. Averaged spatially, the mean annual recharge in Okanagan valley is estimated at 65 mm/a, but varies spatially (from 0 to 186 mm/a) due to the strong precipitation gradient along the valley and differences in the soil properties throughout the valley. At a local scale, in the south Okanagan near Oliver (Fig. 14.2), Toews and Allen (2009a) obtained similar estimates of recharge using HELP; average annual recharge varied from 0 to 199 mm/a with a spatial mean of 42 mm/a, representing 12% of the annual precipitation. In the north Okanagan near Vernon (local scale; Fig. 14.2), Smerdon et al. (2009) estimated recharge using the MIKE SHE code (DHI, 2007). Average annual recharge varied from −8 mm/a (i.e., upwards capillary flux of water, no recharge) to 135 mm/a, with a spatial mean of 6 mm/a, representing 1.3% of the annual precipitation. Minor differences in the recharge simulated at the local and regional scales are attributed to the choice of software (Liggett and Allen 2010); HELP is a water-balance based code and MIKE SHE solves Richards' equation. In this semi-arid region, HELP tended to overestimate recharge (by underestimating ET) compared to MIKE SHE – a problem noted by other researchers (e.g., Scanlon et al. 2002). Nevertheless, all models show peak recharge in March and April coincident with snowmelt. Recharge declines through the summer months (July to October) and attains the lowest values in January and February (when the ground is frozen and covered by snow). Overall, mean annual recharge is less than 12% of annual precipitation (and likely closer to 0%) throughout much of the Okanagan valley bottom, suggesting that diffuse recharge to valley-bottom aquifers in these types of settings is, in general, very low.

Toews and Allen (2009a) examined the sensitivity of recharge to the valley-bottom aquifer in the south Okanagan near Oliver (Fig. 14.2). Three GCMs were considered (CGCM1 GHG +A1, CGCM3.1 A2, and HadCM3 A2). The authors considered irrigation return flow estimated through the HELP recharge model using seasonal crop water demand based on growing degree days, ET and precipitation. Thus, 'effective' precipitation was used to drive the recharge model for both current and future climate periods. For future periods, shifts in growing degree days and overall climate were simulated. Increases in demand due to growth in agriculture were not considered – only those related directly to climate change were considered in order to separate out such effects. No changes in irrigation practices were considered, such as improvement in irrigation systems or changes in crop type. Changes to recharge in future time periods for each GCM resulted in modest increases of annual recharge, with the peak recharge shifting from March to February. For the summer months, lower recharge rates and higher potential ET rates were predicted by all three models. All models showed that the potential growing season will expand between 3 and 4 weeks due to increases in temperature. However, the magnitude of the change varied considerably between models. CGCM3.1 had the largest increases of recharge rates, CGCM1 had very minor increases, and HadCM3 was relatively stable (as indicated by the near-zero changes between climate periods). The significant difference between these three models indicates that prediction of future recharge is highly dependent on the model selected (Fig. 14.3). The minor increase of annual recharge in future predicted climate periods is due the shift of peak recharge from increased temperature affecting earlier valley-bottom snowmelt and ground thaw. Similar sensitivity of groundwater recharge to choice of GCM was observed by Allen et al. (2010b) for a coastal aquifer in BC. In that study, recharge was simulated using climate data series derived from the TreeGen downscaling model (Cannon 2008) for three future time periods: 2020s, 2050s, and 2080s for each of four GCMs (CGCM3.1, ECHAM5, PCM1, and CM2.1). By the 2080s, the range of model predictions spanned -10.5% to $+23.2\%$ relative to historical recharge. This variability in recharge predictions suggested that the seasonal performance of the downscaling tool is important and that a range of GCMs should be considered for water management planning.

Using the recharge predictions above, Toews and Allen (2009b) simulated the impacts of future climate change on groundwater levels in the Oliver aquifer. The total amount of recharge, including irrigation return flow, was applied to the top of a groundwater flow

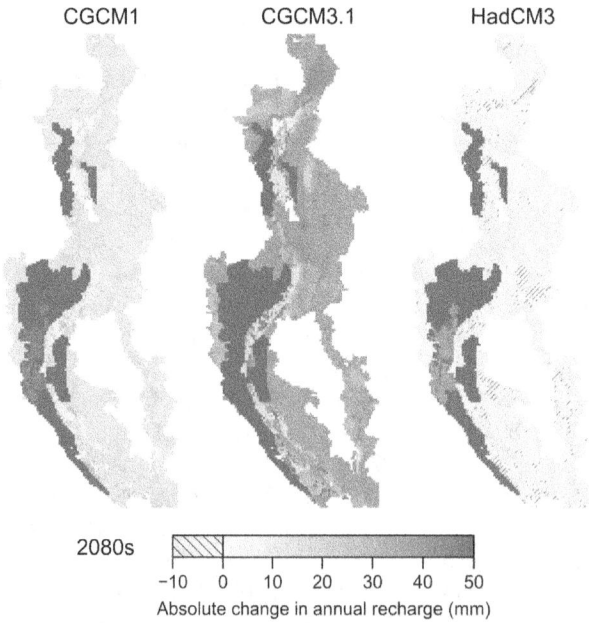

Figure 14.3.   Absolute changes to spatial recharge rates for the 2080s for three GCMs (CGCM1, CGCM3.1 and HadCM3) Modified from Toews and Allen (2009a).

model implemented in GMS (version 6.0; EMRL 2005). In future time periods (the 2050s and 2080s), the most noticeable change was the increased contribution of recharge to the annual water budget, estimated at 1.2% (2050s) and 1.4% (2080s) of the total annual budget relative to the current conditions. This increase is related primarily to increased irrigation return flow resulting from greater summer irrigation due to warmer temperatures and a longer growing season. The increased recharge and irrigation return flow result in higher water tables under future climate conditions, particularly in the irrigation districts. Median value increases in groundwater level of up to 0.7 m by the 2080s were estimated from the groundwater model.

14.2.2   Grand Forks

Scibek and Allen (2006) simulated changes to diffuse valley-bottom recharge in the Grand Forks valley (Fig. 14.2). Sixty-four unique recharge zones were defined considering combinations of soil permeability, soil hydraulic conductivity, and depth to water table. Recharge was simulated for each recharge zone using the HELP recharge model (Schroeder et al. 1994). Simulation results suggest that mean annual recharge varies considerably across the 64 recharge zones, ranging from slightly less than 30 mm/a to over 120 mm/a (6% and 24% of mean annual precipitation) (Scibek and Allen 2006). In winter the ground is frozen recharge is low (< 2 mm/month). Most of the recharge is received in late spring and summer through snowmelt and summer rainstorm events, respectively (up to ~12 mm/month). The autumn season has moderate recharge (~8 mm/month).

The climate data series used to drive the HELP model was altered for future climate conditions based on downscaled output from CGCM1. Recharge in the 2010–2039 period is predicted to increase by 2 to 7% from historical mean annual recharge, while recharge in the 2040–2069 period is predicted to increase by 11 to 26% (Fig. 14.4). Although not shown, changes in seasonal recharge follow a pattern consistent with what is predicted annually, with April to June showing the greatest increases in monthly recharge. To consider the effect of irrigation return flow, the recharge zones were modified by superimposing estimated irrigation return flow to the aquifer. Generalized estimates of return

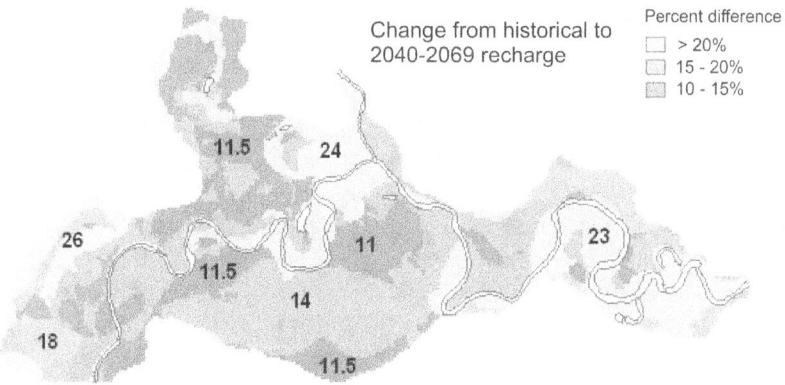

Figure 14.4.   Percent increase in recharge (range 11 to 26%) to the Grand Forks aquifer by the 2040–2069 climate period relative to historical mean annual recharge. Modified from Scibek and Allen (2006).

flow were obtained through consultation with experts in irrigation practices (roughly 25% of the amount of irrigation for the types of crops present). A constant irrigation return flow was assumed for irrigated fields for each month (June to August only) in the historical and future climate periods. Recharge maps for each climate time period were applied to a transient groundwater flow model implemented in Visual MODFLOW (Waterloo Hydrogeologic Inc. 2004) to quantify the effect of changes in recharge on groundwater levels in the aquifer (discussed below). In the MODFLOW model, additional recharge zones were created to represent the modified recharge after addition of return flow from irrigation; pumping was also included.

As discussed above, many valley-bottom aquifers are in close hydraulic connection to surface water bodies. As such, these aquifers are indirectly recharged by surface water and have been observed to respond in a bi-directional fashion – gaining water from the river during the snowmelt season when river stage is high, and losing water to the river when river stage is low. Within the Grand Forks valley, the Kettle River is a meandering gravel-bed river incised into glacial outwash sediments. Aquifer water levels in the surrounding unconfined alluvial aquifer are highly sensitive to water levels in the Kettle River (Allen et al. 2004). At peak flow, during the spring snowmelt period (freshet), river water recharges the aquifer and moves laterally away from the channel, causing groundwater levels to rise over a broad area (Scibek et al. 2007). Within a relatively short period following the peak discharge, when river levels begin to fall, the groundwater direction is reversed and groundwater contributes to baseflow. Figure 14.5 shows a cross plot of groundwater level measured in a provincial observation well (BC Obs. 217) situated approximately 400 m from the Kettle River against the discharge of the Kettle River (logarithmic scale). The cyclic nature of the plot shows that the groundwater system lags behind the river discharge; this type of plot is common for aquifer systems that are in strong hydraulic connection to rivers (Allen et al. 2010a).

In the Kettle River drainage area (Fig. 14.2), the snowpack increases over the winter until early April, and melts between April and the end of June, with the end date of the snowmelt season varying from mid-May to mid-July. The hydrological response of the Kettle River is extremely sensitive to seasonal variations in climate. During years with unusually warm winters, the system shifts from a snowmelt-dominated regime to a regime where there is an increasing number of days of higher flows due to rain, but with a decreasing number of days of high flow due to snowmelt. The predicted warming trends in global, and also regional, climate are expected impact the snowfall amounts and the duration of winter season, shifting the hydrological regime, and potentially affecting hydrologically linked regional aquifers. For example, in a recent study of the Rocky Mountain headwaters regions in eastern BC, Rood et al. (2008) investigated historical changes in seasonal patterns of streamflow by comparing mean monthly flows and analyzing cumulative hydrographs over the periods of record of about a century. They tested predictions of change due to winter and spring warming that would increase the proportion of rain versus snow, and alter snow accumulation and melt. Based on the analysis of records from 14 free-flowing, snowmelt-dominated rivers that drain relatively pristine parks and protected areas they observed: 1) increased winter flows (especially March), 2) earlier spring runoff, 3) earlier peak flows, and most substantially, 4) considerably reduced summer and early autumn flows (July–October). Streams in south-central BC have also shown similar changes in temporal pattern over the past two decades: earlier spring runoff (freshet), lower late summer to early fall flows, and higher early winter flows (Leith and Whitfield

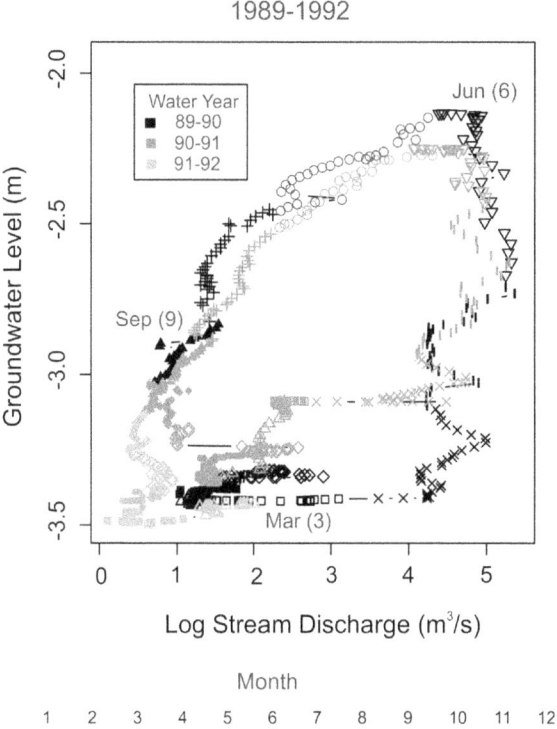

Figure 14.5. Cross plot for water years 1989–1992. Shown is groundwater level in BC Observation Well 217 situated ~400 m from the Kettle River in the Grand Forks valley, and the logarithm of discharge in the Kettle River. The cycling response indicates close hydraulic connection between aquifer and river.

1998). In the Cascade Mountains of the Pacific Northwest of the USA (south of BC), temperature trends over the past half century indicate warmer winters (e.g., Regonda et al. 2005), and declining snowpacks over the past 50 years (Mote et al. 2005). In accordance with earlier spring snowmelt, streamflow timing is earlier by one to four weeks, compared to the middle of the twentieth century (Stewart et al. 2005).

To capture the hydraulic connection between the Grand Forks aquifer and the Kettle River, Scibek et al. (2007) modelled changes in streamflow for future climate periods and used these as boundary conditions for a three-dimensional groundwater flow model. The hydrological changes were integrated in the model with the recharge shifts determined by Scibek and Allen (2006). In order to model the interaction between groundwater and surface water in the valley, stage elevations were required as a function of time for each river node in the groundwater flow model for each climate period (see Scibek et al. (2007) for the detailed approach used to obtain stage values). Hydrological models suggest that in the future climate periods, the hydrograph peak is shifted to an earlier date, although the peak flow remains the same (similar to what has been observed in recent decades) (Fig. 14.6). There is also a measurable increase in winter discharge in the

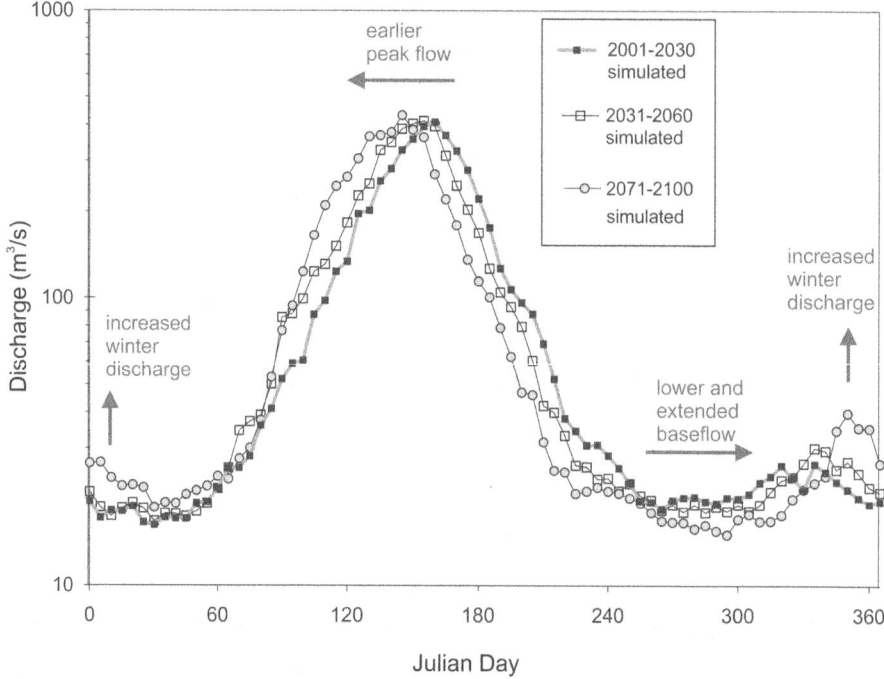

Figure 14.6. Predicted discharge in the Kettle River modelled using statistical downscaling, compared to observed discharge in last 30 years (data from Environment Canada).

future climate scenarios, most likely caused by an increase in rain and snowmelt volumes during the winter under warmer climate scenarios. A longer and lower baseflow is also anticipated (Fig. 14.6).

The simulated shift in the Kettle River hydrograph results in a shift in the timing of peak and low groundwater levels in the aquifer, particularly in the floodplain (Fig. 14.7). By the 2050s, groundwater levels during the winter snowmelt period (freshet) may be as much as 0.5 m higher compared to historical groundwater levels (shown for April 10 in Fig. 14.7). By the 2050s, groundwater levels during the summer low flow period (shown for June 30 in Fig. 14.5) may be up to 0.5 m lower compared to historical groundwater levels. These results highlight the importance of considering not only changes to groundwater recharge, but also shifts in streamflow, particularly in aquifer systems closely connected to surface water. In fact, the groundwater model simulations suggest that variations in recharge to the aquifer under climate change conditions (Scibek and Allen 2006) have a much smaller impact on the groundwater system than changes in the river stage elevation of the Kettle River (Scibek et al. 2007).

14.3   POLICY RECOMMENDATIONS

Mountain areas have complex, coupled hydrogeological and hydrological processes owing to the diverse climatology, variable terrain and complex geology. From a groundwater assessment perspective, it is unreasonable to assume that each small, valley-bottom

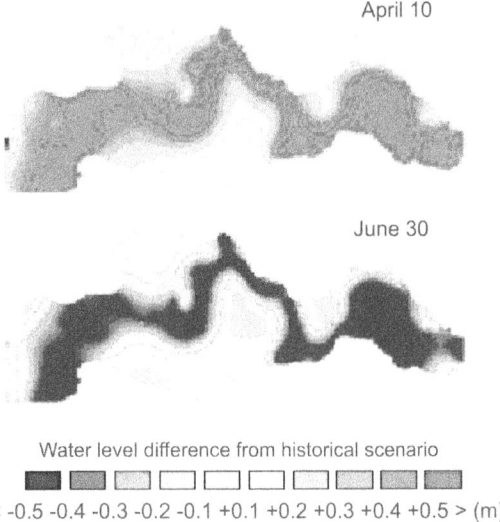

April 10

June 30

Water level difference from historical scenario

< -0.5 -0.4 -0.3 -0.2 -0.1 +0.1 +0.2 +0.3 +0.4 +0.5 > (m)

Figure 14.7.   Groundwater level changes relative to historical groundwater levels in the Grand Forks aquifer for two simulated dates for the 2040–2069 period (2050s). During winter snowmelt (freshet), here shown for April 10, groundwater levels are up to 0.5 m higher in the aquifer, particularly in the floodplain, due to earlier peak flow of the Kettle River on that day. During the summer low flow period (here shown for June 30), groundwater levels are up to 0.5 m lower due to decreased simulated baseflow in the Kettle River on that day (modified from Scibek et al. 2007).

aquifer system can be modelled to the degree as those discussed as case studies in this paper, given the vast number of such small systems in mountain regions and the degree of complexity of the modelling approaches highlighted in this paper. Rather, a broader approach for assessing climate change impacts on mountain groundwater systems is needed. Such a framework should consider two aspects: 1) the current and future hydro-climatic regime; and 2) the mechanisms for groundwater recharge (i.e. diffuse recharge, through connection with surface water, artificial recharge, etc.). Based on this conceptual framework, the likely response of aquifer systems could be evaluated in a preliminary way, with additional prioritized studies targeted to heavily populated areas or agricultural areas where the demand for water is high.

For aquifers where the dominant recharge mechanism is diffuse recharge, including aquifers forming upland areas of mountains, changes to the form of precipitation (snow versus rain), the timing of snowmelt, and evapotranspiration are critical for establishing future seasonal water balances. In aquifers strongly connected to surface water, changes to the hydrology originating from high elevation, remote areas of the watershed need to be considered, and adequate models used to simulate the coupled response of the system. While some attempts have been made to simulate the behaviour of such systems (e.g. Grand Forks; Scibek et al. 2007), more research is needed, particularly using watershed scale integrated groundwater-surface water models. Given the contributions of groundwater to mountain streamflow and mountain block recharge, and the strong interconnection between valley-bottom aquifers and the rivers flowing through them, it seems reasonable that integrated (or coupled) surface water-groundwater models should be used for simulating these complex hydrological processes. However, most hydrological models that are commonly employed to study watershed processes in mountain regions (e.g., Merritt et al. 2006) do not include a groundwater component. Rather groundwater is treated as a linear reservoir, with water exchange to storage adjusted to account for the observed surface water response. Similarly, most groundwater flow models assume fixed boundary conditions (albeit variable in time) to represent surface water bodies (e.g., the approach used by Scibek et al. 2007). Coupled surface water-groundwater codes, such as MIKE SHE (DHI 2007), PARFLOW (e.g., Kollet and Maxwell 2006) and GSFLOW (Markstrom et al. 2008), are perhaps more suitable for predicting possible consequences of climate change to mountain groundwater systems.

14.4   FUTURE WORK

Future work on the impacts of climate change needs to focus on better integrating groundwater and surface water systems. In addition to studies focusing on valley-bottom recharge, more effort is needed to understand the magnitude of groundwater recharge originating in the mountain block and which ultimately discharges along the mountain front into the valley bottom. Hydrological processes in upland watersheds are poorly understood, particularly with respect to groundwater recharge. How changes in snow-pack and snowmelt timing, and evapotranspiration under a warming climate will affect groundwater recharge is not well known, and given that mountains are the dominant source of recharge to the inhabited valley-bottom areas, it seems reasonable that more research be directed at these topics.

In addition, most of the research conducted to date has largely ignored socio-economic factors that will invariably affect water demand, particularly in heavily-populated valley-bottom areas. In the Okanagan Basin, for example, population is expected to increase significantly by the mid twenty-first century. Likewise, longer growing seasons, potential changes in crop types, greater demand for food, etc. are expected to put added stress on water resources for agriculture. These are among the various factors that have not been adequately incorporated into climate change models of groundwater systems in general.

ACKNOWLEDGEMENTS

The author acknowledges several colleagues and former and current students and post doctoral fellows who contributed to the various research projects described in this paper. In particular, Paul Whitfield and Alex Cannon from Environment Canada; Mike Wei from British Columbia Ministry of Environment (BC MoE); Arelia Werner from the Pacific Climate Impacts Consortium; Jacek Scibek, Mike Toews, Jessica Liggett, Laurie Neilson-Welch, and Brian Smerdon. Funding for this research was provided by the Natural Sciences and Engineering Research Council (NSERC) of Canada, Environment Canada, BC MoE, Natural Resources Canada through the Climate Change Action Fund, and the Canadian Water Network (CWN).

REFERENCES

Allen, D.M., Cannon, A.J., Toews, M.W., et al. (2010a) Variability in simulated recharge using different GCMs. *Water Resources Research*. 46:W00F03. : doi:10.1029/2009WR008932

Allen, D.M., Mackie, D.C. & Wei, M. (2004) Groundwater and climate change: A sensitivity analysis for the Grand Forks aquifer, southern British Columbia, Canada. *Hydrogeology Journal*, 12(3), 270–290.

Allen, D.M., Werner, A. & Whitfield, P.H. (2010b) Groundwater level responses in mountainous terrain: Regime classification, and linkages to climate and streamflow. *Hydrological Processes*. [Online] 24(23), 3392–3412. Available from: doi:10.1002/hyp.7757

Back, H., Rosenshein, J.S. & Seaber, P.R. (1988) Hydrogeology. In: *The Geology of North America*, DNAG Vols. O–2. Boulder, CO, Geological Society of America.

Berger, K. (2004) *The hydrologic evaluation of landfill performance (HELP) model-engineering documentation for HELP 3.80D-enhancements compared to HELP 3.07.* Institute of Soil Science, University of Hamburg, Hamburg, Germany. 9 pp.

Cannon, A.J. (2008) Probabilistic multi-site precipitation downscaling by an expanded Bernoulli-gamma density network. *Journal of Hydrometeorology*, 9(6), 1284–1300.

Clague, J.J. (1977) *Quadra sand: A study of the late Pleistocene geology and geomorphic history of coastal southwest British Columbia.* Geological Survey of Canada, Paper number: 77–17.

DHI. (2007) *MIKE-SHE Software*. Hørsholm, Denmark, DHI Water and Environmental.

EMRL. (2005) *Groundwater Modeling System (GMS)*, Version 6.0. Environmental Modeling Research Laboratory, Brigham Young University, Provo, UT.

Feth, J.H. (1964) Hidden recharge. *Ground Water*, 2(4), 14–17.

Holland, S.S. (1976) *Landforms of British Columbia: A Physiographic Outline*. British Columbia Department of Mines and Petroleum Resources, Bulletin 48. 138 pp.

Kollet, S.J. & Maxwell, R.M. (2006) Integrated surface–groundwater flow modeling: A free-surface overland flow boundary condition in a parallel groundwater flow model. *Advances in Water Resources*, (29)7, 945–958.

Leith, R.M.M. & Whitfield, P.H. (1998) Evidence of climate change effects on the hydrology of streams in South-Central BC. *Canadian Water Resources Journal*, 23(3), 219–230.

Liggett, J.E. & Allen, D.M. (2010) Comparing approaches for modeling spatially-distributed direct recharge in a semi-arid region. *Hydrogeology Journal*. [Online] 18(2), 339–357. Available from: doi:10.1007/s10040-009-0512-5

Markstrom, S.L., Niswonger, R.G., Regan, R.S., Prudic, D.E. & Barlow, P.M. (2008) *GSFLOW-coupled ground-water and surface-water FLOW model based on the integration of the Precipitation-Runoff Modeling System (PRMS) and the Modular Ground-Water Flow Model (MODFLOW-2005).* US Geological Survey Techniques and Methods 6-D1. 240 pp.

Merritt, W.S., Alila, Y., Barton, M., Taylor, B., Cohen, S. & Neilsen, D. (2006) Hydrologic response to scenarios of climate change in sub watersheds of the Okanagan Basin, British Columbia. *Journal of Hydrology*, 326, 79–108.

Moore, R.D., Spittlehouse, D.L., Whitfield, P.H. & Stahl, K. (2008) Chapter 3 – weather and climate. In: Pike, R.G., Redding, T.E., Moore, R.D., Winkler, R.D. & Blandon, K.D. (eds.) *Compendium of forest hydrology and geomorphology in British Columbia*. BC Ministry of Forests and Range Research Branch, Victoria, BC and FORREX Forest Research Extension Partnership, Kamloops, BC. Land Management Handbook. 805 pp. [Online] Available from: http://www.forrex.org/program/water/compendium.asp

Mote, P.W., Hamlet, A.F., Clark, M.P. & Lettenmaier, D.P. (2005) Declining mountain snowpack in western North America. *Bulletin of the American Meteorological Society*, 86, 39–49.

Neilson-Welch, L.A. & Allen, D.M. (In Review) Partitioning of regional groundwater flow to streamflow and mountain block recharge (MBR) in mountainous terrain. *Canadian Water Resources Journal.*

Regonda, S.K., Rajagopalan, B., Clark, M. & Pitlick, J. (2005) Seasonal cycle shifts in hydroclimatology over the western United States. *J Climate*, 18, 372–384.

Rood, S.B., Pan, J., Gill, K.M., Franks, C.G., Samuelson, G.M. & Shephard, A. (2008) Declining summer flows of Rocky Mountain rivers: changing seasonal hydrology and probable impacts on floodplain forests. *Journal of Hydrology*, 349, 397–410.

Scanlon, B.R., Christman, M., Reedy, R.C., Porro, I., Simunek, J. & Flerchinger, G.N. (2002) Intercode comparisons for simulating water balance of surficial sediments in semiarid regions. *Water Resources Research.* [Online] 38, 1323. Available from: doi:10.1029/2001WR001233

Schroeder, P.R., Dozier, T.S., Zappi, P.A., McEnroe, B.M., Sjostrom, J.W. & Peyton, R.L. (1994) The Hydrologic Evaluation of Landfill Performance (HELP) Model: Engineering Documentation for Version 3. EPA/600/R-94/168b. Washington, DC, US EPA.

Scibek, J. & Allen, D.M. (2006) Modeled impacts of predicted climate change on recharge and groundwater levels. *Water Resources Research.* [Online] 42, W11405. Available from: doi:10.1029/2005WR004742

Scibek, J., Allen, D.M., Cannon, A.J. & Whitfield, P.H. (2007) Groundwater–surface water interaction under scenarios of climate change using a high-resolution transient groundwater model. *Journal of Hydrology.* [Online] 333, 165–181. Available from: doi:10.1016/j.jhydrol.2006.08.005

Smerdon, B.D., Allen, D.M., Grasby, S.E., et al. (2009) An approach for predicting groundwater recharge in mountainous watersheds. *Journal of Hydrology*, 365, 156–172.

Stewart, I.T., Cayan, D.R. & Dettinger, M.D. (2005) Changes toward earlier streamflow timing across western North America. *The Journal of Climate*, 18, 1136–1155.

Toews, M.W. & Allen, D.M. (2009a) Evaluating different GCMs for predicting spatial recharge in an irrigated arid region. *Journal of Hydrology*, 374, 265–281.

Toews, M.W. & Allen, D.M. (2009b) Simulated response of groundwater to predicted recharge in a semi-arid region using a scenario of modeled climate change. *Environmental Research Letters*, 4, 035003.

Viviroli, D., Weingartner, R. & Messerli, B. (2003) Assessing the hydrological significance of the world's mountains. *Mountain Research and Development*, 23(1), 32–40.

Wade, N.L., Martin, J. & Whitfield, P.H. (2001) Hydrologic and climatic zonation of Georgia Basin, British Columbia. *Canadian Water Resources Journal*, 26, 43–70.

Waterloo Hydrogeologic Inc. (2004) *Visual MODFLOW Version 3.1.84 Software and Documentation.* Waterloo, ON, Waterloo Hydrogeologic Inc.

Wilson, J.L. & Guan, H. (2004) Mountain-block hydrology and mountain-front recharge. In: Phillips, F.M., Hogan, J. & Scanlon, B. (eds.) *Groundwater Recharge in a Desert Environment: The Southwestern United States.* Washington, DC, American Geophysical Union

CHAPTER 15

Possible effects of climate change on groundwater resources in the central region of Santa Fe Province, Argentina

Ofelia Tujchneider, Marta Paris, Marcela Pérez & Mónica D´Elía

ABSTRACT

To evaluate the possible effects of climate change on groundwater resources in the central region of Santa Fe Province, Argentina, available hydrogeological data were analysed to describe the regional-scale system and quantify present (2011) groundwater availability. Future recharge to the aquifer system was estimated based on climate models for the A2 and B2 emission scenarios, and incorporated into a numerical groundwater flow model to assess groundwater availability for drinking and food production at present and during the years 2081–2090. The results indicate that the regional groundwater systems will experience greater stress due to increasing water demand and decreasing recharge rates. Future research should strive to improve the necessary knowledge of the hydrogeological system, both at local and regional scales, to fully develop an appropriate, adaptive, and dynamic groundwater management model.

15.1 INTRODUCTION

Groundwater has a fundamental importance in the centre of Santa Fe province, Argentina. In this region, more than 250,000 inhabitants are supplied with groundwater, as it is the only source of water supply for all regional demands. The aquifer used is commonly known as "Puelche" aquifer. The estimated daily average flow from pumping wells in the aquifer is 51 million liters, of which 26 million liters are consumed in the cities of Esperanza and Rafaela, the two most important towns in the area (Figure 15.1).

The supply to these urban centres is from a pumping field located in the rural area of Esperanza, from where a flow rate of more than one million cubic meters per hour is driven by an aqueduct to Rafaela (Paris et al. 2010). Regional economic activity is diverse with prevalence of agriculture, livestock rearing, and industry. The development of the area has had a sustained growth, which has lead to an increased demand of groundwater resources.

15.1.1 Purpose

Based on these considerations, the purpose of this study is to evaluate the possible effects of climate change on the groundwater system in the study area. This system is naturally sensitive to the recharge variation and the exploitation flow rates (Tujchneider et al., 2005b; D´Elía et al., 2007), which generates negative impacts on groundwater quality and quantity, such as depleted processes, upconing or unwanted chemical compound input. For these reasons it is extremely important to estimate the variations in groundwater storage that occur due to changes in recharge from both local and regional flows.

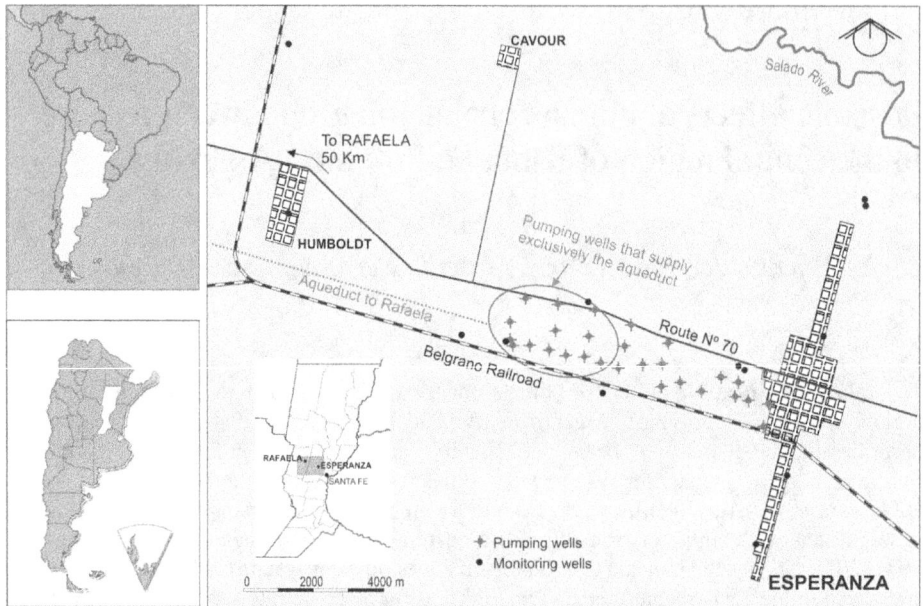

Figure 15.1.   Location of the study area.

15.1.2   Description of the area: the central region of Santa Fe Province

Previous research by the authors characterized the groundwater system (Tujchneider et al. 1998; Fili et al. 1999). The climate of the region is temperate humid, with an average annual temperature of 18°C, an average annual rainfall of 930 mm, and a mean real evapotranspiration (ETR) of 890 mm. The latter was estimated by the Thornthwaite-Matter method and adopting storage values based on the moisture content of the different soil types mainly present in the area.

The hydrogeological system is described by Tujchneider et al. (2000) as follows (Figure 15.2):

–   The upper layer (15–20 m average thickness) is composed by aeolian sedimentary deposits of silt, clay, and loess from the Pampa Group (Holocene). The unconfined aquifer located in this sequence is used mainly in the rural area, but it has low yield and quality restrictions. At the base of the upper layer is an aquitard (3–5 m average thickness) with regional discontinuities.
–   Beneath the upper layer is a semi-confined aquifer (24 m average thickness) of very good quality and yield, that consists of thin and medium sands of fluvial origin (also called "Puelches" sands) belonging to the Ituzaingó Formation (Pliocene). Currently, this is the main source of water supply.
–   Underlying these fluvial sediments, grey sands, sandy clays and green clays of marine origin (Paraná Formation, Miocene) have high salinity water of continental origin. For this study, this layer is considered the aquiclude basement of the hydrogeological column.

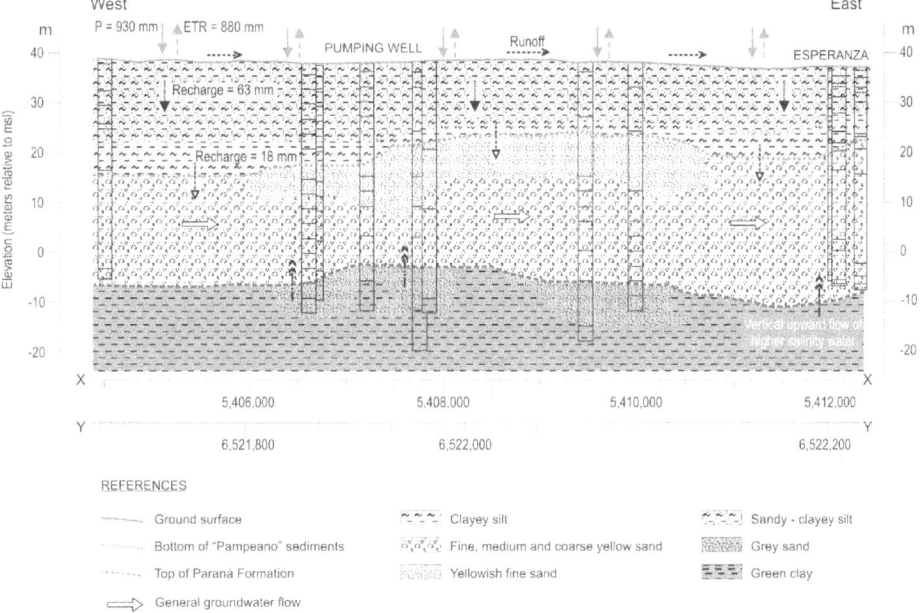

Figure 15.2.   Scheme of the hydrogeological section.

Hydraulically, this sedimentary package behaves as a multiunit aquifer. According to the prevailing relations of hydraulic heads there is the possibility of a downward-flow gradient across the aquitard layer and an upward-flow gradient from the underlying sands of marine origin (Fili et al. 1999).

In order to quantify the natural recharge to the aquifer system, the outputs of the average water balance were analyzed. For the period 1936–2001, this balance indicates an excess of 40 mm/a. In a hydrologic budget the excess is considered as an outflow from the soil zone available to runoff and seepage. For the same period, a serial monthly water balance gives an average potential recharge of 28 mm/a. The estimation of this value involved the different soil types present in the area and the associated hydrologic soil group as well as the particular ground slope and land uses. The precipitation and temperature series were stochastically processed (Paris et al. 2003).

Further investigations (2002–2006) showed that depth to the water table ranged from 4 to more than 9 m and the lag time between precipitation events and subsequent recharge events is approximately one or two months. The use of water-table fluctuation and chloride mass balance methods indicate average annual recharge to the unconfined aquifer layer of 65 mm and 63 mm, respectively. Results from the chloride mass balance method also indicate annual recharge rates of 18 mm to the semi-confined aquifer layer (D´Elía et al. 2007).

The general direction of the groundwater flow is from west to east. The existence of local and regional flows has been corroborated with isotope hydrology techniques (D´Elía et al. 2005).

Groundwater is predominantly sodium bicarbonate type water (Tujchneider et al. 1998). In some cases, the unconfined aquifer has important arsenic and fluoride contents (Nicolli et al. 2009). Changes in the chemical composition of groundwater in the semi-confined aquifer have been noticed in some specific area under uncontrolled exploitation, as well as a significant decline in the groundwater levels (Tujchneider et al. 2005a).

For quantifying the annual average excess in the area, studies carried out by FICH (2006) were taken into account. The FICH's study was a regional study to determine the vulnerability of the water resources in an area that covers the Litoral and Mesopotamia regions, Argentina (Figure 15.3), the richest hydrological region of the country. This study includes an assessment of the current situation of the water resources and the pose of future scenarios for the (2080–2090) period. Issues related to socioeconomic conditions, climate, water resources and soil uses in the region were analyzed for the conditions specified in the two climate scenarios selected (A2 and B2).

The current situation of the water resources (Environmental Base Line) involved climatologic studies, surface and groundwater hydrology, the assessment of available structural and nonstructural measures for flood control, appraisal of environmental stress by water deficits and surplus, the characteristics of the social subsystem and the identification of soils and their diverse uses. In this study, the period (1981–1990) was adopted as a reference because of the available information.

15.1.3   Methods

Using the IPCC (2001) forecasts for anthropogenic emission of greenhouse gases and the most probable future climatic scenarios (A2 and B2), the water balance for the study area was re-calculated for A2 and B2 scenarios during the years 2081–2090. These results were used as inputs (recharge boundary condition) in a 3-D numerical model to evaluate the temporal and areal evolution of the groundwater levels.

The 3-D numerical model was implemented and calibrated using Visual MODFLOW v4.1 (Waterloo Hydrogeologic Inc. 2005). This calibrated model (Tujchneider et al. 2000) constituted the fundamental tool to evaluate the influence of the proposed climate

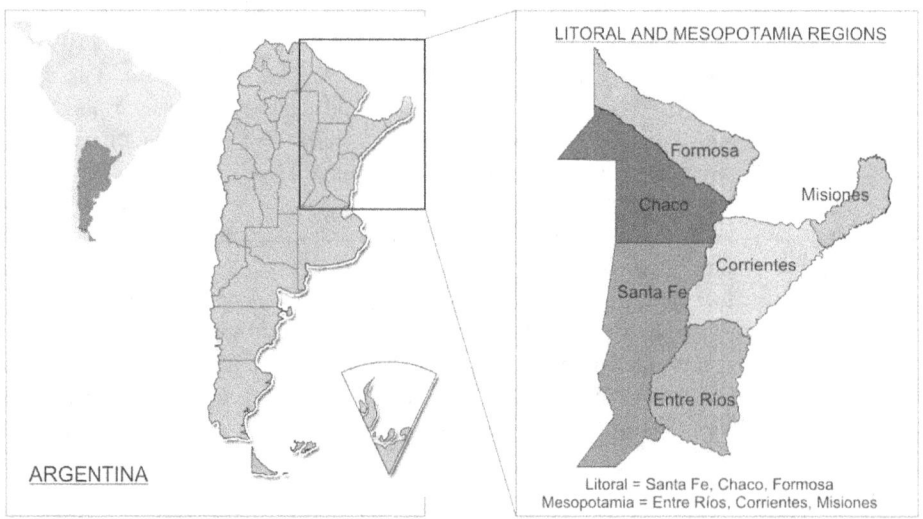

Figure 15.3.   Littoral and Mesopotamia regions, Argentina.

change scenarios on the aquifer system. The model was verified using the groundwater levels measured between 1999 and 2003 (Tujchneider et al. 2005b).

The horizontal grid of the model domain consists of rectangular elements, with varying widths of 200, 250, and 500 m. The vertical discretization includes three layers, according to the conceptual hydrogeological characterization of the area described above. The layers were modelled as Type 3 – Confined/Unconfined. Layer 1 represents the unconfined aquifer, layer 2 is the aquitard unit, and layer 3 is the semi-confining aquifer that is being used as the only freshwater supply in the area.

The finite difference grid and the boundary conditions are illustrated in Figure 15.4. As the model domain did not coincide with the rectangular shape of the model grid, inactive cells were designated at the north-eastern region. Specified flux and head dependant boundary conditions were assigned along the western and eastern limits, respectively. Recharge as described above and river boundary conditions were also used at layer 1. No-flow boundaries were defined at the remaining northern and western cells.

The hydraulic parameters were estimated from pumping tests and major lithologic characteristics of the geologic formation. As the water needs for domestic, industrial and agriculture uses are supplied by pumping wells, they were added to the model and were completed in layers 1 and 3.

15.1.4 Relevance for GRAPHIC

It is well known that in many countries, groundwater is often the main source for domestic water supply and thus of vital importance for the livelihood and health of the people. Groundwater is also widely used for irrigated agriculture and industry. This is

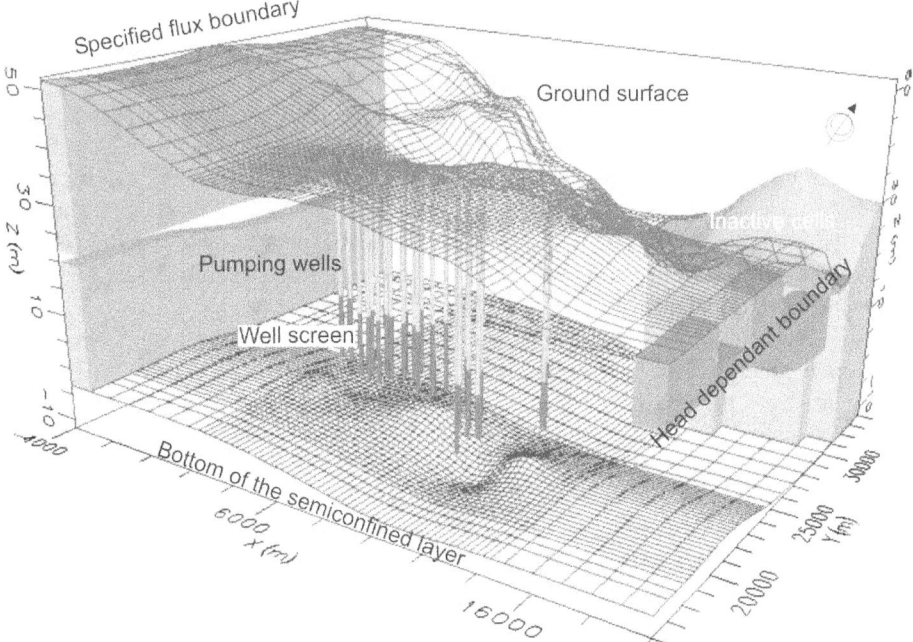

Figure 15.4.   Finite difference grid and boundary conditions used.

particularly true in arid and semiarid regions where surface water is scarce or seasonal, in rural areas with dispersed populations, and in those regions where surface-water bodies are not available or require expensive works for collecting. Climate change is likely to lead to a greater dependence on groundwater as a cushion against drought and increasing uncertainty in surface-water availability (Taylor et al. 2010).

On the other hand, it is very interesting the point of view given by Green et al. (2011) about the subject. They analyze the possible effects of the climate variability and change on groundwater and the evidence of greater needs of the resource in the future. They also provide a very accurate methodological approach for exploring subsurface global change, facing the challenge to adapt and perform management tools.

The case study presented in this chapter is a typical situation where the increased development leads to a growing water demands and where groundwater is the only and current water supply. In such cases, uncontrolled and unplanned groundwater extraction is usual. This is a frequently consequence of the lack of knowledge of the hydrogeologic characteristics and an inappropriate legal and institutional framework and/or policies.

In 1999, the Geohydrological Research Group (GIG) of the Faculty of Engineering and Water Sciences (National University of El Litoral) formulated a management model for the groundwater resources (Fili et al. 1999; Tujchneider et al. 2000). In this model, the location of a single pumping field to supply the cities of Esperanza and Rafaela was proposed in an area away from the influence of urban and industrial pollution. Also, the well design, the distance among wells, their operation schedule, and monitoring network were determined. This new pumping field was designed using a comprehensive understanding of the regional groundwater resources (physical flux and state variables) and the interaction of this system with the climate and population pressures.

It is important to remark that the set of proposed management strategies had the consensus of the local government, the water utility and the water supply control agency, and the different local sectors interested in groundwater. The operation of this unified pumping field began at the end of 2004, and is currently producing 14 millions m^3/a.

To evaluate the future impact on groundwater resources, some scenarios were considered according to different alternatives for the location of the pumping field. This assessment was based on the conceptual and numerical model implemented to represent the aquifer system. These tasks were carried out after a great effort to integrate and analyze a large amount of data. It is indispensable to understand the site specific controls that govern groundwater flow in order to create models and simulations that are useful management tools in decision making.

Groundwater is a part of the "blue and green component" of the water cycle and is in dynamic equilibrium with the other reservoirs of the cycle. Falkenmark (2003) refers as "green water" the part of the precipitation over an area which is consumed in plant production and evaporation from the moist surfaces. It sustains the terrestrial ecosystem. The surplus flow goes to recharge aquifers and rivers ("blue water"), available for societal use. In this sense, the natural water cycle, or budget, balances the global fluxes with the ecosystems demands. However, it is necessary to consider a more realistic and comprehensive vision of the global water cycle that takes into account human influence that alters the natural dynamic equilibrium. This "environmental water cycle" includes the human pressure and natural processes that affect water resources.

In addition to the research results presented here, the study area represents several other lines of investigation that assesses the sustainable development of groundwater resources under human and climate change pressures. These lines focus on: a) quantifying local and regional recharge, b) evaluating groundwater discharges, c) estimating the water requirements for the phreatophyte vegetation, d) identifying the presence and diversity of invertebrates in the groundwater environment like biological indicators of the water quality conditions, and e) indentifying indicators that best describe the behaviour toward sustainable use of the groundwater system.

The different lines of research can be carried out thanks to the design, installation, and operation of a monitoring network that collects time varying hydro-meteorological, hydrogeological, and environmental data within the pilot area. This is a true "laboratory in the Nature", where different sectors and institutions were involved to achieve its implementation.

The approach of this case study could be applied and adapted to other regions where groundwater is subject to multiple uses, usually with little or no planning, and where hydrological events (dry and wet periods) are often poorly estimated.

15.2 RESULTS AND DISCUSSION

As expected, the climatic model outputs used by FICH (2006) indicate a greater impact from the A2 scenario than from the B2 during the decade 2081–2090. Figure 15.5 shows the areal distribution of the average annual precipitation and temperature for the reference period (1981–1990). Figure 15.6 shows the difference in the average annual rainfall amounts for the A2 and B2 scenarios between the periods (1981–1990) and (2081–2090). A decrease of 100 mm can be expected in the study area (Santa Fe province) for the A2 scenario (Figure 15.6a). Annual average rainfall under the B2 scenario remains the same between the periods (1981–1990) and (2081–2090) (Figure 15.6b).

For the reference period (1981–1990) the annual distribution of the excess indicates that they occur from the beginning of autumn to the end of spring. During this

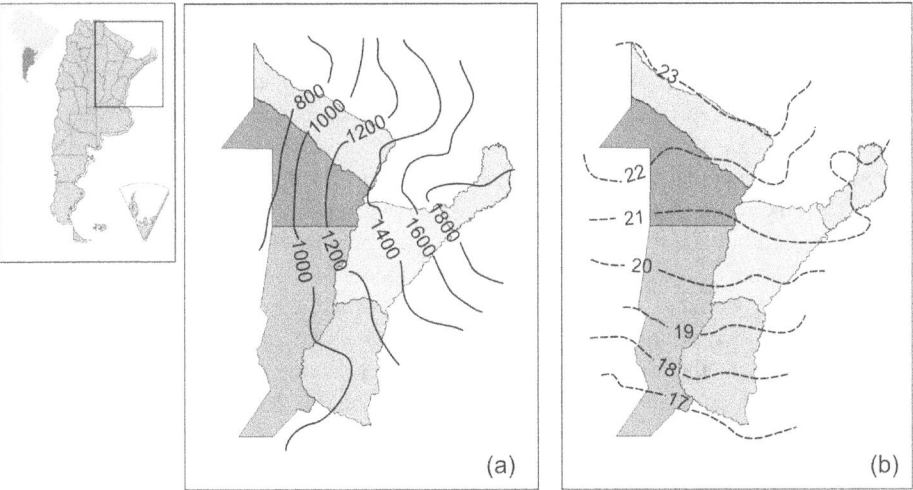

Figure 15.5. (a) Average annual precipitation and (b) average annual temperature. Period 1981–90 [Adapted from FICH (2006)].

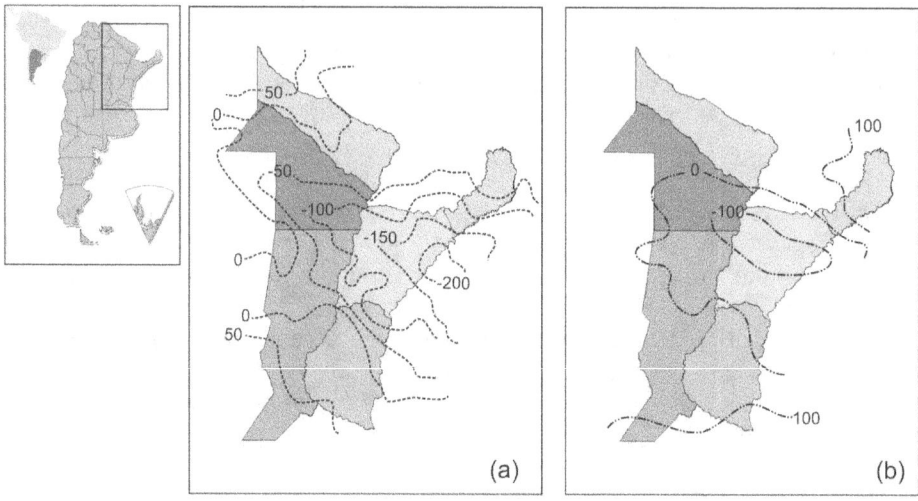

Figure 15.6.   (a) Difference of annual precipitation rates for the A2 scenario and (b) Difference of annual precipitation rates for the B2 scenario. The differences are between (1981–90) and (2081–90) periods [Adapted from FICH (2006)].

period, recharge to the unconfined aquifer occurs. The recharge rate is also dependent on the moisture content of the vadose zone and the water table depth. Average annual deficits, for the same period, occur from mid-spring to the end of summer (Southern Hemisphere).

For the A2 and B2 climate scenarios, values of mean annual recharge were used in the numerical simulations. These values were estimated as a percentage of the average annual excesses, considering a lineal variation between the estimated recharge values for the years 2001 and the period 2081–2090. For the groundwater level evolution, aspects such as: a) increase of the future groundwater demand, b) reduction of the local recharge rate, and c) decrease of the regional groundwater flow were also taken into account.

For the scenario A2, the areal distribution of the excesses has an increasing gradient towards the Northwest (NW), as it can be seen in Figure 15.6a. The annual amounts of these excesses are significantly smaller than the ones calculated for the reference period 1981–1990. Along the average year 2081–2090, the excesses would generally occur in autumn while the deficits would occur from mid-spring through the end of summer. As it can be seen the recharge rates and their seasonal occurrence would be modified.

In the scenario B2, the areal distribution of the excesses is reduced and their annual amounts are lower (Figure 15.6b). Excesses become null to the west of the study area. These excesses would occur from the beginning of autumn until the end of winter. During the summer months, the excesses are null. Average annual deficits would occur from mid-spring to the end of summer. This scenario shows an intermediate condition between the Environmental Base Line and the A2 scenario. However, the recharge possibilities to the aquifer systems would decrease significantly.

Figures 15.7 and 15.8 show the outputs of the groundwater flow model for the simulated scenarios and for the semi-confined aquifer, at present and in the period 2081–2090, respectively. The general groundwater direction is from west to east, but it is clear

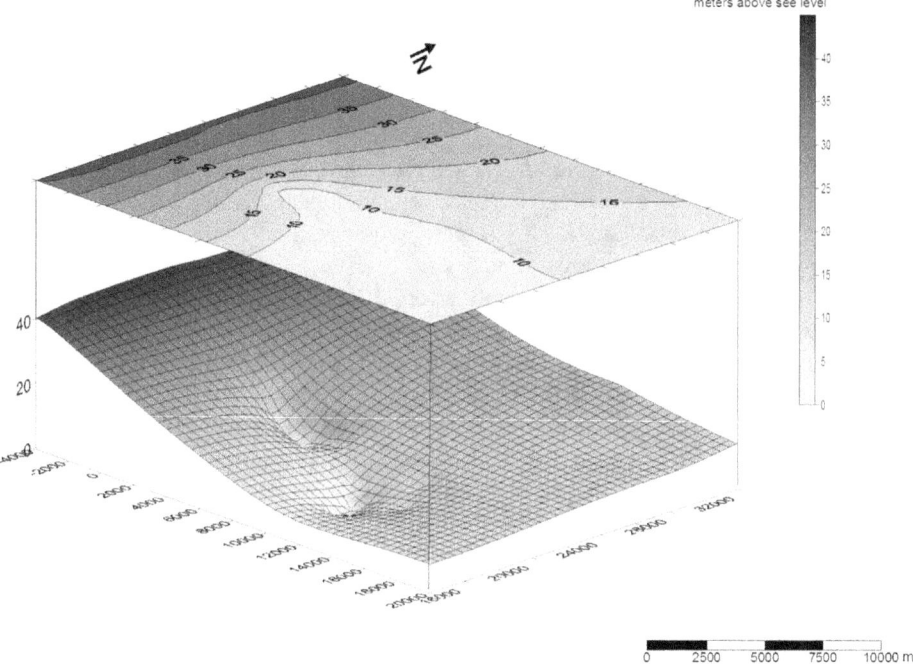

Figures 15.7.   Groundwater level contour maps for the semi-confined, at present.

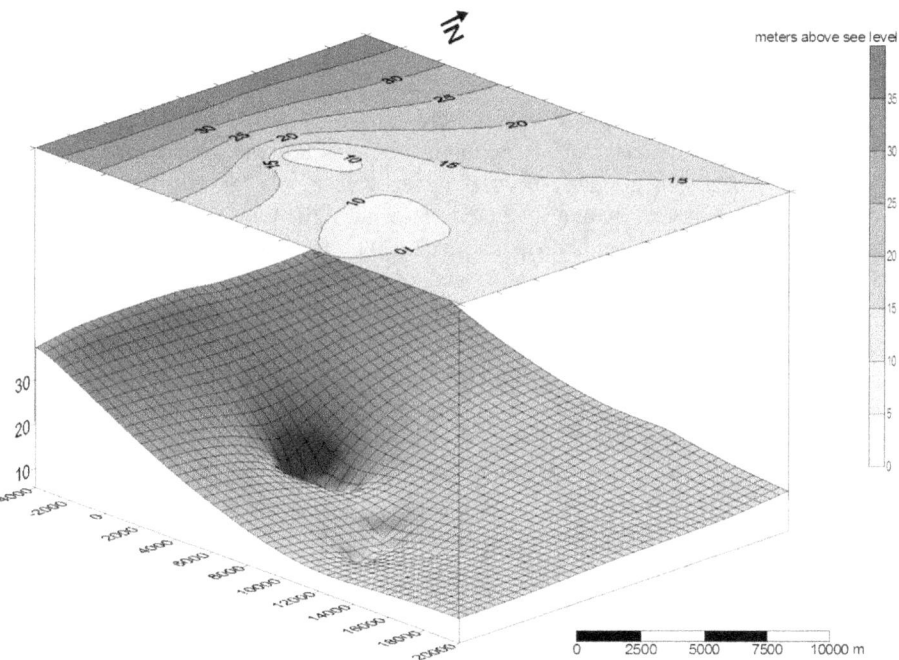

Figures 15.8.   Groundwater level contour maps for the semi-confined, at period 2081–2090.

the formation of a cone of depression around the pumping field. When comparing these figures it can be observed an average decrease in the groundwater levels of 5 m. This fact is most evident at the western boundary where the hydraulic head decreased from 40 m (Figure 15.7) to 35 m (Figure 15.8) and in the pumping field where the cone of depression is deeper (Figure 15.8).

Figures 15.9 and 15.10 represent very well the impacts exerted on the groundwater system as a consequence of the human pressures and climate variability for a particular groundwater level (15 m). Figure 15.9 represents the aquifer volume bounded by the 15 m water table elevation at present time.

Because in the A2 climate scenario for the period 2081–2090 the local aquifer recharge possibilities are null the groundwater levels would drop below the top of the semi-confined aquifer, possibly creating an unconfined aquifer. Figure 10 illustrates this situation where the top of the well screen would be above the groundwater level. The semi-confined aquifer would only be fed by the regional flow. Taking into account the present upconing process (vertical rise of underlying saline water) identified in the area (Tujchneider et al. 2005a), the aquifer could suffer an irreversible salinization process, as well.

For A2 and B2 scenarios, the lowering of the recharge rates would also cause a decreased base flow contribution to surface water bodies (rivers, wetlands, lagoons, etc.). Also, this would alter their water dynamic and groundwater dependant ecosystems, involving important losses of biodiversity.

It is necessary to consider that in this area the semi-confined groundwater level is the combined result of local and regional flow. So, if there is not enough water to recharge the unconfined aquifer, this will be depleted, and only the regional flow will maintain the groundwater system.

15.3   POLICY RECOMMENDATIONS

In this study case, A2 and B2 scenarios were used as input to numerical groundwater flow simulations as the basis for developing a future water resources management plan. Keeping local issues in mind, a successful management plan should articulate actions includes the reducing the volume of extraction; achieve the joint use of groundwater and surface water to satisfy potential demands; establish protection zones around water-supply wells; improve the legal framework; and promote awareness and education at all levels.

In this regard it should be noted that the government of Santa Fe province plans to build an aqueduct to carry surface water from the Paraná River toward the west. This aqueduct will reduce the stress on groundwater resource. Although the environmental effects of imported water must be taken into account. Moreover, the provincial government is also promoting a water bill that includes, among other things, the delineation of well-protection zones and environmental education.

15.4   FUTURE WORK

Based on the findings of this study, the aquifer system will be substantially stressed by the climate change as a consequence of higher demands, decreasing recharge rates, and a significant increase of anthropogenic activities. Therefore, to improve the knowledge

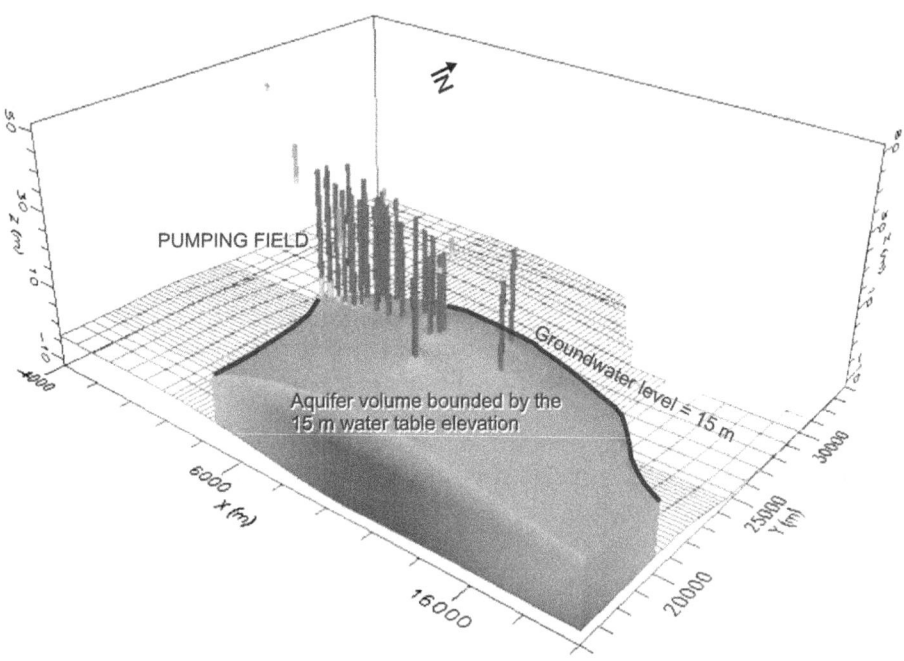

Figures 15.9.   Aquifer volume bounded by the 15 m water table elevation, at present.

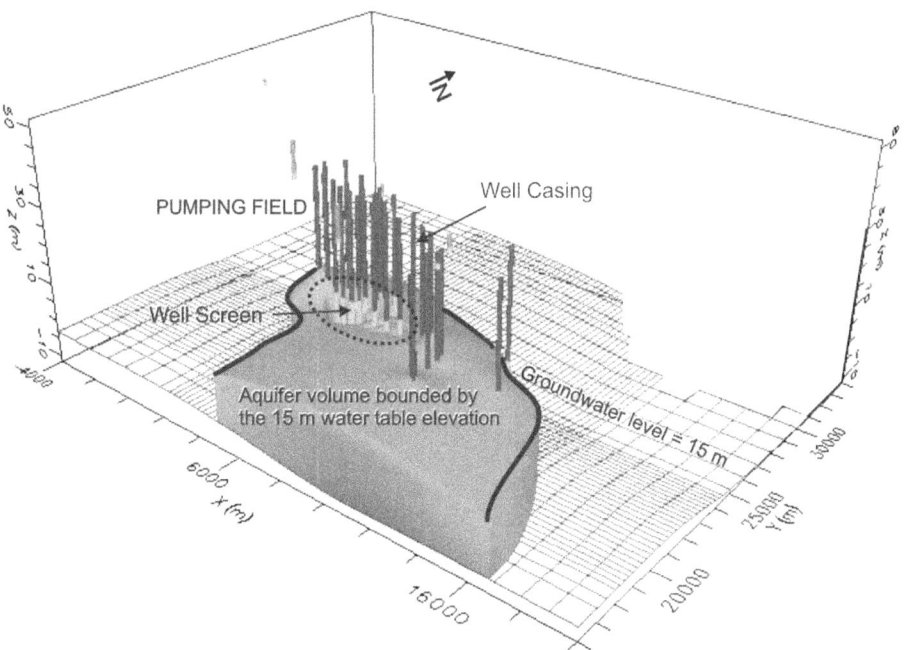

Figure 15.10.   Aquifer volume bounded by the 15 m water table elevation, at 2081–2090 period for A2 scenario.

of the hydrogeological conditions and to minimize the uncertainty level some actions should be addressed. These actions include, among others, the systematic monitoring of the aquifer system behaviour.

There are key aspects where additional scientific research would greatly increase understanding regarding the potential for groundwater contamination related to society and development in the area. Specifically, assessing aquifer pollution vulnerability, mapping of groundwater pollution hazards, establishing appropriate protection strategies for the groundwater supply sources, and monitoring the aquifer response will be important aspects of future studies.

The groundwater management model implemented in the area should be able to cope with the variability of the hydrogeological system attributes (depth, recharge flux, discharge flux, water quality). Therefore, all the actions described here could be used to reach an appropriate, adaptive, and dynamic groundwater management plan according to the challenges and goals proposed by the government and society considering and mitigating the probable effects of climate change.

ACKNOWLEDGEMENTS

The authors would like to deeply acknowledge the anonymous reviewer for the detailed and respectful revision of the draft. His/her comments improved the original manuscript. They would also thank the National University of El Litoral (UNL) and the National Agency of Scientific and Technological Promotions (ANPCYT) for the financial support.

REFERENCES

D´Elia, M., Tujchneider, O., Paris, M. & Perez, M. (2007) The importance of groundwater recharge quantification for the sustainability of ecosystems in plains of Argentina. In: *Proceedings of XXXV IAH Congress*: Groundwater and Ecosystems, September 17–21, 2007, International Association of Hydrogeologists (IAH), Lisbon, Portugal, ISBN 978-989-95297-3-1, variously paginated.

D´Elia, M., Tujchneider, O., Paris, M., Perez, M. & Aravena, R. (2005) Técnicas isotópicas en la caracterización de sistemas de flujo subterráneo [Isotopic techniques for characterizing groundwater flow systems] *Revista Latino-Americana de Hidrogeología (ALHSUD)*, 5, 31–38.

Falkenmark, M. (2003) Water Management and Ecosystems: Living with Change. TEC, Background Paper number: 9.

FICH. (2006) *Actividades Habilitantes para la 2ª Comunicación Nacional del Gobierno de la República Argentina a las partes de la Convención Marco de las Naciones Unidas sobre Cambio Climático. Vulnerabilidad de los Recursos Hídricos del Litoral – Mesopotamia. Argentina* [Enabling activities for the 2nd National Communication of the Argentina Government to the parties of the Framework Convention on Climate Change. Water resources vulnerability in the Litoral – Mesopotamia region. Argentina]. Facultad de Ingeniería y Ciencias Hídricas. Technical Report.

Fili, M., Tujchneider, O., Paris, M., Perez, M. & D´Elía, M. (1999) *Estudio del sistema de aguas subterráneas en el área de Esperanza-Humboldt y zona de influencia* [Groundwater system study in Esperanza – Humboldt region and surrounding area]. Facultad de Ingeniería y Ciencias Hídricas. Technical Report.

Green, T.R., Taniguchi, M., Kooi, H., Gurdak, J.J., Allen, D.M., Hiscock, K.M., Treidel, H. & Aureli, A. (2011) Beneath the surface of global change: Impacts of climate change on groundwater. *Journal of Hydrology*: doi: 10.1016/j.jhydrol.2011.05.002

IPCC, (2001) *Climate Change 2001: The Scientific Basis. Contribution of Working Group I to the Third Assessment Report of the Intergovernmental Panel on Climate Change* [Houghton, J.T.,Y. Ding, D.J. Griggs, M. Noguer, P.J. van der Linden, X. Dai, K.Maskell, & C.A. Johnson (eds.)]. Cambridge University Press, Cambridge, United Kingdom and New York, NY, USA, 881 pp

Nicolli, H., Tujchneider O, Paris M, Blanco M, Barros A (2009) Movilidad del arsénico y oligoelementos asociados en aguas subterráneas del centro – norte de la provincia de Santa Fe, Argentina [Mobility of arsenic and oligoelements associated to groundwaters in the north centre of Santa Fe province, Argentina]. VI Congreso Argentino de Hidrogeología. Seminario: Presencia de flúor y arsénico en aguas subterráneas. Santa Rosa, La Pampa, Argentina. pp. 81–90.

Paris, M., D´Elía, M., Pagliano, M., Pusineri, G., Gualini, S., Tujchneider, O. & Perez, M. (2010) Mapa de vulnerabilidad a la contaminación de acuíferos en Esperanza (Santa Fe, Argentina). Consideración de su dinámica temporal [Map of vulnerability to aquifer contamination (Esperanza city, Santa Fe, Argentina) Considerations about its temporal dynamic]. *X Congreso Latinoamericano de Hidrología Subterránea – ALHSUD*. Libro de Resúmenes más CD. Ref. 10 Comisión T1. Caracas, Venezuela.

Paris, M., D´Elía, M., Perez, M. & Tujchneider, O. (2003) Análisis estocástico de variables hidrometeorológicas para la estimación de recarga de acuíferos. [Stochastic analysis of hydrometeorological variables to estimate groundwater recharge]. *Boletín Geológico y Minero*, 114-2, 193–201.

Taylor, P., Owen, R. & Tuinhof, A. (2010) *Groundwater Management in IWRM Training Manual*, International Network for Capacity Building in Integrated Water Resources Management (Cap-Net), Rietfontein, Pretoria, South Africa, Available on the world wide web: http://www.cap-net.org/sites/cap-net.org/files/Cap-Net%20Groundwater%20(web%20res).pdf (Accessed November 7, 2011), 166 p.

Tujchneider, O., Paris, M., D´Elía, M., Perez, M. & Fili, M. (2000) *Modelo de gestión de los recursos hídricos subterráneos en el centro-oeste de la provincia de Santa Fe, Argentina* [Model of Groundwater Management for the west-centre of Santa Fe Province, Argentina]. *1ST Joint World Congress on Groundwater*. XI Congreso de ABAS (Asociación Brasilera de Aguas Subterráneas) y V Congreso de ALHSUD. Trabajo completo en CD Rom. Fortaleza, Brasil.

Tujchneider, O., Paris, M., Fili, M., D´Elía, M. & Perez, M. (1998) Protección de aguas subterráneas. Caso de estudio: ciudad de Esperanza (República Argentina). Primera fase: Diagnóstico del sistema [Groundwater protection. Case of study: Esperanza City (Argentina). Preliminar diagnostic of the system]. *IV Congreso Latinoamericano de Hidrología Subterránea*. Vol. 2, páginas 805 a 821. Editor: ALHSUD. Montevideo, Uruguay.

Tujchneider, O., Paris, M., Perez, M. & D´Elía, M. (2005a) Singularidad constitutiva de sistemas geohidrológicos de llanura y gestión de los recursos hídricos subterráneos [Constitutive singularity of hydrogeological systems in plain areas and the management of the associeted groundwaters]. *Revista Latino-Americana de Hidrogeología (ALHSUD)*, 5, 117–121.

Tujchneider, O., Perez, M., Paris, M. & D´Elia, M. (2005b) Deterioro de fuentes de agua subterránea por ascenso de agua salada [Groundwater supply deterioration due to an upconing process]. *IV Congreso Argentino de Hidrogeología y II Seminario Hispano – Latinoamericano sobre temas actuales de la Hidrología Subterránea*. Río Cuarto, Córdoba, Argentina. Tomo II, pp. 217–226.

Waterloo Hydrogeologic Inc. (2005) *Visual Modflow Professional Edition*. User´s Manual, Ontario, Canada, variously paginated.

Continental Climates

CHAPTER 16

Impacts of drought on groundwater depletion in the Beijing Plain, China

Yangxiao Zhou, Liya Wang, Jiurong Liu & Chao Ye

ABSTRACT

Beijing, the capital of China, has suffered a prolonged drought of 12 years from 1999 to 2010. The regime shift detection methods have identified the increase of mean annual temperature, the decrease of annual evaporation, and the decrease of annual precipitation. The Standardized Precipitation Index clearly shows a downward shift from more wet years in 1960s and 1970s to more dry years in 2000s. The observed total decrease of groundwater levels from 1999 to 2010 ranges from 20 to 40 m, indicating rapid depletion of groundwater resources. Water shortage during the recent long dry spell was simply met with the construction of six new emergency well fields since 2003. The drought has a direct impact on groundwater resources by reducing natural recharge. The drought also triggered the increase of groundwater abstraction for emergency water supply. These direct and indirect impacts of the drought on the depletion of groundwater storage have been analysed with a transient groundwater flow model. The drought-triggered increase of groundwater abstraction has accelerated the depletion of groundwater storage, especially in the Chaobai River catchment. To reverse the trend of further groundwater depletion, a combination of the reduction of groundwater abstraction and the artificially enhanced groundwater recharge was simulated with the model and was found as the most feasible scenario. A drought management plan is urgently needed to shift from crisis management to drought preparedness and risk management. The groundwater depletion situation in the Beijing Plain is very grim and extraordinary actions must be taken to achieve sustainable groundwater use.

16.1 INTRODUCTION

16.1.1 Purpose and scope

Beijing City, the capital of China, has experienced rapid population growth, urbanization, and industrialization in last 30 years. The total population was nearly 20 million and the GDP was around 10,000 US$ per capita in 2010 (Beijing Statistics Bureau 2010). With a temperate climate, water resources are limited. The total water resources were estimated around $38 \times 10^8 m^3$/annum (a), only $190 m^3$/a per capita, which is much lower than the world average and is categorized as absolute water scarcity (Yang and Zehnder 2001). The inflows of surface water have been substantially decreased due to the construction of reservoirs and water diversions in upstream areas. Water demand for rapid social and economical developments has been met with the overexploitation of groundwater since the 1980s. Groundwater supply accounts for two-thirds of the water supply for Beijing municipality: roughly 60% for irrigation, 25% for drinking water, and 15% for industrial water supply (Beijing Water Authority 2006). Beijing has experienced prolonged consecutive droughts since 1999. The average annual precipitation from 1999 to 2010 is only 78% of the mean annual precipitation from 1959 to 1998. The water shortage during this severe dry spell was met with the installation of six new emergency groundwater well fields. The combined effects of decreasing natural recharge and increasing abstraction have caused the rapid depletion of groundwater storage. The consequences are serious: drying shallow wells, drying streams, degraded ecosystems, and land subsidence (Beijing Engineering Geological Survey Institute 2007).

This paper analyses climate changes with regime shift detection methods (Taylor 2000; Rodionov 2004). The recent severe dry spell was identified from 1999 to 2010 by a Standardized Precipitation Index (McKee et al. 1993). The rapid decline of groundwater levels in recent dry spell was detected with linear regression analysis. Groundwater depletion was simulated with a transient groundwater flow model. Options for mitigating further groundwater depletion were proposed and their effectiveness was simulated with the same model. Management issues to mitigate drought and to achieve sustainable groundwater resources development in the Beijing Plain were also discussed.

This case study is consistent with the objectives of the UNESCO-IHP programme on Groundwater Resources Assessment under the Pressures of Humanity and Climate Changes (GRAPHIC) (UNESCO-IHP 2008). The case study provides a unique example of accelerated depletion of groundwater storage under the consecutive dry years and drought-triggered increase of groundwater abstraction for emergency water supply. The case also demonstrates the lack of preparedness for drought management in Beijing city.

16.1.2   Description of the study area: the Beijing Plain

The Beijing Plain is located in the northwest corner of the North China Plain (Fig. 16.1). Surrounded by mountains to the west and north and bounded by the Hebei province and Tianjin city in the east and south, the area of the Beijing plain is around 6,000 km². The Beijing plain is formed by alluvial fans and plains of two large rivers and several small rivers (Fig. 16.2). Land elevation is high in the northwest and gradually decreases towards the southeast. The capital city of China, Beijing, is located in the middle west of the Beijing Plain. The data used for the analysis in this study includes monthly mean temperature and monthly precipitations from 11 meteorological stations from 1959 to 2009, monthly pan evaporation rates from five meteorological stations from 1959 to 2009, and monthly groundwater level measurements from three groundwater observation wells for multiple years. The locations of the stations are shown in Figure 16.2.

The Beijing Plain is located in the warm temperature zone and has a continental climate, with cold and dry winters, hot and humid summers, and moderate temperatures in spring and autumn. The long-term (1959–2010) average annual precipitation from 11 rainfall stations is around 590 mm. The long-term (1959–2009) average annual evaporation measured in evaporation pan from five stations in Beijing Plain is around 1,836 mm. The pan water evaporation is more than 3 times the precipitation. The precipitation occurs mainly in the summer months from June to September, accounting for 81% of the annual total precipitation. The peak evaporation occurs in April, May, and June (Fig. 16.3). Irrigation in the spring season is necessary for most crops because it is very dry.

Five rivers flow across the Beijing Plain. The combined discharge of two large rivers, Yongding and Chaobai, accounts for 90% of the total surface water inflow to the

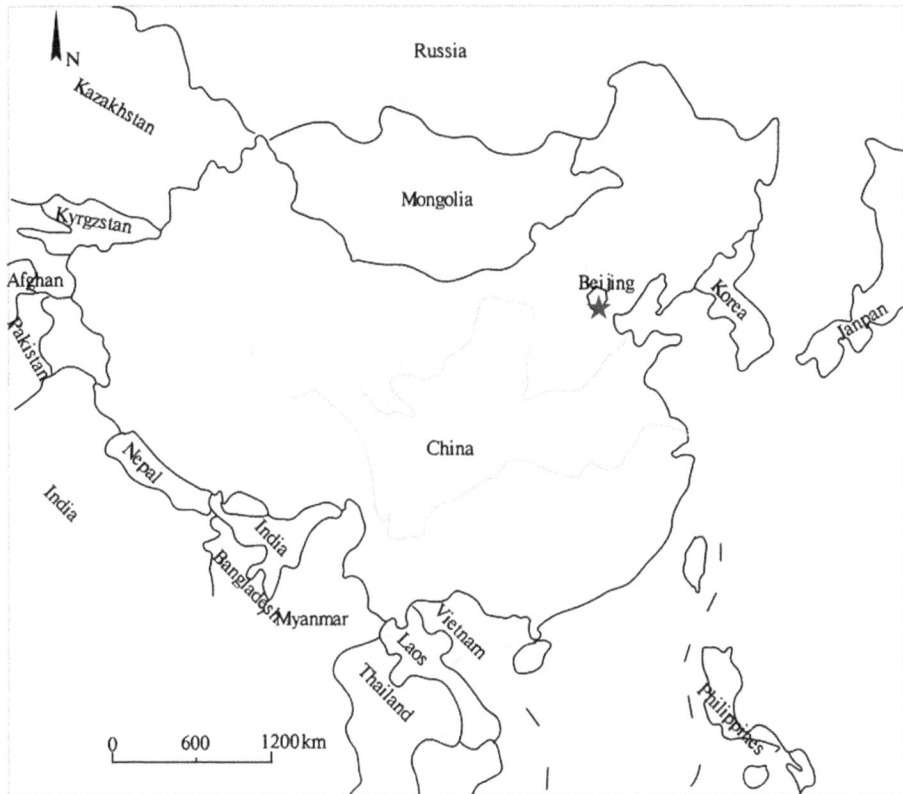

Figure 16.1.   Location map of the Beijing Plain.

Beijing Plain. These two large rivers are dammed by the Guanting Reservoir and Miyun Reservoir, respectively. River inflows have been drastically decreased since 1970's. The consecutive droughts since 1999 had significant impact on further reduction of river inflows (Fig. 16.4). The river discharge during the 2000s is only around 10% of natural discharge in the 1950s.

The Beijing Plain consists of Quaternary deposits formed by alluvial fans and plains. The thickness of Quaternary deposits varies from tens of meters to more than 500 meters. The aquifer system is complex with variable sediment thickness and lithology. Five hydrogeological sub-systems may be distinguished (Fig. 16.5). The Yongding River alluvial fan-plain and Chaobai River alluvial fan-plain occupy the majority area of the Beijing Plain. Nankou-Wenyu River alluvial fan is located in between these two large systems. Daqing River alluvial fan is located in the southwest corner while Jiyun River alluvial fan is located in the northeast corner. From the alluvial fans to the plain, the sediment thickness increases and grain size decreases, aquifer systems change from a single gravel aquifer to multiple aquifer systems of sand layers separated by silt and clay layers. The water bearing layers within a depth of around 50–150 m is called the shallow aquifer. The large part of the shallow aquifer is unconfined. It receives all natural groundwater recharge. The majority of agricultural wells are located in the shallow aquifer. Some water supply well fields are installed in

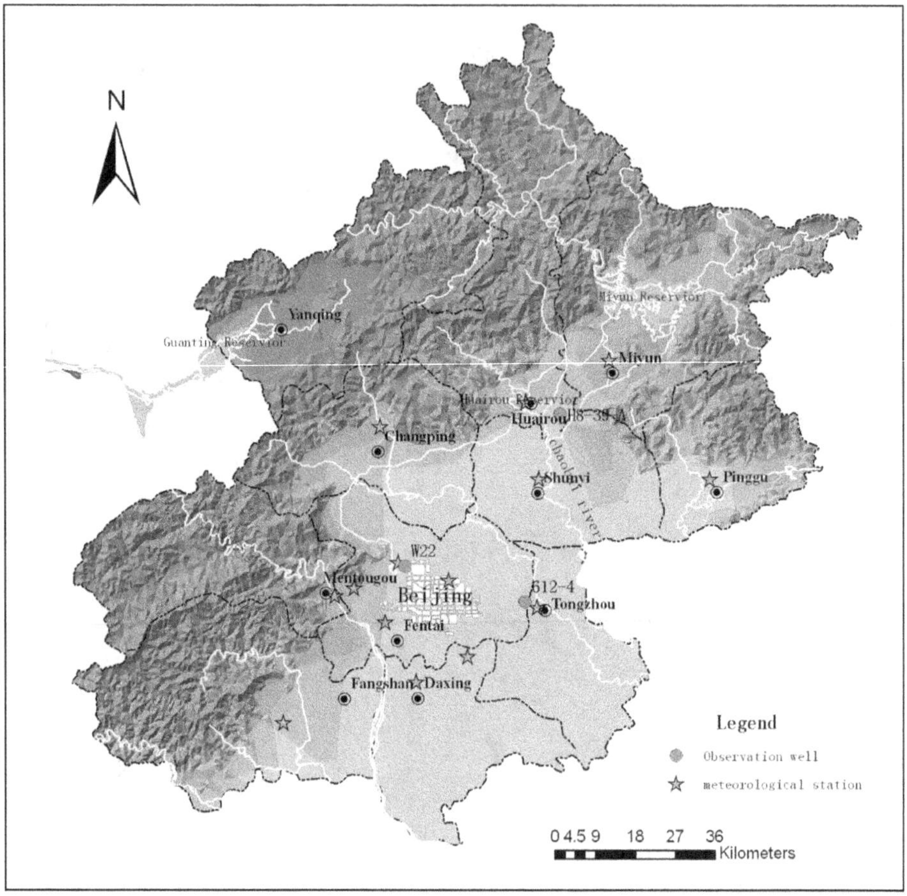

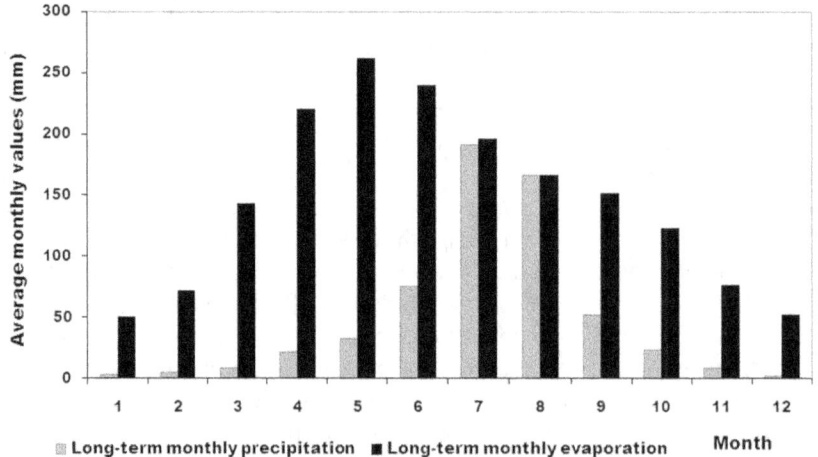

Figure 16.2.   Locations of meteorological stations and groundwater observation wells.

Figure 16.3.   Monthly distribution of precipitation and evaporation.

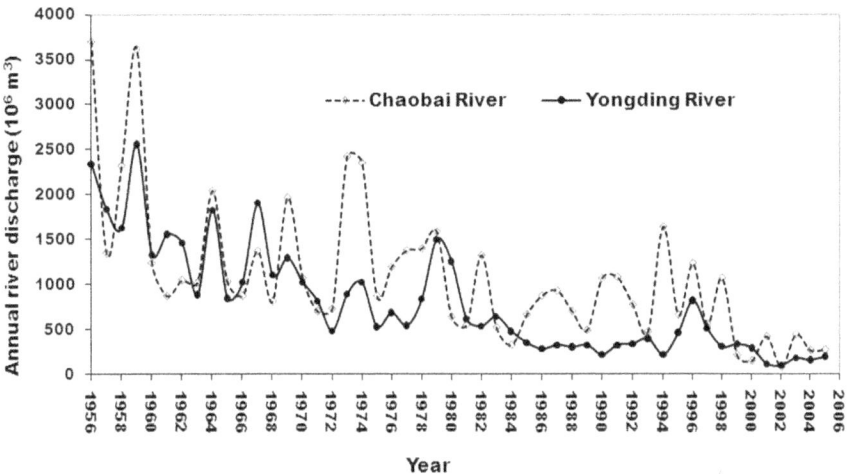

Figure 16.4.   Annual discharges of the Yongding and Chaobai rivers.

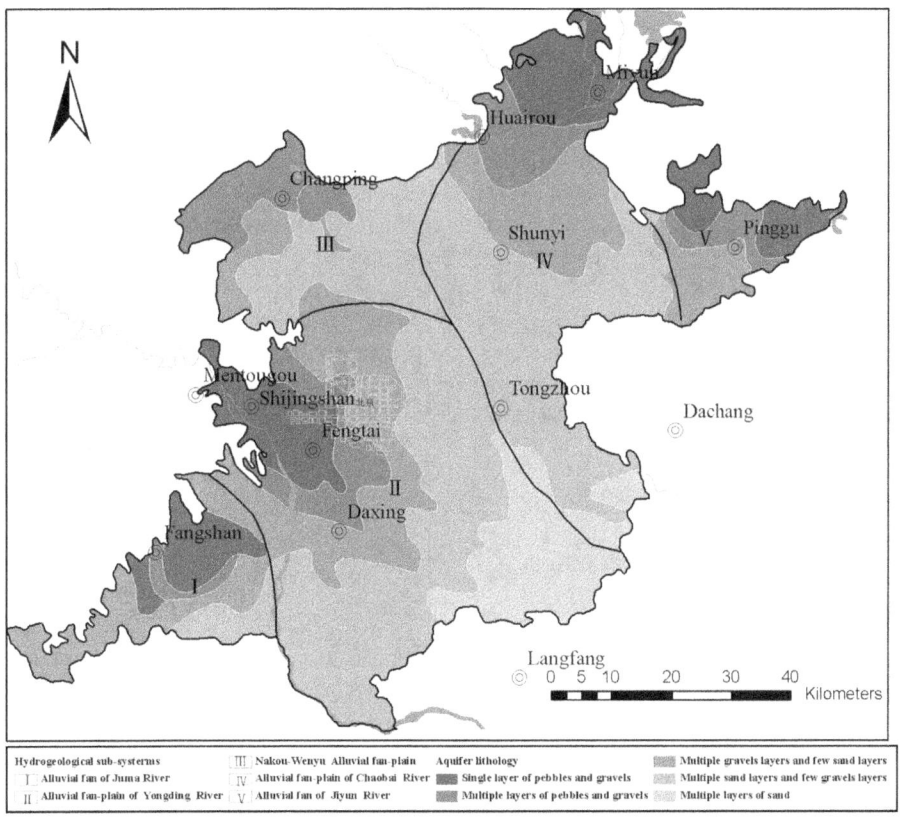

Figure 16.5.   Hydrogeology and aquifer lithology of the Beijing Plain.

Table 16.1   Groundwater balance components of Beijing Plain in 2000.

Components		Annual rate ($10^6\,m^3$)	Percentage
Recharge	Areal recharge	1,140	65%
	Lateral inflow and river leakage	625	35%
	Total	1,765	100%
Discharge	Abstraction	2,559	96%
	Outflow to downstream	49	2%
	Evapotranspiration	42	2%
	Total	2,650	100%
Change of storage		−885	50% of recharge

the shallow aquifer on the top of alluvial fans. The deep aquifers are confined. The recharge to the deep aquifers comes mainly from the leakage of the shallow aquifer. The majority of industrial water supply wells and some drinking water well fields are drilled into the deep aquifer. Groundwater recharge is mainly derived from infiltration of precipitation, lateral inflow from the mountain front, and river leakage. Groundwater discharge is dominated by abstraction, seepage to rivers, and underground outflow to downstream areas.

A groundwater resources assessment study (Beijing Geo-environmental Monitoring Station 2002) has compiled a groundwater balance table (Table 16.1) of the Beijing Plain in 2000. Areal recharge from precipitation infiltration is the major source of groundwater recharge. Lateral inflow from surrounding mountains is an important source of natural groundwater recharge, but river leakage is substantially reduced because rivers are dry during most time of the year. Abstraction for irrigation and urban water supply dominates groundwater discharge. Seepage to rivers has almost ceased. Underground groundwater outflow to downstream is also reduced. The evapotranspiration loss is low since the water table is deep. It is very clear that groundwater is severely overexploited, the depletion of groundwater storage in 2000 amounts to 50% of the total recharge.

16.2   RESULTS AND DISCUSSION

16.2.1   Detection of climate changes

Change of temperature

Shifts in the average mean annual temperature from ten meteorological stations were detected in 1988 and 1994 by the regime shift detection methods (Fig. 16.6). The mean annual temperature was increased by 1.3°C from the mean annual temperature of 11.6°C in the period 1959 to 1987 to the mean of 12.9°C in the period 1994–2009. Further analysis reveals that the mean winter and spring temperatures were increased by 1.7°C while the mean summer and autumn temperatures were increased by around 1°C. A much hotter spring means even higher demand of irrigation water for crops.

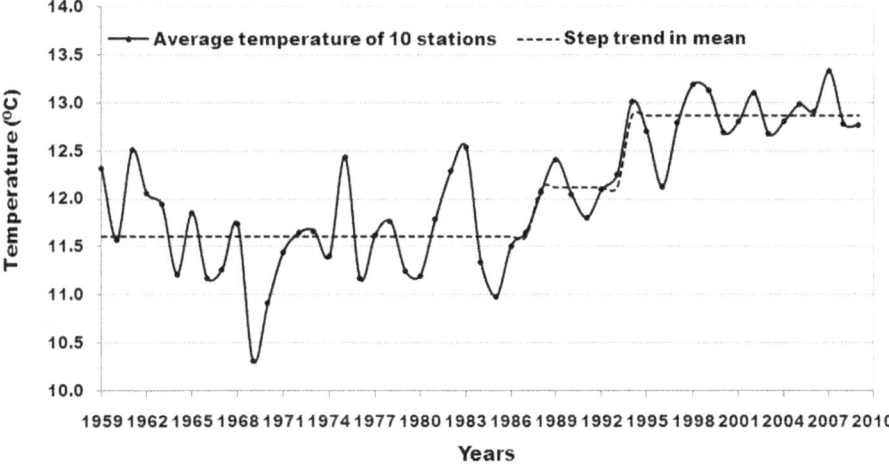

Figure 16.6.   Step trend of mean annual temperature increase detected by the regime shift detection methods.

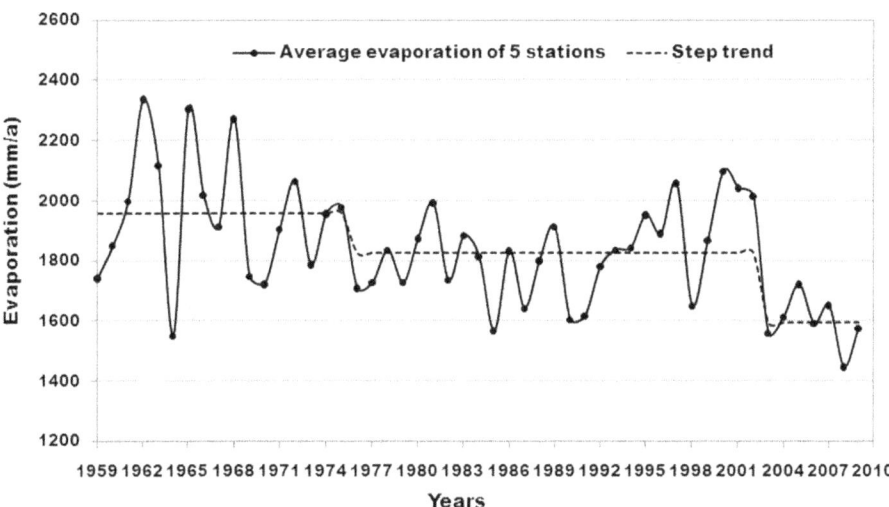

Figure 16.7.   Step trend of average annual evaporation decrease detected by the regime shift detection methods.

Change of evaporation

Shifts in average annual evaporation (measured with evaporation pan) of five meteorological stations with the long-term measurements were detected in 1976 and 2002 (Fig. 16.7). Contrary to the temperature increase, evaporation shows a steep trend of decrease. The mean annual evaporation was 1,955 mm/a in the period 1959–1975, 1,824 mm/a in the period 1976–2001, and 1,593 mm/a in the period 2002–2009, respectively, with a total decrease of 363 mm/a. The reason for this decrease might be attributed to global dimming (Liu et al. 2004).

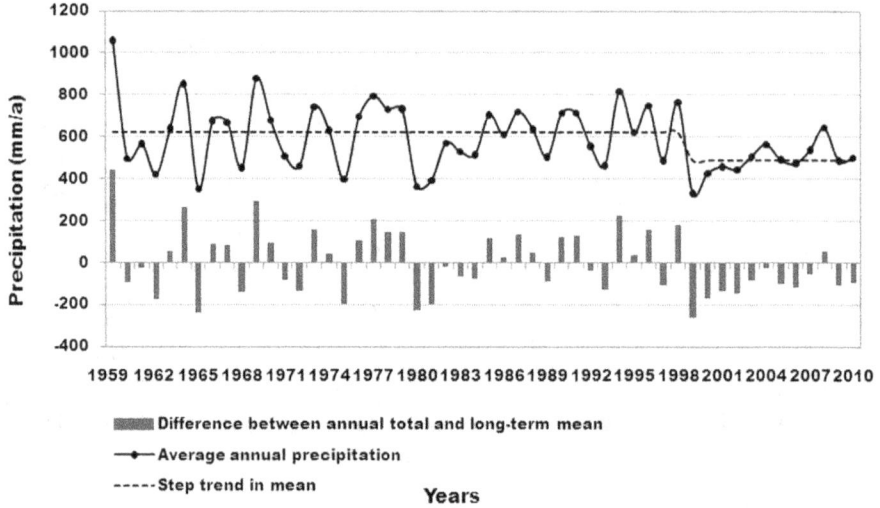

Figure 16.8.   Step trend of average annual precipitation detected by the regime shift detection methods.

Change of precipitation

Only one change point was detected at 1999 in the average annual precipitation of 11 meteorological stations (Fig. 16.8). The mean annual average precipitation was 621 mm/a in the period 1959–1998, and 488 mm/a in the period 1999–2010, only 78% of the previous mean. The difference between annual total precipitation and long-term mean (590 mm/a from 1959 to 2010) shows that annual precipitation in 11 years out of 12 years from 1999 to 2010 was below the long-term average, only the precipitation in the year 2008 was around the average. The other consecutive below-normal precipitation period occurred in 1980–1984.

The Beijing observatory has the longest continuous precipitation record in the country, with data series since 1914 (Fig. 16.9). The long-term average annual precipitation from 1914 to 2010 is 595 mm/a. Four periods of consecutive low precipitation can be found: four years in 1918–1921, five years in 1941–1945, five years in 1980–1984, and 12 years in 1999–2010. Clearly the recent drought spell from 1999 to 2010 is the longest and the severest in the recorded history. The only regime shift in the annual precipitation in this station was found also in 1999.

SPI drought index

The Standardized Precipitation Index (SPI) was calculated for average annual total rainfall of 11 meteorological stations (Fig. 16.10). A value of SPI zero indicates the normal year, higher than $+1$ indicates a wet year, and lower than -1 indicates a dry year. Two dry spells can be clearly seen in SPI index: one from 1980 to 1984, and the other from 1999 to 2010. It is also interesting to notice that the first period from 1959 to 1979 had 8 wet years and no dry years; followed by the first dry spell from 1980 to 1984. The period 1985 to 1998 was very normal, neither wet nor dry years; followed by the second dry spell from 1999 to 2010 which had no wet years, only one year (2008) above the normal,

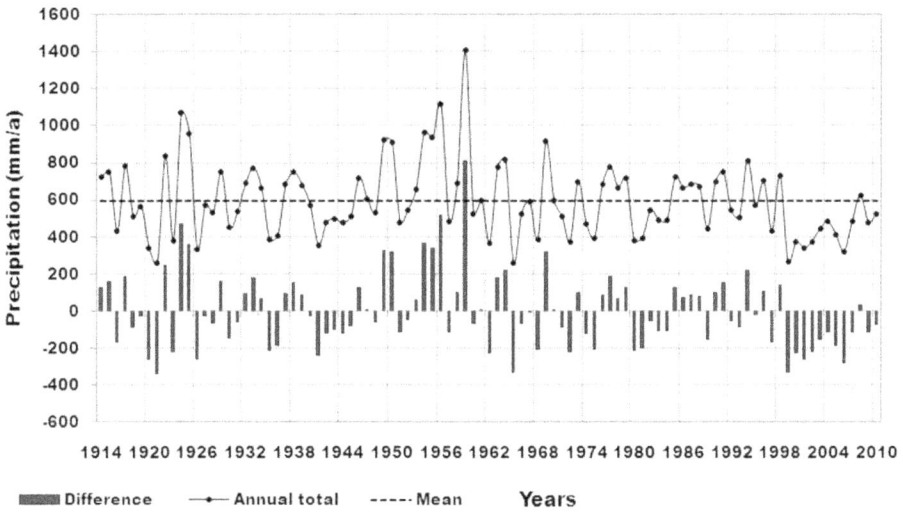

Figure 16.9.   Variations of annual precipitation at Beijing Observatory.

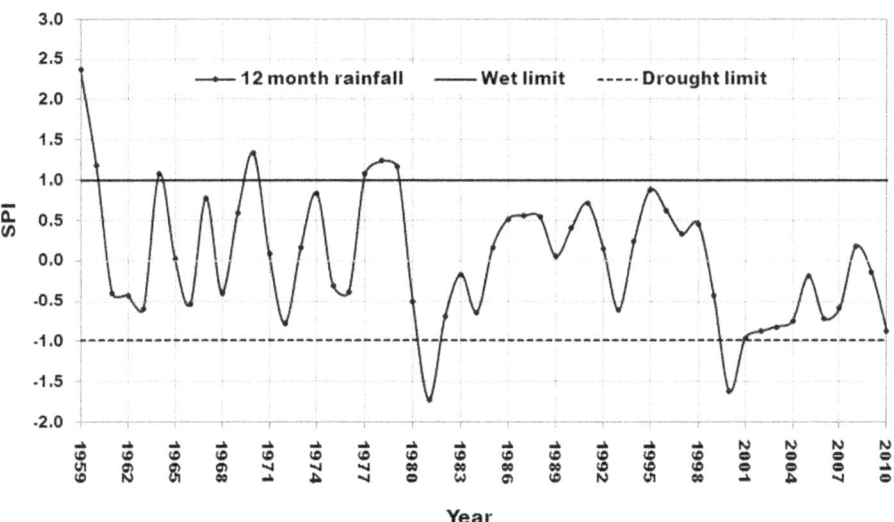

Figure 16.10.   SPI index for average annual total rainfall of 11 meteorological stations.

three dry years, and the rest below the normal. The SPI clearly shows a downward shift from more wet years in the 1960s and 70s to more dry years in the 2000s.

16.2.2   Analysis of rapid decline of groundwater levels

Three representative observation wells were used to analyze changes of monthly groundwater levels (Fig. 16.2). The observation well W22 is screened in the unconfined aquifer and located in the recharge area northwest of Beijing City, a part of the Yongding River

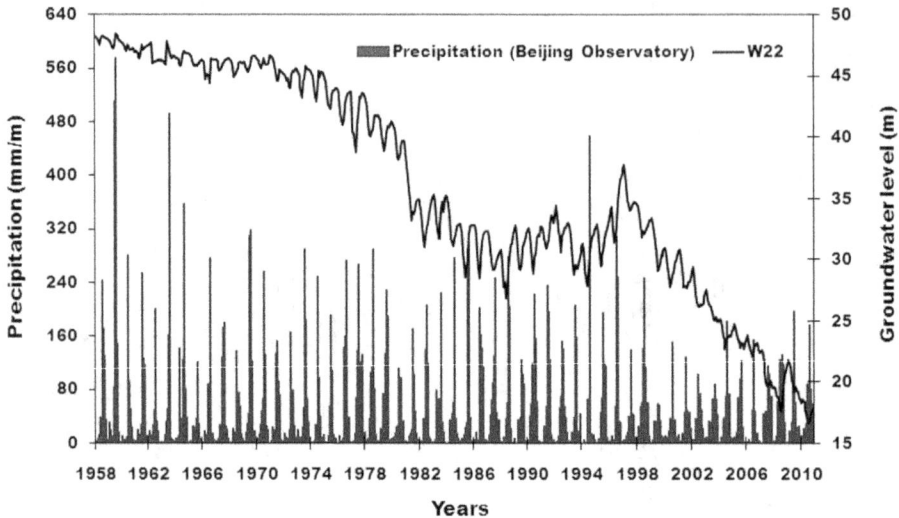

Figure 16.11.   Changes of monthly groundwater levels in W22 in relation to monthly precipitation.

alluvial fan. The observation well H8–39–A is also screened in the unconfined aquifer and located in the recharge area at the Chaobai River alluvial fan. The observation well 612–4 is screened in the deep confined aquifer and located at the cone of depression in the central area.

Figure 16.11 plots monthly groundwater level series at W22 in relation to monthly rainfall from Beijing Observatory station. It shows that the first period of rapid decline of groundwater levels occurred during the drought spell of five consecutive years from 1980 to 1984. The second period of rapid decline of groundwater levels occurred during the recent dry spell from 1999 to 2010. The linear trend analysis indicated a significant trend of decline of groundwater levels (Fig. 16.12). Groundwater levels have decreased by 14 m in 12 years from 1999 to 2010.

Figure 16.13 plots monthly groundwater levels at H8–39–A. Groundwater levels were in stationary variations from 1993 to 1998. Rapid decline of groundwater levels occurred since 1999. The linear trend detection indicated that groundwater levels have decreased by 40 m in 11 years from 1999 to 2009 (Fig. 16.14). The magnitude of groundwater level decrease at H8–39–A is much more than that the decrease at W22. This is due to the fact that a new emergency well field was constructed in the Chaobai River alluvial fan since 2003 (Li et al. 2006).

Figure 16.15 plots groundwater levels at the observation well 612–4 in the deep confined aquifer. It shows that groundwater levels continuously decrease since 1989. The decrease of groundwater levels from 1999 to 2010 is 22 m in 12 years (Fig. 16.16).

16.2.3   Simulation of groundwater depletion under recent droughts

A transient groundwater flow model was constructed using MODFLOW (McDonald and Harbaugh 1988) with the conceptual model approach (Brigham Young University – Environmental Modeling Research Laboratory 2000). The model consists of nine model layers. The model grids consist of 116 rows and 138 columns with a uniform grid size

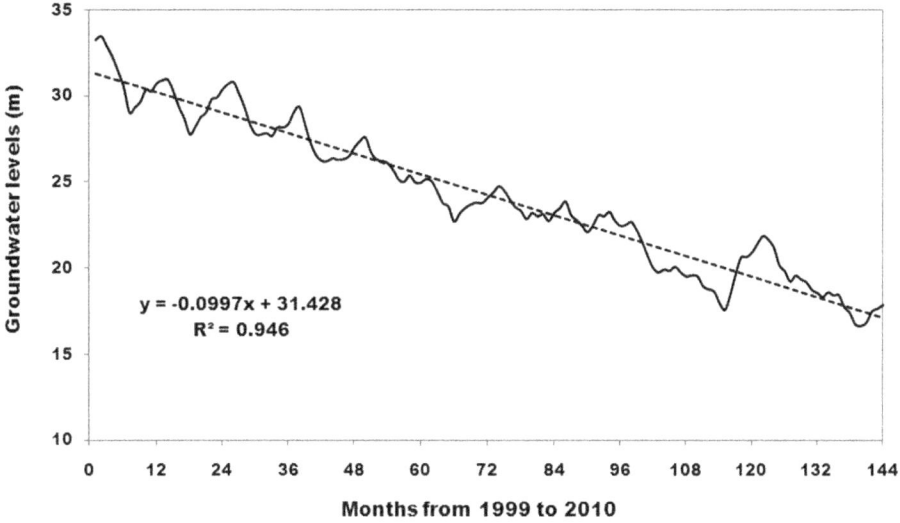

Figure 16.12.   Trend of decline of groundwater levels at W22 from 1999 to 2010.

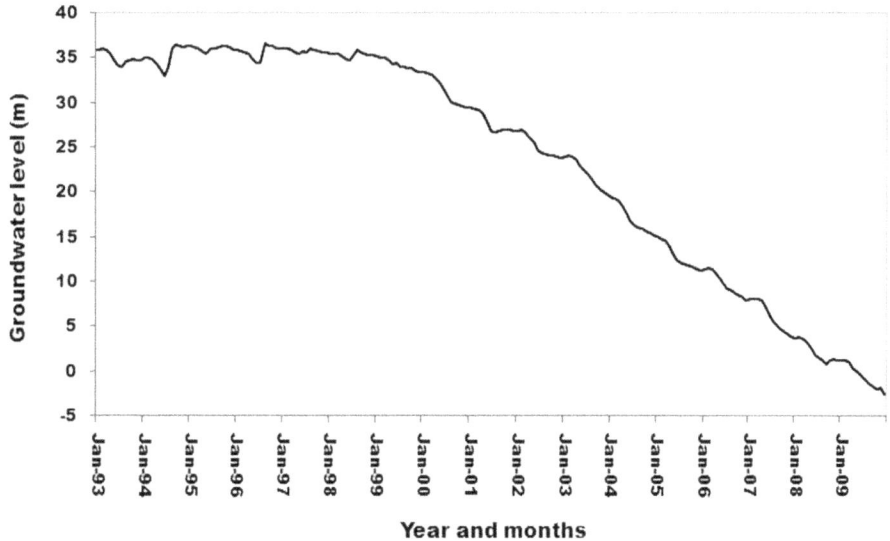

Figure 16.13.   Change of monthly groundwater levels at H8–39–A in the Chaobai River alluvial fan.

of $1,000\,\text{m} \times 1,000\,\text{m}$. The calibration time was 11 years from 1995 to 2005 divided into monthly stress periods resulting in 132 stress periods (Wang et al. 2008). The transient model was used to analyse options of sustainable groundwater resources development in the Beijing Plain (Zhou and Li 2011). In this study, the transient model was extended up to 2009 and used to analyse the impacts of drought on groundwater resources in the Beijing Plain.

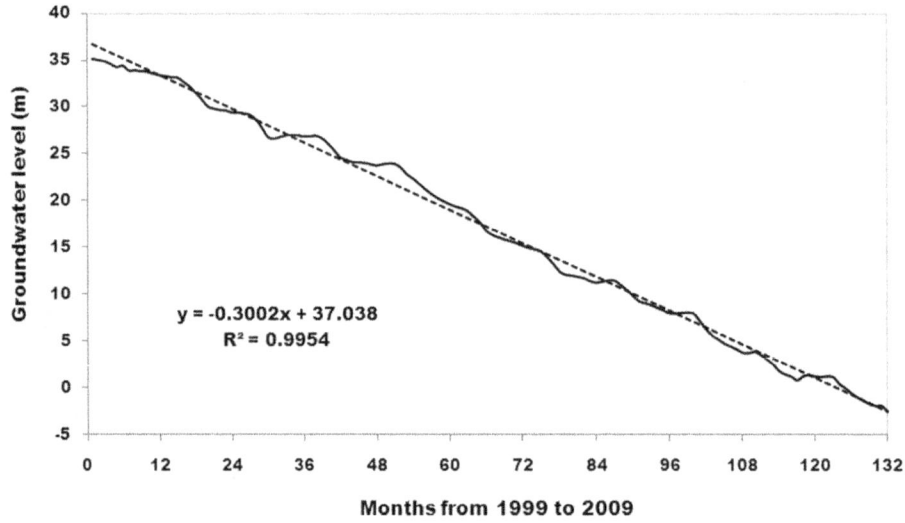

Figure 16.14.   Trend of decline of groundwater levels at H8–39–A from 1999 to 2009.

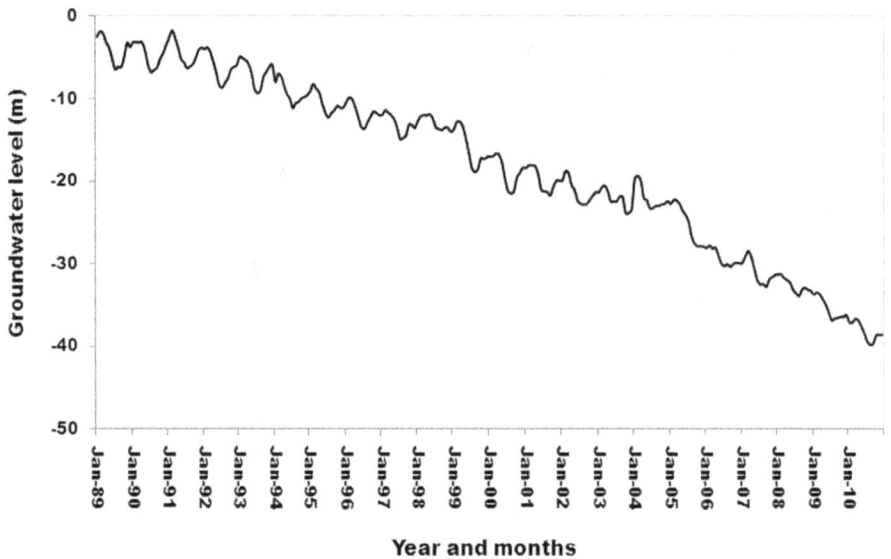

Figure 16.15.   Change of monthly groundwater levels at 612–4 in the cone of depression.

The drought is conceptualized to have a direct and an indirect effect on the depletion of groundwater storage. The direct impact is due to the decrease of groundwater recharge. The indirect impact is due to the increase of groundwater abstraction to compensate for the shortage of surface-water supply during the drought. Indicators for the assessment of climate impact on groundwater resources used in this study are the depletion of groundwater storage and the decrease of groundwater levels. In order to assess the direct impact, the transient groundwater flow model was run from

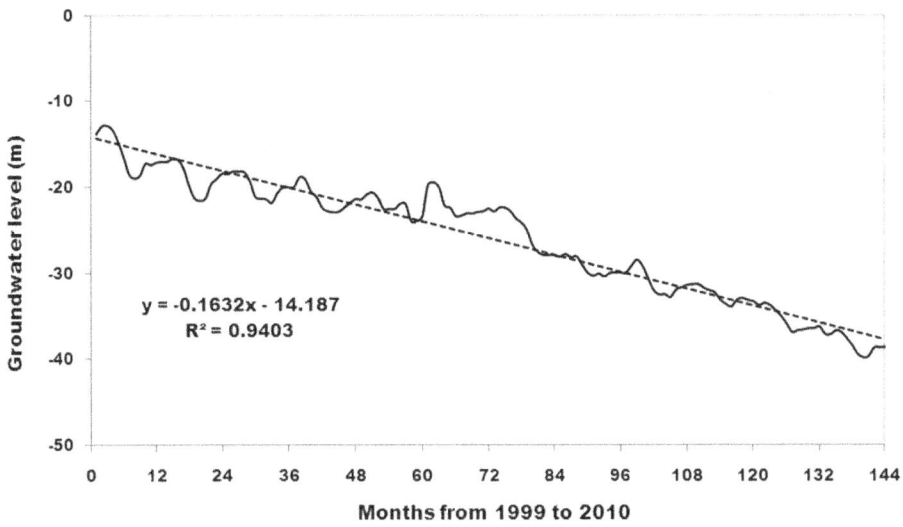

Figure 16.16.   Trend of decline of groundwater levels at 612–4 from 1999 to 2010.

1999 to 2009 under two situations. The first situation uses the actual groundwater recharge from monthly precipitation from 1999 to 2009, while groundwater abstraction is kept the same amount of $2.56 \times 10^9 \, m^3/a$ (which is from the well inventory data in 2000). The groundwater flow model for this first situation is called the direct climate model. The second situation uses groundwater recharge that was calculated from average precipitation from 1959 to 1998 before the recent drought, while groundwater abstraction is the same. The groundwater flow model for the second situation is called the benchmark model. The change in groundwater storage and water levels between the benchmark and direct climate models indicates the direct impact of the drought. For the assessment of the indirect impact of the drought, the transient groundwater flow model was re-calibrated with actual data of precipitation and abstraction from 1999 to 2009. This calibrated model represents the actual situation during the drought. The differences between the actual model and the benchmark model show the total impact of the drought on groundwater resources. The differences between the total impact and the direct impact indicate the indirect impact.

Figure 16.17 shows the accumulated depletion of groundwater storage in the Beijing Plain during 1999–2009. The direct impact of the reduction of groundwater recharge during the drought has caused a continuous depletion of groundwater storage. The drought-triggered increase of groundwater abstraction has accelerated the groundwater storage depletion.

The depletion of groundwater storage from 1999 to 2009 has caused a rapid decline of groundwater levels in the Beijing Plain. Figure 16.18 shows the spatial distribution of the accumulated decreases of groundwater levels at the end of 2009 as the total impact of the drought. The largest impact occurs in the Chaobai River catchment with a decrease in groundwater levels ranging from 8 to 34 m. The decrease in groundwater levels is primarily caused by increased abstraction from a large emergency well field in 2003. In

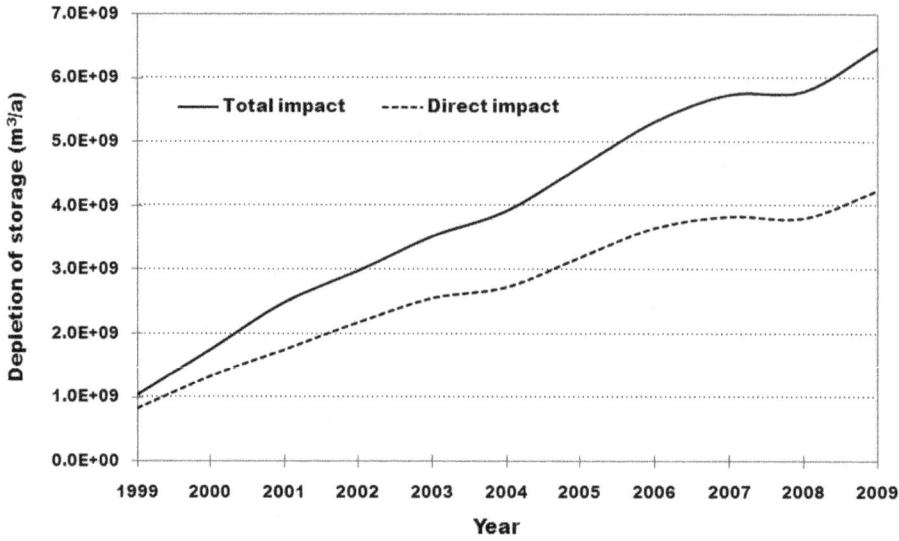

Figure 16.17.   Impact of the drought on the depletion of groundwater storage in the Beijing.

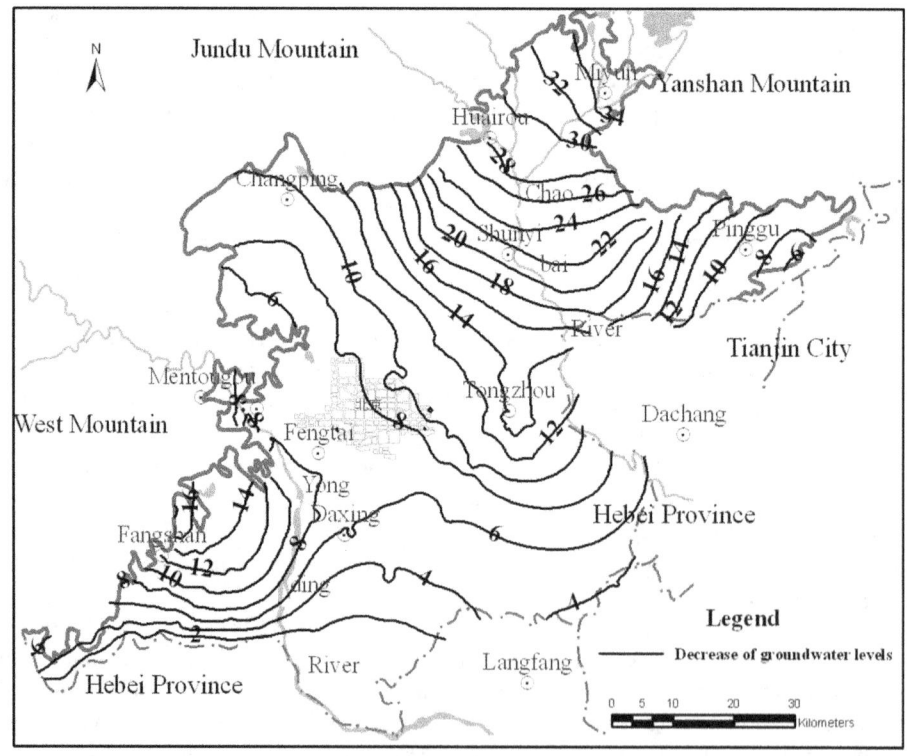

Figure 16.18.   Total impact of the drought on the decrease of groundwater levels from 1999 to 2009.

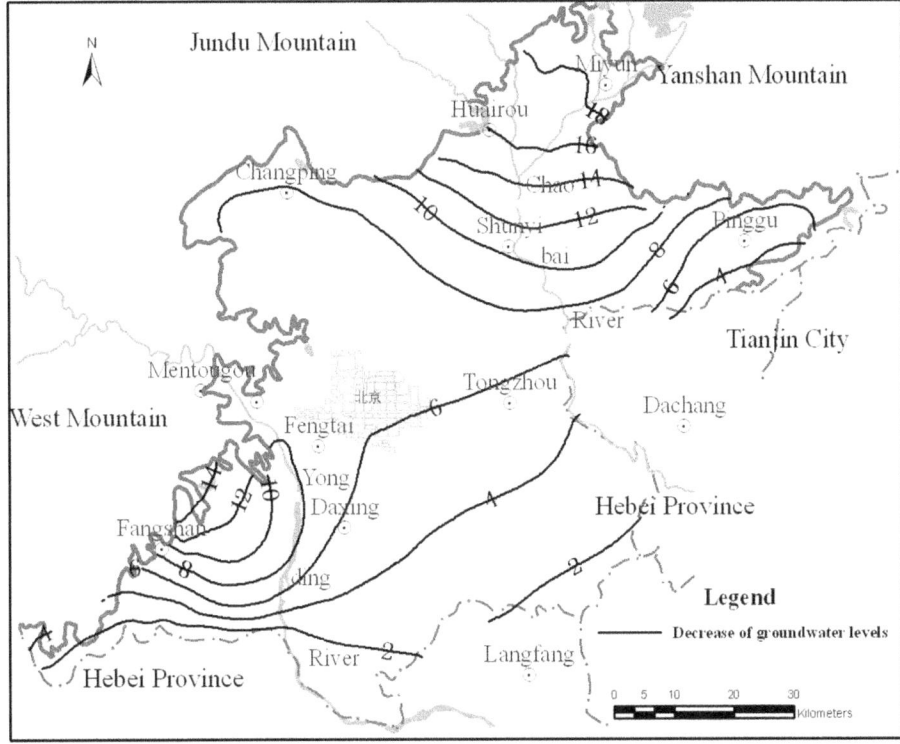

Figure 16.19. Direct impact of the drought on the decrease of groundwater levels from 1999 to 2009.

the Yongding River catchment, the total decrease of groundwater levels ranges from 6 to 16 m between 1999 and 2009.

The direct impact of the drought on groundwater levels is shown in Figure 16.19. Groundwater levels decline the most in the Chaobai River and Yongding River catchments because of the impact of drought on the reduction of recharge. Comparison of the contour maps of Figures 16.17 and 16.18 indicate that the total impact is much larger than the direct impact, indicating that the drought-triggered increase of groundwater abstraction has accelerated the depletion of groundwater storage and the decline of regional groundwater levels. The indirect impact of drought on the depletion of groundwater resources is most pronounced in the Chaobai River catchment because of abstraction from the newly constructed emergency well field.

Figure 16.20 shows the impact of the drought on groundwater levels at the observation well H8–39–A in the Chaobai River catchment. It is clear that the accumulated decrease of groundwater levels caused by direct impact of the drought is around 15 m while the total impact decreases groundwater levels by 30 m. The drought-triggered groundwater abstraction for emergency water supply has a profound impact on the depletion of groundwater resources in the Chaobai River catchment. This is not sustainable and the trend of depletion must be reversed.

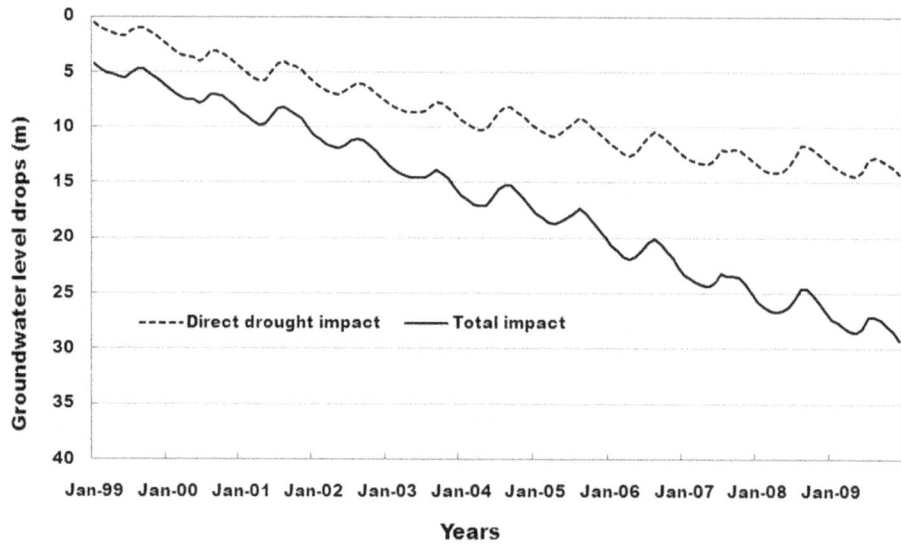

Figure 16.20.   Impact of the drought on the decrease of groundwater levels at H8-39-A.

16.2.4   Options for mitigating further groundwater depletion

Groundwater development in the Beijing Plain is not sustainable. Urgent measures are required to reverse the further depletion of groundwater resources. In the 11th Five Year Plan (from 2006 to 2010) for the protection and utilization of water resources in Beijing (Beijing Water Authority 2006), various measures were introduced to reduce the water demand and to increase water supply. On the water demand side, by the year 2010, the total water demand would be controlled at $4,000 \times 0^6 \, \text{m}^3/\text{a}$, of which urban water demand as $1,500 \times 10^6 \, \text{m}^3/\text{a}$; industrial as $700 \, 10^6 \, \text{m}^3/\text{a}$; agricultural as $1,300 \times 10^6 \, \text{m}^3/\text{a}$; and environmental water demand as $500 \times 10^6 \, \text{m}^3/\text{a}$. Compared to the previous situation, the major changes were to reduce agricultural water use by $400 \times 10^6 \, \text{m}^3/\text{a}$ and to allocate more water for environmental water requirement. On the water supply side, three major actions have been taken to argument the supply sources. The first action is to expand the wastewater reuse to $600 \times 10^6 \, \text{m}^3/\text{a}$. By 2008, a total of 14 wastewater treatment plants were operated for the treatment of 90% of urban wastewater; 30 wastewater treatment plants were constructed in townships with a treatment capacity of 50% of total wastewater. The second action is the completion of the middle route of the South to North water transfer project to transfer water from the Yangtze River to Beijing. By 2010 (it is now delayed to 2014), the transferred water to Beijing would reach $1,000 \times 10^6 \, \text{m}^3/\text{a}$. In a normal hydrological year, local surface-water supply is around $400 \times 10^6 \, \text{m}^3/\text{a}$, and groundwater abstraction can be controlled at $2,200 \times 10^6 \, \text{m}^3/\text{a}$. The combined water supply source amounts to $4,200 \times 10^6 \, \text{m}^3/\text{a}$. The third action is the rehabilitation of Yongding and Chaobai rivers. Both rivers have been designated as ecological corridors. The planned Yongding river ecological corridor has an area of $1,500 \, \text{km}^2$ and a total river length of 170 km: mountainous reach is 92 km and alluvial river reach in the plain area is 78 km. According to the plan (Beijing Water Authority 2009), two constructed wetlands and four lakes will be linked by flowing streams. The wetlands serve for inflow water treatment from pre-treated domestic wastewater and rainfall harvesting. The lakes are for water storage

Table 16.2   Groundwater development scenarios under a normal hydrological year (unit: $10^6 \, m^3/a$).

Scenarios	Recharge	Abstraction	Balance
Business as usual	2,300	2,600	−300
Reduction of abstraction	2,300	2,000	+300
Increase of recharge	2,900	2,600	+300
Combined	2,500	2,200	+300

and enhanced groundwater recharge. The total water surface area is around 1,000 ha and estimated groundwater recharge is around $100 \times 10^6 \, m^3/a$. The construction has started in 2010 and will be completed by 2012.

The implementation of the 11th five year water resources plan for Beijing City provides the opportunity to restore groundwater resources. The necessary condition to achieve sustainable groundwater resources development is to sustain groundwater abstraction with the increased recharge and decreased discharge (Zhou 2009). In the Beijing Plain, all natural groundwater discharges are ceased and all rivers are dry, there are no options to induce natural recharge and to reduce natural discharge. Artificially enhanced groundwater recharge methods are to be used to increase groundwater recharge, and reduction of groundwater abstraction is the only way to reduce groundwater discharge. Therefore, four basic scenarios were formulated to investigate options to achieve sustainable groundwater resources development in Beijing Plain (Table 16.2). In the normal hydrological year, the total groundwater recharge is estimated at $2,300 \times 10^6 \, m^3/a$. Under the business as usual scenario, the total groundwater abstraction of $2600 \times 10^6 \, m^3/a$ would continue. The reduction of abstraction and the increase of recharge scenarios will bring the groundwater in a positive balance to recover the depleted resources. The more realistic scenario is the combined increasing recharge and decreasing abstraction. Under this scenario, the excess of $200 \times 10^6 \, m^3/a$ supply source will be used for artificially enhanced groundwater recharge, and groundwater abstraction is reduced by $400 \times 10^6 \, m^3/a$. The effectiveness of these scenarios in restoring groundwater resources were analysed quantitatively with the transient groundwater flow model.

The calibrated transient groundwater model from 1995 to 2005 was used to simulate effects of proposed development scenarios. The simulation time was taken 50 years from 2006 to 2055. The calculated groundwater levels at December 2005 were used as the initial conditions. Natural groundwater recharge and boundary inflows under a normal hydrological year were used for all simulations. The current abstraction pattern and rates would continue for the business as usual scenario. For the reduction of abstraction scenario, three emergency well fields in the plain and most deep self-supply wells would be stopped, and pumping rates in another two well fields would be reduced. The total reduction of annual abstraction rates amounts to $600 \times 10^6 \, m^3$. To implement the increase of recharge scenario, reclaimed wastewater and transferred surface water would be artificially recharged to groundwater in river alluvial fans. The annual total artificial recharge reaches $600 \times 10^6 \, m^3$. The combined scenario could be more realistically implemented: closure of three emergency well fields and most self-supply wells would reduce the annual abstraction rate of $400 \times 10^6 \, m^3$, an annual total of $200 \times 10^6 \, m^3$ would be

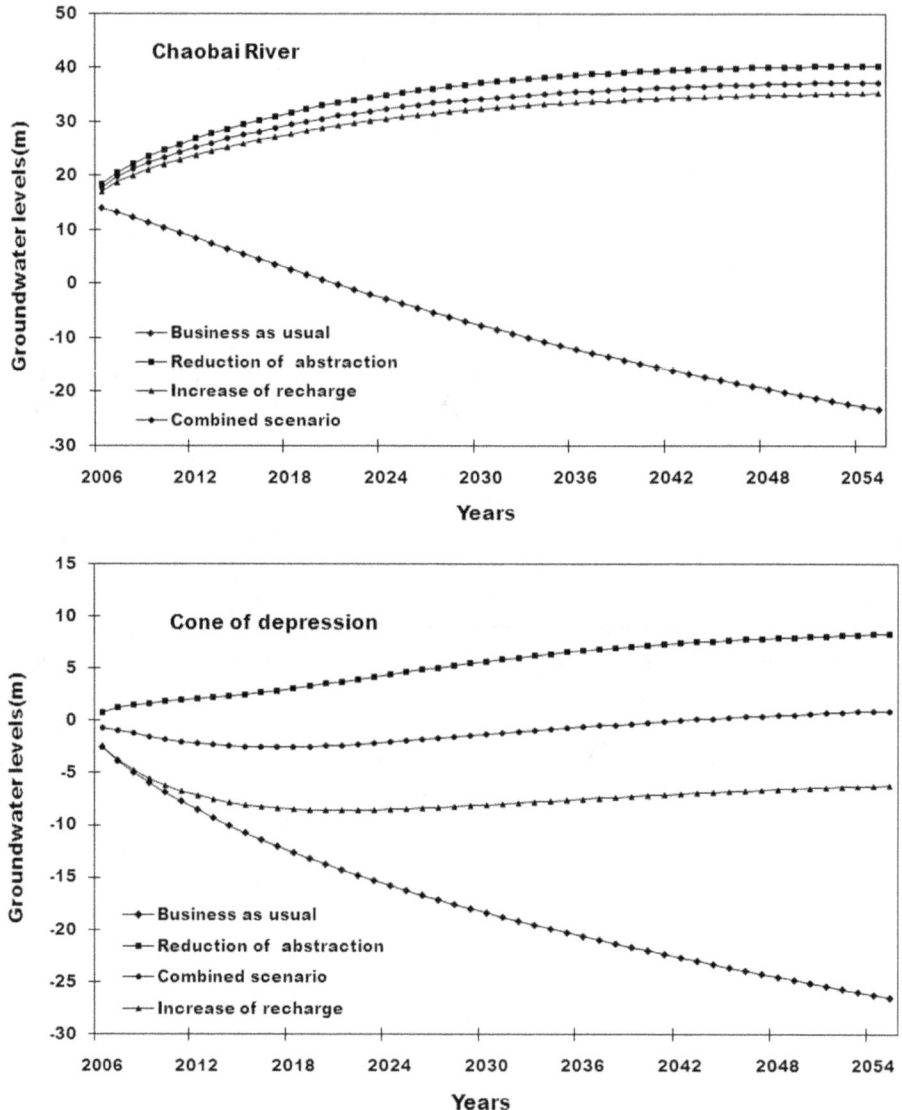

Figure 16.21.   Responses of groundwater levels to the development scenarios.

artificially recharged in the alluvial fans of the Yongding and Chaobai rivers. To compare different scenarios, the annual total recharge would exceed the total abstraction by $300 \times 10^6\,\mathrm{m}^3$ in the last three scenarios.

The responses of groundwater levels and storage changes were simulated to assess the effects. Figure 16.21 shows the change of groundwater levels at the Chaobai alluvial fan and the cones of depression under four development scenarios. It is clear that groundwater levels would continue to decline under the business as usual scenario. The rate of decline is much larger in the Chaobai river fan because of the operation of the emergency well field. The reduction of abstraction scenario would have immediate and fast recovery

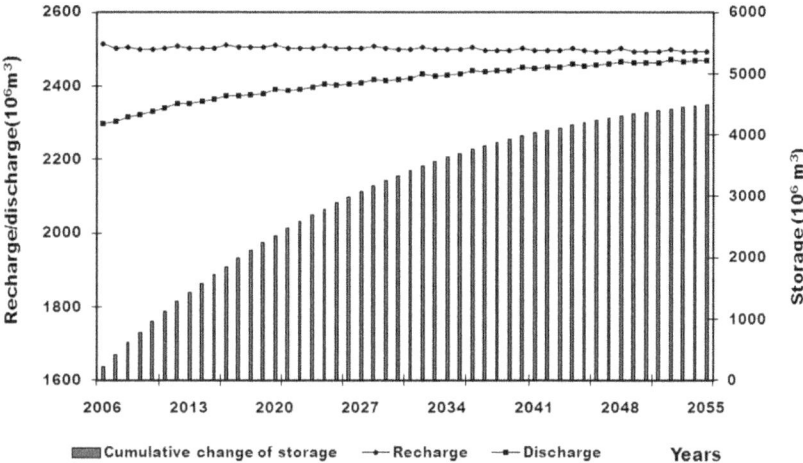

Figure 16.22.   Change of groundwater balance under the combined scenario.

of groundwater levels, especially at the cone of depression. The increase of recharge scenario would also reverse the trend of groundwater level decrease. However, it would take longer time for recharged water to have significant impacts on the recovery of groundwater levels, especially at the cone of depression. The combined scenario is less effective in recovering groundwater levels compared to the reduction of abstraction, but it is better than the increase of recharge scenario.

Figure 16.22 shows the change of groundwater balance under the combined scenario. Since the total groundwater recharge is larger than the total discharge, groundwater storage is gradually increasing. The rising groundwater levels result in the increase of discharge to the rivers and evapotranspiration rate, so that total groundwater discharge is gradually increasing until a new equilibrium state is reached. Under the new equilibrium condition, total groundwater recharge equals total discharge and the net change of groundwater storage becomes zero. The time to reach the approximate equilibrium state under the combined scenario is around 50 years. The increase of groundwater levels range between 10 to 35 m in the Chaobai river catchment, and 10 to 25 m in the Yongding river catchment (Fig. 16.22). A sustainable groundwater resources development could be achieved under the combined scenario.

16.3   MANAGEMENT ISSUES

16.3.1   Legal aspects

The new comprehensive Water Law of China was issued in 2002 (National People's Congress 2002). The law specifies the water resources planning, utilization, protection, allocation, and water saving aspects. Water use permit application procedures were detailed by the Ministry of Water Resources of China (Ministry of Water Resources 2006). For the extraction of groundwater with a well yield of more than 30 m^3/day, a permit can be only given with a feasibility study report conducted by an independent and qualified institution.

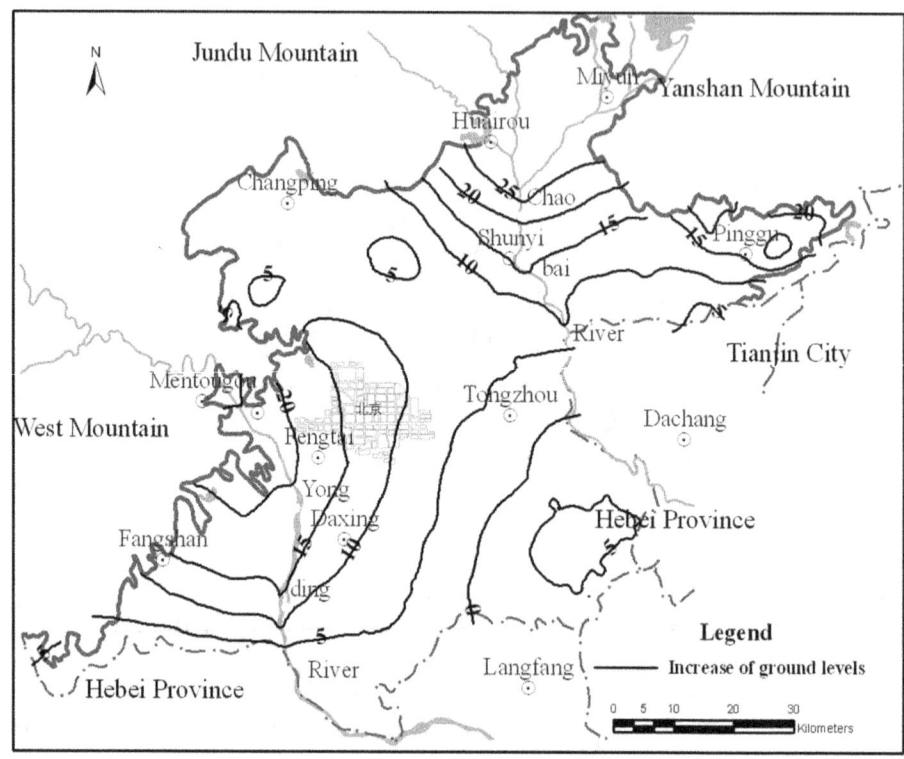

Figure 16.23   Contour map of the increase of groundwater levels when a new equilibrium state reached under the combined scenario compared to the groundwater levels in December, 2005.

Beijing municipality issued a Water Resources Management Act and a Water Saving Act, respectively, in 1991. These acts were further integrated into an implementation policy of the new Water Law of China in 2004 (Beijing People's Congress 2004). Water saving and control of groundwater abstraction were clearly specified. In areas of groundwater overexploitation and source areas of existing well fields, drilling of new wells is strictly controlled and needs to be approved by the Beijing Water Authority. In 2011, Beijing Municipality issued the Water Pollution Prevention Act (Beijing People's Congress 2011). Protection of groundwater pollution, re-use of treated wastewater, monitoring of groundwater quality and pollution are included.

16.3.2   Institutional aspects

The Beijing Water Affairs Department is the water authority in charge of the planning and management of water resources in Beijing City. Departments of Land Resources, Infrastructure, and Environmental Protection are also involved in water resources development, management and protection. In recent years, Beijing Water Affairs Department is implementing projects of the rehabilitation of the Yongding River and the Beijing part of the middle route of South-North water transfer. In cooperation with Environmental

Protection Department, the Department of Land Resources has established a network of groundwater quality and pollution monitoring in the Beijing Plain. The Infrastructure Department was responsible for constructing six emergency well fields since 2003.

Despite the control of well drilling by regulations started as early as in 1991, groundwater overexploitation continues. The responsibility for groundwater overexploitation is with the municipal government itself. Especially during the recent consecutive drought, six new emergency well fields were constructed with a combined capacity of $400 \times 10^6 \, \text{m}^3/\text{a}$, which attributes to rapid depletion of groundwater resources in the Beijing Plain.

16.3.3 A drought management plan

China suffers from drought hazards every year. Only in 2009, the State Council issued Directives on combating drought (State Council 2009). It includes clauses on prevention, mitigation, and recovery of drought hazards.

Despite the fact that Beijing was struck several times in the past by prolonged droughts, Beijing municipality has not yet developed a drought management plan. The drought crisis were handled ad hoc by drilling new wells. The last consecutive drought has forced Beijing municipal government to implement the strict water saving measures in 2011 in conjunction with the 12th five year plan (2011–2016). The objectives are to control total water use, to increase water use efficiency, and to prevent water pollution. However, the absence of a drought management plan does not only accelerate groundwater depletion, but also causes severe economic, ecological, and social damages. Many countries have developed drought management plans (White et al. 1993; Wilhite 1997; European Commission 2007). The Directives on combating drought (State Council 2009) require provincial and municipal governments to develop drought management plans, but don't specify what should be in this plan. A proper drought management plan should consist of three key components (Wilhite and Svoboda 2000): (1) monitoring and early warning; (2) risk and impact assessment; and (3) mitigation and response.

A technical drought mitigation strategy for Beijing should consider the storage of water in wet years for the use during the dry years. Aquifers are the best natural reservoir for water storage. Water demand must be reduced first in order to have excess water in wet years to be stored in the aquifer by artificially enhanced groundwater recharge. The control of water demand should include measures of water permit, increase of water use efficiency, closure of water-intensive industries, reduction of agricultural water use, and cyclic use of urban waters. Because of the depletion, aquifers in the Beijing Plain have very large storage capacity. High permeability at alluvial fans provide ideal locations for artificial groundwater recharge. Recovery of groundwater storage in the coming years will provide reserves to combat next drought.

16.4 CONCLUSIONS AND RECOMMENDATIONS

There is no doubt that climate is changing in the Beijing Plain: temperature is increasing, evaporation is decreasing, but most importantly, Beijing is suffering a prolonged drought of 12 years from 1999 to 2010. The impacts of the recent drought on groundwater are two-folds: the direct impact due to the reduction of natural groundwater recharge, and the indirect impact due to the drought-triggered increase of groundwater abstraction for emergency water supply. The combined impacts have caused rapid depletion of ground-water storage. The observed total decreases of groundwater levels in 12 years amount to 20 to 40 m. The drought-triggered groundwater abstraction has a profound impact on the depletion of groundwater storage in the Chaobai River catchment. This is due to the fact that the Beijing municipal government reacts to the drought with crisis management. The shortage of water supply during the drought was simply met with the construction of new emergency well fields. A drought management plan is urgently required to shift from cri-sis management to drought preparedness and risk management.

Beijing faces unprecedented challenges in managing scarce water resources to meet the increasing demand for rapid urban expansion and population increase. The situation of groundwater depletion is seriously grim. To achieve water and environmental secu-rity with 190 m^3/a per capita water resources is very difficult. The long-awaited South to North water transfer project may provide some relief to thirsting aquifers and ecosystems in the near future. In the long run, Beijing has to further reduce agricultural and indus-trial water uses if it continues developing into an international centre of politics, culture, and tourism. The importance, limitations, and grim status of groundwater resources must be clearly recognised and communicated to decision-makers and public. There is still an opportunity to reverse the trend of groundwater storage depletion, if emergency well fields were closed and groundwater storage was recovered with the artificially enhanced groundwater recharge during wet years. Only then, aquifers could be used as storage res-ervoirs to combat the next drought in the Beijing Plain.

ACKNOWLEDGEMENTS

The authors are grateful for the constructive comments made by Dr Neno Kukuric and Dr Jianyao Chen which were useful for the improvement of the manuscript.

REFERENCES

Beijing Engineering Geological Survey Institute. (2007) *Investigation and monitoring of land sub-sidence in Beijing* (internal report in Chinese), Eds. Jia, S., Zhao,S., Luo,Y., et al.

Beijing Geo-environmental Monitoring Station. (2002) *Second round groundwater resources assessment in Beijing* (internal report in Chinese), Eds. Sun, Y., Miao, L., Xie, Z., et al.

Beijing Municipal People's Congress. (2004) Implementation policy of the Water Law of China in Beijing (in Chinese), [Online] Available from: http://www.cnki.com.cn/Article/CJFDTotal-BMZB200417004.htm [Cited 6th June 2011].

Beijing People's Congress. (2011) *Water Pollution Prevention Act in Beijing* (in Chinese).

Beijing Statistics Bureau. (2010) *Beijing Statistical Year Book* (in Chinese). China Statistics Press, Beijing, China

Beijing Water Authority. (2006) *The 11th 5 year plan of water resources development and protec-tion in Beijing* (internal report in Chinese), Beijing, China.

Beijing Water Authority. (2009) *Plan for the rehabilitation and ecological corridor construction of the Yongding River* (internal report in Chinese).

Brigham Young University – Environmental Modelling Research Laboratory. (2000) *Groundwater Modeling System (GMS)*, GMS 3.1 Tutorials.

European Commission. (2007) *Drought management plan report, including agricultural, drought indicators and climate change aspects*. Technical Report-2008-023.

Li, W., Hao, A., Ye, C. & Zheng, Y. (2006) Emergency plan for water supply in consecutive drought and sustainable water resources management in Beijing. In: Vrba, J. & Salamat, A.R. (eds.) *Proceedings of the International Workshop*, October 2006, Tehran. IHP-VI, Series on Groundwater No. 15.

Liu, B., Xu, M., Henderson, M. & Gong, W. (2004) A spatial analysis of pan evaporation trends in China. 1955-2000. *Journal of Geophysical Research*. [Online] 109 (D15), D15102. Available from: doi:10.1029/2004JD004511[Cited 6th June 2011].

McDonald, M.G. & Harbaugh, A.W. (1988) *A Modular Three-dimensional Finite-difference Ground-water Flow Model*. U.S. Geological Survey Open-File Report 83-875, Book 6.

McKee, T.B., Doesken, N.J. & Kleist, J. (1993) The relationship of drought frequency and duration to time scales. In: *Preprints, 8th Conference on Applied Climatology*, 17–22 January, Anaheim, CA. pp. 179–184.

Ministry of Water Resources. (2006) *Regulations on Water Use Permit and Charges* (in Chinese). [Online] Available from: http://www.mwr.gov.cn/zwzc/zcfg/xzfghfgxwj/200602/t20060221_155923.html [Cited 6th June 2011].

National People's Congress. (2002) *Water Law of The People's Republic of China* (in Chinese). [Online] Available from: http://www.mwr.gov.cn/zwzc/zcfg/fl/200210/t20021001_155904.html [Cited 6th June 2011]. Rodionov, S.N. (2004) A sequential algorithm for testing climate regime shifts. *Geophysical Research Letter*, 31, L09204.

State Council. (2009) *Directives of combating drought in The People's Republic of China* (in Chinese). [Online] Amiable from: http://www.mwr.gov.cn/zwzc/zcfg/xzfghfgxwj/200903/t20090308_155927.html [Cited 6th June 2011].

Taylor, W. (2000) *Change-point analysis: A powerful new tool for detecting changes*. [Online] Available from: http://www.variation.com/cpa/tech/changepoint [Cited 6th June 2011].

UNESCO-IHP. (2008) *GRAPHIC Framework Document*. [Online] Available from: http://www.unesco.org/new/en/natural-sciences/environment/water/ihp/ihp-programmes/graphic/publications [Cited 6th June 2011].

Wang, L., Han, J., Liu, J., Ye, C., Zheng, Y., Wan, L., Li, W. & Zhou, Y. (2009) Modelling of regional groundwater flow in the Beijing Plain. *Chinese Journal of Hydrogeology and Engineering Geology*, 36(1), 11–19.

White, D.H., Collins, D. & Howden, M. (1993) Drought in Australia: Prediction, Monitoring, Management, and Policy. In: Wilhite, D.A. (ed.) *Drought Assessment, Management, and Planning: Theory and Case Studies*. Boston, MA, Kluwer Academic Publishers. 320 pp.

Wilhite, D.A. (1997) *Improving Drought Management in the West: The Role of Mitigation and Preparedness*. Report to the Western Water Policy Review Advisory Commission.

Wilhite, D.A. & Svoboda, M.D. (2000) Drought early warning systems in the context of drought preparedness and mitigation. In: *Proceedings of an Expert Group Meeting*, 5–7 September 2000, Lisbon, Portugal.

Yang, H. & Zehnder, A. (2001) China's regional water scarcity and implications for grain supply and trade. *Environment and Planning*, 33, 79–95.

Zhou, Y. (2009) A critical review of groundwater budget myth, safe yield and sustainability. *Journal of Hydrology*. [Online] Available from: doi:10.1016/j.hydrol.2009.03.009

Zhou, Y. & Li, W. (2011) *Groundwater Monitoring, Information System, Modelling and Sustainable Development*. China Sciences Publishing (in Chinese), Beijing, China. 372 pp.

CHAPTER 17

Possible effects of climate change on hydrogeological systems: results from research on Esker aquifers in northern Finland

Bjørn Kløve, Pertti Ala-aho, Jarkko Okkonen & Pekka Rossi

ABSTRACT

Eskers are important aquifers in Fenno-Scandia and many other Northern regions where the crystalline bedrock is covered by glaciofluvial sand and gravel deposits. Esker ridges stretches in the direction of past ice melt. In coastal regions eskers are covered by marine deposits. Inland systems have unconfined recharge areas and sometimes confined discharge areas. Eskers are often connected to important ecosystems such as lakes, rivers, springs and peatlands. Kettle lakes are located within the esker and are usually groundwater-dependent. In Finland, groundwater in eskers is threatened by peatland drainage, roads and other land-uses. In this paper we review previous results on climate change in Finland, present results from our previous studies on eskers and discuss the impacts of climate change and land-use in eskers. Climate models predict an increase in precipitation but this is uncertain. Analysis of past annual climate variability based on 100-year data show a clear cyclical variation in climate, without any clear trend in precipitation or temperature in Central Finland. Periods of dry and wet years seem to occur at regular intervals. Assuming current predictions on changes of climate in Finland, the period of permanent snow cover will decrease. In central Finland, recharge will with assumed warmer climate occur in areas with permanent winter snow cover. Consequently, the minimum groundwater level in winter will increase in winter as recharge occurs all year. Modelling also show a risk of reduced summer groundwater levels as the snow melt recharge occur earlier in spring. As changes in winter climate seems to be the main impact of climate warming, groundwater models should include winter processes such as snow accumulation, melt and impact of frost on soil permeability.

17.1   INTRODUCTION

Eskers are a typical landscape element of the boreal zone in the lowlands of Fenno-Scandia, Russia and Canada, where the landscape is typically covered also by lakes, peatlands, bed outcrops and glacial till. Eskers aquifers were formed during the last deglaciation some 9000–12000 years ago (Tikkanen 2002). Long and narrow formations of sand and gravel are associated with the retreat of the ice (Svendsen et al. 2004), which in Fenno-Scandinavia had its centre just north of the Bay of Bothnia (Fig. 17.1). Eskers form important aquifers in the Fenno-Scandinavian shield and are also common in other regions covered by the last glaciation.

Besides being the main source of potable groundwater in Finland, eskers support many important ecosystem services, including wildlife habitats, recreation, hunting,

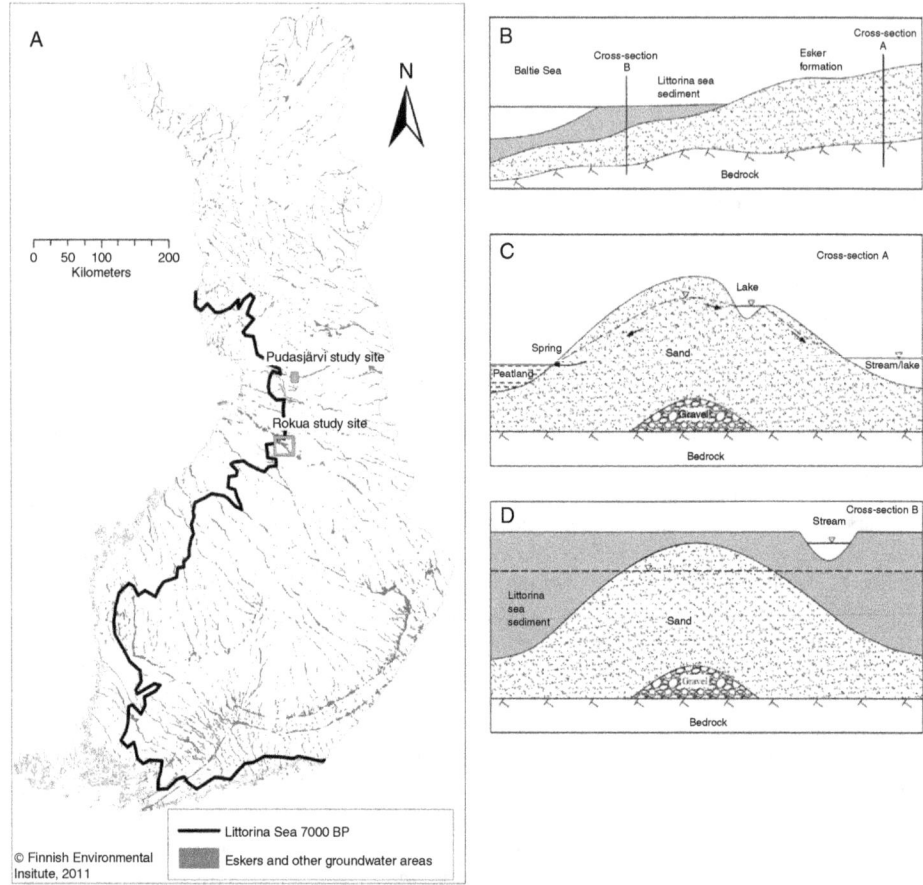

Figure 17.1.   **a** Map of Finnish eskers (SYKE) and the past Litorina sea shore line with location of Esker case sites Pudasjärvi and Rokua, **b** cross-section showing the Litorina sea sediment deposits and glacifluvial sand and gravel deposits, **c** cross-section and main flow lines of unconfined inland eskers, and **d** cross-section of costal esker confined by former sea sediments.

lichens, berry and mushroom picking. Eskers are often connected to rivers, lakes and wetlands, and such groundwater-dependent systems are of high ecological value (Kløve et al. 2011a; 2011b). Groundwater-fed lakes are often clear and clean and therefore have a high recreational value. Land use poses a threat to esker groundwater. Roads and their associated pollution risk pose a major threat to esker groundwater. Roads are often laid on eskers as these sand deposits are suitable for construction. Gravel and sand mining is a threat locally. Wide-scale peatland drainage for forestry, peat harvesting, and agriculture occurs in the esker discharge or recharge zones and this impact on groundwater is not well understood.

The impact of climate change and global warming in the cold region is predicted to occur as increased temperature and precipitation (IPCC 2007). In Finland, a linear trend analysis shows 0.76°C warming in the 20th century with most warming in the spring

(Tuomenvirta 2004). The mean annual runoff shows no trend for the period 1912–2004, but the seasonal distribution of streamflow has changed with increases in river discharge in winter and spring months (Korhonen and Kuusisto 2010). The timing of the spring has changed and it now occurs earlier in some catchments (Korhonen and Kuusisto 2010). Climate change scenarios show that in Northern and Central Finland the snowmelt floods decrease or remain unchanged (Veijalainen et al. 2010) due to decreasing snow accumulation (Venäläinen et al. 2001). In southern Finland, the river runoff may increase showing the importance for local predictions and hydrological knowledge (Veijalainen et al. 2010). So far, climate changes have mainly been observed and simulated for surface waters. Changes are expected for groundwater as a result of changes in snow accumulation and melt, precipitation and evapotranspiration (Okkonen et al. 2010). Recharge patterns in Southern Finland with melt in winter will be more likely also in Central Finland (Mäkinen et al. 2008). Future simulations with a warming climate by Okkonen and Kløve (2010) show a more irregular recharge pattern and groundwater level variation after a potential climate warming in central Finland due to melt in winter months. As eskers are connected to surface water, potential changes in surface water can also affect groundwater. The impact of climate change on the groundwater-dependent ecosystems is not known, and not included in water industry planning.

In order to better understand the impacts of climate change and land-use on groundwater, several studies have been initiated on eskers in the Oulu region by the University of Oulu in the past decade. The aim of these studies is to determine the impacts of climate change and climate variability. As eskers have been little studied in the past, a particular aim is to examine the hydrology of these systems. Studies on groundwater flowpaths, recharge, and discharge patterns have been initiated and several sites have been studied. As land-use changes can have similar impacts to climate change (Prowse et al. 2006), these impacts must be included when climate changes are being analysed. The impact of peatland drainage for forestry is being studied in the Rokua esker (Fig. 17.1), where restoration of peatland ditches is also being carried out. A key environmental consequence of changes in groundwater level is changes in groundwater-dependent ecosystems. The impacts of groundwater-surface water interaction have been studied in the Pudasjärvi and Rokua eskers. In Rokua, detailed studies have been conducted on groundwater interaction in the kettle lakes that are commonly found in eskers. The results show that the water level in these valuable kettle lakes has declined, which has caused concern to holiday home owners and local authorities (Ala-aho 2010). This paper reviews past research and summarizes the research that has been carried out at the two eskers and their groundwater dependent ecosystems. In particular, it evaluates the impact of climate change on groundwater based on previous work on foreseen changes to Finnish climate in Central Finland where the studied eskers are located.

17.1.1   Study area description: esker aquifers, northern Finland

Groundwater in eskers can be unconfined or confined. Confined eskers are typically found in the coastal region, where eskers are covered by marine clay from the former Baltic Sea stages, such as the Littorina Sea (e.g. Tikkanen 2002; Miettinen 2008; Sohlenius et al. 1996). These eskers can receive water from marine deposits or inland eskers if a hydraulic connection exists. Sometimes a chain of eskers stretches inland (Fig. 17.1), but the contact between individual deposits is not well known. For inland eskers (Fig. 17.1c),

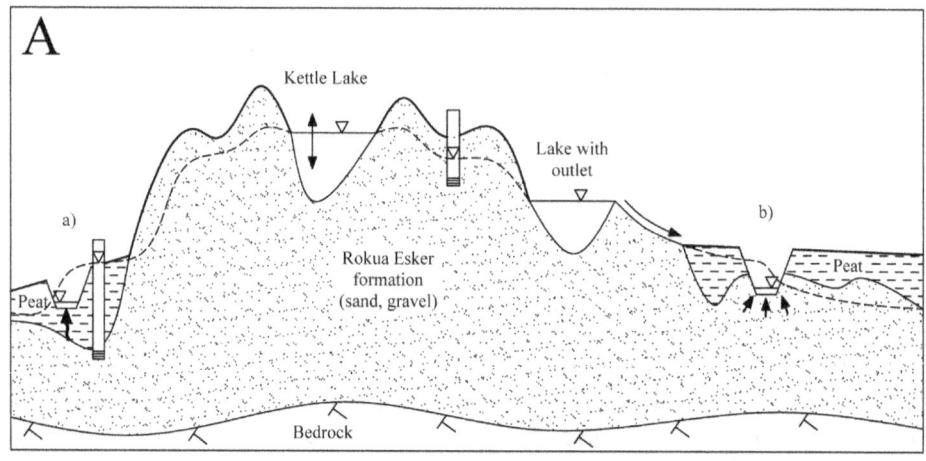

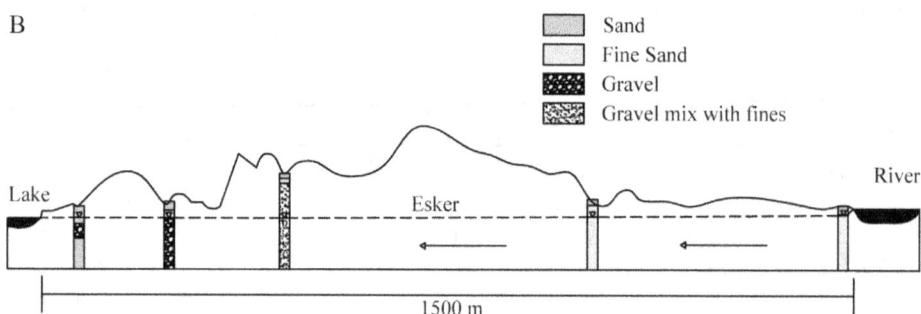

Figure 17.2.   **a** Cross-section of Rokua esker with location of groundwater table, potentiometric surfaces and indication of water level variation and flow patterns in ecosystems (arrows), and **b** Pudasjärvi esker with adjacent surface water and groundwater dependent ecosystems.

the unconfined part normally forms the esker recharge area. On the edges the eskers are typically confined by peatlands and the aquifers can be artesian (Fig. 17.2a). Some eskers discharge directly to lakes and rivers, e.g. the Pudasjärvi esker (Fig. 17.2b).

The Rokua esker area forms part of a long esker ridge stretching inland from the North Ostrobothnian coast (Aartolahti 1974; Ala-aho 2010). It is situated 100 km inland from the coast, has an area of 90 km^2 and rises at its highest point about 80 m above the surrounding peatlands. It is clearly visible in an otherwise flat landscape. The esker material consists mainly of sand with layers varying in thickness from 30 m to more than 100 m above the bedrock. A deposit of gravel has also been found (Fig. 17.2a). Rokua has a rolling terrain because of kettle hole, wave action and aolian dunes (Aartolahti 1974). In contrast, the surrounding peatlands started to form some 8000 years ago between the sand deposits and in some kettle holes (Pajunen 1995). These peat layers have grown to be in some locations more than 5 m thick and have a low permeability (Fig. 17.2a).

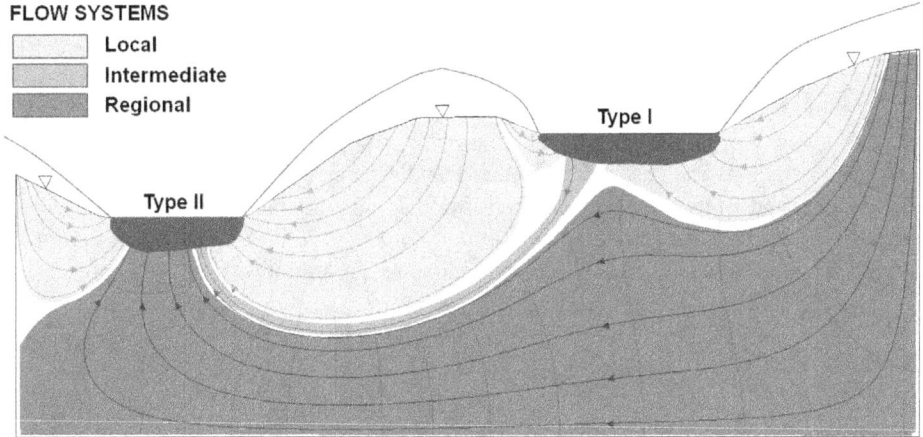

Figure 17.3. Conceptual drawing of an oligotrophic kettle lakes type I receiving local flow system, and eutrophic lake type II receiving recharge also from the regional groundwater.

The central part of the Pudasjärvi aquifer consists of gravel extending from the bedrock to the ground surface. In the northern part of the aquifer, there is a thick layer of silt that creates a perched water table 6 m higher than the actual groundwater. Fine sand and glacial till border the gravel, but extend to the ground surface in only a few parts of the aquifer. Sand, ranging from coarse to very fine, occurs in various parts of the aquifer, in layers varying in thickness from 1 m to 20 m. The ground surface is from 110 to 135 m above sea level. The surface of the bedrock varies between 94 and 96 m above sea level and the bedrock consists of gneiss and granite (Fig. 17.2b). The esker is in contact with surrounding surface water as seen in Figure 17.2b and 17.3.

17.1.2   Importance of esker aquifers in climate change studies

As eskers are a key aquifer type in northern regions, studies from Finland can be used to increase existing knowledge of groundwater in cold climate. This is important in climate change studies, as most changes are expected in the northern climate, particularly in snow accumulation and melt processes. Other human pressures are present in eskers, posing complex and coupled pressures of human activity and climate change. This includes roads, as they are commonly laid on sandy soils such as eskers. Peatland use for forestry and energy is also a major threat. Future increases in demand for biofuel can result in increased pressure on eskers as more peatlands are drained for forestry.

Groundwater from eskers is commonly used for potable water in Finland and elsewhere. The use of groundwater has increased over recent decades and will continue to increase in the future as cities seek multiple water sources to ensure a high level of water supply security. A number of ecosystem services are related to the role of groundwater in lakes and therefore tourism and recreation. These are sometimes under threat when drinking water is extracted. The current studies in Finland are focusing on groundwater hydrogeology, geochemistry, ecology and socio-economic systems for better groundwater management (Kløve et al. 2011a; 2011b). Valuable ecosystems such as springs exist where groundwater esker discharges. These springs are threatened by forestry and agriculture.

The Rokua esker, the largest aquifer in Finland, forms the main case site for studies at the University of Oulu. Water levels in groundwater and lakes are monitored continuously. A weather station has been installed. The discharge from the esker is monitored at many points and water geochemistry and isotopic composition are also measured regularly. The site is currently the most closely monitored aquifer in Finland. Long-term data exist from some wells and from nearby climate observation stations.

17.2   RESULTS AND DISCUSSION

17.2.1   How should we assess climate change and land-use changes?

The effects of global warming on water resources, and especially on groundwater, will depend on the groundwater system, its geographical location, and changes in hydrological variables (Alley 2001; Sophocleous 2004; Huntington 2006). Climate change can be analysed by statistical methods (Chen et al. 2002; Okkonen and Kløve 2010), numerical models (e.g. Rosenberg et al. 1999: Scibek and Allen 2006) or expert evaluation based on past research and judgement (Okkonen et al. 2010). In all approaches, a good conceptual model of the aquifer is needed as a starting point. Communication of result from one aquifer to a broader audience is a way also to assess impacts (Green et al. 2011). Sometimes we need to clearly understand the impacts, not only on the aquifer but also on nearby ecosystems (Kløve et al. 2011a). The impact on the socio-economic system and ecosystem services is also crucial and must be understood and accounted for (Kløve et al. 2011b). The methods used depend on the focus of the analysis and the available data.

In order to understand the impacts of climate change for the next 40 or 100 years, we must examine changes in both the climate and land-use. We must also determine how the climate has changed in the past in order to put present and future changes into perspective. We need to understand trends in climate such as increasing or decreasing temperatures. It is important to analyse past variability in climate for a considerable period, e.g. at least 50–100 years. This will reveal climate variability patterns such as periods of high and low precipitation. These patterns are typically cyclic and related to events in atmosphere and oceans, such as the NAO-index for Scandinavia (Tuomenvirta 2004). These signals depend on local climate drivers and can be clear or more unpredictable. In some areas rapid climate shifts can occur (Swanson and Tsonis 2009). A shift in mean values of e.g. precipitation can also arise from changes in measurement devices and location of weather stations, so the analysis must be carried out carefully. After climate variability patterns are excluded, the remaining trend can be observed and its significance statistically tested. Once we know past trends, future trends can be analysed with downscaled climate data predicted using climate models. An ensemble of climate model outputs is normally used to account for the uncertainty in future climate predictions (e.g. Lind and Kjellström 2008).

For groundwater, the key factor to be analysed is the change in recharge caused by changes in precipitation and evapotranspiration. The most common method of predicting changes in evapotranspiration is to use FAO56 Penman–Monteith based method to calculate reference ET (e.g. Dinpashoh et al. 2011) or temperature-based empirical models, as future scenarios for temperature are available. The impact of global warming on ET is uncertain. Increased carbon assimilation results in more biomass (LaMarche et al. 1984), higher leaf area, and thereby potentially higher evapotranspiration. However, as CO_2 increase, the stomata opening has been found to decrease (Penuelas and Matamala 1990) resulting in less water loss from the plant counteracting the influence on plant growth on transpiration. Betts et al. (2007) discuss that runoff will on average increase due to reduced transportation caused by reduced stomata opening.

For cold climate it is essential to separate precipitation into rain and snow and account for snow accumulation and melt. Any model used for recharge must account for winter hydrology. As most changes are expected in winter months (Okkonen and Kløve 2010; Veijalainen et al. 2010), for cold climates it is essential to study changes in the timing of melt and the melt peak rather than changes in average values alone. The timing of the melt peak can influence summer low groundwater levels and droughts in ecosystems.

To understand past and future changes in esker hydrology and flow patterns, detailed information is required on climate, hydrogeology, water use and land-use. Impacts of land-use must be included if climate changes in flow patterns or water levels are to be understood, so precise land-use records are needed. In Finland, intensive land drainage and water construction has occurred since the 1950s. Construction of a network of forest roads also began at that time. In many cases information on land-use changes such as drainage is not easily available. Forest cutting and growth will have an impact on spatial variability in evapotranspiration and recharge, but is not likely to affect the overall water budget or discharge patterns from eskers, as forestry and clear-cutting are usually only carried out on one part of the esker at a time.

17.2.2   Models used and our experiences from modelling

In northern, snow-dominated regions, it is important that any models used to assess the responses of climate variability and change on groundwater levels are able to simulate seasonal snow accumulation and snowmelt. As ice usually reduces soil hydraulic conductivity, it is also important to include soil freezing-thawing and its impact on soils in models. In addition, if changes to nearby ecosystems depending on groundwater are to be predicted, the models need to incorporate the interaction between surface water and groundwater. Models such as HELP (Hydrologic Evaluation of Landfill Performance) and CoupModel (coupled heat and mass transfer model for soil-plant-atmosphere systems, Jansson and Karlberg 2004) are able to simulate snow accumulation, snowmelt and soil frost. The CoupModel can also be used to estimate water infiltration and groundwater recharge during the soil frost period (partially frozen soil), which can be important in highly permeable esker aquifers (Sutinen et al. 2007). Unsaturated models can be coupled to Modflow (Harbaugh 1990) in a sequential manner or by a fully coupled approach which allows feedback from Modflow to the unsaturated model.

Although numerical modelling is a convenient tool to assess the potential impacts of climate variability and change on groundwater levels, the hydrogeology of aquifers is often less well-known, even when groundwater level variations and key climate variables are well monitored. In addition, physically-based models require a large amount of field work to collect the data needed to simulate and calibrate the large watersheds. Esker aquifers in Northern conditions are usually small and numerous and a detailed investigation would therefore be costly. Empirical statistical models (Chen et al. 2002; Okkonen and Kløve 2010) could be used in areas where more physically-based models would be difficult to implement.

17.2.3   Impact of future climate change on hydrology and recharge

In cold, snow-dominated regions, the long term mean groundwater level fluctuation pattern shows a clear seasonal variation (Mäkinen et al. 2008; Okkonen et al. 2010). In shallow unconfined esker aquifers of Central Finland, two maxima and two minima usually occur annually. The spring maximum occurs due to snowmelt and the winter minimum due to permanent snow cover and lack of percolation. The summer minimum occurs due to high evapotranspiration and the autumn maximum due to low evapotranspiration and high rainfall. At present the spring water table maximum is large in Northern and Central Finland, but in Southern Finland the spring and fall maximum are about equal (Mäkinen et al. 2008). It is expected that along with changes due to global warming, the intra-annual groundwater fluctuation pattern will change and the southern conditions move to central Finland (Mäkinen et al. 2008).

A future increase in air temperature is expected to increase evapotranspiration in summer and could lead to an increase in soil moisture deficit compared with the present climate if precipitation does not increase as well. The future precipitation is expected to increase, but estimation of future climate change remains uncertain (Green et al. 2011). For Central Finland the 100 year record in Kajaani does not show increasing trends in temperature or precipitation (Irannezhad et al 2011, in preparation). Future scenarios of climate warming impacts in central Finland show that the length of dry periods may increase as spring melt occurs earlier and the summer ends later (Okkonen and Kløve 2010). This may increase the soil moisture deficit and, together with enhanced transpiration through the vegetation canopy, eventually lower groundwater levels in summer. The duration of the low groundwater period may increase with earlier onset of summer and a shift of the minimum groundwater level to autumn due to increased soil moisture deficit.

It is expected that along with a warmer climate, winter rainfall will increase and snow accumulation decrease (Venäläinen et al. 2001; Jylhä et al. 2004). An increase in winter temperature would potentially increase the amount of precipitation as shown by climate model simulations in Sweden (Lind and Kjellström 2008). This can partly be confirmed by past observations (Hansson et al. 2011) that show a positive past trend between river flow and temperature for Northern part of the Baltic sea and the Gulf of Finland, but negative trend in southern Baltic with more dry future conditions indicating large regional differences. Changes in snowmelt are likely to have significant impacts on groundwater recharge. During the snowmelt period, the evapotranspiration is low and all the melt can recharge the aquifer. The snow melt is of particular importance because the evapotranspiration is lacking or insignificant, which means that the snow (given as mm precipitation) gives a larger recharge than rain (mm). The snowmelt is important because it gives a fast recharge and a spring maximum in the groundwater level. The increase in winter precipitation (Jylhä et al. 2004) and shorter period of snow accumulation (Venäläinen et al. 2001) will most likely increase the winter minimum groundwater level as recharge in a warmer climate occur all winter at Pudajärvi and Rokua Eskers. The spring maximum will decrease as the maximum snow depth decreases.

Higher temperatures in autumn may delay the beginning of snow accumulation and in spring may advance snowmelt. In the future temperate climate, where there is a decrease in the depth of protective snow cover (Venäläinen et al. 2008), the depth of soil frost is expected to decrease rather than increase due to the higher winter temperatures. Frost days are expected to be fewer in cold regions of Finland such as the Pudasjärvi region and the frost season will contract as a result of the rise in daily winter minimum air temperature. A decrease in the depth of soil frost and an increase in winter rainfall and snowmelt will potentially enhance the groundwater recharge and water levels in temperate climates. This results in larger recharge during the winter and weakens the winter minimum (Okkonen et al. 2010).

It is expected that warm winters may shift and extend snowmelt and runoff (Veijalainen et al. 2010) earlier in the year. Ultimately this will also increase recharge and may raise the winter groundwater levels, shift the date of the spring maximum earlier in the year and reduce the response of the groundwater level to the snowmelt water. The summer groundwater levels are expected to decrease and the spring maximum groundwater levels will arrive earlier (Okkonnen et al. 2010). The autumn maximum is caused by increased rainfall induced recharge when evapotranspiration is low. This is expected to remain unchanged.

The changes in frost cover duration and depth can influence recharge. The effects of soil frost on the groundwater are not well understood. In snowy landscapes the soil frost depth is usually lower than in snow-free areas (Venäläinen et al. 2001). In recent years, particular attention has been devoted to estimating water infiltration through partially frozen soil and the impacts of frozen soil on hydraulic conductivity (Stähli et al. 1996). Soil frost may impede groundwater recharge and also cause an upward water flow from the groundwater. Gravelly esker aquifers promote the infiltration of almost all melt water, while other materials such as till may resist percolation and promote surface runoff instead (Gray et al. 2001). The rate of infiltration depends on the soil ice content, the soil and ice structure, and the snowmelt rate (Stähli et al. 1999). Simulations with numerical models for Pudajärvi esker show that it is sensitive to the impact of ice on soil permeability and the amount of recharge (Okkonen and Kløve 2011).

With global warming, groundwater quality may also change. Groundwater quality in northern regions changes seasonally (Soveri et al. 2001). In unconfined esker aquifers, the concentration maximum usually occurs in summer and late autumn and the concentration minimum after spring melt as a result of the infiltrating oxygen-rich water. However, these seasonal changes are site-specific and the impacts of climate variability and change in groundwater quality are usually poorly understood. In agricultural areas, the seasonal changes in recharge due to global warming might accentuate the seasonal variation in pesticide concentrations in groundwater, and an increase in storm events might increase the maximum concentration of pesticides (Bloomfield et al. 2006). Shallow unconfined esker aquifers may also be at risk of contamination. Increased flooding poses a risk to groundwater by increasing surface water intrusion. An increase in the winter recharge, on the other hand, may contribute more oxygen-rich water to the groundwater system and improve groundwater quality (Silander et al. 2006). If the groundwater level falls in summer and increases in dry periods as expected, concentrations of contaminants may increase due to a decrease in dissolved oxygen in groundwater (Silander et al. 2006). However, the amount and seasonal changes in recharge are very site-specific and affect groundwater quality in different ways.

As stated above the most dominant effect of climate change will be increased winter temperature. An increase in precipitation and evapotranspiration can occur and the net recharge can also increase but this seems to be a minor effect. It is expected that the impacts of climate change will in Pudasjärvi esker depend on future winter climate and surface water levels that surround the esker (Okkonen et al. 2010; Okkonen and Kløve

2011). The winter minimum groundwater table will decrease as winter water recharge the aquifer. The summer minimum will most likely decrease as the snow melt occurs earlier. A more random groundwater level will result due to more recharge in winter (Okkonen and Kløve 2011). From the water balance conceptual Figure 17.4 it the impacts of climate can be roughly assessed when changes in precipitation, evapotranspiration and snow accumulation and melt is known. In Rokua esker the main impacts will be changes in lake levels, surface-groundwater-interaction and stream runoff. Warmer climate will increase lake evaporation even if transpiration remains at present level. In headwaters of the esker, the role of lake evaporation is considerable part of the water budget (Ala-aho 2010). This will also yield less water to the discharge zone. It is likely that the sub-catchments (Fig. 17.5) with less base flow will suffer most relative water loss. Lakes of type II will be less impacted by changes in recharge as they receive a large portion of water at present. However, it is likely that peatland drainage will have even a higher impact on the esker lakes than climate change. This is confirmed by the preliminary results that show a declining trend in groundwater table despite an increasing trend in the potential recharge (P-ET).

17.2.4   Surface water-groundwater interaction in lakes

In the Rokua esker, a number of important crystal-clear lakes are valuable for tourists and summer house owners. As lakes on the Rokua esker are embedded in the aquifer, both lake water levels and water quality are highly dependent on the aquifer system. The highly dynamic relationship between lake systems and aquifers is largely determined by geology, climate and topology of the area. Interactions between lakes and groundwater can be divided into three basic types. The lake can have: 1) groundwater inflow from the entire lake bed (groundwater discharge); 2) groundwater outflow from the entire lake bed (groundwater recharge); or 3) both situations occurring at the same time in different parts of the lake (Ala-aho 2010, Ala-aho et al. in preparation). Groundwater exchange not only affects lake water levels and water chemistry, but also lake ecosystems by providing nutrients, inorganic ions and stable water temperature (Hayashi and Rosenberry 2002).

In glacial terrain such as esker formations, special attention should be paid to interactions between lakes and aquifers. As the geology of an esker consists mostly of sandy deposits, the terrain is highly permeable and land areas often do not contribute runoff to the surface stream network. Instead, water ponds in landscape depressions, forming 'closed' lakes and wetlands (Fig. 17.3). Due to lack of stream outlets and inlets, groundwater can have a pronounced role in the water budget of such lakes. The contribution of groundwater to the lake water budget varies between individual lakes even in the same area, depending on the underlying hydrogeology of the lake and its position with respect to local and regional groundwater flow systems (Toth 1963). A common conception is that lakes situated at high topographical elevations recharge groundwater (lakes in *local* groundwater flow systems) and that lakes in low areas serve as groundwater discharge areas (lakes in *regional* groundwater flow systems) (Winter et al. 1998, Fig. 17.3).

Several lakes on the Rokua esker have suffered from periodical groundwater fluctuations in recent years and a long-term trend of declining groundwater and lake levels has been suggested by local inhabitants and officials (Ala-aho 2010). The lakes in question are closed basin lakes situated in high topographical areas, and presumably located in local groundwater flow systems. The water quality of the lakes is generally excellent and the recreational and ecological value high. A pilot lake, Ahveroinen, was chosen for detailed studies of groundwater-lake interactions (Ala-aho 2010). Water balance calculations

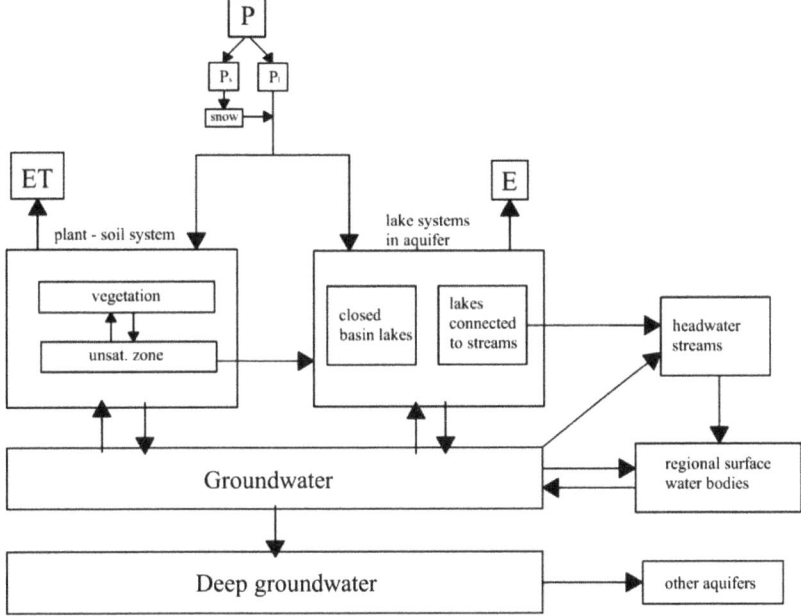

Figure 17.4. Flow diagram of the main water balance components of an inland esker connected to surface waters and wetlands depending on groundwater.

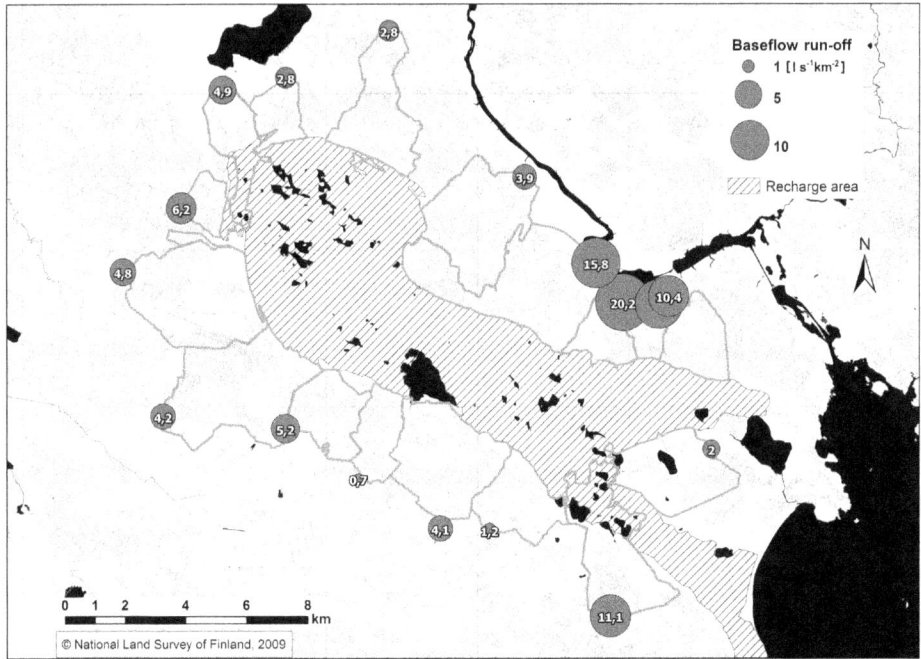

Figure 17.5. Rokua esker recharge area, groundwater dependent lakes and discharge areas and local catchment areas shown with the amount of groundwater discharge approximately shown.

showed that groundwater inflow to the lake was equal to lake evaporation and groundwater outflow was equal to precipitation during the one-year study period. These results indicate significant exchange and dependence between lake and groundwater. Seepage meters applied to directly measure flow rates and direction between lake and groundwater (Lee 1977) revealed distinct and permanent groundwater recharge and discharge areas in the lake. Spatial variability of flow was observed outside more permanent recharge/discharge areas. Changes in inflow and outflow rates during ice-free periods showed some temporal variability. Groundwater inflow and outflow was concentrated near the shoreline, partly due to organic lake sediments deposited in deeper parts of the lake.

An analysis of long-term climate data suggests that consecutive years of below average precipitation can result in temporary lowering of water levels (Ala-aho et al. 2010), especially in lakes located at high topographical elevations (Fig. 17.2a). However, a long-term declining trend in lake water levels can only be explained by an overall lowering of the groundwater level. Some tentative data suggest that water quality in the lakes is partly determined by their position in the groundwater flow system. Lakes situated at lower topographical elevations (and presumably in regional flow systems) are eutrophic and enriched with nutrients and inorganic ions from weathering and groundwater of long residence time (Fig. 17.3). In comparison, lakes located at higher elevations and fed by local groundwater flow systems are in general oligotrophic. The complicated groundwater flow systems controlling lakes of the area need to be better understood in order to draw final conclusions about the interdependent relationship between Rokua lakes and the aquifer.

17.2.5   Impact of peatland drainage

Peatlands are typically located in esker discharge areas. As in many other regions, most of the peatlands surrounding the Rokua esker were drained in the period 1950–1990, mainly for forestry, but also for agriculture and peat extraction. Besides climate change, peatland drainage, road construction and gravel mining are the greatest environmental threats to esker groundwater. Drainage ditches in peatland are usually 1 m deep to ensure sufficient root growth. The potential impact of drainage on esker groundwater was not well known, so due to declining lake levels at Rokua, groundwater intrusion into drains was studied in 2009 and 2010.

A detailed survey of peatland ditches revealed that groundwater leakage to drainage canals occurred in two distinctly different ways near the esker (Kupiainen 2010). In areas with a thick peat layer, the piezometric pressure difference between the confined sand layer beneath the peat and ditch water level was highest, more than 1 m, and artesian. This high water pressure against a peat layer of low conductivity seemed to erode a channel or pipe that discharged groundwater (compare with karst) through peat to the bottom (Fig. 17.2b, point a) and was evident as spring-like formations. On the other hand, in areas of shallow peat, parts of the ditch were dug through to the sand layer and in other parts the peat layer beneath were shallow. In these conditions groundwater was able to seep through the whole ditch bottom (Fig. 17.2b, point b). This caused a gradual flow increase along the whole length of the drain. Both conditions can also occur naturally in the area but drainage has changed the leakage points and possibly increased the groundwater discharge from the esker. A discharge increase is indicated by the fact that in some areas of the Rokua esker, the mean groundwater level has gradually declined since the 1980s and this decrease is not caused by climate conditions.

A study was conducted to restore part of the drainage system by elevating the water levels (Rossi et al. in preparation). From the surveyed peatland drainage system, a 200 m long ditch was selected in order to study whether groundwater leakage can be reduced by increasing the water level with dams. The selected stretch of ditch had spring-like formations in the ditch bottom and the groundwater discharge from these formations was $1000–1500 \, m^3/d$. By damming the ditch, the hydraulic gradient difference between ditch water level and the piezometric pressure in the underlying sand was decreased by 1 m. This gradient difference decreased the discharge rates from the spring-like formations by about $500 \, m^3/d$. The damming also increased the piezometric pressure level in the sand by 20 cm and the effects on groundwater level could be seen at 150 m distance from the dammed drainage ditch. The pilot study showed that the effects of drainage on lowering the groundwater can be reversed. More studies are needed to clarify whether restoration by damming can be used on a large scale to reverse the decline in Rokua groundwater level.

17.3 POLICY RECOMMENDATIONS

For future policy we need to understand the impacts of climate on groundwater and the services groundwater provides for ecosystems and human. It is important to possess good data records and a good understanding of groundwater in order to provide support for correct decision making. This requires increased monitoring efforts and specific research on aquifers and ecosystems related to aquifers.

Besides the impact of climate change, we need to understand the impact of land-use and find methods to remediate and restore potential negative impacts. Understanding the impacts of land-use is crucial so that all negative impacts and benefits are identified. The impacts of land-use are complex and interwoven with climate change impacts, so the associated uncertainty must be included when future scenarios are predicted. Peatland drainage can have negative impacts on other landowners and land, which is much more valuable for recreation than the peatlands. Drainage for marginal and low economic profit with huge damage to other properties and values must be avoided. The link between drainage and groundwater needs to be better understood. Methods and funding mechanisms for potential restoration of negative impacts must be established.

Land-use regulation and groundwater protection zones need to be carefully drawn, or in some cases extended from the present, to better account for activities in the groundwater discharge zone. Forestry practices and other land-use and water use practices should give more consideration to groundwater. For example, at present drainage is allowed on deep peat, as in theory peat layers have low hydraulic conductivity and prevent groundwater exfiltration. However, our studies show that groundwater does discharge from shallow ditches dug into deep peat layers. Therefore care must be taken and all practices recommended tested in practice to ensure that damage will not occur. Where possible, restoration should be considered as an alternative to increased forestry in order to maintain land property and ecosystem values.

17.4 FUTURE WORK

The analysis of climate change in Northern esker needs more research. At the moment we have projections for future temperature. For the evaluation of climate change impacts we need reliable estimates of future precipitation and evapotranspiration. Until now these are very uncertain and the scientific literature predicts both increases and decreases in transpiration with increased CO_2. In order to assess impacts locally we need reliable long term data on past recharge. Variation in recharge needs to be better linked to cyclic weather patterns such as NAO, and we should better include cycles and trends in future recharge scenarios.

For groundwater eskers we need better conceptual models on hydrogeology and groundwater dependent ecosystems. The schematic presented in this paper are one of the first presented for eskers in Finland. These models needs to be further developed and tested for other sites. The present models used need to include realistic components that take into account also ecosystems. The models should be run for transient situations and not for steady state to better include the impact of climate variability. It is also important that realistic land-use scenarios are included in such modelling.

The assessment of climate change and land-use impacts on eskers will continue at Rokua esker, in particular. Future work at the Rokua esker will include detailed research on hydrogeology, including a detailed analysis of recharge and discharge patterns. As a result, the conceptual model will be further improved. Tracer studies using natural geochemical concentrations of groundwater and stable isotopes will be carried out. The hydrology will be modelled and past and future climate simulated. The modelling needs to include dominant land-uses such as forestry and peatland drainage. Also the impact of vegetation (trees and lichen) need to be better included. The scenarios modelled will be based on analysis of the past climate and on well-known scenarios for future climate change generated by an ensemble of climate models for future predictions. Rokua will also be used as a test site for socio-economic studies in the GENESIS project (www.thegenesisproject.eu). The monitoring of climate, groundwater and discharge from the esker will continue and the results will be used in future research and education at the University of Oulu.

ACKNOWLEDGEMENTS

The work was carried out as part of the GENESIS project on groundwater systems (http://www.thegenesisproject.eu) financed by the European Commission 7FP contract 226536. We thank the referees for valuable comments that improved the manuscript.

REFERENCES

Aartolahti, T. (1973). Morphology, vegetation and development of Rokuanvaara, an esker and dune complex in Finland. *Fennia*, 127, 1–53.
Ala-aho, P. (2010) Water budget and interaction between surface water and groundwater in kettle lake Ahveroinen at Rokua esker formation. *MSc Thesis*. University of Oulu, Oulu, Finland, 104 pp.
Ala-aho, P., Rossi, P. & Kløve, B. (2010) Hydrology of a groundwater dependent esker lake. In: *XXXVIII IAH Congress*, 12–17 September 2010, Krakow, Poland.
Alley, W.M. (2001) Ground water & climate. *Ground Water*, 39(2), 161–161.
Betts, R.A., Boucher, O., Collins, M., Cox, P.M., Falloon, P.D., Gedney, N., Hemming, D.L., Huntingford, C., Jones, C.D., Sexton, D.M.H. & Webb, M.J. (2007) Projected increase in continental runoff due to plant responses to increasing carbon dioxide. *Nature*, 448, 1037–1041.
Bloomfield, J.P., Williams, R.J., Gooddy, D.C., Cape, J.N. & Guha, Pm. (2006) Impacts of climate change on the fate and behaviour of pesticides in surface and groundwater – a UK perspective. *Science of the Total Environment*, 369(1–3), 163–177.

Chen, Z., Grasby, S. & Osadets, M.K. (2002) Prediction of average annual groundwater levels from climate variables: An empirical model. *Journal of Hydrology*, 260(1–4), 102–117.

Dinpashoh, Y., Jhajharia, D., Fakheri-Fard, A., Singh, V.P. & Kahya, E. (2011) Trends in reference crop evapotranspiration over Iran. *Journal of Hydrology*, 399(3–4), 422–433.

Gray, D.M., Brenda, T., Zhao, L., Pomeroy, J.W. & Granger, R.J. (2001) Estimating areal snow-melt infiltration into frozen soil. *Hydrological Processes*, 15(16), 3095–3111.

Green, T.R., Taniguchi, M., Kooi, H., Gurdak, J.J., Allen, D.M., Hiscock, K.M., Treidel, H. & Aureli, A. (2011) Beneath the surface of global change: Impacts of climate change on ground-water. *Journal of Hydrology*. : doi: 10.1016/j.jhydrol.2011.05.002

Harbaugh, A.W. (1990) *A computer program for calculating subregional water budgets using results from USGS MODFLOW model*. U.S. Geological Survey Open File Report, 90-392. 46 pp.

Hansson, H., Eriksson, C., Omstedt, A. & Chen, D. (2011) Reconstruction of river runoff to the Baltic Sea, AD 1500–1995. *International Journal of Climatology*, 31, 696–703.

Hayashi, M. & Rosenberry, D.O. (2002) Effects of ground water exchange on the hydrology and ecology of surface water. *Ground Water*, 40, 309–316.

Huntington, T.G. (2006) Evidence for intensification of the global water cycle: Review and synthesis. *Journal of Hydrology*, 319(1–4), 83–95.

IPCC. (2007) In: Parry, M.L., Canziani, O.F., Palutikof, J.P., van der Linden, P.J. & Hanson, C.E. (eds.) *Climate Change 2007: Impacts, Adaptation and Vulnerability. Contribution of Working Group II to the Fourth Assessment Report of the Intergovernmental Panel on Climate Change*. Cambridge, UK, Cambridge University Press, 976 pp.

Jansson, P.-E. & Karlberg, L. (2004) *Coupled heat and mass transfer model for soil–plant–atmosphere systems*. Royal Institute of Technology, Stockholm. 435 pp.

Jylhä, K., Tuomenvirta, H. & Ruosteenoja, K. (2004) Climate change projections for Finland during the 21st century. *Boreal Environment Research*, 9, 127–152.

Kløve, B., Ala-aho, P., Guillaume, B., et al. (2011a) Groundwater Dependent Ecosystems Part I: Hydroecological status and trends. *Environmental Science and Policy, doi:10.1016/j.envsci.2011.04.002*.

Kløve, B., Allan, A., Bertrand ,G., et al. (2011b) Groundwater Dependent Ecosystems: Part II – Ecosystem services and management under risk of climate Change and Land-Use Management. *Environmental Sciences and Policies* (accepted).

Korhonen, J. & Kuusisto, E. (2010) Long-term changes in the discharge regime in Finland. *Hydrology Research*, 41(3–4), 253–268.

Kupiainen, V.H. (2010) Groundwater discharge to forest ditches at Rokua esker area and restoration by ditch dams. *MSc Thesis*. University of Oulu, Oulu, Finland. 68 pp.

LaMarche, Jr., V.C., Graybill, D.A., Fritts, H.C. & Rose, M.R. (1984) Increasing atmospheric carbon dioxide: Tree-ring evidence for growth enhancement in natural vegetation. *Science*, 225, 1019–1021.

Lee, D.R. (1977) A device for measuring seepage flux in lakes and esturies. *Limnology and Oceanography*, 22, 140–147.

Lind, P. & Kjellström, E. (2008) *Temperature and precipitation changes in Sweden: A wide range of model-based projections for the 21st century*. Swedish Meteorological and Hydrological Institute, RMK No. 113. 50 pp.

Mäkinen, R., Orvomaa, M., Veijalainen, N. & Huttunen, I. (2008) The climate change and ground-water regimes in Finland. In: *Proceedings 11th International Specialized Conference on Watershed and River Basin Management*, Budapest, Hungary. ISBN 978-963-06-5689-4.

Miettinen, A. (2008) Holocene sea-level changes and glacio-isostasy in the Gulf of Finland, Baltic Sea. *Quaternary International*, 120, 91–104.

Okkonen, J., Jyrkama, M. & Kløve, B. (2010) A climate change assessment framework for ground water. *Hydrogeology Journal*. [Online] Available from: doi:0.1007/s10040-009-0529-9

Okkonen, J. & Kløve, B. (2010) A conceptual and statistical approach for the analysis of climate impact on ground water table fluctuation patterns in cold conditions. *Journal of Hydrology*, 388(1–2), 1–12.

Okkonen, J. & Kløve, B. (2011) Coupling of surface water and groundwater flow models to assess climate variability on groundwater resources in a cold climate: Scenario studies of water resources in an unconfined esker aquifer in Finland (personal communications).

Pajunen, H. (1995) Holocene accumulation of peat in the area of an esker and dune complex, Rokuanvaara, central Finland. *Geological Survey of Finland*, Special Paper 20, 125–133.

Penuelas, J. & Matamala, R. (1990) Changes in N and S leaf content, stomatal density and specific leaf area of 14 plant species during the last three centuries of CO_2 increase. *Journal of Experimental Botany*, 41(9), 1119–1124.

Prowse, T.D., Belta, S., Gardner, J.T., et al. (2006) Climate change, flow regulation and land-use effects on the hydrology of the peace-athabasca-slave System; findings from the northern rivers Ecosystem initiative. *Environmental Monitoring and Assessment.* [Online] Available from: doi:10.1007/s10661-005-9080-x

Rosenberg, N.J., Epstein, D.J., Wang, D., Vail, L., Srinivasan, R. & Arnold, J.G. (1999) Possible impacts of global warming on the hydrology of the Ogallala aquifer region. *Climate Change*, 42(4), 677–692.

Scibek, J. & Allen, D.M. (2006) Comparing modelled responses of two high-permeability, unconfined aquifers to predicted climate change. *Global Planet Change*, 50(1–2), 50–62.

Silander, J., Vehviläinen, B., Niemi, J., Arosilta, A., Dubrovin, T., Jormola, J., Keskisarja, V., Keto, A., Lepistö, A., Mäkinen, R., Ollila, M., Pajula, H., Pitkänen, H., Sammalkorpi, I., Suomalainen, M. & Veijalainen, N. (2006) *Climate change adaptation for hydrology and water resources.* FINADAPT Working Paper 6. Finnish Environmental Institute Mimeographs 336, Helsinki.

Sohlenius, G., Sternbeck, J. & Westman, P. (1996) Holocene history of the Baltic Sea as recorded in a sediment core from the Gotland Deep. *Marine Geology*, 134, 183–201.

Sophocleous, M. (2004) Climate change: Why should water professionals care? *Ground Water*, 42(5), 637–637.

Soveri, J., Mäkinen, R. & Peltonen, K. (2001) *Changes in Groundwater Levels and Quality in Finland 1975–1999*. Helsinki, Finland, Tummavuoren kirjapaino Oy. 382 pp.

Stähli, M., Janson, P.-E. & Lundin, L.-C. (1996) Preferential flow in a frozen soil–a two-domain model approach. *Hydrological Processes*, 10(10), 1305–1316.

Stähli, M., Janson, P.-E. & Lundin, L.-C. (1999) Soil moisture redistribution and infiltration in frozen sandy soils. *Water Resources Research*, 35(1), 95–103.

Sutinen, R., Hänninen, P. & Venäläinen, A. (2007) Effect of mild winter events on soil water content beneath snowpack. *Cold Regions Science and Technology.* [Online] Available from: doi:10.1016/2007.05.014

Svendsen, J.I., Alexanderson, H., Astakhov, V.I., et al. (2004) Late quaternary ice sheet history of northern Eurasia. *Quaternary Science Reviews*, 23, 1229–1271.

Swanson, K.L. & Tsonis, A.A. (2009) Has the climate recently shifted? *Geophysical Research Letters.* [Online] 36, L06711. Available from: doi:10.1029/2008GL037022

Tikkanen, M. (2002) The changing landforms of Finland. *Fennia*, 180(1–2), 21–30.

Toth, J. (1963) A theoretical analysis of groundwater flow in small drainage basins. *Journal of Geophysical Research*, 68, 4795–4812.

Tuomenvirta, H. (2004) Reliable estimation of climatic variations in Finland. *PhD Thesis.* Finnish Meteorological Institute, Helsinki, Finland, Contributions 43. 82 pp.

Veijalainen, N., Lotsari, E., Alho, P., Vehviläinen, B. & Käyhkö, J. (2010) National scale assessment of climate change impacts on flooding in Finland. *Journal of Hydrology*, 391(3–4), 333–350.

Venäläinen, A., Tuomenvirta, H., Heikinheimo, M., Kellomäki, S., Peltola, H., Strandman, H. & Väisänen, H. (2001) Impact of climate change on soil frost under snow cover in a forested landscape. *Climate Research*, 17(1), 63–72.

Winter, T.C., Harvey, J.W., Franke, O.L. & Alley, W.M. (1998) *Ground water and surface water: A single resource.* US Geological Survey Circular 1139. Denver, CO, USGS. 79 pp.

Polar Climates

CHAPTER 18

Impacts of climate change on groundwater in permafrost areas: case study from Svalbard, Norway

Sylvi Haldorsen, Michael Heim & Martine van der Ploeg

ABSTRACT

Polar areas experience a temperature increase due to global warming, with decrease of permafrost thickness. Permafrost degradation may give dramatic changes in the interaction between groundwater and surface water. In areas where permafrost is thick and continuous no changes in the subpermafrost groundwater flow-pattern is reported. However, here changes of recharge and discharge conditions may be significant. On the high arctic peninsula of Svalbard, Norway, groundwater discharge has decreased after the end of the Little Ice Age from around 1930 due to a gradual melting of the glaciers where groundwater recharge takes place. A future decrease in the discharge as a result of human-induced global warming may result in the termination of groundwater springs. Large-scale circum-polar monitoring of the groundwater systems and numerical groundwater simulation models for the data-poor subpermafrost aquifers, which are exposed to a global climate change may be important international research tasks.

18.1 INTRODUCTION

18.1.1 Purpose and scope

In the northern hemisphere permafrost underlies approximately 24% of the exposed land and extends from 26°N in the Himalayas to 84°N in Canada (Zhang et al. 1999). In Antarctica permafrost can be found throughout the continent. Geographically permafrost regions are divided into areas of continuous permafrost (permafrost underlying 90–100% of the area) and discontinuous permafrost (permafrost underlying 50–90% of the area) (Fig. 18.1). Although permafrost areas are relatively sparsely populated, they include towns, industrial sites and settlements for research groups. Important habitats for vulnerable polar ecosystems are found both in the northern and southern hemisphere.

Unconfined groundwater occurs in seasonally thawed sediments above the permafrost ('the active layer', Fig. 18.2, van Everdingen 1990). Its recharge is by direct infiltration of precipitation and annual melting of snow. Groundwater below the permafrost is partly or entirely (Weaver 2003) confined, with the permafrost forming the confining layer (Fig. 18.2, Haldorsen et al. 1997; van Everdingen 1990; Williams 1970). Recharge and discharge require unfrozen conduits through the permafrost (called 'talik'). Taliks are commonly associated with the bottom of lakes, rivers, major snow fields and glaciers (Fig. 18.2, van Everdingen 1990). Although water in the active layer is very important for fauna and flora and a challenge for human construction work, this review paper will only focus on subpermafrost groundwater.

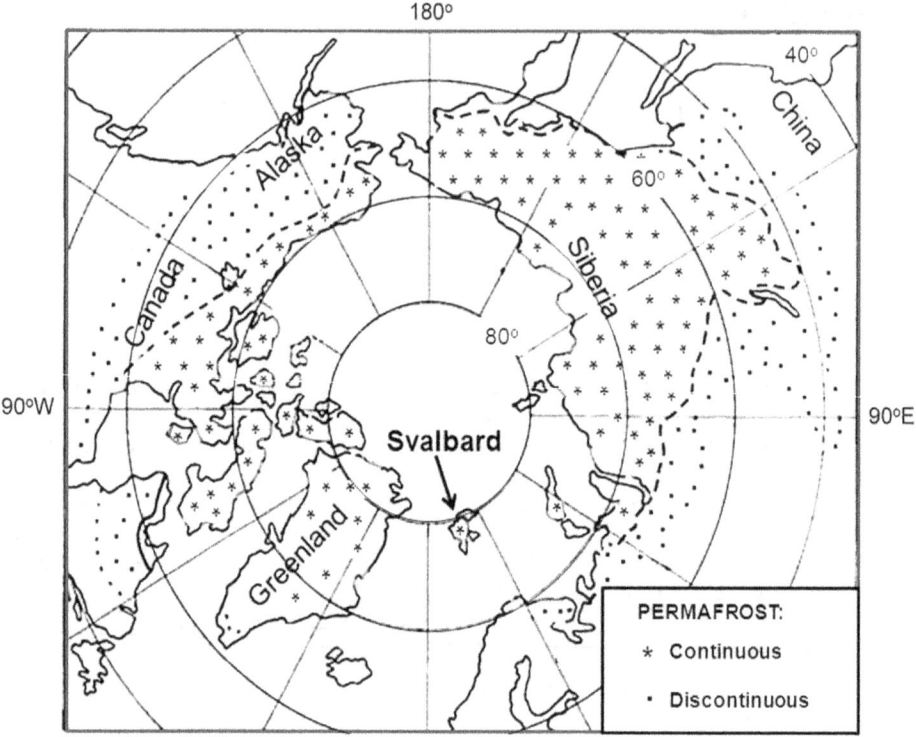

Figure 18.1.   The distribution of permafrost in the northern hemisphere, modified from Williams (1970).

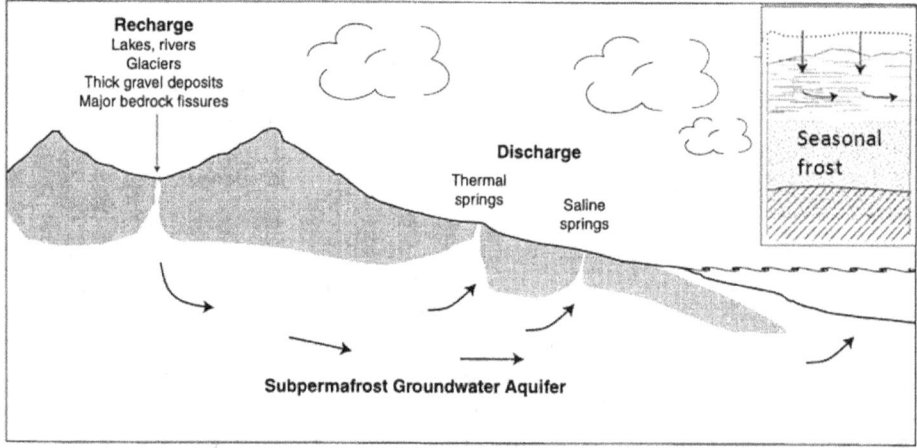

Figure 18.2.   A subpermafrost groundwater system, with recharge in the upland areas, confined flow under the permafrost, and supraaquatic discharge in major open springs or open-end pingos. Inset: groundwater in the active layer above the permafrost. (From Haldorsen et al. 1997).

Some arctic settlements in areas of thin or discontinuous permafrost apply sub-permafrost groundwater as drinking water (Ford and Bedford, 1987; Prowse et al. 2006a; van Everdingen 1973, 1974; Wang 1990). In particular groundwater stored in permeable sediments beneath lake and river beds can form reliable and economically affordable water sources (e.g. van Everdingen 1973; Williams 1970). The restricted discharge of groundwater in areas of thick and continuous permafrost may be of particular importance for ecosystems that require open water during the winter.

The paper first gives a review of some general studies of climate change and subpermafrost aquifers. Thereafter a case study from Svalbard, Norway is used as an example of long-time groundwater dynamic in an area of continuous permafrost.

18.1.2   Area description

In areas of thin and/or discontinuous permafrost, groundwater flows above, below and between separate permafrost bodies (see description in e.g. van Everdingen 1990). This creates a complex flow pattern. Groundwater interacts with water in rivers and lakes as a part of the hydrological regime (Ford and Bedford 1987; Haldorsen et al. 2002; Kane and Slaughter 1973; Prowse et al. 2006a; Williams 1970; Woo et al. 2000). The groundwater may regulate surface water temperatures (e.g. prevent a full freezing of the water in the winter) and have an impact on hydrochemistry and biochemistry (Frey et al. 2009; MacLean et al. 1999; Prowse et al. 2006a).

Groundwater systems below continuous permafrost are characterized by low recharge and long residence times, and the water is often salty (see Fig. 18.2). The deep confined flow pattern does not change much over time. The spatially restricted discharge requires that the flow is large enough and the temperature high enough to prevent freezing when the water passes up through site-specific taliks. Despite of this restriction discharge nevertheless occurs in major springs (e.g. Haldorsen and Heim 1999; Heldmann et al. 2005; Pollard 2005) particularly related to karst rocks (Clark and Lauriol 1997; Ford 1987; Lauritzen 2006; Lauritzen and Bottrell 1994; Michel and van Everdingen1988; Weaver 2003). There are some few direct studies of deep subpermafrost aquifers connected to arctic oil and gas fields (Cramer et al. 1999; Deming et al. 1992; Hanor et al. 2004; Kharaka and Carothers 1989; Nunn et al. 2005) and arctic mining (Haldorsen et al. 1996; Haldorsen and Heim, 1999; Orvin, 1934, 1944; Stotler et al. 2009; Weaver 2003), but aquifers beneath deep and continuous permafrost are probably the least studied groundwater systems in the world.

Svalbard (Fig. 18.1) is a high-arctic Norwegian archipelago situated between latitudes $74°$ and $81°$N. The annual mean temperature varies between -2 and $-7°$C and climate is dry with an annual precipitation between 400 and 600 mm (Førland and Bauer-Hanssen 2003). The continuous permafrost is generally from 150 to 450 m thick (Liestøl 1980). Approximately 60% of the area is covered by glaciers and all upland depressions are filled with ice. Exposed ground is found along the coasts or inland where mountain peaks rise above the ice surface. A large part of the archipelago is declared a national park.

18.1.3   Methodology

The groundwater studies in Svalbard included water sampling (temperature, chemical composition, stable and radioactive isotopes) and measurement of discharge at spring outlets. In addition, rock cores from the coal mining period in Ny-Ålesund were used to measure permeability and porosity of the Permian and Tertiary bedrock aquifers. The results are presented in a series of papers, referred to in the text below. The data were also used in numerical groundwater simulation models. The modelling was done in two steps: A 2D model using SUTRA software code (Booij et al. 1998) and a 3D model using the METROPOL and METROHEAT software codes (van der Ploeg 2002).

18.1.4   Relevance to GRAPHIC

A temperature rise of 2–4°C is predicted to occur in the polar areas during the 21st century (IPCC 2007). The related permafrost degradation was discussed by Anisimov and Nelson already in 1996 and has later been confirmed by field observations (Hinzman et al. 2005; Smith et al. 2005; Grosse and Romanovsky 2011). Some melting may be natural. From air photos Jorgensson et al. (2001) found that 83% of the total permafrost degradation in the Tanana Flats in Alaska took place before 1950 and palaeo-data showed that degradation must have started as early as in the mid-1700. The impact of climate change on permafrost is not equal all over the northern hemisphere. While permafrost temperature has increased during the last decades in parts of Alaska and Russia, it has decreased some places in Canada (Serreze et al. 2000).

The thick permafrost in the high arctic responds slowly to climate changes (e.g. Gregersen and Eidsmoen, 1988). In Siberia permafrost older than 100,000 yr survived the last Interglacial as well as the mildest part of the Holocene (11,000–6,000 years ago) (Andreev et al. 2004; Grosse and Romanovsky 2011). Warming of continuous permafrost has been most significant along coasts, while no inland thawing of cold continuous permafrost has been observed during the last six decades (Grosse and Romanovsky 2011).

As recharge and discharge of subpermafrost groundwater occur via taliks through the climate-controlled permafrost aquiclude (Fig. 18.2), subpermafrost groundwater is very climate dependent and probably more vulnerable to climate change than most other groundwater types. This makes subpermafrost groundwater a very relevant theme for GRAPHIC.

In high-latitude areas in North America, Europe and Asia it is a big concern that the polar areas will experience a dramatic climate change in the near future (IPCC 2007). This is important in the international discussions about greenhouse gas emissions. Large parts of Antarctica and the Arctic consist of pristine nature. Generally, there is no or very limited information about permafrost groundwater in these remote areas. There is very little knowledge about the change of the groundwater systems during a global warming. The impact of groundwater on the total hydrology is not known. Considering the fast climate change in the polar areas, studies of permafrost groundwater will indeed fit the long-term goals defined by GRAPHIC.

18.2   RESULTS AND DISCUSSION: SUBPERMAFROST GROUNDWATER

18.2.1   Discontinuous permafrost

From the 1990s and onwards it has been reported that a milder climate in permafrost regions increases the thickness of the active layer, gives shorter seasonal snow cover, later freeze-up and earlier breakup of lake and river ice, accelerates the wasting of glacier mass, and increases plant growth and shrub cover (Prowse et al. 2006b; Jorgenson et al. 2001; Serreze et al. 2000; White et al. 2007; Woo et al. 2008). The studies have from the very start included groundwater (Michel and van Everdingen 1994; Rouse et al. 1997; White et al. 2007).

Thawing of permafrost can alter the hydrological regime drastically when aquifers change from confined to unconfined (e.g. Rivière et al. 2010). Satellite images showed that from the 1970s until 2005 there was a widespread decline in the number of lakes in the Siberian discontinuous permafrost areas. This was related to formation of new taliks and drainage down into sub-permafrost aquifers (Smith et al. 2005). Similar results are obtained from local field studies (Frey and McClelland 2009 and Yoshikawa and Hinzman 2003) and by numerical simulation models (Bense et al. 2009). On a scale of decades to centuries degrading of permafrost has increased the amount of shallow groundwater and given a larger contribution of groundwater to streams (Bense et al. 2009). This is associated with higher winter base flow and lower summer flow peaks (Alessa et al. 2008a, 2008b; Walvoord and Strieg 2007). Climate-dependent changes in vegetation may influence the recharge and discharge conditions. An increased vegetation cover in some cases prevents recharge of groundwater (Alessa et al. 2008b; Jorgenson et al. 2001). A study by Muskett and Romanovsky (2009) was based on the GRACE model combined with field studies in western North America and eastern Eurasia. They found that for the period 1990 to 2009 the amount of groundwater had increased in the Lena and Yenisei watersheds, decreased in Mackenzie watershed, while they found no changes in the Ob watershed. Their study illustrates that the changes in the groundwater storage cannot be easily predicted without a special knowledge about permafrost conditions in the individual catchments.

18.2.2 Continuous permafrost, case study Svalbard: results and discussion of previous work

There is a number of large groundwater discharge springs in Svalbard (Fig. 18.3). Already 150 years ago hot springs (20 – 30°C) were observed in Bockfjorden, northern Svalbard (Fig. 18.3, Banks et al. 1998; Banks et al. 1999; Hoel 1914; Hoel and Holtedahl 1911). The warmest springs are habitats for algae, which are not identified elsewhere in the world (Langangen 1979, 2000). Most groundwater springs in Svalbard, many of them related to karst rocks (Lauritzen and Bottrell 1994; Salvigsen et al. 1983; Salvigsen and Elgersma 1985; Smart 1984), are concentrated in the west, where the geothermal gradient is higher than elsewhere on the archipelago (Liestøl 1976; Haldorsen et al. 2010). There are also pingos in Svalbard (Liestøl 1976; Liestøl 1996). Pingos are sediment-covered ice mounds. It is assumed that the ice cores of the pingos in Svalbard are formed by the freezing of artesian groundwater flowing up from deep sub-permafrost aquifers.

In order to understand the dynamics of deep sub-permafrost groundwater systems, it is useful to analyze their long-time history:

i) What controls the recharge and discharge today?
ii) How and when was the present recharge and discharge established?
iii) What happened to the system from the time it was established up to present?
iv) Which role does the permafrost history play for the present groundwater system?

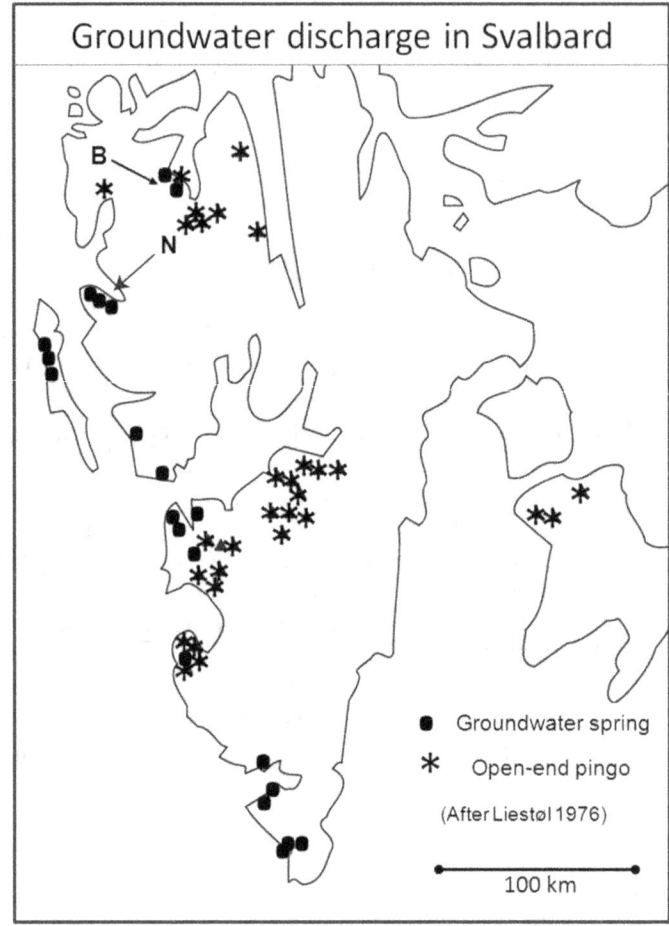

Figure 18.3.   Groundwater discharge in Svalbard (for location, see Fig. 18.1). B: Bockfjorden, N: Ny-Ålesund.

i) It was early understood that groundwater recharge in Svalbard is related to the glaciers as there are no other potential taliks where groundwater recharge can occur (Fig. 18.4, Liestøl 1976; Orvin 1934; Orvin 1944). However, the relation between groundwater recharge and glaciers was not fully understood before it was found that the bottom of the glacier can be either 'cold-based' (temperature below the pressure freezing point) or 'warm-based' (temperatures at the pressure freezing point) (Hagen et al. 2003a, 2003b; Liestøl 1976; Paterson 1994). Most glaciers in Svalbard as well as in many other polar regions are polythermal, i.e. a part of the glacier is cold-based while another part of the same glacier is warm-based (Fig. 18.4, Paterson 1994). The groundwater recharge can only take place along zones of warm-based ice. Such ice requires one of the following two conditions (Fig. 18.4, Haldorsen et al. 2010):

a. A part of the glacier is thick enough to insulate the bottom from the cold air and thereby prevents the ice from freezing to the geological substratum.

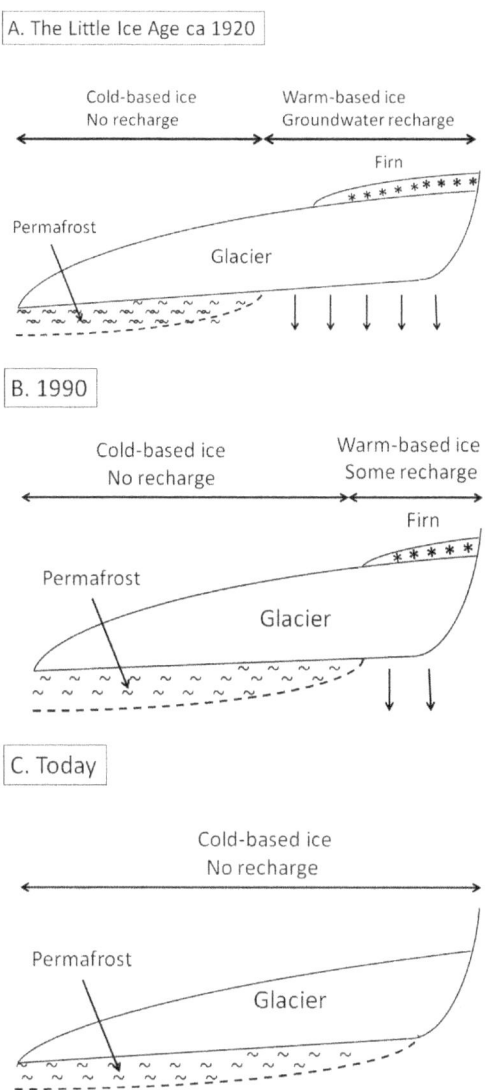

Figure 18.4. The relation between the thermal regime of a polythermal glacier and the groundwater discharge during three phases of melting. Based on the study of the Vestre Lovénbreen glacier with groundwater discharge in Ester Spring: A: the Little Ice Age, B: 1990, C: today. For location: see Fig. 18.5.

b. The glacier has a significant accumulation zone covered by several meters of firn (coarse-grained snow accumulated through several seasons). During the summer season rain and meltwater is stored in the very permeable firn. Heat is released when this water freezes in the winter. This heat prevents freezing along the bottom of the ice (Paterson 1994).

When a glacier decreases in size, its surface is lowered and its accumulation area shrinks (Fig. 18.4b). Both of these factors will decrease the area of warm-based ice and thereby reduce the groundwater recharge area. In the end, the entire glacier may become cold-based and groundwater recharge terminates completely (Fig. 18.4c).

ii) and iii) During the coal mining in Ny-Ålesund, western Svalbard in 1921–1927 and in 1947–1963 there was a large flow of groundwater into the mines (Orvin 1944; Haldorsen and Heim 1999) where the mines crossed major faults or passed through a porous Permian sandstone aquifer (Fig. 18.5). The amount of water pumped out of the mines was continuously measured and on average showed a drastic drop from the first (ca. $3 \cdot 10^{-4}$ m^3/s) to the second mining period (ca. $1.5 \cdot 10^{-2}$ m^3/s). After the mining was terminated in 1963 the mines became water-filled and water was flowing out of the entrance to the now abandoned Ester Mine. When the natural discharge was measured in 1993–1995 and in 2000 it was lower (ca. 10^{-2} m^3/s) than during the mining periods. In 2007 the discharge terminated abruptly. The water to Ny-Ålesund is supplied from the small lake Tvillingvatn. During a drilling near the lake in 1928 artesian groundwater was reached 17 m below the lake bottom (Orvin 1934). Detailed water balance studies in the 1990s showed that there is no longer any groundwater discharge into the lake (Haldorsen et al. 2002).

Photos from the hot springs in Bockfjorden, northern Svalbard taken in 1911 (Hoel 1914; Hoel and Holtedahl 1911) were compared with observations during visits to the hot springs in 1993 and 1995. There had been a significant reduction in the groundwater discharge compared to the situation in 1911 (Haldorsen et al. 2010).

At the start of the 20th century the glaciers in Svalbard had their largest extension since the last glaciation. This represents the end of the Little Ice Age, a phase with a particularly cold climate (Dowdeswell et al. 1997; Hagen 1989), also observed in other Arctic areas (Grosse and Romanovsky 2011). Around 1930 the climate in Svalbard became remarkably milder (Hagen 1989). The glaciers retreated as a response to this natural warming (Førland and Bauer-Hanssen 2003), and this was conducted by a decrease in the areas of warm-based ice (Haldorsen and Heim 1999; Hagen 1989). This applies to all the glaciers around Ny-Ålesund, including Vestre Lovénbreen, which was the recharge area of the water flowing out of Ester Mine. The termination of the discharge out of Ester Mine took place when the entire Vestre Lovénbreen glacier became cold based. Very simplified one can conclude that the milder the arctic climate becomes the less groundwater is recharged. It is difficult to know if the fast melting observed in Svalbard during the last few decades is still natural or now mainly related to a human-induced global warming (Dowdeswell et al. 1997; Pohjola 2007).

iv) During the Late Glacial Maximum glaciers covered most of Svalbard, with their fronts terminating in the sea (Landvik et al. 1998). Thick warm-based ice streams filled the valleys (Landvik et al. 2003, 2005), and the subglacial recharge was not restricted by permafrost Haldorsen et al. 2010). During the first part of the deglaciation the recharge was still considerable. Before new permafrost was formed the groundwater discharge was controlled by the bedrocks and more widespread than today. The recharge decreased when the glaciers retreated. At the same time the discharge gradually became more climate-controlled when new permafrost developed and gradually thickened outside the retreating glacier fronts. During the mildest part of the Holocene (8,700–7,800 radiocarbon years ago (Salvigsen 2002)) the temperatures in Svalbard were somewhat higher than today but still below the permafrost temperature of $-3 - -4$°C

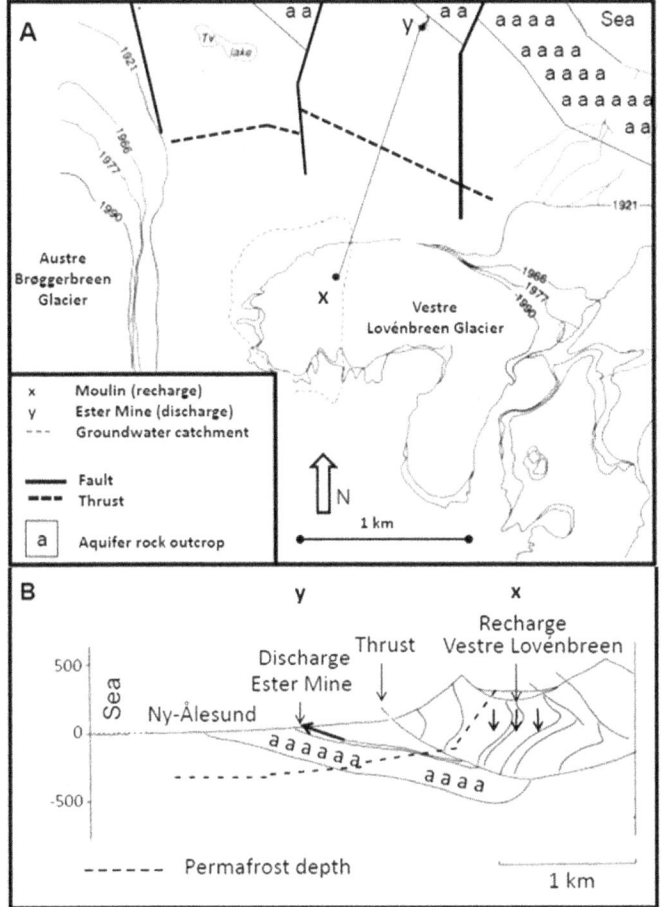

Figure 18.5. A. Simplified hydrogeological map of the Vestre Lovénbreen glacier – Ester Mine groundwater system. Moulin: channel where surface water flows down to the bottom of the ice. Year-lines show the outline of the glacier from 1921 to 1990.
B. Simplified section x–y (location shown on map A). The major groundwater aquifer is the Permian sandstone marked with 'a' symbols, with faults forming high-flow systems. Thrusts (overthrusting of older bedrocks upon younger) are not of major importance for the groundwater flow. Based on data from Haldorsen and Heim (1999).

on an annual basis (Humlum et al. 2003; Landvik et al. 1988). The glaciers shrank and some disappeared (Svendsen and Mangerud 1992) or became cold based. Groundwater recharge areas consequently decreased or disappeared. In Bockfjorden the groundwater discharge was significantly reduced during the Holocene (Haldorsen et al. 2010). Also a number of fossil pingos in Svalbard (Liestøl 1996) indicate a significant reduction of groundwater discharge some time during the Holocene. Later the climate cooled down again and the glaciers increased in size. Based on these observations it is speculated if there was a minimum groundwater discharge situation during the mildest part of the Holocene.

The water temperature is close to 0°C when it is recharged below the glacier. Geothermal heat increases the groundwater temperature during its transit through the aquifer, how much depends on depth, geothermal gradient and residence time. The water cools down again when it passes up through the talik in the permafrost to the ground surface. The heat balance will control if the water remains unfrozen up to the surface. Simulation of the heat transfer in the groundwater aquifer in Ny-Ålesund (van der Ploeg 2002) indicated that a discharge would persist under decreasing recharge-conditions until the recharge terminated almost completely. A lower recharge would result in a higher water temperature when it entered the permafrost and thereby prevent the water from freezing. Theoretically the present springs in Svalbard should survive a continued global warming, as long as the climate remains colder than it was during the mildest part of the Holocene. However, if a discharge spring terminates and new permafrost is formed, the spring will not be reactivated before the permafrost melts (Haldorsen et al. 2010). Under natural climate conditions this will probably not happen before the next glaciation. However, it may perhaps happen much earlier if the global warming continues into the future and results in a degradation of the permafrost along the coast of Svalbard.

In conclusion, it is the climate-dependent glacier dynamic and not the permafrost thickness that controls the sub-permafrost groundwater regime in Svalbard. Parallel groundwater recharge and discharge conditions are probably typical also in other areas where polythermal glaciers occur (for distribution, see Paterson 1994).

Precipitation is a factor not considered in the conceptual model for Svalbard. Svalbard is a relatively dry area and hitherto the glacier mass budget has been temperature controlled. Warmer summers result in a faster melting of the glaciers. As long as this is not compensated by higher winter precipitation the glacier retreat will continue. A change to more snow could give the opposite effect; the glacier mass balance could become positive and give a new advance of most glaciers. However, such a shift in precipitation has not been experienced up to now.

Also other types of taliks are climate dependent. For instance, where recharge takes place through taliks at the bottom of lakes a shift in pressure due to a changed lake volume may be critical for the sustainability of the recharge along the bottom of the lake (Smith et al. 2005). The response on the groundwater system may be different from what was experienced in Svalbard, even if the climate variability is the same.

For areas of discontinuous permafrost it has been argued that the situation today is the same as during the mildest part of the Holocene. Much of the permafrost, which degrades today, was probably formed as late as during the Little Ice Age a couple of hundred years ago (Grosse and Romanovsky 2011). Natural climate-dependent processes may have been repeated several times over the past ten thousand years. However, there is one great difference between the past and the present situation: Today societies in the permafrost regions require a modern infrastructure including a safe water supply. At the same time these societies have to cope with a fast changing environment, which is probably induced by human release of climate gases.

18.3   POLICY-RELEVANT RECOMMENDATIONS

Although permafrost areas are not densely populated, important human activities take place there. The impact of local management actions can add to or slow down the impact of global warming. Alessa et al. (2008a, 2008b) developed a tool to evaluate the weight of different factors that influence the water budget and water quality in permafrost areas in Alaska. A vulnerability index was used as an integrated assessment tool for resilience and vulnerability with respect to freshwater. In addition to the effect of climate change Alessa et al. (2008a, 2008b) found that local management some places had a significant impact on the freshwater systems. Similar tools may be of great value for managers in many other permafrost areas. Water extraction, dam- and road construction as well as human changes of the vegetation (Jorgenson et al. 2001) including the impact of forest fires (Prokushkin et al. 2011) should form an integrated part of such management assessments.

Industrial activity and mining takes place in many areas of discontinuous permafrost. A change from confined to open aquifers due to disintegration of the permafrost will increase the risk of groundwater pollution, especially where frozen waste deposits melt. In many cases the only action would be to remove the waste physically and deposit it under controlled conditions. This will be an important task for the local politicians.

Increased petroleum activity, not least in the Barents Sea region and offshore Siberia, will result in population growth and increased water demand, implicating a need for groundwater. The first part of the planning should include a monitoring of available water resources. In the high arctic the amount of unfrozen water may be very limited during the cold season and groundwater may then supply settlements with water. A calculation of the recharge is critical and groundwater samples from discharge springs or from drilled wells have to be analyzed to see if the water is drinkable. Also, the technology connected to the drilling of deep wells in permafrost has sometimes been a challenge (see e.g. examples from northeast China (Wang 1990)) and well construction may be too costly in smaller settlements in areas with deep permafrost. In the water planning it is also necessary to account for the impact of a continued global warming. This is obvious for areas of discontinuous permafrost where the hydrology changes so fast. However, it may also be needed where the permafrost is thick and continuous, but where the recharge-conditions change relatively fast, as shown in the case study from Svalbard.

The restricted discharge in areas of continuous permafrost may, as mentioned above, be important for the ecology in lakes and springs. Water managers must judge the need for special protection plans. As a result of the studies of hot springs in Bockfjorden the authorities in Svalbard established a protection plan for the springs, where, among other regulations, it is now illegal to approach the springs with snow scooters. Global change is not an aspect in these regulations, but may result in a termination of these springs.

18.4   FUTURE WORK

There is a great need for more knowledge about the polar groundwater systems. This includes basic studies of recharge and discharge. The impact of global change cannot be realistically accounted for if the groundwater flow system is not fully understood. In areas of discontinuous permafrost a future continuous monitoring is critical. An international net of model catchments should be established as an analogue to the International Hydrological decade of the 1970s. The catchments should be adequately monitored and followed over a long time. It could be an important future task to apply GRACE and other remote sensing methods for a circumpolar inspection of the groundwater changes, i.e. a continuation of the work by Muskett and Romanovsky (2009) and

Smith et al. (2005). Numerical modelling will strongly gain from continuous monitoring and measurements in selected catchments.

For deep subpermafrost groundwater systems a continuous and frequent monitoring may not be so important. However, the main weakness today is that the subpermafrost systems in the high Arctic and in Antarctica are almost completely unknown. As shown in Bockfjorden, some discharge springs may form habitats for ecosystems of particular international importance. When important groundwater flow systems are identified the first step should be to construct a conceptual model for the recharge and discharge conditions. Thereafter the impact of a global change on the recharge and discharge conditions should be targeted. The study from Svalbard shows that the long-term groundwater balance may be better understood when climate history of the past is known, as a change over the last decades may be linked to natural processes that took place a long time ago. The validity of hydrogeological modelling in the remote polar areas may be debated, because the data material is usually so limited. However, also for these remote areas a future application of GRACE and other remote sensing methods may bring the understanding of the groundwater systems a large step forward.

REFERENCES

Alessa, L., Kliskey, A., Busey, R., et al. (2008a) Freshwater vulnerabilities and resilience on the Seaward Peninsula: Integrating multiple dimensions of landscape change. *Global Environmental Change*, 18, 256–270.

Alessa, L., Kliskey, A., Lammeres, R., et al. (2008b) The Arctic water resource vulnerability index: An integrated assessment tool for community resilience and vulnerability with respect to freshwater. *Environmental Management*, 42, 523–541.

Andreev, A.A., Grosse, G., Schirrmeister, L., et al. (2004) Late Saalian and Eemian palaeoenvironmental history of the Bol'shoy Lyakhovsky Island (Laptev Sea region, Arctic Siberia). *Boreas*, 33, 319–348.

Anisimov, O.A. & Nelson, F.E. (1996) Permafrost distribution in the Northern Hemisphere under scenarios of climatic change. *Global and Planetary Change*, 14, 59–72.

Banks, D., Siewers, U., Sletten, R.S., et al. (1999) The thermal springs of Bockfjorden, Svalbard: II: Selected aspects of trace element hydrochemistry. *Geothermics*, 28, 713–728.

Banks, D., Sletten, R., Haldorsen, S., et al. (1998) The thermal springs of Bockfjord, Svalbard: Occurrence and major ion hydrochemsitry. *Geothermics*, 27, 445–467.

Bense, V.F., Ferguson, G. & Kooi, H. (2009) Evolution of shallow groundwater flow systems in areas of degrading permafrost. *Geophysical Research Letters*. [Online] 36, L22401. Available from: doi:10.1029/2009GL039225, September 29, 2011.

Booij, M., Leijnse, J., Haldorsen, S., et al. (1998) Subpermafrost groundwater modelling in Ny-Ålesund, Svalbard. *Nordic Hydrology*, 29, 385–396.

Clark, I.D. & Lauriol, B. (1997) Aufeis of the Firth River Basin, Northern Yukon, Canada: Insights into permafrost hydrogeology and karst. *Arctic and Alpine Research*, 29, 240–252.

Cramer, B., Poelchau, H.S., Gerling, P., et al. (1999) Methan released from groundwater: The source of natural gas accumulations in northern West Siberia. *Marine and Petroleum Geology*, 16, 225–244.

Deming, D., Sass, J.R., Lachenbruch, A.H., et al. (1992) Heat flow and subsurface temperature as evidence for basin-scale ground-water flow, North Slope of Alaska. *Geological Society of America Bulletin*, 104, 528–542.

Dowdeswell, J.A., Hagen, J.O., Björnsson, H., et al. (1997) The mass balance of Circum-Arctic glaciers and recent climate change. *Quaternary Research*, 48, 1–14.

Ford, D. (1987) Effects of glaciations and permafrost upon the development of karst in Canada. *Earth Surface Processes and Landforms*, 12, 507–521.

Ford, J. & Bedford, B.L. (1987) The hydrology of Alaskan wetlands, U.S.A.: A review. *Arctic and Alpine Research*, 19, 209–229.

Frey, K.E. & McClelland, J.W. (2009) Impacts of permafrost degradation on arctic river biochemistry. *Hydrological Processes*, 23, 169–182.

Førland, E.J. & Bauer-Hanssen, I. (2003) Past and future climate variations in the Norwegian Arctic: Overview and novel analyses. *Polar Research*, 22, 113–124.

Frey, K.E., Siegel, D.I. & Smith, L.C. (2007) Geochemistry of west Siberian streams and their potential response to permafrost degradation. *Water Resources Research*. [Online] 43, W03406. Available from: doi: 10.1029/2006WR004902, September 29, 2011.

Gregersen, O. & Eidsmoen, T. (1988) Permafrost conditions in the shore area at Svalbard. In: *VInternational Conference on Permafrost*. 2-5 August, Trondheim, Norway, Tapir Publishers. pp. 933–936.

Grosse, G. & Romanovsky, V. (2011) Vulnerability and feedbacks of permafrost to climate change. *EOS*, 92, 73–74.

Hagen, J.O. (1989) Isbreer og permafrost som klimaindikatorer [Glaciers and permafrost as climate indicators]. In: Orheim, O. & Brekke, A. (eds.) *Hva skjer med klimaet i polarområdene* [What Happens with the Climate in the Polar Areas]. Norwegian Polar Institute Report, 53. 7 pp.

Hagen, J.O., Kohler, J. & Melvold, K., et al. (2003a) Glaciers in Svalbard: Mass balance, runoff and freshwater flux. *Polar Research*, 22, 145–159.

Hagen, J.O., Melvold, K., Pinglot, F., et al. (2003b) On the net mass balance of the glaciers and ice caps in Svalbard, Norwegian Arctic. *Antarctic Alpine Research*, 25, 264–270.

Haldorsen, S. & Heim, M. (1999) An Arctic groundwater system and its dependence upon climatic change: An example from Svalbard. *Permafrost and Periglacial Processess*, 10, 137–149.

Haldorsen, S., Heim, M., Dale, B., et al. (2002) The water balance of an arctic lake and its dependence on climate change: Tvillingvatnet in Ny-Ålseund, Svalbard. *Norwegian Journal of Geography*, 56, 146–151.

Haldorsen, S., Heim, M., Dale, B., et al. (2010) Sensitivity to long-term climate change of subpermafrost groundwater systems in Svalbard. *Quaternary Research*, 73, 393–402.

Haldorsen, S., Heim, M. & Lauritzen, S.-E. (1996) Subpermafrost groundwater, western Svalbard. *Nordic Hydrology*, 27, 57–68.

Haldorsen, S., Leibman, M., Nelson, G., et al. (1997) *State-of-the-Art Report on Saturated Water Movement in Permafrost Areas. Results of the Northern Research Basins Task Force on Water Management in Frozen Ground IHP*. Department of Soil and Water Sciences, Rep 4/97. Norwegian University of Life Sciences, Aas, Norway.

Hanor, J.S., Nunn, J.A. & Lee, Y. (2004) Salinity structure of the central North Slope foreland basin, Alaska, USA: Implications for pathways of past and present topographically driven regional fluid flow. *Geofluids*, 4, 152–168.

Heldmann, J.L., Pollard, W.H., Mckay, C.P., et al. (2005) Annual development cycle of an icing deposit and associated perennial spring activity on Axel Heiberg Island, Canadian high Arctic. *Arctic, Antarctic, and Alpine Research*, 37, 127–135.

Hinzman, L.D., Bettez, N.D., Bolten, W.R., et al. (2005) Evidence and implications of recent climate change in northern Alaska and other regions. *Climatic Change*, 72, 251–298.

Hoel, A. (1914) *Nouvelles observations sur la district volcanique du Spitsberg du nord* [New observations in the volcanic district of northern Spitsbergen]. Videnskapsselskapets skrifter (Christiania), Matematikk, Naturvidenskapelig klasse 9.

Hoel, A. & Holtedahl, O. (1911) *Les nappes de lave, les volcans et les sources thermals dans les environs de la baie Wood au Spitsberg* [The lava formations, the volcanos and the thermal spring in the area of the Wood fjord in Spitsbergen]. Videnskapselsk skr (Christiania), Matematikk, Naturvid kl 8, Nor Acad Sci, Oslo.

Humlum, O., Instanes, A. & Sollid, J.L. (2003) Permafrost in Svalbard: A review of research history, climatic background and engineering challenges. *Polar Research*, 22, 191–215.

IPCC. (2007) Solomon, S., Qin, D. & Manning, M., et al. (eds.) *Climate Change (2007): The Physical Science Basis. Contribution of Working Group I to the Fourth Assessment Report of the Intergovernmental Panel on Climate Change.* Cambridge and New York, Cambridge University Press. 1056 pp.

Jorgenson, T.M., Racine, C.H., Walters, J.C., et al. (2001) Permafrost degradation and ecological changes associated with a warming climate in Central Alaska. *Climatic Change*, 48, 551–579.

Kane, D.L. & Slaughter, C.W. (1973) Recharge of a central Alaska lake by subpermafrost groundwater. In: *Permafrost: The North American Contribution to the Second International Conference.* 13-28 July, Yakuts, USSR. Washington, DC, National Academy of Science. pp. 458–462.

Kharaka, Y.K. & Carothers, W.W. (1989) *Geochemistry of Oil-field Water from the North Slope.* U.S. Geological Survey of Professional Papers 1399. p. 551–561.

Landvik, J.Y., Bondevik, S., Elverhøi, A., et al. (1998) The last glacial maximum of the Barents Sea and Svalbard area: Ice sheet extent and configuration. *Quaternary Science Reviews*, 17, 43–75.

Landvik, J.Y., Brook, E.J., Gualtieri, L., et al. (2003) Northwest Svalbard during the last glaciation: Ice free areas existed. *Geology*, 31, 905–908.

Landvik, J.Y., Ingólfsson, Ó., Mienert, J., et al. (2005) Rethinking Later Weichselian ice-sheet dynamics in coastal NW Svalbard. *Boreas*, 34, 7–24.

Landvik, J.Y., Mangerud, J. & Salvigsen, O. (1988) Glacial history and permafrost development in the Svalbard area. In: *Permafrost: Fifth International Conference Proceedings.* Vol. 1. Trondheim, Norway, Tapir Publishers. pp. 194–198.

Langangen, A. (1979) *Chara canenscens* reported from Spitsbergen. *Phycologia*, 18, 436–437.

Langangen, A. (2000) Charophytes from the warm springs of Svalbard. *Polar Research*, 19, 143–153.

Lauritzen, S.-E. (2006) Caves and speleogenesis at Blomstrandsøya, Kongsfjord, W. Spitsbergen. *International Journal of Speleology*, 35, 37–58.

Lauritzen, S.-E. & Bottrell, S. (1994) Microbiological activity in thermoglacial karst springs, South Spitsbergen. *Geomicrobiology*, 12, 161–173.

Liestøl, O. (1976) Pingos, springs and permafrost in Spitsbergen. *Norsk Polarinstitutt Årbok*, 1975, 7–29.

Liestøl, O. (1980) Permafrost conditions in Spitsbergen. *Frost actions in soils*21, 23-28..

Liestøl, O. (1996) Open-system pingos in Spitsbergen. *Norwegian Journal of Geography*, 50, 81–84.

MacLean, R., Oswood, M.W., Irons, J.G., et al. (1999) The effect of permafrost on stream biogeochemistry: A case study of two streams in the Alaskan (U.S.A.) taiga. *Biogeochemistry*, 47, 239–267.

Michel, F.A. & van Everdingen, R.O. (1988) Karst development in permafrost regions of northern Canada. IAHS 21 Congress. In: *Karst Hydrogeology and Karst Environment Protection*, 10–15 October, Guiun, China. pp. 249–254.

Michel, F.A. & van Everdingen, R.O. (1994) Changes in hydrogeologic regimes in permafrost due to climate change. *Permafrost and Periglacial Processes*, 5, 191–195.

Muskett, R.R. & Romanovsky, V.E. (2009) Groundwater storage changes in arctic permafrost watersheds from GRACE and in situ measurements. *Environmental Research Letters*. [Online] 4, 04009. Available from: doi:10.1088/1748-9326/4/4/045009, September 29, 2011.

Nunn, J.A., Hanor, J.S. & Lee, Y. (2005) Migration pathways in the Central North Slope foreland basin, Alaska USA: Solute and thermal constraints on fluid flow simulations. *Basin Research*, 17, 403–416.

Orvin, A. (1934) Geology of the Kings Bay region, Spitsbergen. *Skrifter om Svalbard og Ishavet*, 57.

Orvin, A. (1944) Litt om kilder på Svalbard [Notes about springs in Svalbard]. *Norwegian Journal of Geography*, 10, 16–38.

Paterson, W.S.B. (1994) *The Physics of Glaciers.* London, Pergamon. 480 pp.

Pohjola, V. (2007) Arctic warming – a perspective from Svalbard. *Global Change Newsletter*, 69, 9–12.

Pollard, W.H. (2005) Icing processes associated with high arctic perennial springs, Axel Heiberg Island, Nunavut, Canada. *Permafrost and Periglacial Processes*, 16, 51–68.

Prokushkin, A., Pokrovsky, O. & Viers, J. (2011) Wildfire effects on stream hydrochemistry within the zone of continuous permafrost distribution in Central Siberia. *Geophysical Research Abstracts*, 13, EGU2011-10524.

Prowse, T.D., Wrona, F.J. & Reist, J.D., et al. (2006a) General features of the arctic relevant to climate change in freshwater ecosystems. *Ambio*, 35, 330–338.

Prowse, T.D., Wrona, F.J., Reist, J.D., et al. (2006b) Climate change effects on hydroecology of Arctic freshwater ecosystems. *Ambio*, 35, 347–358.

Rivière, A., Goncalves, J., Jost, A., et al. (2010) Pore water pressure variations in subpermafrost groundwater: Numerical modeling compared with experimental modeling. *Geophysical Research Abstracts*, 12, EGU 2010-10970-3.

Rouse, W.R., Douglas, M.S.V., Hecky, R.E., et al. (1997) Effects of climate change on the freshwaters of arctic and subarctic North America. *Hydrological Processes*, 11, 873–902.

Salvigsen, O. (2002) Radiocarbon-dated *Mytilus edulis* and *Modiolus modiolus* from northern Svalbard: Climatic implications. *Norwegian Journal of Geography*, 56, 56–61.

Salvigsen, O. & Elgersmaa, A. (1985) Large-scale karst features and open taliks at Vardeborgsletta, outer Isfjorden, Svalbard. *Polar Research*, 3, 145–153.

Salvigsen, O., Lauritzen, Ø. & Mangerud, J. (1983) Karst and kartification in gypsiferous beds in Mathiesondalen, Central Spitsbergen, Svalbard. *Polar Research*, 1, 83–88.

Serreze, M.C., Walsh, J.E., Chapin III, F.S., et al. (2000) Observational evidence of recent change in the northern high-latitude environment. *Climatic Change*, 46, 159–207.

Smart, C.C. (1984) Glacier hydrology and the potential for subglacial karstification. *Norwegian Journal of Geography*, 38, 157–161.

Smith, L.C., Sheng, Y., MacDonald, G.M., et al. (2005) Disappearing Arctic lakes. *Science*, 308, 1429.

Stotler, R.L., Frape, S.K., Ruskeeniemi, T., et al. (2009) Hydrogeochemistry of groundwaters in and below the base of thick permafrost at Lupin, Nunavut, Canada. *Journal of Hydrology*, 373, 80–95.

Svendsen, J.I. & Mangerud, J. (1992) Palaeoclimatic inferences from glacial fluctuations on Svalbard during the last 20 000 years. *Climate Dynamics*, 6, 213–220.

Van der Ploeg, M.J. (2002) Simulation of coupled groundwater flow and transport of heat in the groundwater system under Vestre Lovénbreen, with the model METROHEAT. A surveying study. *MSc Thesis*. Wageningen University, Wageningen, The Netherlands.

Van Everdingen, R.O. (1973) Perennial discharge of subpermafrost groundwater in two small drainage basins, Yukon, Canada. In: *The North American Contribution to the Second International Conference*. 13-28 July, Yakuts, USSR. Washington, DC, National Academy of Science. pp. 639–643.

Van Everdingen, R.O. (1974) Groundwater in permafrost regions in Canada. In: *Permafrost Hydrology, Proceedings of Workshop Seminar 1974, Canadian National Committee for the International Hydrological Decade*, Ottawa, Candada. pp. 83–93.

Van Everdingen, R.O. (1990) Ground water hydrology. In: Prowse, T.D., Ommanney, C.S.L. (eds.) *Canadian Perspectives*. NHRI Science Report 1. pp. 77–101.

Walvoord, M.A. & Striegl, R.G. (2007) Increased groundwater to stream discharge from permafrost thawing in the Yukon River basin: Potential impact on lateral export of carbon and nitrogen. *Geophysical Research Letters*. [Online] 34, L12402. Available from: doi:10.1029/2007GL030216, September 29, 2011.

Wang, B. (1990) Permafrost and groundwater conditions, Huola Basin, northeast China. *Permafrost and Periglacial Procesess*, 1 45–52.

Weaver, J. (2003) Assessment of sub-permafrost groundwater conditions at the Red Dog Mine, Alaska. In: Phillips, M., Springman, S.M. & Arenson, L.U. (eds.) *Permafrost*. Lisse, Swets & Zeitlinger pp. 1223–1228.

White, D., Hinzman, L., Alessa, L., et al. (2007) The arctic freshwater system: Changes and impacts. *Journal of Geophysical Research*. [Online] 112, G04S54. Available from: doi:10.1029/2006JG000353, September 29, 2011.

Williams, J.R. (1970) Groundwater in the permafrost regions of Alaska. *Geological Survey of Professional Paper*, 696, 83 pp.

Woo, M., Marsh, P. & Pomeroy, J.W. (2000) Snow, frozen soils and permafrost hydrology in Canada, 1995–1998. *Hydrological Process*, 14, 1591–1611.

Woo, M.-K., Kane, D.L., Carey, S.K., et al. (2008) Progress in permafrost hydrology in the new millennium. *Permafr Periglac Process*, 19, 237–254.

Yoshikawa, K. & Hinzman, L.D. (2003) Shrinking thermokarst ponds and groundwater dynamics in discontinuous rermafrost near Council, Alaska. *Permafrost and Periglacial Processes*, 14, 151–160.

Zhang, T., Barry, R.G., Knowles, K., et al. (1999) Statistics and characteristics of permafrost and ground-ice distribution in the northern hemisphere. *Polar Geography*, 23, 132–154.

Various Climates

CHAPTER 19

Groundwater management in Asian cities under the pressures of human impacts and climate change

Makoto Taniguchi

ABSTRACT

This research assesses the effects of climate change and human activities on urban subsurface environments including groundwater, which is an important but largely unexamined field of human-environmental interactions. Subsurface conditions merit particular attention in Asian coastal cities including Tokyo, Osaka, Seoul, Taipei, Bangkok, Jakarta and Manila, where population numbers, urban density and use of subsurface environments have expanded rapidly. The goals of this research were to evaluate the subsurface environments of Asian coastal cities for such problems as water shortage, land subsidence, groundwater contamination and thermal anomalies, and to suggest how they can be addressed or adapted. The effects of climate change were examined in particular for changes in water mass including groundwater by satellite GRACE, capacity analyses of groundwater, and subsurface warming due to global warming and heat island effect of urbanization.

19.1 INTRODUCTION

More than one third of the people in the world currently depend on groundwater as a water resource. According to the global distribution of groundwater dependency (UNESCO-IHP, 2008), the dependence on groundwater is greater in the arid and semi-arid regions where the availability of surface water is low. This is not exceptional even for developed countries, such as the EU region with groundwater dependency of 65% (EU, 2000). The level of dependence on groundwater can be higher in the sectors such as agriculture with non-point water use. In contrast to sectors such as metropolises and industries where concentrated and focused intensive water use (point water use) is needed with the use of surface water playing a comparatively central role.

The dependency on groundwater varies with the region/usage in this manner, however, it causes a reduction of the groundwater storage occurs whenever the water consumption exceeds the groundwater recharge rate. It may be pointed out that the cause of this reduction in groundwater storage is not only due to increased water demand due to increase in human population, but also from sociological factors based on globalized market-oriented economic principles which lead to the deep groundwater usage that accompanies with agricultural activities and/or virtual water trade. There are reports that the reduction in global groundwater storage is about 200 billion tons per year (IAH, 2008), which corresponds to one-sixth of the river flow over the world, and is equivalent to a reduction of the depth of global groundwater of about 1.2 mm/a.

To evaluate the global change of land water storage, satellite gravity missions are a new technique of the 21st Century. GRACE (Gravity Recovery And Climate Experiment) launched in March 2002 is a dedicated satellite gravity mission and provides new global geopotential models. The GRACE mission is much more important in terms of temporal variation of gravity fields. It has been shown that GRACE data provided temporal variations of the Earth's gravity fields due to land water including groundwater storage changes in the Earth (Tapley et al., 2004). Even though it may be difficult to employ GRACE data directly for studies of local scale phenomena, it is important to see more large-scale gravity changes as background. There is no doubt that GRACE data contributes to study of global/regional scale water circulations and/or larger scale gravity variations.

In terms of climate change effects on the earth, the recent global warming is considered a global environmental issue only above the ground. However, subsurface temperatures are also affected (Pollack et al., 1998, Huang et al., 2000, Taniguchi et al., 2003). In addition to this, the heat island effect due to urbanization creates subsurface thermal anomalies in many cities (Taniguchi et al., 2003). The combined effects of heat island and global warming reach more than 100 meters below the surface (Taniguchi et al., 1999).

For adaptation to the climate change in terms of sustainable use of groundwater, capacity analyses are necessary. the use of safe and inexpensive groundwater has continued from the initial development stages of the major cities of Tokyo, Osaka and others situated on the coastal alluvial plain in Asia, and the excessive pumping of it has resulted in progressive ground subsidence since the 1960s. Pumping regulations to control land subsidence were enacted in the latter half of the 1960s, and the groundwater levels recovered relatively quickly due to adequate groundwater recharge rate. However, the replenished groundwater led to a new problem with underground structures such as subway stations due to buoyancy of groundwater water level rising. This illustrates how policies and measures which failed to take into consideration. In general speaking for global environmental problems, once humans act to nature then have to keep managing it. In those regions where there is adequate replenishment of groundwater, it should be possible to utilize groundwater sustainably as a part of water circulation by managing it adequately while keeping in mind its limits of capacities.

The research methods being used by this research are as follows; (1) Changes in land water including groundwater were analyzed to compare in-situ satellite-GRACE gravity data and numerical groundwater modeling, (2) Subsurface thermal anomalies due to the global warming and heat island effects in urban areas were evaluated by reconstruction of surface temperature history and urban meteorological analyses, and (3) Capacity analyses of groundwater in seven Asian coastal cities for adaptation to the climate change.

Target study areas are basins including the cities of Tokyo, Osaka, Bangkok, Jakarta, Manila, Taipei and Seoul. In addition to the methods mentioned above, numerical modelling of the subsurface environment and data compilation has been made for social, economic, land use/cover change, and observed subsurface environment based on Geographical Information System (GIS) analyses. Integrated indices on changing society/environment and natural capacity have been developed.

19.1.1   Relevance for GRAPHIC

Climate change causes changes in the hydrological cycle such as precipitation patterns, and therefore causes temporal and spatial changes in groundwater recharge, groundwater storage, and groundwater discharge. Climate change such as global warming causes increases in not only surface temperature but also subsurface temperature because surface warming penetrates into the subsurface environment. Under projected climate change conditions, integrated water management including both surface water and groundwater are necessary, and capacity analyses of groundwater are important for sustainable use of groundwater.

In this study, we show the following three subjects which are deeply related to GRAPHIC theme;

1. Satellite GRACE data for regional/basin scales can be compared with hydrogeological data and results from numerical modelling to evaluate the change in land water storage,
2. Subsurface warming due to global warming and heat island effects may cause subsurface environmental change such as effects on micro biomass, and energy efficiency for heat pump, and
3. Capacity analysis of subsurface environment is a key for adaptation to the climate change.

The reasons why this research fits in GRAPHIC theme are;

1. Development of methodology is important for GRAPHIC theme for evaluating change in land water storage including groundwater, which is not established yet for global and regional/basin scale,
2. Subsurface warming is not recognized yet widely, however, it is very important for global environmental issue, and
3. Capacity analyses of groundwater are extremely important because limitation of water resources under the climate change and integrated water management.

The results from this case study in Asian mega cities can be used to model adaptation and management of groundwater resources under the climate change. In particular, the GRACE satellite methods can be applicable to the basin scale analyses of water budget including groundwater, and it can be good tool for not only global scale but also regional/basin scale of water management.

Subsurface warming shown in this study can contribute to the long-term goals of GRAPHIC, because the effects and impacts of increase in subsurface temperature due to global warming and heat island effects are not evaluated yet, but as same as global surface warming, global subsurface warming with longer residence times is also key for global environmental change as water and heat cycle changes in the earth.

For GRAPHIC long-term goals, the sustainable use of groundwater is the most important issue. Therefore for adaptation to the climate change, resources capacity analyses including groundwater recharge and groundwater storage are important to evaluate the limitation of groundwater uses. This study shows some results for these three issues for long-term GRAPIC goals.

19.2 RESULTS AND DISCUSSION

19.2.1 Satellite GRACE

In order to maintain a sustainable groundwater use, integrated water management of land water storage monitoring including groundwater is important at basin to continental scales. However, traditional methods cannot evaluate large-scale groundwater variations. Previous in situ and satellite techniques can evaluate individual water balance components such as soil water and surface water, however, they cannot evaluate groundwater or total land water storage changes (Taniguchi et al., 2011). Numerical models for groundwater such as MODFLOW can be used to evaluate the land water storage changes, however they have limitations such as insufficient information of geological structure, process description or parameterization as well as errors in model input data, which can all cause large uncertainty (Hasegawa et al., 2008).

The GRACE satellite was launched in March 2002 to solve this problem, and GRACE has been providing monthly global gravity field data with more accurate than before (Tapley et al., 2004). GRACE can recover temporal variations of the Earth's gravity fields due to mass redistribution in and on the Earth (Wahr et al., 1998). Many previous studies concluded that GRACE-inferred annual changes of land water storage agree reasonably well with hydrological model estimates in certain large river basins, for instance, the Amazon (Tapley et al., 2004), Mississippi (Chen et al., 2005), Indochina (four major combined basins) (Yamamoto et al., 2007), and India (Rodell et al., 2009).

Using updated GRACE data, we have succeeded to reveal seasonal variations and secular trend of the land water mass variations in the Chao Phraya river basin (Figure 19.1). The result showed that the total mass change after 2002 was decreasing in the downstream section of the Chao Phraya (Bangkok) while it was increasing in the upstream portions. Figure 19.1d shows temporal variations of averaged terrestrial water storage (the total amount of water existing over the land area, including river water, soil moisture, groundwater, snow, etc.) in Asia from 2002 to 2008, which were calculated from satellite gravity mission GRACE data. GRACE satellite observes temporal variations of the Earth's gravity field precisely. The gravity field variations correspond to the mass variations on and beneath the Earth surface. Because many geophysical phenomena occurring on the Earth, e.g. land water movements, ocean flow, ice sheet mass changes, mass changes associated with earthquakes, etc., accompany mass movement, we can recover those variations using GRACE data. Figure 19.1 was made from CNES/GRGS 10-day gravity field solutions, version2. (Yamamoto, 2011).

The satellite GRACE data were compared with 3D groundwater simulation (MODFLOW) which shows spatial change of the groundwater recharge area, the major recharge area of the pumped aquifer (Taniguchi, 2011). This spatial change of the groundwater potential was strongly affected by regional groundwater pumping regulations, and the success or failure of those regulations are mostly affected by the availability of alternative water resources for the city area and the legal aspect of the groundwater resources. MODFLOW model results show the increasing trend of groundwater storage due to regulation of pumping during this period. Therefore the reason for decreasing land water storage obtained from GRACE must be decreasing natural condition such as decreasing groundwater recharge rate due to decrease of precipitation, but not anthropogenic effect such as groundwater pumping.

19.2.2   Subsurface warming

Global warming causes not only increases in air temperature but also subsurface temperature. In addition to this, the heat island effect due to urbanization also causes increases in air and subsurface temperature. With regard to the cities of Tokyo, Osaka, Seoul, Bangkok, Jakarta and Taipei, a comparison based on the measured average values for

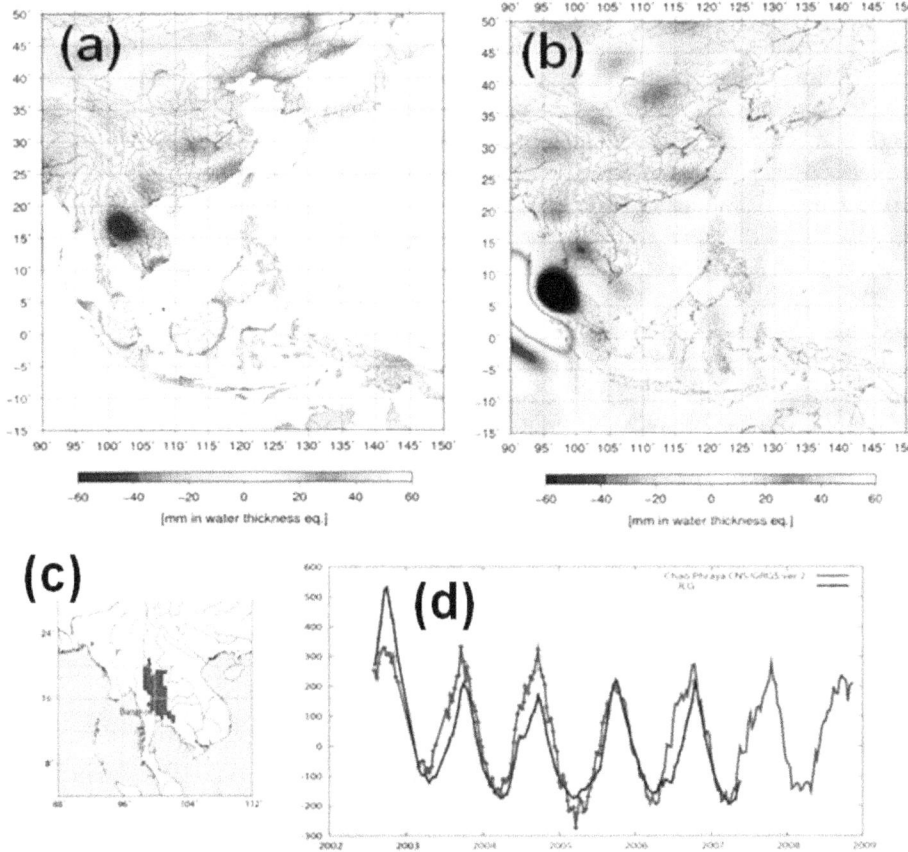

Figure 19.1.   Top: Changes in land water storage during 2002 & 2008 by (a) Model (JRA) and (b) satellite GRACE. Bottom: (c) Location of the basin, and (d) Comparisons of LWS between models (blue) and GRACE (red) (Yamamoto et al. 2009).

the vertical distribution in groundwater temperatures between the geothermal gradient in the natural state (determined by the terrestrial heat flow for the region and the average temperature together with the heat characteristics of the geological features) found that in contrast to the influence of an increase in surface temperature causing an increases in underground temperatures at a depth of more than 120 m in Tokyo, the influence in Osaka was up to the depth of about 80 m and to about 50 m depth in Seoul and Bangkok.

In contrast, an examination of the periods when the air temperatures increase began were evaluated based on an analysis of the increase in average air temperatures in each city for the past 100 years. The results showed larger increase in air temperature began around 1910 in Tokyo, near 1940 in Osaka, in the period from 1970–1980 at Bangkok and Seoul. Consideration of these rises in air temperature and the increase in subsurface temperatures makes it clear that the increases in air temperatures due to the effect of heat islands by urbanization and the global warming are increasing the underground temperatures under the cities in Asia. Furthermore, the periods of sharp rises in air temperature are related to the expansionary times of urbanization with the influence of the heat islands accompanying the city building evincing the great influence exerted on the subterranean temperature environment. This is also evident from an analysis that takes the periods of the beginning of the rise in temperatures (land surface temperatures). The amount of rise in temperatures as parameters and subterranean stored heat together with the differences in depth from geothermal gradients (120 m for Tokyo, 80 m for Osaka and 50 m for Seoul and Bangkok) subsequently proposing them as indices for the degree of urbanization. It has been pointed out that the subterranean heat accumulated in these 7 Asian cities during the past 100 years corresponds to 2 to 6 times of the average subterranean temperature rises worldwide in the heat storage resultant from global warming (Taniguchi et al., 2007, 2009a, b).

Analysis of temperature profiles measured in boreholes provided information on the history of past ground surface temperature (GST). Increased subsurface thermal storage showed the starting time and degree of surface warming due to the combined effects of global warming and urbanization. In the Bangkok area, the amount of the GST increase is larger in the city center than that in the suburban and rural areas, reflecting the degree of urbanization (Figure 19.2). Subsurface heat storage, the amount of heat accumulated under the ground as a result of surface warming, is a useful indicator of the subsurface thermal environment. For example, we can compare the heat storage values at different times in the past with other parameters representing urban subsurface environment obtained through various approaches.

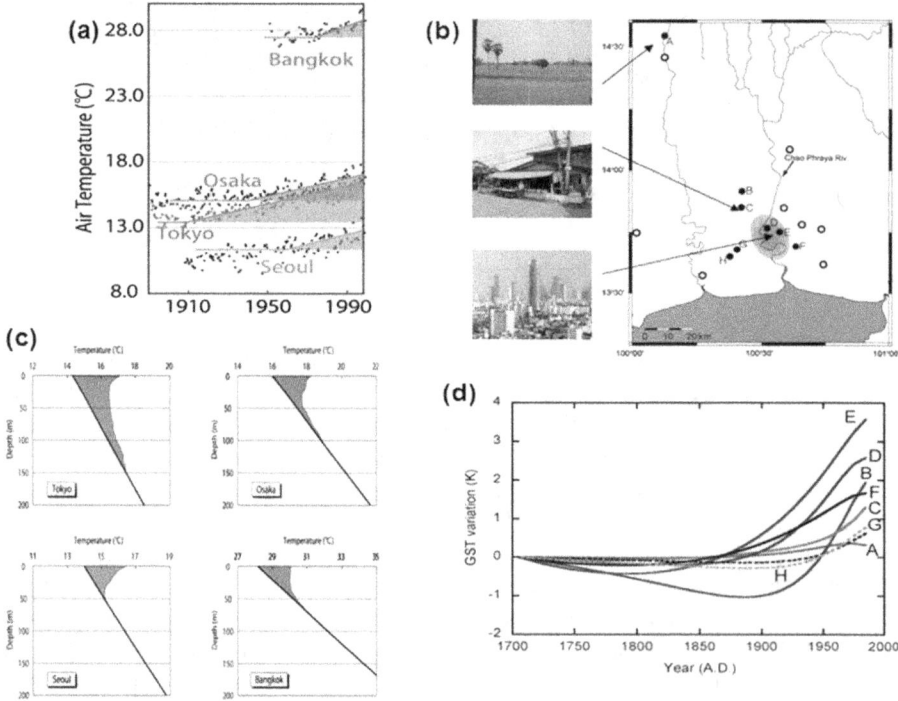

Figure 19.2.   Changes in air temperature (b) Locations of boreholes in BKK (c) Increased subsurface temperature (Taniguchi et al., 2007), and (d) Reconstruction of surface temperature history in Bangkok (Yamano et al., 2009).

19.2.3   Groundwater assessment as natural capacity

In each of the regions, there are different water fluxes such as precipitation, evapotranspiration, river discharge and groundwater recharge. In addition to this, the groundwater storage in aquifers depends on a region's geomorphological and geological factors. These become the critical factors for the sustainable usage of groundwater for the point of view of "natural resources capacity" of the region (Taniguchi 2011). For a instance, it becomes un-sustainable when the groundwater consumption exceed the groundwater recharge rate as natural resources capacity.

Land subsidence due to the excessive pumping of water is being repeated among the coastal metropolises of Asia with the same problem recurring even at present although there is some staggering corresponding to the stage of development of the city. This problem is caused by a failure to understand the "natural resources capacity" for the groundwater in the region and the use of groundwater which exceeds it. Here, natural capacity for groundwater will be considered in two parts: the capacity as volume and the capacity in terms of quality.

The permissible amount for groundwater as seen from the standpoint of volume is the groundwater storage and the groundwater recharge rate which comprise the stock and flow of groundwater. Table 19.1 lists the groundwater storage in aquifers and the potential groundwater recharge rates for 7 Asian metropolises (Tokyo, Osaka, Seoul, Taipei, Bangkok, Jakarta and Manila). It can be seen that Tokyo, Osaka and Bangkok have large volumes of storage from the standpoint of stock, while Manila and Taipei enjoy large volumes of recharge rates from the viewpoint of flow.

Natural resource capacity and social/environmental development indices allowed integration of findings. The indices in five-stage of city development model and DPSIR (Driving force, Pressure, State, Impacts, and Response) model described patterns of water shortage, land subsidence, groundwater contamination, and subsurface thermal anomaly. We suggest a range of suitable policy approaches, taking account of latecomer's benefits, patterns of development, and natural resource capacities (Taniguchi, 2011). In total, our project findings highlight the importance of careful public cross-boundary surface-subsurface environmental management (Taniguchi, 2011).

Table 19.1   Characteristics of aquifer in seven Asian cities.

	Aquifer thickness m	Area km^2	Storage M ton	Potential recharge mm/year	Turnover time Year
Tokyo	600	620	75	350	1,700
Osaka	1,200	220	53	430	2,800
Seoul	100	610	12	340	290
Taipei	100	270	5	640	160
Bangkok	300	1,570	94	450	670
Manila	300	630	38	820	370
Jakarta	100	740	15	640	160

19.3   POLICY RECOMMENDATIONS

The ground subsidence repeated in the coastal cities of Asia indicates that "a trade-off" between groundwater as an environmental base of land and groundwater as a water resource has taken place. As long as water remains in the geological stratum as a "reservoir (stock)" it does not merely exist as a subterranean non-entity, but can be understood afresh as playing a role as a part of the composition of the environmental base comprising the stratum after recognizing ground subsidence as the outcome of the "Tragedy of the Commons". Hardin (1968) called the use of unmanaged common land owned by all yet used by individuals for each person's selfish gain with detrimental results to the benefits of all of the owners "the Tragedy of the Commons". This tragedy of common property (natural resources) is undoubtedly being played out with groundwater and land subsidence as the key actors.

Societal changes such as urbanization that accompany increased human activity influence the subterranean environment through changes in the forms of water usage, material/heat loads and accumulation. With respect to the "natural resources capacities" index group, which is based on groundwater recharge rate, groundwater storage, oxidation-reduction state, heat gradient and other items. Another important index group is the "changing environment and society" that is based on groundwater pumping volumes, socioeconomic indices and other factors. Although there are various factors entailed in the indexing of a "changing society and environment", a stage model corresponding to the development stage of the city (early period of urbanization, industrialization period, urban region expansion period, regulatory and monitoring period, and problem-solving) was formulated based upon: D (Driving force); P (Pressure); S (State); I (Impact); and R (Response). Population, income, industrial structure, and urbanization ratio were selected for the D (Driving force) index; water usage volume, groundwater pumping volume and the degree of dependence on groundwater for the P (Pressure) index; groundwater level for the S (State) index; amount of land subsidence for the I (Impact) index; and pumping regulations for the R (Response) index; and a data set covering the past 100 year period for the 7 cities based on measured, restored, statistical, and estimated data was completed (Taniguchi, 2011).

An analysis of land subsidence, groundwater pollution and subterranean heat pollution using these index groups clearly showed that with regard to land subsidence, Osaka fit the "Tokyo Follower Model" when Tokyo was used as a standard; Bangkok came under the "Follower's Benefit Model" which learned from preceding problems; while Jakarta and Manila fell within the "Excessive Development Model" and Taipei fit the "Benefit of Natural Resources Capacity" (Taniguchi, 2011).

The groundwater storage, groundwater recharge rate and other items in the "natural resources capacity" possess fixed values within the range of their respective regions as natural changes. In other words, this means that the use of values which exceed them is not sustainable. On the one hand, the capability of society with regard to change is an aspect that is difficult to quantify and yet which provides an important element for solving the problem presented by the "latecomer's benefit" generalized by environmental Kuznets curve (Vogel, 1999) and the traditional knowledge present in the respective regions.

19.4   CONCLUSION AND FUTURE WORK

We conclude that subsurface environmental processes can be successfully managed under the changing climate and society, especially in their critical capacity in provision of water, if policies correspond to actual material flows across surface-subsurface and land-ocean boundaries. In regard to water quality, human societies should pay closer attention to the subsurface accumulation of contaminants and heat, especially as these loads can often be controlled or managed from the surface. Designing such policies, however, depends on accurate assessment of the stage of urban growth in relation to natural capacities and social capabilities. In particular for the effects of climate change on groundwater, the following three conclusions were made;

1.   Satellite GRACE is good tool for evaluating land water changes including ground-water, which was shown in Chao Praya basin in this study,
2.   Capacity analyses of groundwater such as groundwater recharge rate and storage as well as adaptation to the subsurface environmental changes are necessary for sustainable use of groundwater under the pressures of climate change and human impacts, and
3.   Subsurface warming due to global warming and heat island effect of urbanization should be analyzed in detail for the future works.

REFERENCES

Chen, J.L., Rodel, M., Wilson, C.R. & Famiglietti, J.S. (2005) Low degree spherical harmonic influences on Gravity Recovery and Climate Experiment (GRACE) water storage estimates, Geophysical Research Letters. Vol. 32, No. 14, L14405. doi:10.1029/2005GL022964

European Union. (2000) Directive 2000/60/EC of the European Parliament and of the Council of 23 October 2000 establishing a framework for Community action in the field of water policy.

Hardin, G. (1968) The Tragedy of the Commons. *Science*, 162(1968), 1243–1248.

Hasegawa, T., Fukuda, Y., Yamamoto, K. & Nakaegawa, T. (2008) The 2006 Australian drought detected by GRACE. In: Taniguchi, M. et al. (eds.) *From Headwaters to the Ocean*. London, Taylor & Francis. pp. 363–367.

Huang, S., Pollack, H.N. & Shen, P.Y. (2000) Temperature trends over the past five centuries reconstructed from borehole temperatures. *Nature*, 403, 756–758.

Pollack, H.N., Huang, S. & Shen, P.Y. (1998) Climate change record in subsurface temperatures: A global perspective. *Science*, 282, 279–281.

Rodell, M., Houser, P.R., Jambor, U., Gottschalck, J., Mitchell, K., Meng, C.J., Arsenault, K., Cosgrove, B., Radakovic, J., Bosilovich, M., Entin, J.K., Walker, J.P., Lohmann, D. & Toll, D. (2004) The global land data assimilation system. *Bulletin of the American Meteorological Society*, 85(3), 381–394.

Taniguchi, M. (ed.) (2011) Groundwater and Subsurface Environment – Human Impacts in Asian Coastal Cities. Springer, Tokyo. 312 pp.

Taniguchi, M., Burnett, W.C. & Ness, G. (2009a) Integrated research on subsurface environments in Asian urban areas. *Science of the Total Environment*, 407, 3076–3088.

Taniguchi M, Shimada J, Fukuda Y, Yamano M, Onodera S, Buapeng S, Delinom R, Fernando S, Wang CH, Lee B, Yasumoto J, Yamamoto K (2009b) Degradation of subsurface environment due to human activities and climate variability in Asian cities, IAHS Publication, 329, 124–129.

Taniguchi, M., Shimada, J., Tanaka, T., Kayane, I., Sakura, Y., Shimano, Y., Dapaah-Siakwan, S. & Kawashima, S. (1999) Disturbances of temperature-depth profiles due to surface climate-change and subsurface water flow: (1) An effect of linear increase in surface temperature caused by global warming and urbanization in Tokyo metropolitan area, Japan. *Water Resources Research*, 35, 1507–1517.

Taniguchi, M., Shimada, J. & Uemura, T. (2003) Transient effects of surface temperature and groundwater flow on subsurface temperature in Kumamoto Plain, Japan. *Physics and Chemistry of the Earth*, 28, 477–486.

Taniguchi, M., Uemura, T. & Jago-on, K. (2007) Combined effects of urbanization and global warming on subsurface temperature in four Asian cities. Vadose Zone Journal, 6(3), 591–596.

Taniguchi, M., Yamamoto, K. & Sarukkalige, P.R. (2011) Groundwater Resources Assessment based on Satellite GRACE and Hydrogeology in Western Australia. IAHS Publication, 343, 3-8

Tapley, B.D., Bettadpur, S., Watkins, M.M. & Reigber, C. (2004) The gravity recovery and climate experiment: Mission overview and early results. *Geophysical Research Letters*. [Online] 31, L09607. Available from: doi:10.1029/2004GL019920

UNESCO-IHP. (2008) *GRAPHIC Framework Document (GRAPHIC Series No. 2)*. 30 pp.

Vogel, M.P. (1999) Environmental Kuznets Curves: A Study on the Economic Theory and Political Economy of Environmental Quality Improvements in the Course of Economic Growth. Springer - Verlag. Berlin, Heidelberg, and New York, 197 pp.

Wahr, J., Molenaar, M. & Bryan, F. (1998) Time-variability of the Earth's gravity field: Hydrological and oceanic effects and their possible detection using GRACE. Journal of Geophysical Research, 103(B12), 30205–30230.

Yamamoto, K. (2011) Human impacts on Asian Subsurface Environment. CD-ROM. Research Institute for Humanity and Nature, Kyoto, Japan.

Yamamoto, K., Fukuda, Y., Nakaegawa, T. & Nishijima, J. (2007) Landwater variation in four major river basins of the Indochina peninsula as revealed by GRACE. Earth, Planets and Space, 59, 193–200.

Yamano, M., Goto, S., Miyakoshi, A., Hamamoto, H., Lubis, R.F., Monyrath, V. & Taniguchi, M. (2009) Reconstruction of the thermal environment evolution in urban areas from underground temperature distribution. Science of the Total Environment, 407, 3120–3128.

CHAPTER 20

Evaluation of future climate change impacts on European groundwater resources

Kevin Hiscock, Robert Sparkes & Alan Hodgson

ABSTRACT

In this study, climatological, geological and land use data were used to characterise five study areas in northern and southern Europe, centred on the Å, Medway, Seine, Guadalquivir and Po river basins. To analyse the impacts of climate change on groundwater resources in these areas, four Global Circulation Models (GCMs) were used to predict future precipitation and temperature trends based on a 'high' (SRES A1FI) gas emissions scenario for the 2020s, 2050s and 2080s with these values then used in a soil moisture balance model to calculate future potential groundwater recharge. Most GCMs predict that, by the end of this century, northern Europe will receive more winter rainfall, leading to increased groundwater recharge but during a shorter time period, and that summers will be drier with a longer period of limited or no groundwater recharge. In order to adapt to these conditions, it will be necessary to capture the winter recharge and use it efficiently to meet the summer demand for water. Southern Europe is expected to experience lower groundwater recharge overall and may become more water stressed than present, with any increase in winter recharge unable to compensate for reduced autumn groundwater recharge. Southern Spain is predicted to be one of the worst affected regions with almost the total disappearance of groundwater recharge. In this and similar water-stressed regions, future demand for drinking water and irrigation will need to be managed sustainably with supplies potentially supplemented by non-conventional water resources such as waste water reclamation and reuse.

20.1 INTRODUCTION

At the European level, groundwater is a significant economic resource. As shown in Table 20.1, large quantities of groundwater are abstracted in France, Germany, Italy and Spain (all in excess of $5 \times 10^9 \, m^3/a$) comprising 16% of the total water abstracted in these four countries. Average annual water abstraction from groundwater accounts for 20% of the total in Europe, ranging from greater than 50% in Austria, Belgium, Denmark and Luxembourg to only 10% and 12%, respectively, in Finland and Ireland. The data given in Table 20.1 probably underestimate the contribution made by groundwater to municipal water supplies since, according to a report commissioned for the European Commission (RIVM and RIZA 1991), about 75% of the inhabitants of Europe depend on groundwater for their water supply. Additionally, in Mediterranean regions, groundwater is a valuable resource in meeting the high agricultural irrigation demand (Krinner et al. 1999; UNEP 2003).

In common with other areas of the world, European groundwater resources are threatened by over-abstraction and contamination from surface-derived pollutants (IYPE 2005), with these pressures potentially exacerbated by the additional anthropogenically-driven

Table 20.1   Average annual municipal water supply abstractions in European Union member states by type for the period 1980–1995. The data are ordered in terms of the percentage groundwater contributes to the total abstraction (Source: European Environment Agency Data Service).

Country	Surface water $(\times 10^6 \, m^3)$	Groundwater $(\times 10^6 \, m^3)$	Total	% Groundwater
Denmark	9	907	916	99
Belgium	2385	4630	7015	66
Austria	1038	1322	2360	56
Luxembourg	28	29	57	51
Portugal	4233	3065	7298	42
Greece	3470	1570	5040	31
Italy	40000	12000	52000	23
United Kingdom	9344	2709	12053	22
Sweden	2121	588	2709	22
Spain	29901	5422	35323	15
Germany	51151	7711	58862	13
France	35195	5446	40641	13
Netherlands	10965	1711	12676	13
Ireland	945	125	1070	12
Finland	3011	335	3346	10
Total	193796	47570	241366	20

Note: Data given in this table should be treated with caution due to the lack of a common European procedure to estimate water resources.

threat of climate change. The interpretation of climate change on these important groundwater resources is difficult to predict but will be dependent on regional hydrogeological characteristics, as well as socio-economic conditions that will determine future water supply demand (Holman et al. 2005; Holman 2006). Climatic warming during this century is expected to lead to changes in the global hydrological cycle that will impact on regional water resources. In Europe, records show that, over the last century (1901–2001), the average temperature has risen by 0.95°C and that climate change has caused a steepening of precipitation and temperature gradients resulting in wetter conditions in northern regions and drier conditions in southern areas (IPCC AR4 WGII, Alcamo et al. 2007).

To explore the potential response of European groundwater resources to future climate change, this study presents an analysis of the application of four Global Circulation Models (GCMs) together with a soil moisture balance model to predict current and future potential groundwater recharge until 2100.

Five river basin areas were selected from northern and southern Europe on the basis of the importance of their groundwater resources. Comparisons are made between current and future recharge to deduce changes in groundwater resources with data grouped into the 2020s, 2050s and 2080s time periods under a 'high' greenhouse gas emissions scenario. Changes in potential groundwater recharge are compared to the observed 1961–1990 period in order to analyse for changing severity and frequency of hydrological periods. The results from the modelled scenarios are considered in terms of the general hydrological response of European groundwater resources to climate change and, based on these results, recommendations are made for future adaptation of groundwater resources management.

20.1.1 Description of the areas: aquifer units in northern and southern Europe

The choice of study areas for analysis of climate change impacts on European groundwater resources was determined by the presence of important aquifer units and, in part, by the availability of reliable information. Data for land use, aquifer types, river flows and water resources use were taken from a range of sources including the Global Runoff Data Centre (www.gewex.org/grdc.html), the UK National Water Archive (www.nwl.ac.uk/ih/nwa/index.htm), the Corine Land Cover 2000 map (www.eea.europa.eu/data-and-maps/data/corine-land-cover-2000), the Digital Dataset of European Groundwater Resources (eusoils.jrc.ec.europa.eu/ESDB_Archive/groundwater/gw.html; Hollis et al. 2002) and the International Groundwater Resources Assessment Centre (www.igrac.net).

Five river drainage basins were chosen, centred on the locations shown in Figure 20.1. The five areas were characterised in terms of land use, major aquifer units and the degree of stress on water resources. In northern Europe, the River Å catchment in Denmark is an area of arable land and industrial/commercial development. The major groundwater resources are contained in glacial sands and gravels, with water abstractions equal to less than 10% of available recharge. The River Medway catchment in South East England is an area of mainly non-irrigated arable land, pasture and urban development. Major aquifers include the Cretaceous Chalk and Lower Cretaceous Sands that supply London and the South East, with water abstractions equal to about 25% of available recharge. The River Seine catchment in France is an area of arable agriculture together with urban and semi-urban development. The region is underlain by the Cretaceous Chalk and Lower Cretaceous Sands that are important for supplying Paris and the northwest of the basin, with water abstractions equal to less than 10% of available recharge, but with some aquifer units that are over-abstracted. In southern Europe, the River Guadalquivir in central-southern Spain is an important agricultural region, mainly planted with cotton, asparagus, maize and peppers, grown from April to September (Lorite-Herrera et al. 2009), with groundwater resources found in dolomitic limestone and alluvial deposits. Water supplies are under stress with abstractions up to 85% of available recharge, making the Guadalquivir one of the most exploited rivers in Europe. Of the groundwater abstractions, up to 45% are withdrawn illegally for crop irrigation from unlicensed boreholes, a practice that started after the prolonged drought of the 1980s (WWF 2006). The River Po in northern Italy is a region of irrigated arable land as well as having urban and industrial/commercial development (Morari et al. 2004). Major aquifers are contained in alluvial sediments that are recharged twice each year with snow melt in the spring and rainfall in the autumn, with abstractions equal to about one third of available recharge.

20.2 METHODOLOGY

In the assessment of climate change impacts on groundwater recharge, an ensemble of four Global Circulation Models (GCMs) was chosen (HadCM3, CGCM2, CSIRO2, PCM) to provide scenarios of climate until 2100, with the scenarios used as input to a soil moisture balance model in order to estimate current and future potential groundwater recharge rates.

Figure 20.1.   Map of Europe showing locations of the five study areas centered on the: River Å, Denmark; River Medway, England; Seine, France; River Guadaiquivir, Spain; and River Po, Italy.

The simulation of future potential groundwater recharge values adopted the SRES A1FI gas emissions scenario (IPCC 2000) in which the world remains reliant on fossil fuels leading to a best estimate temperature rise of 4.0°C from 1990 levels by 2100 (IPCC 2007). Climate change data were made available from the Climatic Research Unit, University of East Anglia (www.cru.uea.ac.uk) as a set of high-resolution grids of monthly climate for Europe at a spatial resolution of 10 minutes for the observed record (1901–2000) and 16 transient scenarios (2001–2100) as presented by Mitchell et al. (2004). The calculation of potential groundwater recharge was based on the Penman-Grindley method as described by Lerner et al. (1990). In this study, potential groundwater recharge is considered to equate to drainage below the soil zone and is equivalent to the hydrological excess water, or effective precipitation, that can generate both surface runoff and groundwater recharge. In the presence of underlying permeable strata found in major aquifer areas, the surface runoff component via interflow is considered in this study to be of limited importance. Values of potential evapotranspiration were estimated using the Hargreaves method (Hargreaves and Allen 2003). Land use data were taken from the Corine Land Cover 2000 map and values for soil properties and crop coefficients were based on Allen et al. (1998).

The methodology for calculating current potential groundwater recharge was validated by comparison with other studies. For example, for the current baseline period, a modelled mean annual potential groundwater recharge value of 287 mm for South East England (Table 20.2) is comparable to a mean annual effective precipitation value of 285 mm obtained for the UK Meteorological Office MORECS square 172 (Hough and Jones 1997) and within the range of 148 mm to 300 mm for average annual recharge calculated by Cross et al. (1995) and Jones and Robins (1999). For southern Spain, the long-term average discharge for the River Guadalquivir at Puente de Palmas for the period 1912–1989 is 80.356 m^3/s and this equates to a total potential groundwater recharge in support of river flow of 52.2 mm for a catchment area of 48,515 km^2, slightly more than the baseline recharge value of 30.6 mm calculated in this study (Table 20.2), which itself is slightly more than the recharge value of 13.5 mm presented by Lorite-Herrera et al.

Table 20.2 Percentage change in mean annual potential groundwater recharge values (in mm) calculated using four GCMs (HadCM3, CGCM2, CSIRO2 and PCM) for the 2020s, 2050s and 2080s 'high' gas emissions scenarios compared with the baseline period, 1961–1990, for the five study areas (Denmark, England, France, Spain and Italy) shown in Figure 20.1. Negative percentage changes indicating a decrease in annual groundwater recharge are shown in bold.

	Baseline	2020s	% change	2050s	% change	2080s	% change
HadCM3							
Denmark	279.8	312.0	11.5	333.6	19.2	302.3	8.0
England	286.8	301.5	5.1	284.8	**−0.7**	347.2	21.0
France	140.7	159.0	13.1	175.3	24.6	235.5	67.4
Spain	30.6	24.5	**−20.1**	30.2	**−1.4**	5.1	**−83.4**
Italy	494.0	370.6	**−25.0**	330.3	**−33.1**	346.2	**−29.9**
CGCM2							
Denmark	279.8	333.5	19.2	384.3	37.3	376.3	34.5
England	286.8	303.6	5.8	289.0	0.8	340.4	18.7
France	140.7	148.2	5.4	147.5	4.9	184.4	31.1
Spain	30.6	33.4	9.0	46.9	53.0	12.1	**−60.6**
Italy	494.0	392.1	**−20.6**	362.0	**−26.7**	383.4	**−22.4**
CSIRO2							
Denmark	279.8	330.9	18.3	376.3	34.5	369.3	32.0
England	286.8	308.6	7.6	294.4	2.7	349.2	21.7
France	140.7	162.0	15.1	173.1	23.0	232.9	65.6
Spain	30.6	43.2	41.0	70.6	130.5	35.8	16.9
Italy	494.0	420.6	**−14.9**	424.6	**−14.1**	493.9	0.0
PCM							
Denmark	279.8	318.4	13.8	348.3	24.5	318.5	13.8
England	286.8	297.2	3.6	268.1	**−6.5**	302.7	5.5
France	140.7	138.9	**−1.2**	117.7	**−16.3**	138.1	**−1.8**
Spain	30.6	37.5	22.4	57.0	86.1	22.0	**−28.1**
Italy	494.0	406.6	**−17.7**	390.3	**−21.0**	438.4	**−11.3**

(2009). However, it should be recognised that the actual groundwater recharge rates in irrigated areas of southern Spain can, according to Jiménez-Martínez et al. (2010), be as much as 397 mm for annual row crops, equal to 31% of the total applied water, as a result of irrigation return flow, even under drip irrigation which is used widely due to water scarcity and the need for water conservation.

Once validated, the methodology for computing potential groundwater recharge for all five study areas was executed for the future climate change scenarios. To facilitate the analysis of results, comparisons were made between current baseline and future recharge to infer changes in groundwater resources, with data grouped into the 2020s (2011–2040), 2050s (2041–2070) and 2080s (2071–2100) 30-year time periods and compared to the observed baseline period 1961–1990.

20.3   RESULTS AND DISCUSSION

Values of percentage change in mean annual potential groundwater recharge compared with the baseline period, 1961–1990, calculated using the above methodology for the four GCMs and the 2020s, 2050s and 2080s 'high' gas emissions scenarios for the five study areas are shown in Table 20.2. Three GCMs (HadCM3, CGCM2 and PCM) agree in predicting negative percentage changes in groundwater recharge by the 2080s for southern Spain and northern Italy, with the greatest losses in groundwater recharge of 83% and 30%, respectively, predicted by the HadCM3 GCM. In contrast to these results, the CSIRO2 GCM predicts a small increase (17%) in recharge for southern Spain and no change in mean annual recharge for northern Italy. Percentage changes are generally positive by the 2080s for the northern European countries, Denmark, England and France with, for example, predicted gains in recharge of 8%, 21% and 67%, respectively, based on the HadCM3 model. However, in contradiction, some scenarios for the HadCM3 and PCM GCMs predict small losses of groundwater recharge for England and France. For the most northern study area, Denmark, all scenarios for the four GCMs show positive changes in groundwater recharge, ranging from 8% for the HadCM3 2080s scenario to 37% for the 2050s CGCM2 scenario.

Changes in mean monthly potential groundwater recharge values compared to the observed baseline period 1961–1990 for the five study areas for the HadCM3 2080s 'high' gas emissions scenario are presented in Figure 20.2. Inspecting these results, and considering that the HadCM3 GCM tends to predict drier climatic conditions (see table of groundwater recharge values given in Table 20.2), regions of northern Europe represented by the Rivers Å, Medway and Seine show an increase in both precipitation and evapotranspiration values leading to increased groundwater recharge in the winter months. In contrast, these areas demonstrate decreases in groundwater recharge in all months from April to October and, for the River Seine, also in November. As shown in Figures 20.3 to 20.5, the decrease in recharge in the autumn months represents both a reduction and delay in the replenishment of groundwater resources in the northern European study areas by the 2080s. For southern Europe, the Rivers Guadalquivir and Po demonstrate a small increase in evapotranspiration and a marked decrease in precipitation leading to a reduction in groundwater recharge values in most months by the 2080s compared with the historic baseline period. This dramatic loss of recharge in nearly all months by the 2080s is clearly shown for the Rivers Guadalquivir and Po in Figures 20.6 and 20.7, respectively, Therefore, these results confirm those of the IPCC (Alcamo et al. 2007) of a steepening precipitation gradient from southern to northern latitudes, generally drier spring and summer periods across the whole European continent and increasingly arid conditions in southern European regions leading to a loss of groundwater resources.

The results presented here for northern European areas can be compared to those presented by Herrera-Pantoja and Hiscock (2008) who used climate data derived from

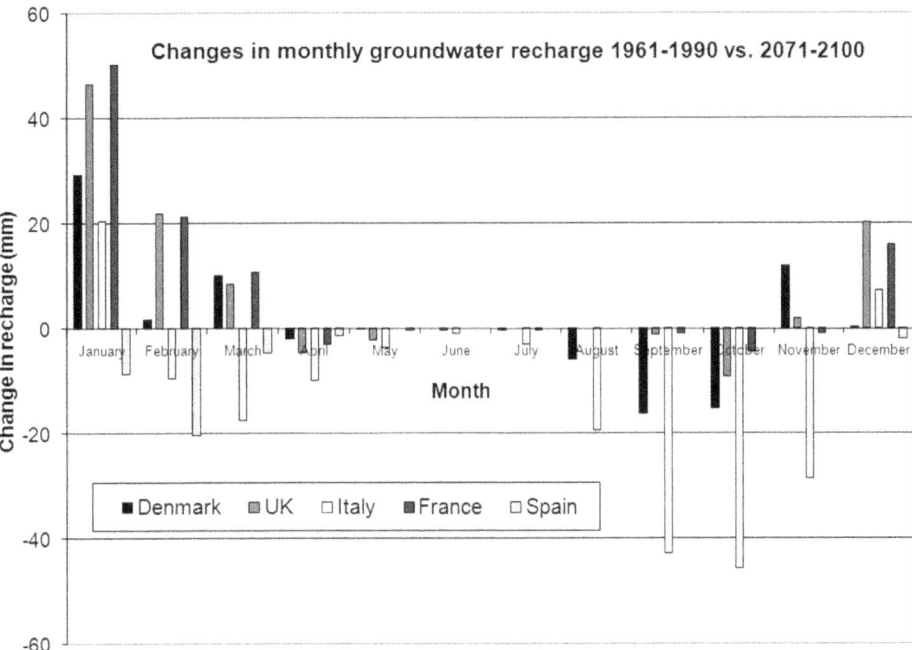

Figure 20.2. Changes in mean monthly potential groundwater recharge (mm) predicted by the HadCM3 GCM for the 2080s 'high' gas emissions scenario compared with the baseline period, 1961–1990, for the five study areas shown in Figure 20.1. Most regions show wetter winters and drier summer recharge periods.

the HadRM3H regional climate model to simulate potential groundwater recharge time-series for the baseline period 1961–1990 and for a future 'high' gas emissions scenario for the 2020s, 2050s and 2080s time periods for East Anglia in eastern England, Gatwick in southern England and Paisley in west Scotland. Unlike the results presented in Table 20.2, Herrera-Pantoja and Hiscock (2008) predicted a decrease of 20% in potential groundwater recharge for East Anglia, 40% for Gatwick and 7% for Paisley by the end of this century, although there is agreement in predicting an increase in the persistence of dry periods in both England and Scotland during the 2050s and 2080s. The difference in findings between this and the current studies is considered to be due to the choice of climate change projections. Herrera-Pantoja and Hiscock (2008) used an approach based on a stochastic weather generator in contrast to the method used in this study which is based on Mitchell et al. (2004) who combined time-series of global warming and patterns of change from GCMs with the baseline climate and sub-centennial variability from the observed record.

In better comparison, similar findings to the current study are presented by Jackson et al. (2011) for an investigation of Chalk groundwater resources in central-southern England in which precipitation and temperature change factors for the 2080s, derived from 13 GCMs run under a medium-high (A2) gas emissions scenario, were applied to a distributed recharge model to yield a spread of results ranging from a 26% decrease to a 31% increase in annual potential groundwater recharge, with ten GCMs predicting a decrease and three an increase. Jackson et al. (2011) also found that the seasonal

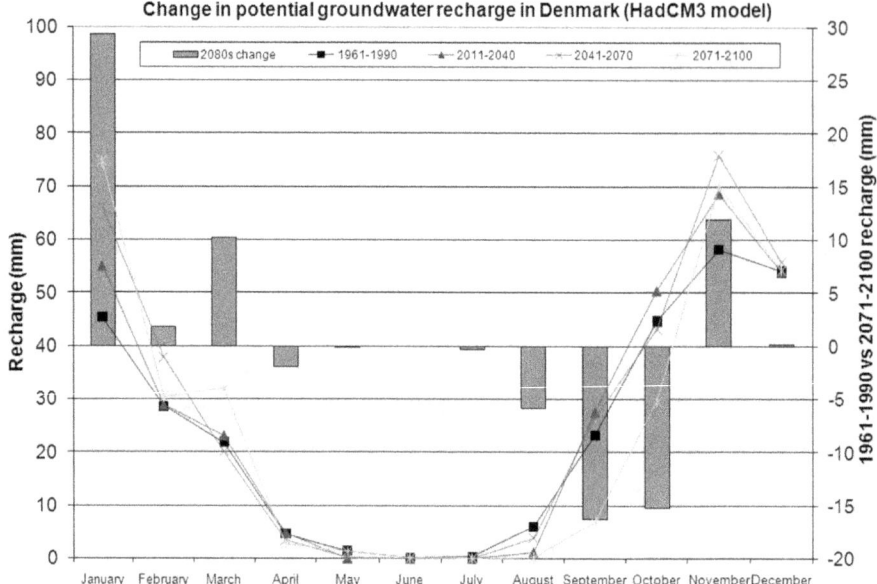

Figure 20.3. Graphs of mean monthly potential groundwater recharge for the baseline period (1961–1990) and three HadCM3 future 'high' gas emissions scenarios (2020s, 2050s and 2080s) for the River Å catchment, northern Denmark. Also shown as bars are mean monthly changes in groundwater recharge comparing the 1961–1990 baseline period with the future 2080s 'high' gas emissions scenario.

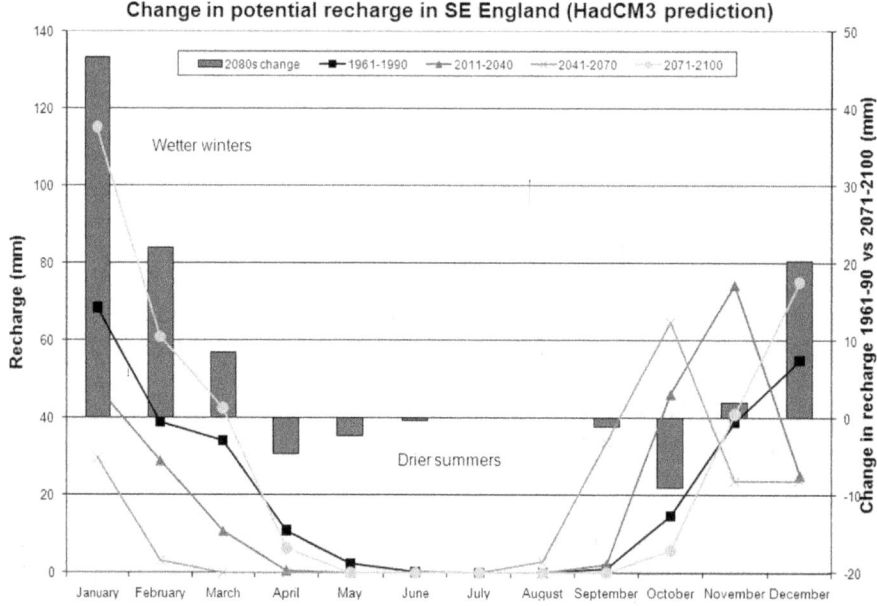

Figure 20.4. Graphs of mean monthly potential groundwater recharge for the baseline period (1961–1990) and three HadCM3 future 'high' gas emissions scenarios (2020s, 2050s and 2080s) for the River Medway catchment, southern England. Also shown as bars are mean monthly changes in groundwater recharge comparing the 1961–1990 baseline period with the future 2080s 'high' gas emissions scenario.

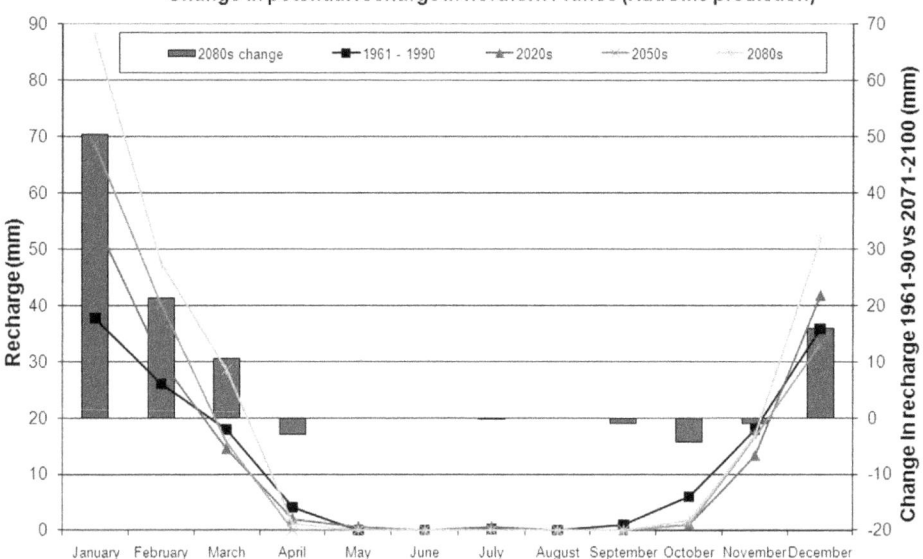

Figure 20.5. Graphs of mean monthly potential groundwater recharge for the baseline period (1961–1990) and three HadCM3 future 'high' gas emissions scenarios (2020s, 2050s and 2080s) for the River Seine catchment, northern France. Also shown as bars are mean monthly changes in groundwater recharge comparing the 1961–1990 baseline period with the future 2080s 'high' gas emissions scenario.

variation in groundwater recharge will be greater in the future, with higher recharge rates during a reduced time period in the winter season, with the effects of climate change shown to depend significantly on the type of land use.

Also for northern Europe, Brouyère et al. (2004) used an integrated distributed model to show the potential impact of climate change on the Chalk aquifer of the Geer Basin in Belgium in which most scenarios tested predicted a decrease in groundwater levels and storage. In extending this study in the Geer Basin, Goderniaux et al. (2009) showed that for the climate scenarios considered, simulations exhibited significant decreases of up to 8 m in groundwater levels and decreases of between 9% and 33% in groundwater-supported surface flows by 2080. In Denmark, climate change impacts on groundwater-river interaction have also been modelled by Van Roosmalen et al. (2007) using a regional climate model comparing the periods 1961–1990 and the 2080s run under SRES gas emissions scenarios A2 and B2, with the climate data then input to a physically-based, distributed hydrological model. The results showed significant increases in annual net precipitation, but with decreases in the summer months, that produced increased groundwater recharge, higher groundwater levels and increased groundwater-river interaction. Stream discharges were simulated to increase by 50% in winter months and decrease by 50% in summer months.

In southern Europe, Polemio and Casarano (2008) statistically analysed available hydrometeorological data from 1921–2001 for southern Italy and showed a widespread decreasing trend of annual rainfall, with the spatial mean values ranging from −0.8 mm/a in Apulia to −2.9 mm/a in Calabria, predominantly concentrated in the winter season.

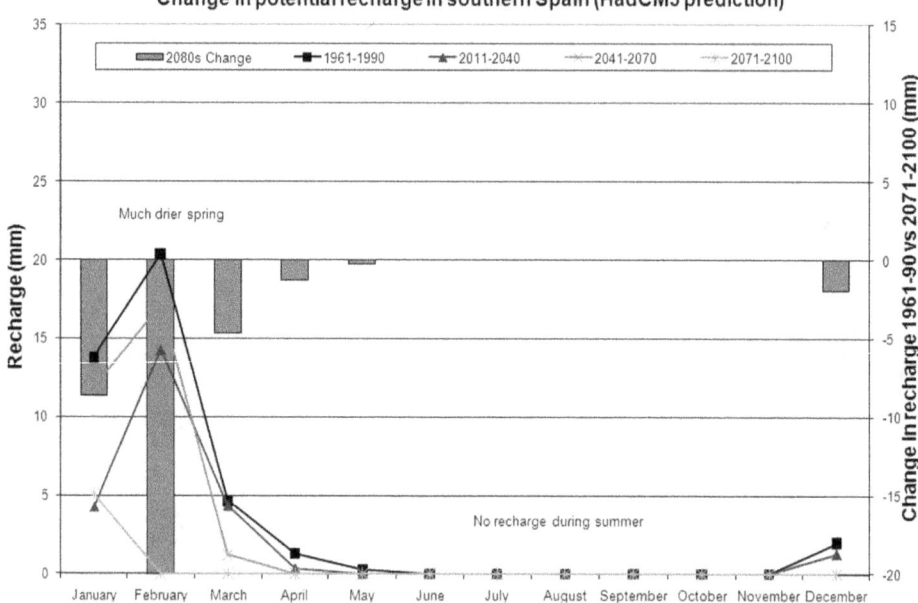

Figure 20.6.   Graphs of mean monthly potential groundwater recharge for the baseline period (1961–1990) and three HadCM3 future 'high' gas emissions scenarios (2020s, 2050s and 2080s) for the River Guadalquivir catchment, southern Spain. Also shown as bars are mean monthly changes in groundwater recharge comparing the 1961–1990 baseline period with the future 2080s 'high' gas emissions scenario.

The decrease in rainfall is notable after 1980 with the droughts of 1988–1992 and 1999–2001 appearing to be exceptional. Polemio and Casarano (2008) also analysed 239 well records and showed a widespread negative trend in groundwater availability in southern Italy causing a serious water supply situation as a result of a decrease in groundwater recharge, exacerbated by an increase in groundwater abstractions to compensate for the non-availability of surface water reservoirs during drought periods. Similar conclusions were obtained by Ducci and Tranfaglia (2008) for the region of Campania in southern Italy, with many parts of the region experiencing a regional temperature increase of about 0.3°C in the last 20 years (1981–1999) and a mean decrease of 30% in average infiltration under the present climate. Ducci and Tranfaglia (2008) found that carbonate aquifers are the most affected hydrogeological regions, especially in the mountainous areas in the southern and northern areas of Campania, with spring flows significantly reduced by about 70 m^3/s. In Portugal, Stigter et al. (2009) applied a groundwater modelling approach and showed an enhanced risk under climate change of increased water demand and decreasing rainfall that will exacerbate groundwater salinisation and ecosystem degradation in the Algarve. In northern Portugal, Oliveira et al. (2007) presented a daily sequential water balance model to study the impact of changes in precipitation on groundwater recharge and found that for a scenario of 70% of actual precipitation, recharge would decrease to 45% of the estimated actual recharge.

In eastern Europe, in a study of the karst aquifers in Bulgaria, Benderev et al. (2008) showed that the influence of climate change is especially evident for karst springs.

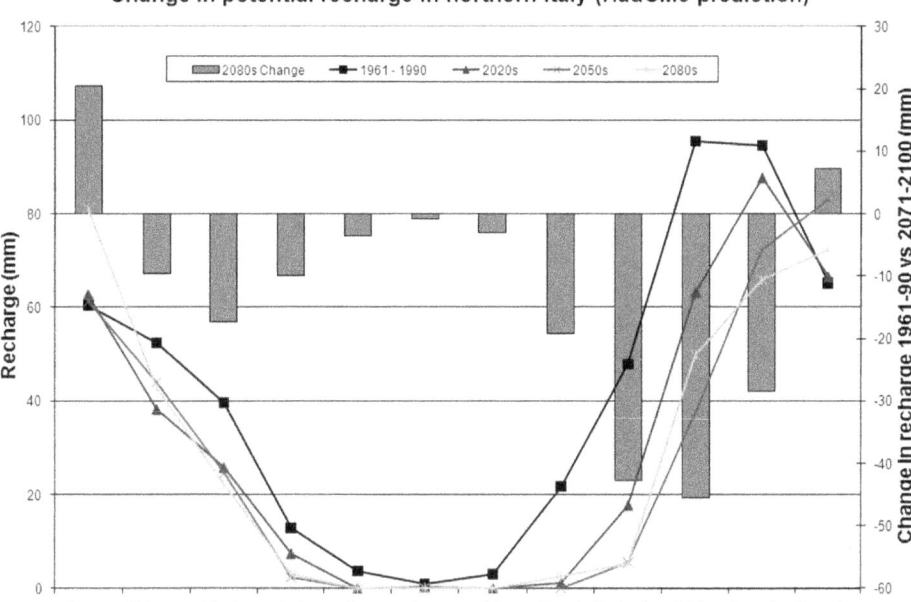

Figure 20.7. Graphs of mean monthly potential groundwater recharge for the baseline period (1961–1990) and three HadCM3 future 'high' gas emissions scenarios (2020s, 2050s and 2080s) for the River Po catchment, northern Italy. Also shown as bars are mean monthly changes in groundwater recharge comparing the 1961–1990 baseline period with the future 2080s 'high' gas emissions scenario.

A time series analysis of spring discharges for selected karst basins showed the impact of the 1982–1994 drought period on the groundwater regime which led to limited resource availability, in particular for springs that drain unconfined mountainous karst. This case study illustrates the important influence of geological structure in determining the effects of climate variability on groundwater resource availability.

In central Europe, in a study of the Upper Danube catchment located in parts of Germany, Austria, Switzerland and Italy, Barthel (2011) presented an integrated modelling approach for assessing the availability of groundwater under several different climate scenarios (2011–2060) under the SRES gas emissions scenario A1B, downscaled using a stochastic weather generator. The results showed that responses of the groundwater system to climate change in the Upper Danube are very different, again demonstrating the important role that aquifer geometries and hydraulic properties exert in the assessment of future conditions. In general, a steady decrease in groundwater recharge is simulated which results in decreasing groundwater levels over the 50-year simulation period. It is observed that conditions remain stable throughout the first two decades of the scenario with only the northern border of the Alps showing any deterioration due to the combination of decreasing precipitation in this area and unfavourable hydrogeological conditions (moraines). This spatial trend is then observed to develop more widely towards the end of the simulation period.

20.4   CONCLUSIONS

Based on the methodology and results presented here, for northern Europe, where present groundwater supplies are generally good with high abstraction rates sustained by high recharge rates, the impacts of climate change will affect summer recharge conditions but potentially increase winter recharge, albeit for a shorter period, potentially leading to an increased risk of groundwater flooding (Ireson and Butler 2011). Hence, the key to climate change adaptation in northern Europe in order to cope with longer, drier summer periods is to capture the winter recharge and use it efficiently through good demand management in which surface water and groundwater resources are used conjunctively, especially in the densely populated areas in southern England and northern France.

In southern Europe, groundwater supplies are already heavily stressed with climate change expected to further reduce potential groundwater recharge and available groundwater resources throughout the year, with groundwater recharge virtually disappearing in some areas by the 2080s. Hence, in combination with water requirements that are projected to increase under a drier climate, severe water shortages are a distinct possibility in southern Europe by the end of this century. Under these conditions, adaptation responses to climate change are to conserve groundwater resources by controlling the irrigation of crops in the most water-stressed areas in order that domestic water supplies can be met. Furthermore, in Mediterranean coastal areas where water scarcity is aggravated by high population densities and intense economic activities leading to acute seasonal water demands, Salgot and Torrens (2008) recommend the increasing use of non-conventional water resources of lower quality, such as waste water reclamation and reuse that can augment groundwater resources and so meet additional demands (other than for drinking water) for urban, industrial and agricultural applications. Ultimately, the achievement of long-term sustainability of groundwater resources in water-stressed regions, can only be met by effective integration of regional water and agricultural policies that, for example, control illegal groundwater abstractions, establish water banking, and promote cropping diversification and modern irrigation methods, to ensure ecological protection and human development at acceptable social cost (Varela-Ortega et al. 2011).

20.5   FUTURE WORK AND RELEVANCE TO GRAPHIC

The research presented in this paper applied a consistent methodology to five different areas of Europe which enabled the calculation of potential groundwater recharge using standard equations. More detailed analysis is challenging given the variety of hydrogeological conditions and variability in data sources encountered in a region as geologically and geographically diverse as Europe. Hence, and in keeping with the primary objectives of the GRAPHIC Project to understand the impacts of human activities and climatic changes on groundwater availability and sustainability, further integrated assessments, supported by consistent regional monitoring efforts, are encouraged that combine climate, land and hydrological models to better represent recharge processes in the soil and unsaturated zones in all regions of Europe. Further work is also required to improve the representation of the mechanisms and timing of snow melt runoff and recharge that are of significance in the mountainous areas of central and eastern Europe. An example of an improved modelling approach is presented by Goderniaux et al. (2009) who used a coupled physically-based surface-subsurface flow model combined with advanced future climate scenarios to assess climate change impacts on groundwater resources. Similarly, Jackson et al. (2011) emphasised the need for further research to quantify the influence on groundwater recharge of the effects of different vegetation types and thickness of soil cover and their response to a changing climate.

In recommending further, more sophisticated modelling studies to provide more case study examples, consideration should also be given to the uncertainty and importance of socio-economic scenarios in exploring future climate change. As argued by Holman (2006), to focus on the direct impacts of climate change is to neglect the potentially important role of policy, societal values and economic processes in determining the landscape above aquifers, and so requiring a multidisciplinary approach to assessing the impacts of climate change on groundwater resources. As an example, Maia and Schumann (2007) developed a decision support system for a region of southern Portugal that aimed to optimise domestic and irrigation water demand as well as determine the water pricing increase necessary to achieve economic sustainability, with cost recovery goals set in accordance with the EU Water Framework Directive.

ACKNOWLEDGEMENTS

The authors are grateful to the Nuffield Foundation for the award of a research bursary to Robert Sparkes and also to the University of East Anglia in supporting the research presented here. This paper is based on a presentation given at the 2008 meeting of the European Geosciences Union with funding support from the UNESCO GRAPHIC (Groundwater Resources Assessment under the Pressures of Humanity and Climate Change) Project. The following organisations are also thanked for providing help and advice with data collection: the Climatic Research Unit, University of East Anglia, Norwich, UK; the Global Runoff Data Centre, Koblenz, Germany; the International Groundwater Resources Assessment Centre, The Netherlands; and the European Environment Agency, Copenhagen.

REFERENCES

Alcamo, J., Moreno, J.M., Nováky, B., Bindi, M., Corobov, R., Devoy, R., Giannakopoulos, C., Martin, E., Olesen, J.E. & Shvidenko, A. (2007) In: Parry, M.L., Canziani, O.F., Palutikof, J.P., van der Linden, P.J. & Hanson, C.E. (eds.) *Europe. Climate Change 2007: Impacts, Adaptation and Vulnerability. Contribution of Working Group II to the Fourth Assessment Report of the Intergovernmental Panel on Climate Change.* Cambridge, Cambridge University Press. 541–580.

Allen, R.G., Pereira, L.S., Raes, D. & Smith, M. (1998) *Crop evapotranspiration – guidelines for computing crop water requirements.* FAO Irrigation and Drainage Paper 56. Rome, Italy, Food and Agriculture Organization of the United Nations.

Barthel, R. (2011) An indicator approach to assessing and predicting the quantitative state of groundwater bodies on the regional scale with a special focus on the impacts of climate change. *Hydrogeology Journal*, 19, 525–546.

Benderev, A.D., Orehova, T.V. & Bojilova, E.K. (2008) Some aspects of groundwater regime in Bulgaria with respect to climate variability. In: Dragoni, W. & Sukhija, B.S. (eds.) *Climate Change and Groundwater. Geological Society Special Publications (London)*, 288, 13–24.

Brouyère, S., Carabin, G. & Dassargues, A. (2004) Climate change impacts on groundwater resources: Modelled deficits in a chalky aquifer, Geer basin, Belgium. *Hydrogeology Journal*, 12, 123–134.

Cross, G.A., Rushton, K.R. & Tomlinson, L.M. (1995) The East Kent Chalk Aquifer during the 1988–92 drought. *Journal of Water and Environmental Management*, 9, 1–37.

Ducci, D. & Tranfaglia, G. (2008) Effects of climate change on groundwater resources in Campania (southern Italy). In: Dragoni, W. & Sukhija, B.S. (eds.) *Climate Change and Groundwater. Geological Society Special Publications (London)*, 288, 25–38.

Goderniaux, P., Brouyère, S., Fowler, H.J., Blenkinsop, S., Therrien, R., Orban, P. & Dassargues, A. (2009) Large scale surface-subsurface hydrological model to assess climate change impacts on groundwater reserves. *Journal of Hydrology*, 373, 122–138.

Hargreaves, G.H. & Allen, R.G. (2003) History and evaluation of the Hargreaves evapotranspiration equation. *Journal of Irrigation and Drainage Engineering-ASCE*, 129, 53–63.

Herrera-Pantoja, M. & Hiscock, K.M. (2007) The effects of climate change on potential groundwater recharge in Great Britain. *Hydrological Processes*, 22, 73–86.

Hollis, J.M., Holman, I.P. & Burton, R.G.O. (2002) *A Digital Dataset of European Groundwater Resources at 1:500,000*. Vol. 1.0. Brussels, Belgium, European Crop Protection Association.

Holman, I.P. (2006) Climate change impacts on groundwater recharge-uncertainty, shortcomings, and the way forward? *Hydrogeology Journal*, 14, 637–647.

Holman, I.P., Nicholls, R.J., Berry, P.M., Harrison, P.A., Audsley, E., Shackley, S. & Rounsevell, M.D.A. (2005) A regional, multi-sectoral and integrated assessment of the impacts of climate and socio-economic change in the UK: Part 2 results. *Climatic Change*, 71, 43–73.

Hough, M.N. & Jones, R.J.A. (1997) The United Kingdom Meteorological Office Rainfall and Evaporation Calculation System: MORECS version 2.0 – an overview. *Hydrology and Earth System Sciences*, 1, 227–239.

IPCC. (2000) *Special report on emissions scenarios*. Geneva, Intergovernmental Panel on Climate Change.

IPCC. (2007) *Climate Change 2007: Synthesis Report. Contribution of Working Groups I, II and III to the Fourth Assessment Report of the Intergovernmental Panel on Climate Change*. Geneva, Switzerland.

Ireson, A.M. & Butler, A.P. (2011) Controls on preferential recharge to Chalk aquifers. *Journal of Hydrology*, 398, 109–123.

IYPE. (2005) *Groundwater – Reservoir for a Thirsty Planet? International Year of Planet Earth*. Leiden, The Netherlands, Earth Sciences for Society Foundation.

Jackson, C.R., Meister, R. & Prudhomme, C. (2011) Modelling the effects of climate change and its uncertainty on UK Chalk groundwater resources from an ensemble of global climate model projections. *Journal of Hydrology*, 399, 12–28.

Jiménez-Martínez, J., Candela, L., Molinero, J. & Tamoh, K. (2010) Groundwater recharge in irrigated semi-arid areas: Quantitative hydrological modelling and sensitivity analysis. *Hydrogeology Journal*, 18, 1811–1824.

Jones, H.K. & Robins, N.S. (eds.) (1999) *The Chalk Aquifer of the South Downs*. Hydrogeological Report Series. Keyworth, Nottingham, British Geological Survey.

Krinner, W., Lallana, C., Estrela, T., Nixon, S., Zabel, T., Laffon, L., Rees, G. & Cole, G. (1999) *Sustainable water use in Europe – Part 1: Sectoral Use of Water*. Copenhagen, Denmark, European Environment Agency.

Lerner, D.N., Issar, A.S. & Simmers, I. (1990) *Groundwater Recharge: A Guide to Understanding and Estimating Natural Recharge*. Vol. 8. International Contributions to Hydrogeology. Hannover, Germany, Verlag Heinz Heise.

Lorite-Herrera, M., Hiscock, K. & Jiménez-Espinosa, R. (2009) Distribution of dissolved inorganic and organic nitrogen in river water and groundwater in an agriculturally-dominated catchment, south-east Spain. *Water, Air, & Soil Pollution*, 198, 335–346.

Maia, R. & Schumann, A.H. (2007) DSS application to the development of water management strategies in Ribeiras do Algarve River Basin. *Water Resources Management*, 21, 897–907.

Mitchell, T.D., Carter, T.R., Jones, P.D., Hulme, M. & New, M. (2004) *A comprehensive set of high-resolution grids of monthly climate for Europe and the globe: The observed record (1901–2000) and 16 scenarios (2001–2100)*. Tyndall Centre Working Paper Number: 55. Tyndall Centre for Climate Change Research, University of East Anglia, Norwich.

Morari, F., Lugato, E. & Borin, M. (2004) An integrated non-point source model-GIS system for selecting criteria of best management practices in the Po Valley, North Italy. *Agriculture Ecosystems and Environment*, 102, 247–262.

Oliveira, M.M., Novo, M.E. & Lobo Ferreira, J.P. (2007) Models to predict the impact of the climate changes on aquifer recharge. *IAHS-AISH Publication*, 310, 102–110.

Polemio, M. & Casarano, D. (2008) Climate change, drought and groundwater availability in southern Italy. In: Dragoni, W. & Sukhija, B.S. (eds.) *Climate Change and Groundwater. Geological Society, Special Publications (London)* 288, 39–51.

RIVM and RIZA. (1991) *Sustainable use of groundwater: Problems and threats in the European Communities*. Bilthoven, The Netherlands, National Institute of Public Health and Environmental Protection and Institute for Inland Water Management and Waste Water Treatment. Report Number: 600025001.

Salgot, M. & Torrens, A. (2008) Impacts of climatic change on water resources: The future of groundwater recharge with reclaimed water in the south of Europe. In: Dragoni, W. & Sukhija, B.S. (eds.) *Climate Change and Groundwater. Geological Society Special Publications (London)*, 288, 145–168.

Stigter, T.Y., Monteiro, J.P., Nunes, L.M., Vieira, J., Cunha, M.C., Ribeiro, L., Nascimento, J. & Lucas, H. (2009) Screening of sustainable groundwater sources for integration into a regional drought-prone water supply system. *Hydrology and Earth System Sciences*, 6, 85–120.

UNEP. (2003) *Groundwater and its susceptibility to degradation: A global assessment of the problem and options for management*. Early Warning and Assessment Report Series, RS 03-3. Nairobi, Kenya, United Nations Environment Programme.

Van Roosmalen, L., Christensen, B.S.B. & Sonnenborg, T.O. (2007) Regional differences in climate change impacts on groundwater and stream discharge in Denmark. *Vadose Zone Journal*, 6, 554–571.

Varela-Ortega, C., Blanco-Gutiérrez, I., Swartz, C.H. & Downing, T.E. (2011) Balancing groundwater conservation and rural livelihoods under water and climate uncertainties: An integrated hydro-economic modelling framework. *Global Environmental Change*, 21, 604–619.

WWF. (2006) *Illegal Water Use in Spain – Causes, Effects and Solutions*. Madrid, Spain, WWF.

CHAPTER 21

Sustainable groundwater management for large aquifer systems: tracking depletion rates from space

Sean Swenson & James Famiglietti

ABSTRACT

Nearly 2 billion people worldwide rely on groundwater as their primary source of water for domestic and agricultural use. However, in many regions around the globe, groundwater supplies are being stressed to their limits by overuse, population growth and climate change. Likewise, groundwater quality is also threatened by contamination, land use change, salinization, and in coastal regions, seawater intrusion. In the developed world, water management efforts to minimize the impacts of these multiple stressors on groundwater quantity and quality are complicated by a number of institutional, political and socioeconomic barriers. These include the lack of sufficiently dense monitoring well networks, disparate sampling, data reporting and archiving strategies, inadequate data sharing policies, and in many regions, the lack of groundwater use reporting requirements. In the developing world, such management strategies, if they exist at all, are rudimentary or in their infancy. In this paper, a satellite-based method for characterizing groundwater storage changes in large aquifer systems is described that provides new opportunities for water resources monitoring, particularly in data sparse regions. Two case studies of groundwater depletion are presented, one in the relatively data-rich Central Valley aquifer of California (USA) and in the other in more data-poor northern India. The approach uses observations from the NASA Gravity Recovery and Climate Experiment (GRACE) mission, along with varying amounts of supplementary hydrologic data as available. This method can be applied to large aquifers ($>200,000\,\text{km}^2$) at monthly and longer timescales with sufficient accuracy to inform regional water management decisions. The approach could also be implemented globally to track changes in the availability of the world's fresh groundwater resources.

21.1 INTRODUCTION

Freshwater is vital to human existence. While water is abundant on Earth, much of it is saline, and unsuitable for human consumption. Of the only 2.5% of the Earth's water that is fresh, roughly two-thirds is stored as ice, and almost a third is stored deep beneath the surface as groundwater. The most readily accessible stocks of freshwater are those at the land surface, including for example, rivers, lakes, and in human-constructed reservoirs. These account for perhaps a quarter of a percent of all fresh water (*Shiklomanov* 2000). Fortunately, these stocks are renewed on timescales of days to years, as fresh water is evaporated from the oceans and precipitates over land, before returning to the ocean as river discharge. Over the course of a year, about $36,000\,\text{km}^3$ of fresh water will pass through terrestrial fresh water stores (Syed et al. 2010).

However, not all land reservoirs are renewed on these timescales. Water stored as groundwater may have residence times of tens to thousands of years, making these freshwater stores effectively non-renewable on the shorter timescales of typical human interest (Gleeson et al. 2010). Although the total amount of renewable water resources is currently greater than global human demand, it is distributed unevenly both in space and time. Thus, in many locations, for example in the mid-latitude arid and semi-arid regions of the world, groundwater is the primary source of fresh water: at present, it provides domestic, agricultural and industrial supply to over 2 billion people worldwide. Clearly then, effective water management and environmental decision making requires that available water resources be used sustainably, in particular in the context of continuous population growth and changing climate. Quantifying groundwater withdrawal and use, however, remains a difficult challenge (*Giordano* 2009).

21.1.1   Purpose and scope

The goal of this paper is to review a new approach for space-based monitoring of groundwater storage changes in large aquifer systems that in part alleviates the problem of data scarcity. In particular, the potential for combining observations of terrestrial water storage changes from the NASA Gravity Recovery and Climate Experiment (GRACE) satellite mission (Tapley et al. 2004) with supplementary *in situ*, remotely sensed, or modelled hydrologic datasets, to track monthly rates of groundwater storage will be demonstrated. The role of this approach in regional and transboundary water management is also discussed.

21.1.2   Description of the study area

Recently, Famiglietti et al. (2011) used GRACE data to track groundwater depletion in California's Central Valley, a hydrologically data-rich region relative to international standards. Similarly, Tiwari et al. (2009), and *Rodell* et al. (2009), used GRACE and other modelled hydrologic datasets to estimate rates of groundwater depletion in northern India, a much more data-sparse region in which water storage changes are difficult to characterize both within and outside of the country. With the exception of data availability, the regions share several defining characteristics. Both are vast agricultural regions that produce enormous quantities of food for their respective nations, but that rely heavily on groundwater supplies to meet their irrigation water demands. Consequently, both regions have been experiencing significant rates of groundwater depletion, and in the case of California's Central Valley, land subsidence, for several decades.

21.1.3   Relevance to GRAPHIC

The work described in this paper can play a key role in the GRAPHIC effort. As described in the following sections, the GRACE-based approach is well-suited to monitoring groundwater storage changes in the world's major aquifer systems. Since many of these aquifers underlie major population centres, cross international political boundaries, and are subject to new recharge regimes resulting from changing climate, they fall well within the purview of the GRAPHIC mission. Moreover, the approach described here provides an aquifer-scale, holistic view of groundwater storage variations that may otherwise be difficult to construct. This is particularly true where hydrologic monitoring is not well implemented, or where access to *in situ* observations is restricted.

21.2   METHODS AND RESULTS

21.2.1   Ground-based well measurements

How are changes in groundwater storage typically measured? Because of its subterranean nature, groundwater is inherently more difficult to measure compared to the level of a lake or reservoir. The oldest and most commonly used method is to measure changes in the water table depth observed in wells or boreholes. By making records of the distance from the ground surface to the top of the water table, one can infer changes in groundwater storage (Taylor and Alley 2001). This is not a direct measure of storage, because the relationship between change in water table and change in groundwater volume depends on the properties (e.g. specific yield) of the water-holding material. Thus the changes in the levels of two wells might have similar amplitudes, yet the changes in storage could differ greatly depending on the relative specific yield. Subsurface soil or rock properties can also exhibit considerable spatial heterogeneity, making uncertain the relationship between water levels in a well and its surroundings. An accurate picture of the state of groundwater storage over a larger region may therefore require a large quantity of well level measurements, as well as detailed estimates of the subsurface properties needed to reconcile well level and storage changes.

Even at the scale of a local aquifer system, such measurements are often made by different entities, are recorded at varying time intervals, and are stored in a variety of formats at different locations. In many regions of the world, measurements of groundwater pumping are not even required, while in other regions, access to available measurements is severely restricted beyond political boundaries. Simply put, characterizing groundwater storage changes, even at smaller spatial scales, requires a concerted, labour-intensive monitoring and data collection effort. For larger-scale, regional-scale aquifers, and at the global scale, it may not be feasible to assemble a well level dataset with sufficient temporal and spatial resolution to conduct viable groundwater resources assessments.

21.2.2   Hydrologic modelling

While there is no substitute for intensive, ground-based groundwater monitoring and the synergistic use of detailed numerical groundwater flow and transport models, as alluded to above, their implementation for large, regional aquifer systems is a labour-intensive endeavour (e.g. Faunt 2009) which may not even be feasible at the global-scale.

A related approach that may be more amenable for use at continental and global scales utilizes water balance models that estimate the difference between recharge and withdrawal. For example, Wada et al. (2010) recently estimated groundwater depletion around the world by estimating recharge to the water table using a global hydrological model and readily-available supporting datasets. The model partitions net precipitation (i.e. precipitation minus evapotranspiration) from a global atmospheric forcing dataset into river discharge, water storage as snow and soil moisture, and infiltration to deeper groundwater layers. Abstraction was estimated from a database maintained by the International Groundwater Resources Assessment Centre (IGRAC; www.igrac.net) of groundwater withdrawal by country. These national level data were then downscaled to the resolution of the modelled recharge dataset (0.5 degree × 0.5 degree) using information on the location of aquifers and human demand. Regions where withdrawals exceed recharge indicate groundwater depletion. Areas of significant ground water depletion estimated by this method include India, Pakistan, north-eastern China, and the western United States.

While model-based approaches such as that of *Wada* et al. (2010) provide valuable information on the magnitude and spatial distribution of groundwater depletion at regional scales, they are of course subject to uncertainty. Estimates of groundwater recharge will contain biases due to errors in both the atmospheric forcing data and the representation of processes within the land model. Withdrawal estimates may be poorly known, or in some cases, are not reported by some countries. Furthermore, the accuracy of the assumptions used to downscale country-level data to finer spatial scales is also unknown. How, then, can we validate or supplement these model data? How can we obtain more detailed information on changing groundwater availability in data-poor or data-restricted regions?

21.2.3 The GRACE-based approach: case studies from the Central Valley of California (USA) and northern India

One answer may come from satellite remote sensing. Regional scale estimates of changes in groundwater storage have recently been inferred from the Gravity Recovery and Climate Experiment (GRACE). Unlike the radiometric measurements of Earth surface made by the majority of remote sensing instruments, GRACE measures changes in Earth's gravity, which are directly related to changes in the distribution of mass in the Earth system, including those in the deeper, subsurface. At interannual and shorter timescales, most changes in Earth's mass occur within the hydrosphere. By modelling and removing estimates of atmospheric mass and the mass of water stored above the water table (i.e. snow and ice, surface water, soil moisture), groundwater storage changes can be isolated and estimated. GRACE has been used to successfully quantify groundwater variations in northern India (Tiwari et al. 2009; Rodell et al. 2009), north-eastern China (Moiwo et al. 2009), and California, USA (Famiglietti et al. 2011). In all of these regions, groundwater is being depleted at alarming high rates. Case studies for California and for northern India are presented here.

The Central Valley of California, USA

The Central Valley is an approximately $50,000 \, km^2$ region that lies within the $\sim 150,000 \, km^2$ Sacramento and San Joaquin River basins of California in the United States. It is one of the most productive agriculture regions in the world, and its underlying aquifer is the second-most pumped in the country. By international standards, the region is well-monitored and its data are openly accessible within and outside of the United States. Consequently, the occurrence of groundwater depletion in the Central Valley is well-known. As such, the Central Valley offers an important test case for GRACE-based estimation of groundwater storage changes. The successful application of the GRACE-based technique in data-rich regions like the U.S. gives confidence to its application in the many data-poor regions of the world.

In order to isolate the groundwater component of total water storage change observed by GRACE, the other major water storage signals (i.e. from snow, surface water and soil moisture) must be estimated and removed from the GRACE data. The upper left panel of Figure 21.1 (all data in mm; taken from Famiglietti et al. 2011) shows the GRACE data for monthly water storage variations for the period October,

2003–March, 2010. The upper right panel shows the snow water equivalent (SWE) (from the U.S. National Operational Hydrologic Remote Sensing Center), the lower left the surface water (complied for the 20 largest reservoirs in the California Department of Water Resources database) and the lower right, soil water content (the average of three different soil moisture simulations taken from the NASA Global Land Data Assimilation System (GLDAS) Rodell et al. (2004) for the corresponding time period).

Subtracting SWE, surface water and soil moisture variations from the total water storage change from GRACE yields the groundwater storage changes shown in Figure 21.2. A clear decreasing trend is evident beginning in 2006, which is consistent with regional drought conditions observed on the ground. Famiglietti et al. (2011) reported that the total water loss for the time period was nearly $30\,km^3$, two-thirds of which was estimated as net groundwater loss (extraction minus recharge). The estimated loss rate is consistent with earlier losses (prior to 2002) estimated by the U. S. Geological Survey (Faunt 2009) in a detailed observational-modelling study of the Central Valley. Famiglietti et al. (2011) also noted that the GRACE-based estimate of the spatial pattern of groundwater depletion matches the pattern observed from the region's network of monitoring wells.

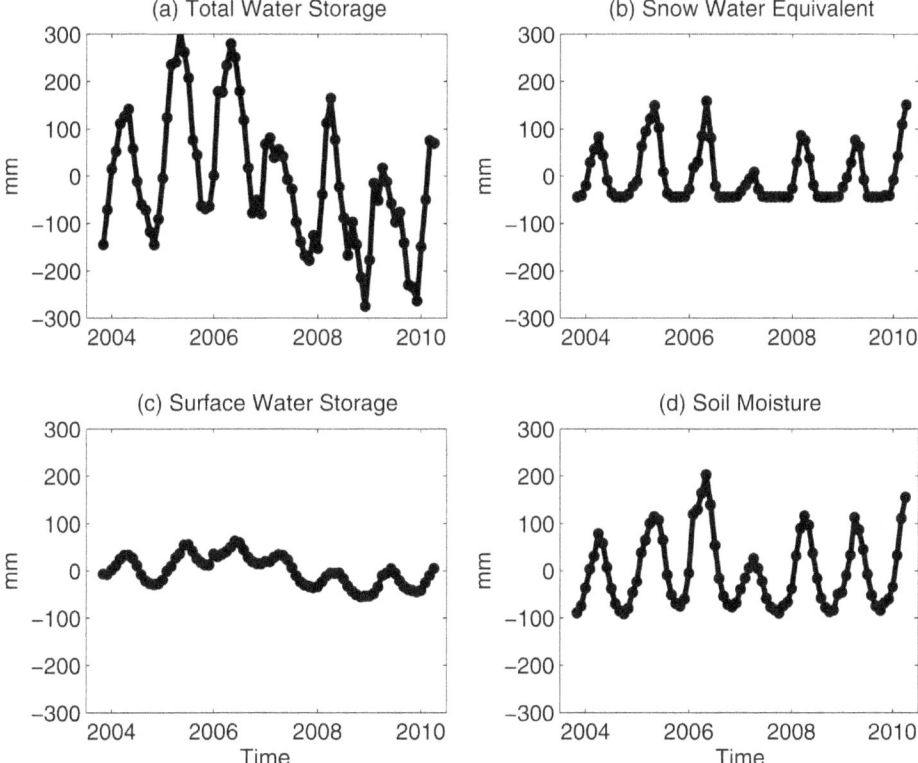

Figure 21.1. Monthly storage variations of a) total water storage from GRACE; b) snow water equivalent; c) surface water storage; and d) soil water content. All stores shown as anomalies with respect to the mean of the study period and reported in mm. From Famiglietti et al. (2011).

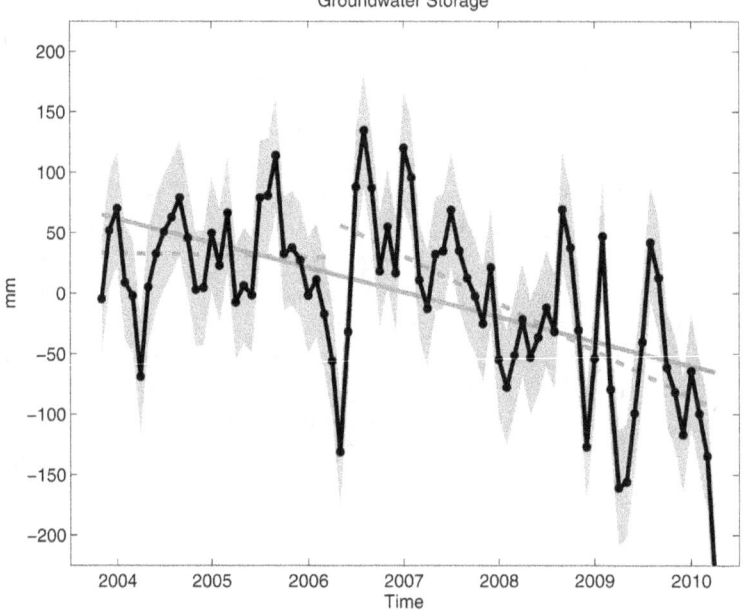

Figure 21.2.   GRACE-derived estimate of monthly groundwater storage variations. Solid black line represents monthly anomalies (mm) of groundwater storage with respect to the study period. Solid gray line is the trend (20.4 ± 3.9 mm/a). Dashed gray lines are piecewise trends before (October, 2003 – March, 2006; 1.4 ± 12.7 mm/a) and during before (April, 2006 – March, 2010; 38.9 ± 9.5 mm/a) the regional drought. Shaded gray area represents monthly errors. From *Famiglietti* et al. (2011).

Northern India

Northern India, like California's Central Valley, is another of the world's most productive agricultural regions. This productivity is derived in part from irrigation, which is supplied by both surface and groundwater. Water tables have been dropping in many parts of Northern India (*CGWB* 2006), but quantifying the total amount of groundwater depletion is challenging because of the highly distributed nature of the extraction, which has rapidly increased in recent times due to the availability of electrical pumps to individual farmers (*Shah* 2007). It is further complicated by limited data collection and access.

A pair of studies (*Tiwari* et al. 2009; *Rodell* et al. 2009) combined GRACE data with land model simulations and demonstrated the efficacy of the GRACE-based approach for estimating groundwater storage changes in regions where data accessibility is relatively poor by international standards. Looking at the entire region of northern India, Tiwari et al. (2009) found an average decrease in groundwater storage of 20 mm/annum (a) or 54 km³/a. Focusing on the north-western states of Rajasthan, Punjab, and Haryana (including Delhi), which show the largest magnitude decreases in water storage, Rodell et al. (2009) estimated losses of 40 mm/a, or 18 km³/a, from these states alone.

Figure 21.3 shows the average groundwater change for the period 2003–2011. The map shown in the top panel reveals a broad groundwater loss across the entire Ganges River basin. This pattern coincides with the highest intensity irrigated areas seen in the Global Map

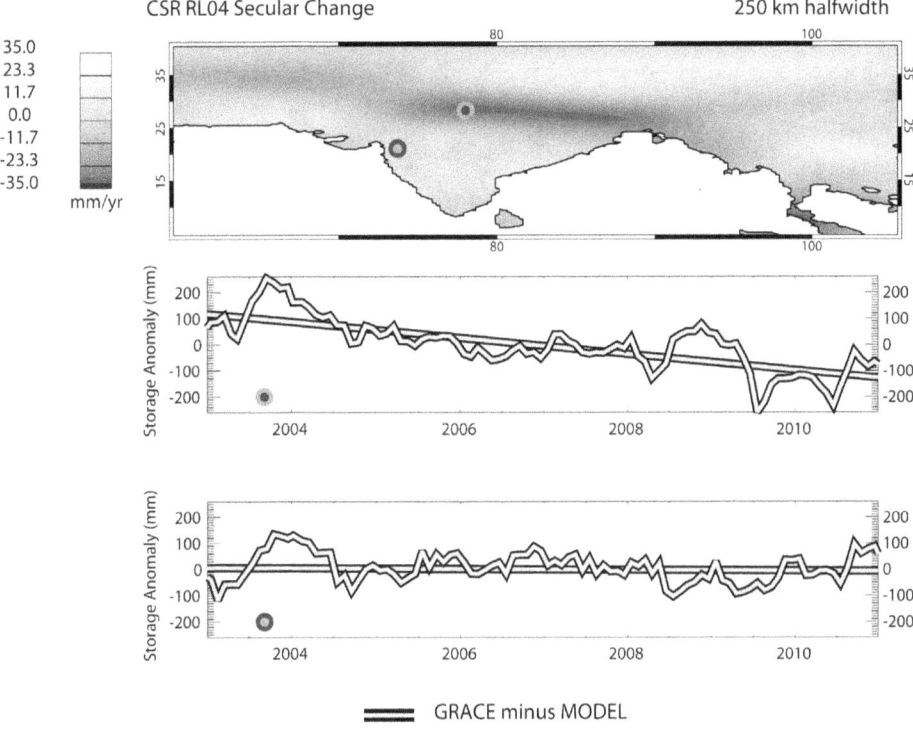

Figure 21.3. (Top) Map of trends in groundwater storage for South Asia (mm/a) derived from GRACE data by removing model-estimated soil water content. (Middle) Time series of groundwater depletion (trend of -31.0 mm/a shown by straight line) for a location in north-western India (shown by the upper dot in top panel), and (Bottom) for a region in western India (shown by the lower dot in top panel) where there is no apparent trend (straight line) in groundwater storage.

of Irrigation Areas (Siebert et al. 2007). A time series near the greatest negative values in north-western India shows a clear decreasing trend (middle panel), while a location further southwest shows little if any long-term groundwater depletion (lower panel).

21.4.   A FRAMEWORK FOR GLOBAL GROUNDWATER MONITORING

Modelled groundwater abstraction combined with satellite observations of water storage independently confirm that the decreasing well levels in numerous parts of the world are not isolated local occurrences, but in fact indicate widespread regional groundwater depletion. This paints a worrisome picture because of the finite nature of groundwater as a resource. Moreover, detrimental effects of groundwater withdrawal occur before abstraction surpasses recharge. First, water pumped from aquifers may come at the expense of discharge to rivers, reducing water availability to others, and second, many ecosystems require minimum groundwater levels to maintain health and function. In some developed countries, the costs associated with these effects have lead to a levelling- off of groundwater pumping, but this is unlikely to occur in the near future in many developing countries (Cole 2004).

The emergence of groundwater depletion as a global phenomenon (Konikow and Kendy 2005) indicates that this precious resource is already under considerable stress. Continued population growth will place additional stress on its limited availability. This scenario will most likely be exacerbated by reduced mid-latitude precipitation in the decades to come (IPCC 2007), which will further reduce groundwater recharge, this in a region that already relies heavily on groundwater resources to meet its domestic and irrigation water demands.

Hence a comprehensive strategy for monitoring the world's groundwater resources is an essential component of sustainable freshwater management. A robust framework should include regional and global networks of *in situ* monitoring wells as its foundation. These would be supplemented by space-based observations of groundwater storage changes from GRACE and its successor missions, and by output from regional aquifer and global hydrological models. Satellite missions such as GRACE are demonstrating that the synergistic use of *in situ* monitoring wells, hydrological models, and spaceborne observations, are providing new insights into the magnitude of regional-to-global groundwater depletion that would not otherwise be possible. Furthermore, the application of the methods described here is may be the only currently-available means for characterizing groundwater storage changes for large aquifer systems in the data-poor or data-restricted regions of the globe.

ACKNOWLEDGEMENTS

This work was supported by grants to the authors from the NASA GRACE Science Team. The contributions of the numerous co-authors to the original research are duly noted.

REFERENCES

Central Ground Water Board of India (CGWB). (2006) *Dynamic groundwater resources of India (as of March 2004)*. [Online] Available from: http://cgwb.gov.in/ (Accessed November 7, 2011).

Cole, M.A. (2004) Economic growth and water use. *Applied Economics Letters.* [Online] 11, 1–4. Available from: doi:10.1080/1350485042000187435 (Accessed November 7, 2011).

Famiglietti, J.S., Lo, M., Ho, S.L., Bethune, J., Anderson, K.J., Syed, T.H., Swenson, S.C., de Linage, C.R. & Rodell, M. (2011) Satellites measure recent rates of groundwater depletion in California's Central Valley, *Geophysical Research Letters.* [Online] 38, L03403. Available from: doi:10.1029/2010GL046442 (Accessed November 7, 2011).

Faunt, C.C. (ed.) (2009) Groundwater availability of the Central Valley Aquifer, California. *U.S. Geological Survey Professional Paper*, 1766, 225.

Giordano, M. (2009) Global Groundwater? Issues and Solutions. *Annual Review of Environmental Resources*, 34, 153–178. Available from: doi: 10.1146/annurev.environ.030308.100251 (Accessed November 7, 2011).

Gleeson, T., Vandersteen, J., Sophocleous, M.A, Taniguchi, M., Alley, W.M., Allen, D.M. & Chou, Y. (2010) Groundwater sustainability strategies. *Nature Geoscience*, 3, 378–379.

IPCC (2007), *Climate Change 2007: The Physical Science Basis. Contribution of Working Group I to the Fourth Assessment Report of the Intergovernmental Panel on Climate Change* [Solomon, S.,

D. Qin, M. Manning, Z. Chen, M. Marquis, K.B. Averyt, M. Tignor & H.L. Miller (eds.)]. Cambridge University Press, Cambridge, United Kingdom and New York, NY, USA, 996 pp.

Konikow, L. & Kendy, E. (2005) Groundwater depletion: A global problem. *Hydrogeology Journal.* [Online] 13, 317–320. Available from: doi: 10.1007/s10040-004-0411-8 (Accessed November 7, 2011).

Moiwo, J.P, Yang, Y., Li, H., Han, S. & Hu, Y. (2009) Comparison of GRACE with in situ hydrological measurement data shows storage depletion in Hai River basin, Northern China. *Water SA*, 35, 663–670.

Rodell, M., et al. (2004) The Global Land Data Assimilation System. *Bulletin of the American Meteorological Society.* [Online] 85(3), 381–394. Available from: doi:10.1175/BAMS-85-3-381 (Accessed November 7, 2011).

Rodell, M., Velicogna, I. & Famiglietti, J.S. (2009) Satellite-based estimates of groundwater depletion in India. *Nature.* [Online] 460, 999–1002. Available from: doi:10.1038/nature08238 (Accessed November 7, 2011).

Shah, T. (2007) The groundwater economy of South Asia: An assessment of size, significance and socio-ecological impacts\u201d. In: Giordano, M. & Villholth, K.G. (eds.) *The Agricultural Groundwater Revolution: Opportunities and Threats to Development.* Wallingford, CT, CAB International. pp. 7–36.

Shiklomanov, I.A. (2000) Appraisal and assessment of world water resources. *Water International*, 25(1), 11–32. Available from: doi:10.1080/02508060008686794 (Accessed November 7, 2011).

Siebert, S., Doell, P., Feick, S., Hoogeveen, J. & Frenken, K. (2007) *Global Map of Irrigation Areas version 4.0.1.* Johann Wolfgang Goethe University, Frankfurt am Main, Germany/Food and Agriculture Organization of the United Nations, Rome, Italy. [Online] Available from: http://www.fao.org/nr/water/aquastat/irrigationmap/index10.stm (Accessed November 7, 2011).

Syed, T.H., Famiglietti, J.S., Chambers, D., Willis, J. & Hilburn, K. (2010) Satellite-Based Global Ocean Mass Balance Estimates of Interannual Variability and Emerging Trends in Continental Freshwater Discharge. *Proceedings of the National Academy of Sciences*, 107, 17916–17921.

Tapley, B.D., Bettadpur, S., Ries, J.C., Thompson, P.F. & Watkins, M.M. (2004) GRACE measurements of mass variability in the Earth system. *Science.* [Online] 305, no. 5683, 503–505. Available from: doi:10.1126/science.1099192 (Accessed November 7, 2011).

Taylor, C.J. & Alley, W.M. (2001) *Ground-Water-Level Monitoring and the Importance of Long-Term Water-Level Data.* Denver, CO, U.S. Geological Survey Circular 1217.

Tiwari, V.M., Wahr, J. & Swenson, S. (2009) Dwindling groundwater resources in northern India, from satellite gravity observations. *Geophysical Research Letters.* [Online] 36, L18401. Available from: doi:10.1029/2009GL039401 (Accessed November 7, 2011).

Wada, Y., can Beek, L.P.H., van Kempen, C.M., Reckman, J.W.T.M., Vasak, S. & Bierkens, M.F.P. (2010) Global depletion of groundwater resources. *Geophysical Research Letters.* [Online] 37, L20402. Available from: doi:10.1029/2010GL044571 (Accessed November 7, 2011).

CHAPTER 22

Major science findings, policy recommendations, and future work

22.1 OVERVIEW

This chapter summarizes the major science findings and policy recommendations that are detailed in the previous chapters, and represents one of the first global-scale and systematic syntheses to address the paucity of data and limited understanding of the effects of climate change and variability and coupled human activities on groundwater resources. The Intergovernmental Panel on Climate Change (IPCC) recently (2008) identified several gaps in knowledge related to climate change and water, and most notably the need to improve understanding of climate change and variability effects on groundwater (Bates et al. 2008). The following findings and recommendations address many of these knowledge gaps.

The findings and recommendations from the 20 studies cases are synthesized here to provide global-scale analogues that may help to guide future studies and inform decision makers where data and detailed studies are not presently available. Findings and recommendations are first presented in order of climate region, from tropical to polar and then summarized across all climate regions. The chapter concludes by listing remaining questions and specific future work that is needed to better understand and manage the effects of climate change and variability on global groundwater sustainability. We use the definition of Alley et al. (1999) that conceptualizes *groundwater sustainability* as development and use of groundwater in a manner that can be maintained for an indefinite time without causing unacceptable environmental, economic, or social consequences.

22.2 TROPICAL CLIMATES

Tropical climate regions are home to approximately 40% of the world's population and have among the highest rates of population growth and poverty. Rapid population growth in the tropics coupled with projected climate change will have potentially severe effects on local water resources (Chapter 2). Sustainable adaptive management of groundwater is critically important, especially in areas entirely dependent on groundwater as the only source of potable water and on small, tropical islands that are among the most vulnerable systems in the world to climate change and variability. Yet, many tropical regions lack the human and institutional capacity and knowledge base to identify the vulnerability of groundwater resources and implement adaptive water management strategies.

22.2.1 Science findings

Climate model simulations for the 21st century project precipitation increases in some tropical-wet regions.

The projected changes in the intensity of rainfall has been rarely considered to date in groundwater sustainability assessments, but may have a substantial influence on the timing and magnitude of groundwater recharge. A shift towards more frequent heavy rainfall events (Bates et al. 2008) may promote rather than restrict recharge in some tropical (Chapter 2) and semiarid regions, including the High Plains aquifer in the United States (Chapter 9). Approaches that use delta factors with historical rainfall time series to simulate climate change impacts may fail to account for changes in the number of rain days, spatial distribution, or changes in rainfall variability and intensity. For example, Taylor and Tindimugaya (Chapter 2) report a substantial (55%) decline in recharge as a result of climate change using delta factors on historical (daily) rainfall distributions, but a substantial (53%) increase in recharge if the rainfall distribution is transformed to account for projected changes in rainfall intensity. Such a wide range in estimated recharge rates illustrates the need to quantify and reduce uncertainty in rainfall projections.

Climate model simulations for the 21st century project precipitation decreases in some tropical-dry regions.

The projected precipitation decreases and air temperature increases (Bates et al. 2008) may result in seasonal increases in evapotranspiration rates and decreases in groundwater recharge rates in some of the tropical-dry regions. In Mali, West Africa, projected recharge may decreases by 8 to 11% depending on soil properties (Chapter 3). The projected decrease in recharge may add pressure on an already stressed groundwater resource.

Rapid development in some tropical regions will likely have a greater effect on groundwater resources than climate change.

Over the coming decades, rapid development is expected in some tropical regions, such as Sub-Saharan Africa (Chapter 2). For example, groundwater abstraction from weathered crystalline rock aquifer systems in Uganda to support domestic and agricultural demands has drawn from the long-term storage and may have a greater negative effect on sustainable groundwater resources than projected climate change over the next two decades (Chapter 2).

Land-use change affects groundwater and groundwater-dependent ecosystems.

In many tropical regions, forested lands were historically cleared for agriculture. However, reforestation and best management practices, for example, in the Pipiripau river basin, central Brazil could substantially increase groundwater base flow to the river during the dry season (Chapter 4). Additionally, a number of tree species on tropical islands directly transpire from groundwater during dry periods and can substantially reduce the available groundwater and exacerbate the effects of drought on groundwater resources (Chapter 6).

Groundwater quantity and quality on small islands are particularly vulnerable to extreme rainfall variations, storm surges, sea-level rise, and other similar effects of climate change and variability.

Groundwater resources on many tropical islands, especially low-lying, carbonate islands are inherently vulnerable because of limited land areas and quantities of groundwater safe for drinking, growing population and demand over supply, surface-water resources that are close to limits of sustainability, over-pumping of fresh groundwater lens, salt-water intrusion, and pollution from human and animal waste (Chapters 5 and 6). Such vulnerabilities are exacerbated by tropical storms and climate change and variability, including rising sea levels, projected precipitation decrease, evapotranspiration increase, and possible decreases in recharge rates. For example, a storm surge from Hurricane Frances in 2004 contaminated the groundwater supply on North Andros Island, Bahamas (Chapter 5). After the storm surge, chloride concentrations in groundwater were nearly 30 times greater and orders of magnitude above than the World Health Organization drinking water guideline of 250 mg/L. Results indicate that the storm surge did not directly compromise the groundwater lens, but rather the trench and conduit system allowed direct infiltration and rapid intrusion of saltwater to the system after it became flooded with sea water. As a result, the trench and conduit systems on North Andros Island were pumped down to encourage recharge and dilution of the brackish groundwater.

The El Niño/Southern Oscillation (ENSO)-induced drought can reduce recharge and the thickness of freshwater lenses beneath low-lying Pacific Islands.

Groundwater resources on low-lying Pacific Islands are especially sensitive to climate variability on interannual to multidecadal timescales. ENSO-related drought can reduce recharge rates and substantially reduce the thickness and extent of freshwater lenses (Chapter 6). Additionally, groundwater salinity on some Pacific Islands is correlated with sea surface temperatures (SST) and the Southern Oscillation Index (SOI). Severe droughts have even forced the abandonment of several very small islands when fresh groundwater was exhausted (Chapter 6). During droughts it is critical to modify groundwater pumping on islands to sustain fresh groundwater lenses.

The current ability to generate climate change scenarios for small tropical islands is limited using ocean-atmosphere general circulation models (GCMs).

Presently, GCMs do not simulate sufficiently fine spatial resolution to generate scenarios for small island countries without using statistical downscaling techniques (Chapter 6). Furthermore, many GCMs have considerable uncertainty for projected precipitation in the tropics because they do not simulate tropical convection well and do not presently reproduce some of the major modes of interannual to multidecadal climate variability, including ENSO (Chapter 6). Improved downscaling techniques, regional climate models, and finer resolution GCMs will improve groundwater sustainability estimates on small islands.

22.2.2   Policy recommendations

To support the growing population, increases in groundwater-fed irrigation will, under some favourable hydrogeologic conditions, help to improve food production and the resilience of agricultural systems.

In some tropical regions, including Sub-Saharan Africa (Chapters 2 and 3), increased development of local aquifers to support irrigated agriculture has been proposed to improve food production and mitigate vulnerabilities in agricultural systems (Chapter 2). Groundwater may provide a partial solution to immediate concerns over food and agricultural systems. However planned development of aquifers should include best understanding of the hydrogeologic system and estimates, including uncertainties, of projected climate change and future socio-economic stresses that may cause unsustainable withdrawals of groundwater for irrigated agriculture, which is now a major concern in many aquifers in dry, temperate, and continental climate regions (e.g., Chapters 9). Sustainable development of groundwater resources in many tropical regions will require much needed groundwater and meteorological monitoring systems at the basin to national scale to assess how groundwater responds to abstraction and climate change and variability.

Continued financial and institutional support is urgently needed for aquifer characterization and groundwater-resource assessment.

Observational data and basic monitoring networks (including water levels, extraction rates, number of wells drilled, hydrogeological map, water quality, etc.) are lacking in many tropical regions, including Sub-Saharan Africa (Chapters 2 and 3) and many small islands in the Caribbean (Chapter 5) and Pacific (Chapter 6). Furthermore, existing data are often incomplete and of questionable reliability (Chapter 3). Some national well inventory databases need improved organization, centralization, and verification for completeness. In Mali and many other countries, each local administrative unit is responsible for its own database and data are not shared among regions. An organized system for data collection and storage is needed at the national level, with decentralized responsibilities for data collection and archiving in each region. Many Pacific Island countries have limited resources, few trained personnel, are transitioning from subsistence cultures to urban settlement, and have inadequate adaptation strategies and mode of coping with the present-day challenges of urban environments (Chapter 6). Thus, there is a strong and urgent need for financial support, capacity building, increased community participation, improved governance, and training opportunities in many developing tropical countries (Chapter 6) towards basic aquifer characterization and groundwater resource assessment before more complex climate change questions can be reliably addressed.

Land-use change may have economic and adaptive benefits for the water sector.

For example, reforestation and best management practices may substantially increase groundwater base flow to the Pipiripau River, central Brazil, which may indirectly help generate additional revenues for local farmers to pay for hydrologic services (Chapter 4). Participation by the farmers would potentially create a positive circle of land conservation and stabilization of groundwater and streamflow in the basin. The economic feasibility of such land conservation might lead to the planning and implementation of other land-conservation projects in similar tropical regions. Land-use policy and management practices that stabilize water availability is an adaptation strategy to increased precipitation variability and drought that may affect parts of the tropics in the 21st century (Bates et al. 2008). Additionally, selective clearing of phreatophytes, such as coconut trees, on small tropical islands may increase recharge and the sustainable yield and decrease groundwater salinity (Chapter 6). However, the potential benefit for groundwater availability must be weighed against the subsistence communities on the islands that depend on the trees.

Vulnerabilities within engineered systems that affect groundwater must be identified and mitigated.

For example, isolating and covering the trench and conduit system on North Andros Island will likely reduce the potential for inundation by future storm surges that are likely to affect The Bahamas (Chapter 5). The installation of rainwater harvesting cistern or large surface storage reservoirs might augment local-scale surface and groundwater supplies (Chapter 6).

22.3 DRY (ARID AND SEMIARID) CLIMATES

The dependence on groundwater is greater in dry (arid and semiarid) regions because the availability of surface water is much lower than in many other regions. These regions are also characterized by high hydroclimatic variability, relatively thick vadose zones, and a strong agricultural dependence on groundwater that creates a spatially and temporally complex relation between climate and groundwater.

22.3.1 Science findings

Shallow, unconfined aquifers are more vulnerable than deeper, confined aquifers.

Groundwater in shallow, unconfined aquifers is cheaper and easier to use, has a more immediate response to climate, shorter residence times, and thus is more sensitive to climate change and human activities than deeper, confined aquifers (Chapter 7 and 8).

Focused or preferential recharge is common in dry climates and may promote increased recharge under projected climate conditions.

In the Iullemmeden basin, West Africa (Chapter 7) and the High Plains aquifer, USA (Chapter 9) focused recharge occurs beneath ponds, gullies, and other topographic depressions where surface runoff concentrates. Because focused recharge is highly dependent upon the intensity of rainfall events, the projected increased precipitation intensity and variability during the 21st century (Bates et al. 2008) may increase recharge rates in some dry regions.

Vadose zone processes strongly affect climate-groundwater interactions in dry climates.

Aquifers in many arid and semiarid climates are characterized by vadose zones that are tens to hundreds of meters thick, which combined with high evapotranspiration rates minimize or restrict diffuse recharge (Chapter 9). Fast pathways for recharge and contaminant transport are facilitated beneath focused and preferential recharge zones (Chapter 7), but also form episodically under wet climate variability scenarios (Chapter 9). Important oceanic-atmospheric phenomenon, such as El Nino/Southern Oscillation partially facilitates temporally variable recharge rates (Chapter 9).

Important differences exist between groundwater and surface water response to climate change and variability.

Findings from of a number of studies in dry regions indicate that groundwater response does not always mirror the surface-water response to climate change and variability. The highly episodic rainfall associated with a small number of large storms often exacerbates drought and surface-water availability but may facilitate episodic recharge events in some dry regions, such as the High Plains aquifer, USA (Chapter 9) and the Murray Groundwater Basin, Australia (Chapter 10). Therefore, even if climate simulations project a drier than average climate, increased rainfall intensity could potentially lead to an increase in groundwater recharge rates (Chapters 9 and 10).

Land-use change affects the direction and magnitude of water level change.

In many dry regions, natural grassland and (or) forest have been converted to agricultural land that is often supported by groundwater. In the High Plains aquifer (Chapter 9), the land-use conversion from natural grassland to irrigated agriculture since the 1950s has indirectly led to substantial declines of water levels because of widespread groundwater pumping and withdrawals to support the irrigated agriculture. In the Iullemmeden basin, West Africa (Chapter 7), land clearing has affected runoff, evapotranspiration, erosion, gullying, silting of lakes, and water levels have risen in parts of the basin. In the Murray Basin, Australia, land clearing has resulted in a rising water table, which has caused widespread salinity problems and degraded agricultural soil and surface water resources. Evidence from the High Plains aquifer and Iullemmeden basin indicate that land-use change, in fact, may currently be a stronger driver of water-level change than climate change. However, the ongoing multi-year drought in the Murray-Darling Basin has, at least temporarily, reversed and surpassed the long-term groundwater trend inherited from land clearance and may induce a temporary halt of secondary salinization processes.

Saltwater intrusion is not the only source of groundwater salinity.

In some arid and semiarid aquifers, salinity in groundwater originates not only from saltwater intrusion, but also from entrapped saline groundwater within the aquifer, salt dissolution, and infiltration of agricultural return flow (Chapter 8). Additionally, pumping by high-capacity wells may also accelerate the upward movement of saline water from underlying formations, such as in the central High Plains aquifer, USA (Chapter 9). Water level declines in the aquifer caused by overpumping could change hydraulic gradients and enhance upwelling of the saline water into fresh groundwater resources (Chapter 9).

22.3.2   Policy recommendations

Adaptive management may benefit from using groundwater vulnerability maps of climate change and coupled human activities.

Vulnerability maps provide a useful tool for decision making and planning regarding groundwater monitoring, protection, and adaptive management (Gurdak 2008). Quantifying, modelling, and mapping groundwater vulnerability to climate change and coupled human activities is a necessary first step before identifying and implementing appropriate adaptive management actions (Chapter 9).

Minimize intensive groundwater extraction in vulnerable coastal areas.

Intensive groundwater pumping in coastal areas increases the potential for saltwater intrusion. As infrastructure allows, move groundwater supply centres from vulnerable coastal areas to more in-land areas of aquifers. For example, in the Souss-Mass basin of Morocco, the groundwater from coastal areas is most heavily used and has salinity problems from saltwater intrusion, while groundwater quantity and quality is better and less vulnerable to saltwater intrusion within in-land alluvial aquifers (Chapter 8). Therefore, groundwater vulnerability to climate change, human activities, and (or) contamination is a relative concept as most aquifers have important spatial differences in vulnerability.

22.4   TEMPERATE CLIMATES

Temperate climates cover large regions in Asia, Australia, Europe, and North and South America, including many important coastal areas, such as the Mediterranean. Coastal areas support approximately one-quarter of the global population, but contain less than 10 percent of the global renewable water supply and are undergoing rapid population growth (Kundzewicz et al. 2007). Thus, the limited groundwater resources in many temperate, coastal regions are highly vulnerable to increased human pressures and the effects of climate change and variability.

22.4.1   Science findings

Vulnerability of coastal groundwater resources to sea-level rise and saltwater intrusion is a topographically relative and high-resolution concept.

Important spatial differences in groundwater vulnerability to climate-induced sea-level rise and saltwater intrusion exist within individual coastal aquifers. Modelling of coastal aquifers in Israel (Chapter 12) indicate that steep coastal topography is an effective hydrologic barrier to saltwater intrusion and is expected to help minimize salinisation due to sea-level rise at local scales within coastal aquifers. Therefore high-resolution methods are needed to effectively integrate small-scale differences in topography of coastal aquifer systems. These findings have important implications for future groundwater monitoring efforts and policy considerations in coastal system, and further highlight the extreme vulnerability of groundwater in low-lying islands, such as in the Caribbean (Chapter 5) and Pacific (Chapter 6), and coastal areas below mean sea level, such as the Netherlands (Chapter 13).

Human-induced land subsidence exacerbates saltwater intrusion in coastal aquifers.

In many coastal regions, especially in large parts of the Netherlands, land subsidence has been a historically strong driver of saltwater intrusion that is likely to be exacerbated by expected climate-induced sea-level rise (Chapter 13). Model simulations indicate that more coastal areas in the Netherlands will experience higher rates of upward groundwater flow and salinisation because the head difference will increase between rising mean sea level and surface water levels of subsiding land (Chapter 13). Findings indicate that the current salinisation of the Dutch groundwater is a slow, continuous, and practically irreversible process. Furthermore, expected future sea-level rise of 0.5 meters per century will likely accelerate the rate of salinisation of groundwater resources in the Netherlands.

Wetlands are vulnerable to reduced groundwater discharge under projected climate change.

Wetlands are important ecosystems in temperate climates and other regions that are often dependent on groundwater discharge to maintain water levels and ecosystem function. Downscaled climate and groundwater model simulations for the first quarter of the 21st century project a 17% reduction in recharge that will likely reduce groundwater discharge into wetlands of Majorca, Spain (Chapter 11). Many other groundwater-dependent ecosystems, such as in the Santa Fe province of Argentina (Chapter 15) may experience substantial losses of biodiversity because of decreasing discharge caused by increasing groundwater demand and decreasing recharge rates. Future preservation of many groundwater-dependent ecosystems, such as the wetlands of Majorca, Spain require adaptive management actions that decrease groundwater abstraction used for irrigated agriculture and (or) re-locate the well(s) that have the most detrimental effects on groundwater discharge to dependent ecosystems (Chapter 11).

22.4.2   Policy recommendations

Groundwater pumping is expected to have more of an effect than climate change and sea-level rise on saltwater intrusion in some coastal aquifers.

Climate-induced sea-level rise will likely increase the vulnerability of many coastal groundwater resources to saltwater intrusion and salinity concerns. However, groundwater pumping is expected to have a stronger effect on the vulnerability of many coastal groundwater resources to saltwater intrusion (Chapter 12). Adaptive groundwater management in coastal aquifer must regulate groundwater pumping and implement or maintain groundwater level and quality monitoring networks (Chapter 12). Successful policy goals will differentiate the relative effects between pumping-induced and sea-level rise vulnerabilities and devise and implement adaptive management plans accordingly.

Mitigating saltwater intrusion and salinisation of groundwater on regional scales requires sound knowledge of sub-surface processes and technical solutions.

Slowing and preventing saltwater intrusion and salinisation of coastal groundwater is an important goal, but is inherently challenging at most spatial scales. Some local-scale approaches, such as land reclamation and inundation, have had positive results in the Netherlands (Chapter 13). Most water management sectors, such as drinking water supply and agriculture will be affected by saltwater intrusion. Therefore, it is necessary for adaptive management to identify and anticipate these threats and explore technical approaches to mitigate the salinisation, such as extraction of saline groundwater that has been successful at some local scales in the Netherlands (Chapter 13). Other approaches may include offshore land reclamation, inundation of polders, injection of fresh surface water, and creation of dykes and subsurface barriers to saltwater intrusion (Chapter 13).

Adaptive management must use a variety of coordinated approaches to mitigate climate- and human-induced stress on groundwater sustainability.

Keeping local issues in mind, a successful management plan should reduce the volume of groundwater extraction, conjunctively manage surface and groundwater resources to meet demand, establish protection zones around water-supply wells, improve the legal framework, and promote awareness and education at all levels (Chapter 15).

22.5   CONTINENTAL CLIMATES

Continental climates are characterized by annual variations in temperature, winter temperatures that support snow, and relatively moderate precipitation and variable evapotranspiration in the summer. Some of Asia, Europe, and North America's most important regional aquifer systems in continental climate regions support extensive irrigated agriculture and drinking-water supply for large populations.

22.5.1   Science Findings

Groundwater is directly and indirectly vulnerable to drought.

Increased precipitation variability is projected to increase drought in many continental climate regions during the 21st century (Bates et al. 2008). Drought directly affects groundwater resources by reducing precipitation and, in turn, recharge, and indirectly triggers increased groundwater withdrawals for emergency water supplies to supplement surface-water resources that are vulnerable to severe and (or) prolonged drought. Near Beijing, China, a prolonged drought from 1999 to 2010 limited recharge and prompted the construction of and withdrawals from emergency well fields, which had the combined effect of decreasing groundwater levels 20 to 40 m during the drought period (Chapter 16).

Climate and human-induced changes of groundwater levels in esker aquifers have important economic, recreation, and ecosystem implications for many lakes, rivers, springs, and peatlands.

Increased temperature and precipitation is projected in some boreal regions in northern Europe, Canada, and Russia (Bates et al. 2008) that will affect groundwater-surface water interactions, recharge, and most notably the discharge of esker aquifers (Chapter 17). Many eskers have strong hydrologic connections to surface water bodies that support recreation, tourism, and sensitive ecosystems. Therefore, climate and human-induced changes to groundwater in eskers will likely have cascading effects in many other human and natural systems (Chapter 17).

22.5.2   Policy recommendations

During wet years, artificially enhanced groundwater recharge and other strategies can help mitigate groundwater vulnerability to drought, climate change, and increased withdrawals.

To reverse the severe groundwater depletion near Beijing, China (Chapter 16) and other similar areas, a combination of reducing groundwater abstraction and implementing artificially enhanced recharge projects during wet years is needed. The use of water permits, improved water-use efficiency, suspending water-intensive industries, reducing agricultural water demand, and recycling urban water may help to control groundwater demand during drought and water crisis (Chapter 16).

Drought management plans must consider groundwater sustainability.

Many countries lack comprehensive and coordinated drought management plans, and often respond to drought ad hoc by drilling emergency wells or rely on unregulated groundwater withdrawals from existing wells. Drought management plans are urgently needed to shift from crises management to drought preparedness and risk management, and must contain monitoring and early warning, risk and impact assessment, mitigation, and planned use of groundwater that maintains long-term sustainability (Chapter 16).

Energy and transportation policy affects groundwater sustainability in eskers and has implications for groundwater-dependent ecosystems, tourism, and drinking-water supply.

In some northern European regions, peatlands drained for forestry and biofuel demand often reduces recharge to esker aquifers, which in turn reduces groundwater discharge into lakes that support ecosystems, recreation, and tourism. Roads are typically constructed on eskers, which can increase the vulnerability of groundwater to nonpoint-source contamination from road salts and other potential contaminants. Improved land-use regulations and groundwater protection zones will benefit groundwater-dependent systems (Chapter 17).

22.6   POLAR CLIMATES

Polar regions are sparsely populated (4 million) (Bogoyavlenskiy and Siggner 2004) and have less demand for groundwater compared to many other regions of the world. However, polar regions are expected to experience some of the earliest and most profound climate-induced change (Bates et al. 2008) that will likely have substantial effects on the surface water, glaciers and ice caps, wetlands, and subsurface water in soil and groundwater. Significant changes to permafrost have been observed in the Arctic over the last 50 years, which will likely affect the seasonal precipitation-runoff response (Bates et al. 2008) and groundwater flow dynamics (Chapter 18). Changing groundwater flow may have negative effects on roads, building foundations, and other infrastructure (Chapter 18). Much of the population in polar regions live in coastal communities that might be vulnerable to sea-level rise and saltwater intrusion of coastal aquifer systems (Warren et al. 2005). Moreover, polar aquifers beneath deep and continuous permafrost are among the least studied ground-water systems in the world (Chapter 18).

22.6.1   Science findings

Climate-induced degradation of permafrost in polar regions will dramatically change groundwater-surface water interactions.

Seasonally available unconfined groundwater above permafrost and subpermafrost groundwater (confined beneath permafrost) are sensitive to climate change that modifies spatiotemporal patterns of permafrost and, in turn, surface water resources (Chapter 18). For example, thawing of permafrost creates new taliks (unfrozen conduits through permafrost) for surface-water drainage that were likely responsible for the decline in the number of Siberian lakes between the 1970s and 2005 (Chapter 18).

Recharge and discharge of subpermafrost groundwater are sensitive to climate change and variability.

Climate-induced changes in vegetation and permafrost extent and thickness will likely affect recharge and discharge of subpermafrost groundwater via taliks. Subpermafrost groundwater is likely among the most vulnerable to climate-induced changes because of the sensitivity of permafrost and taliks to change and variability (Chapter 18). Increased vegetation cover in some areas may reduce recharge, while increased number of taliks caused by degradation of permafrost may enhance recharge to subpermafrost groundwater.

Groundwater-fed surface water and polar ecosystems are highly vulnerable to climate change.

There will likely be substantial effects on polar ecosystems as climate-induced groundwater dynamics change. Groundwater in polar regions can regulate surface water temperatures and affect hydrochemistry and biochemistry of rivers, lakes, and springs (Chapter 18).

Recharge beneath glaciers is sensitive to climate change.

In many polar regions, including Svalbard, Norway (Chapter 18), climate-dependent glacier dynamics and not the permafrost thickness are the primary control on subpermafrost groundwater recharge. Beneath glaciers, groundwater recharge is possible beneath zones of warm-based ice, which is characterized as having a substantial accumulation zone and thick glacial to insulate the bottom from the cold air and thereby prevents the ice from freezing to the geological substratum. Climate-induced melting of glaciers will decrease glacier size, reduce the accumulation area, and lower its surface toward the base geological substratum. These factors will likely decrease the area of warm-based ice and thereby reduce the groundwater recharge beneath glaciers. It is possible that entire glaciers may become cold-based and groundwater recharge beneath these glaciers could terminate completely (Chapter 18).

22.6.2 Policy recommendations

The vulnerability of subpermafrost groundwater to human-induced contamination will likely increase under projected climate change.

Although current human demand for subpermafrost groundwater is relatively low compared to groundwater demand in other regions of the world, important changes to the human-subpermafrost groundwater system may occur under future climate change. Industrial activity and mining takes place in many areas of discontinuous permafrost. A change from confined to unconfined aquifers due to disintegration of the permafrost will increase the risk of groundwater pollution, especially where frozen waste deposits melt (Chapter 18). Moving the waste from permafrost to more controlled base conditions is an important policy and planning action to reduce groundwater contamination.

Protection and management strategies are needed for groundwater-fed springs and ecosystems. Establishing special protection plans for groundwater discharge zones is important to best protect and manage groundwater-fed surface water and ecosystems in many polar regions (Chapter 18).

Large-scale circum-polar monitoring and modelling of data-poor subpermafrost aquifers is needed.

Groundwater in polar regions is among the most vulnerable to climate change and variability, but among the least studied and characterized systems in the world. Many fundamental gaps in knowledge remain, including the location and extent of subpermafrost aquifers, and fundamental processes of recharge and discharge as related to storage and availability of subpermafrost groundwater. Coordinated international efforts to improve monitoring, modelling, and aquifer characterization of subpermafrost groundwater in polar regions are needed (Chapter 18).

22.7 VARIOUS CLIMATES

Many of the following science findings and policy recommendations are derived from the regional case studies presented in Chapters 19–21. Other findings and recommendations are synthesized from important cross-cutting themes of many chapters in this volume.

22.7.1   Science findings

Regional depletion of groundwater resources is a global-scale phenomenon.

Findings from most of the case studies in this volume further support the emerging concern that regional depletion of groundwater resources is a global-scale problem (Konikow and Kendy 2005). This is of particular concern because of the low recharge rates, long (hundreds to thousands of years) groundwater residence times, and large population growth in many regions (Chapter 21). Many groundwater resources are non-renewable on timescales that are meaningful for society and water management. Projected climate change in many regions of the world will only exacerbate these concerns by reducing precipitation and increasing evapotranspiration, which is expected to reduce recharge and increase groundwater withdrawal rates.

Quantifying groundwater withdrawals and use remains a difficult but necessary challenge.

While each chapter in this volume presented scientific findings that addressed, and in most cases, quantified recharge, relatively few chapters addressed or quantified climate change and human effects on discharge, especially groundwater withdrawals (abstractions) and use. Groundwater withdrawals for drinking water, agriculture, and industry has a major effect on most groundwater resources and is one of the few components of the groundwater budget that society can most directly influence by adaptive management practices and policy decisions. Additional scientific studies are needed in most aquifers of the world to quantify spatial and temporal patterns of groundwater discharge, withdrawals, and uses.

Satellite-based methods, such as GRACE, hold promise for assessing groundwater storage.

Data from the Gravity Recovery and Climate Experiment (GRACE) mission was used successfully by a number of studies around the globe (Chapters 3, 10, 19, and 21) to characterize groundwater storage changes in regional aquifers ($>200,000\,km^2$) at monthly and longer timescales. GRACE provides a new and exciting approach to monitor groundwater resources with sufficient accuracy, particularly in data sparse regions, to address a number of pressing scientific questions and to help inform water management at the regional scale (Chapter 21).

Groundwater responds to climate variability on interannual to multidecadal scales.

It is important to analyze climate variability in long-term (50 to 100 years or more) hydrological and meteorological records to identify oceanic-atmospheric climate variability patterns that affect spatiotemporal patterns in precipitation, evapotranspiration, recharge, discharge, groundwater storage, biogeochemical reactions, and contaminant fate and transport as related to the sustainability of groundwater quantity and quality (Chapters 3, 6, 9 and 17). The important global-scale oceanic-atmospheric climate variability patterns that have implications for adaptive management toward groundwater sustainability include, but are not limited to the Atlantic Multidecadal Oscillation (AMO), Arctic Oscillation (AO), El Niño/Southern Oscillation (ENSO), North Atlantic Oscillation (NAO), and Pacific Decadal Oscillation (PDO).

Important vadose zone processes are missing or poorly simulated in many studies.

The vadose zone represents an important link between land surface hydroclimate processes, human activities, and groundwater (Chapter 9), especially in many arid and semiarid climate regions. However, many studies neglect the vadose zone or report that the vadose zone is poorly represented in some numerical simulations (Chapter 3).

Integrate multiple methods of investigation when available.

Groundwater responds to climate change and variability in many complex ways and data availability is often limited in many regions of the world. Therefore, studies that integrate several methods and data types rather than focusing on a single approach may have increased success in making substantial scientific progress. For example in Chapter 3, groundwater level and meteorological data, groundwater modelling, and GRACE satellite data were used to improve understanding of groundwater resources in Mali where groundwater data are limited. Using multiple methods to estimate recharge is especially advantageous because recharge methods provide information over various spatial and temporal scales of interest (Scanlon et al. 2002). For example, global oceanic-atmospheric systems on interannual to multidecadal timescales can modify recharge on temporal scales consistent with water management decisions and planning (see Chapter 9).

Climate model simulations for the 21st century project increases in recharge for northern Europe and decreases in recharge for southern Europe.

Global circulation models (GCMs) used to simulate future precipitation and temperature trends based on a 'high' (SRES A1F1) gas emissions scenario predict that by the end of this century, northern Europe will receive more winter rainfall, leading to increased groundwater recharge but during a shorter time period, and that summers will be drier with longer periods of limited or no groundwater recharge (Chapter 20). In southern Europe, there will be less groundwater recharge overall and the region may become more water stressed than present, with any increase in winter recharge unable to compensate for reduced autumn recharge. Southern Spain is predicted to be among the worst affected regions with almost the total disappearance of recharge.

22.7.2   Policy recommendations

Groundwater is an integral component of the climate-water-society connection.

Water is an important component of the global climate system, which in turn creates feedbacks in the hydrologic cycle that affect surface and groundwater resources over many spatiotemporal scales that are important for society. To varying degrees, groundwater is necessary for many human and natural systems (Chapter 1). Groundwater is a substantial economic resource in most developed and developing countries (Chapter 20). The management of groundwater resources affects many policy implications outside the immediate water sector (Ludwig et al. 2009), including implications for agriculture and food security (Chapter 9), energy, human health and safety (Chapter 6), and the conservation of groundwater dependent ecosystems (Chapters 4 and 17). Many policy and management decisions directly affect groundwater and (or) climate, which in turn further modifies groundwater resources. Thus, policy decisions must carefully assess implications to the climate-water-society connection and the sustainability of groundwater resources.

Community involvement is necessary to establish groundwater sustainability goals.

Because groundwater residence times can often exceed hundreds to thousands of years and thus are effectively non-renewable on human timescales, Gleeson et al. (2010) propose that sustainability goals for groundwater quantity and quality should be set on a multigenerational time horizon of 50 to 100 years, while acknowledging even longer-term impacts. Furthermore, community involvement is critical in establishing sustainability goals for long-term management strategies to succeed (Gleeson et al. 2010). This is especially true in many regions that lack participation of communities in water resource management and planning because of a disconnection between government ministries and communities (Chapter 6).

Inadequate governance is a problem for sustainable use of groundwater resources.

Many groundwater resources are vulnerable to climate change and coupled human activities because of inadequate legislation and regulations, inappropriate national water policies that provide no clear priorities or directions to government agencies of responsibility, and very limited financial and human resources to manage groundwater resources and water supply systems (Chapter 6).

Land ownership may hinder sustainable groundwater resource management.

In Pacific Island countries (PIC) and other regions of the world, land is often owned by traditional owners. This often leads to conflicts between governments and landowners when establishing water reserves on privately owned land. Because customary law in many PICs assigns ownership of groundwater to land owners, governments are often reluctant to enact water legislation specifying that water belongs to all people or the government or banning polluting land uses for fear of infringing on property rights (Chapter 6). As a consequence, in some PICs there is no legal protection of groundwater from over pumping or from contamination.

Adaptive groundwater management must integrate regional water and agricultural policies.

Water-stressed regions can possibly achieve sustainable groundwater resources by effectively integrating regional water and agricultural policies that control illegal groundwater abstraction, create water banking infrastructure and policy, and diversify crops and implement best available water-efficient irrigation (Varela-Ortega et al. 2011) (Chapter 20). Such a management approach must strike a balance between ecological protection, human development, and acceptable socio-economic costs (Chapter 20).

Policy must account for important differences between the developed and developing world in terms of addressing climate and human stresses on groundwater resources.

While groundwater resources in nearly all regions of the world are stressed by overuse (Chapters 2–21), population growth, and climate change, there are important differences in policy and management to minimize stressors on groundwater resources between the developed and developing world (Chapter 21). In the developed world, water management efforts are complicated by a number of institutional, political and socioeconomic barriers, which include the lack of sufficiently dense monitoring well networks, disparate sampling, data reporting and archiving strategies, inadequate data sharing policies, and in many regions, the lack of groundwater use reporting requirements. In the developing world, such management strategies, if they exist at all, are rudimentary or in their infancy (Chapter 21). Policy efforts to implement adaptive management must recognize these important differences and address limitations and challenges appropriately.

Rural, developing, and (or) economically depressed areas that are strongly dependent on groundwater are especially vulnerable to human pressures and climate change.

Communities in rural, developing, and (or) economically depressed areas that rely solely on groundwater resources, especially in dry (arid and semiarid) regions are highly vulnerable to climate change that may enhance drought or further limit stressed surface-water resources. Developed regions may find engineering solutions to mitigate their groundwater-dependence vulnerability. For example, the government of Santa Fe province, Argentina (Chapter 15) plans to build an aqueduct to import surface water from the Paraná River, which will reduce the stress on local groundwater resources. The environmental effects of imported water must, however, be taken into account. Moreover, the provincial government is also promoting a water bill that includes, among other things, the delineation of well-protection zones and environmental education. Developing and economically depressed areas will need external assistance to minimize their groundwater dependent vulnerability.

Nonstationary assumptions must be implemented in groundwater management.

Stationarity (or fluctuations of natural processes within an unchanging envelope of variability) should no longer serve as the central assumption in water-resource management (Milly et al. 2008). Nearly all studies presented in this volume identify nonstationary processes that affect groundwater resources are driven in part by human activities, climate change, and natural low-frequency (interannual to multidecadal) atmospheric and ocean circulation systems, such as El Niño/Southern Oscillation (ENSO), North Atlantic Oscillation (NAO), Pacific Decadal Oscillation (PDO), and Atlantic Multidecadal Oscillation (AMO). Milly et al. (2008) outline a framework to implement nonstationary assumptions, which could be modified for adaptive groundwater management.

Policy is needed to better quantify groundwater withdrawal, use, and best estimates of safe and sustainable yield.

Because there is a temporal lag on the order of months to years in many aquifers before the effects of groundwater withdrawals become evident, there is a tendency to neglect studies that are needed to properly support groundwater management until water-resource crises materialize (Alley 2006). This type of reactionary stance can be improved if there is policy that supports efforts to better monitor, quantify, and regulate groundwater withdrawals and use in most aquifers of the world. Improved estimates of groundwater discharge, including withdrawals, will complement the widespread efforts to quantify recharge, and lead to improved estimates of current and future groundwater storage. Adaptive management planning will require best estimates of safe and sustainable yield, which are not currently available for aquifer system in many regions of the world.

Adaptive groundwater management in Europe must consider projected difference between northern and southern regions.

In northern Europe, adaptation to climate change predictions (Chapter 20) will require groundwater management practices and policies that capture the winter recharge and use it efficiently to meet the summer demand for water. In southern Europe, especially southern Spain and similar regions, the future demand for drinking water and irrigation use will likely require adaptation that strictly controls water supplies and potentially supplements groundwater by non-conventional sources such as waste water reclamation and reuse (Chapter 20).

22.8   FUTURE WORK

Integrate climate change and variability to improve conceptual hydrogeology models.

As demonstrated in this volume, climate change and variability has affected many human and natural processes governing groundwater quantity and quality. The effects of climate change and variability on the hydrologic cycle are expected to intensify over the coming decades (Bates et al. 2008). Therefore, a fundamental first step in defining the problem for new and on-going hydrogeological and groundwater resource studies must include developing conceptual models with appropriate climate change and variability information.

A comprehensive strategy is needed to monitoring global groundwater resources.

Adaptive groundwater management needs a coordinated and comprehensive strategy to monitor global groundwater resources, especially in data sparse regions (Chapter 21). Such a monitoring strategy would supplement monitoring wells with space-based observations of groundwater storage from GRACE and regional to global scale hydrologic models. Additional strategies are needed in data sparse regions that have areas smaller than the current spatial resolution of GRACE and other space-based approaches.

Groundwater quality has received less attention than groundwater quantity.

To date, the majority of studies have addressed questions of climate change effects on groundwater quantity and relatively few have addressed groundwater quality. The limited groundwater quality studies have primarily addressed saltwater intrusion. However, the effect of climate change on air temperature may influence groundwater temperatures and dissolved oxygen concentrations (Chapter 17 and 18), which have important implication for reaction rates and reduction-oxidation (redox) reactions that directly affect the nitrogen and carbon cycle in soil and groundwater, nonpoint- and point-source contamination, and the fate of many groundwater contaminants. The quality of groundwater may be a limiting factor for some intended uses such as drinking- or irrigation-water supply and to the long-term sustainability of many groundwater resources worldwide (Chapter 9), and therefore necessitates additional study.

Mechanisms and timing of snowmelt runoff and recharge needs further study.

Most studies of climate change effects on surface hydrology in alpine, mountainous, and snow-dominated regions do not explore subsurface hydrologic responses (Green et al. 2011).

Further work is needed to improve the representation of the mechanisms and timing of snow melt runoff and recharge that are of significance in the mountainous areas of central and eastern Europe (Chapter 20) and other similar regions. Moreover, the recharge mechanisms, storage capacity, and residence times of high elevation aquifers under climate change and variability is poorly understood (Singleton and Moran 2010). However, recent findings from temperate mountain regions in British Columbia, Canada (Chapter 14) indicate important changes to the rate and timing of recharge. Under projected climate change, peak diffuse recharge will likely coincide with March and April snowmelt, decline through the summer and fall months, and have minor increases in annual recharge due to the shift (earlier) of peak recharge from increased temperature affecting earlier valley-bottom snowmelt and ground thaw (Chapter 14). More effort is needed to understand the magnitude of mountain block recharge and the socio-economic factors that dominate many populated mountain valleys (Chapter 14).

Continue interdisciplinary and multidisciplinary collaboration.

There is considerable prediction uncertainty of groundwater under future climate change and variability. Potential changes in groundwater resources are widely dependent on regional hydrogeological characteristics and, more importantly, socio-economic-political conditions (Holman 2006). Yet, many hydrogeologists lack the training and knowledge to individually and adequately address the human components of society, economics, and policy on groundwater resources. The most substantial advances in groundwater sustainability will likely come as hydrogeologists collaborate with other physical and social sciences that are better skilled in understanding and predicting human stress, specifically abstraction and surface-derived contamination. Additionally, the estimated uncertainty in current projections of climate change impacts on groundwater resources is so large (Chapter 2) to be of limited use to water management. Thus, a concerted effort involving collaborations between groundwater and climate scientists is required to investigate ways of reducing this uncertainty.

Usable groundwater research is needed.

From study design to publication and throughout the scientific process, hydrogeologists must critically evaluate whether their scientific findings are usable by (ground)water managers and leading to better decision making about climate change and variability (Pielke et al. 2010).

REFERENCES

Alley, W.M. (2006) Tracking US groundwater: Reserves for the future? *Environment*, 48(3), 10–25.

Alley, W.M., Reilly, T.E. & Franke, O.L. (1999) *Sustainability of Ground-Water Resources*. U.S. Geological Survey Circular 1186, 79 pp.

Bates, B., Kundzewicz, Z.W., Wu, S. & Palutikof, J.P. (2008) *Climate change and water*. Technical Paper VI of the Intergovernmental Panel on Climate Change. Geneva, Switzerland, Intergovernmental Panel on Climate Change Secretariat. 210 pp.

Bogoyavlenskiy, D. & Siggner, A. (2004) Arctic demography. In: Einarsson, N., Larsen, J.N., Nilsson, A. & Young, O.R. (eds.) *Arctic Human Development Report (AHDR)*. Akureyri, Steffanson Arctic Institute. pp. 27–41.

Gleeson, T., VanderSteen, J., Sophocleous, M.A., Taniguchi, M., Alley, W.M., Allen, D.M. & Zhou, Y. (2010) Groundwater sustainability strategies. *Nature Geoscience*, 3, 378–379.

Green, T.R., Taniguchi, M., Kooi, H., Gurdak, J.J., Allen, D.M., Hiscock, K.M., Treidel, H. & Aureli, A. (2011) Beneath the surface of global change: Impacts of climate change on groundwater. *Journal of Hydrology*. [Online] Available from: doi: 10.1016/j.jhydrol.2011.05.002. accessed 1 October 2011

Gurdak, J.J. (2008) *Ground-Water Vulnerability: Nonpoint-Source Contamination, Climate Variability, and the High Plains Aquifer*. Saarbrucken, Germany, VDM. 223 pp.

Holman, I.P. (2006) Climate change impacts on groundwater recharge-uncertainty, shortcomings, and the way forward? *Hydrogeology Journal*, 14, 637–647.

Konikow, L. & Kendy, E. (2005) Groundwater depletion: A global problem. *Hydrogeology Journal*. [Online] 13, 317–320. Available from: doi: 10.1007/s10040-004-0411-8. Accessed 1 October 2011

Kundzewicz, Z.W., Mata, L.J., Arnell, N.W., Doll, P., Kabat, P., Jimenez, B., Miller, K.A., Oki, T., Sen, Z. & Shiklomanov, I.A. (2007) Freshwater resources and their management. In: Parry, M.L., Canziani, O.F., Palutikof, J.P., van der Linden, P.J. & Hanson, C.E. (eds.) *Climate Change 2007: Impacts, Adaptation and Vulnerability*. Cambridge, Cambridge University Press. pp. 173–210.

Ludwig, F., Kabat, P., van Schaik, H. & van der Valk, M. (2009) *Climate Change Adaptation in the Water Sector*. London, Earthscan. 274 pp.

Milly, P.C.D., Betancourt, J.L., Falkenmark, M., Hirsch, R.M., Kundzewicz, Z.W., Lettenmaier, D.P. & Stouffer, R.J. (2008) Stationarity is dead: Whither water management? *Science*, 319, 573–574.

Pielke, R., Sarewitz, D. & Dilling, L. (2010) *Usable Science: A Handbook for Science Policy Decision Makers*. [Online] Available from: http://cstpr.colorado.edu/sparc/outreach/sparc_handbook/brochure.pdf [Accessed 22nd July 2011].

Scanlon, B.R., Healy, R.W. & Cook, P.G. (2002) Choosing appropriate techniques for quantifying groundwater recharge. *Hydrogeology Journal*, 10, 18–39.

Singleton, M.J. & Moran, J.E. (2010) Dissolved noble gas and isotopic tracers reveal vulnerability of groundwater in a small, high elevation catchment to predicted climate change. *Water Resources Research*. [Online] 46, W00F06. Available from: doi:10.1029/2009WR008718. Accessed 1 October 2011

Varela-Ortega, C., Blanco-Gutiérrez, I., Swartz, C.H. & Downing, T.E. (2011) Balancing groundwater conservation and rural livelihoods under water and climate uncertainties: An integrated hydro-economic modelling framework. *Global Environmental Change*, 21, 604–619.

Warren, J., Berner, J. & Curtis, J. (2005) Climate change and human health: Infrastructure impacts to small remote communities in the North. *International Journal of Circumpolar Health*, 64(5), 487–497.

Contributing Authors and Contact Information

Pertti Ala-aho

Water Resources and Environmental Engineering Laboratory, Department of Process and Environmental Engineering, University of Oulu, Finland

Diana M. Allen*

Department of Earth Sciences, Simon Fraser University, Burnaby, British Columbia, Canada, email: dallen@sfu.ca

Ibrahim Baba Goni

Maiduguri University, Department of Geology, PMB 1069, Maiduguri, Nigeria

Lhoussaine Bouchaou*

Applied Geology and Geo-Environment Laboratory, Ibn Zohr University, BP. 8106 Cite Dakhla, 80000 Agadir, Morocco, email: lbouchaou@yahoo.fr

Saib Boutaleb

Applied Geology and Geo-Environment Laboratory, Ibn Zohr University, BP. 8106 Cite Dakhla, 80000 Agadir, Morocco

John Bowleg*

Water Resources Management Unit, Water & Sewerage Corporation (WSC), Nassau, Bahamas, email: wcjbowleg@wsc.com.bs

Breton W. Bruce

U.S. Geological Survey, Lakewood, Colorado, USA

Ana Paula S. Camelo

School of Technology-ENC, University of Brasilia-UnB, 70910–900, Brasilia-DF, Brazil

Lucila Candela*

Department of Geotechnical Engineering and Geoscience-Universitat Politècnica de Catalunya (UPC), C/Gran Capitán s.n., Barcelona, Spain, email: Lucila.candela@upc.edu

Ian Cartwright

School of Geosciences, Monash University, Melbourne, VIC 3800, Australia

Henrique M.L. Chaves*

School of Technology-EFL, University of Brasilia-UnB, 70910–900, Brasilia-DF Brazil, email: hchaves@unb.br

Mónica D'Elía

Grupo de Investigaciones Geohidrológicas. Facultad de Ingeniería y Ciencias Hídricas. Universidad Nacional del Litoral, Ciudad Universitaria (3000) Santa Fe, Argentina

Harm Demon

Department of Earth Sciences, Simon Fraser University, Burnaby, British Columbia, Canada

---

* Corresponding author

Zine El Abidine El Morjani   Polydisciplinary Faculty of Taroudant, Ibn Zohr University, Hay El Mohammadi (Lastah), BP. 271, 83 000 Taroudant, Morocco

F. Javier Elorza   Universidad Politécnica de Madrid, Ríos Rosas 21, 28003, Madrid, Spain

Adam Fakes   School of Earth and Environmental Sciences, James Cook University, Cairns, QLD 4878, Australia

Tony Falkland   Island Hydrology Services, Canberra, Australia

James Famiglietti*   UC Center for Hydrologic Modeling, Department of Earth System Science, University of California, Irvine, California, USA, email: jfamigli@uci.edu

Guillaume Favreau*   IRD, UMR HydroSciences Montpellier, 276 Av. Maradi, BP 11416, Niamey, Niger, & Université Abdou Moumouni, Faculté des Sciences, département de Géologie, BP 10662, Niamey, Niger, email: Guillaume. Favreau@ird.fr

Frédéric Frappart   GET, GRGS, Observatoire Midi-Pyrénées, 14 Avenue Edouard Belin, 31400 Toulouse Cedex 01, France

Abdou Guéro   Niger Basin Authority, 288 rue du Fleuve Niger, BP 729, Niamey, Niger

Jason J. Gurdak*   Department of Geosciences, San Francisco State University, San Francisco, California, USA, email: jgurdak@sfsu.edu

Sylvi Haldorsen*   Department of Plants and Environmental Science, Norwegian University of Life Sciences, PO Box 5003, N-1432 AAs, Norway, email: sylvi.haldorsen@umb.no

Mohamed Hssaisoune   Applied Geology and Geo-Environment Laboratory, Ibn Zohr University, BP. 8106 Cite Dakhla, 80000 Agadir, Morocco

Michael Heim   Department of Plants and Environmental Science, Norwegian University of Life Sciences, P:O.Box 5003, N-1432 AAs, Norway

Chris M. Henry   Department of Earth Sciences, Simon Fraser University, Burnaby, British Columbia, Canada

Kevin Hiscock*   School of Environmental Sciences, University of East Anglia, Norwich, NR4 7TJ, UK, email: k.hiscock@uea.ac.uk

Alan Hodgson   School of Environmental Sciences, University of East Anglia, Norwich, NR4 7TJ, UK

Joaquín Jiménez-Martínez   Department of Geotecnical Engineering and Geoscience-Universitat Politècnica de Catalunya (UPC), C/Gran Capitán s.n., Barcelona, Spain

Uri Kafri	Geological Survey of Israel, 30 Malkhe Israel, Jerusalem, 95501, Israel
Dirk Kirste	Department of Earth Sciences, Simon Fraser University, Burnaby, British Columbia, Canada
Bjørn Kløve*	Water Resources and Environmental Engineering Laboratory, Department of Process and Environmental Engineering, University of Oulu, Finland email: bjorn.klove@oulu.fi
Henk Kooi	Department of Hydrology and Geo-Environmental Sciences, VU University Amsterdam, De Boelelaan 1085, 1081 HV Amsterdam, The Netherlands
Marc Leblanc*	School of Earth and Environmental Sciences, James Cook University, Cairns, QLD 4878, Australia, email: marc.leblanc@jcu.edu.au
Jiurong Liu	Beijing Geo-environmental Monitoring Station, China
Peter B. McMahon	U.S. Geological Survey, Lakewood, Colorado, USA
Rejane M. Mendes	School of Technology-EFL, University of Brasilia-UnB, 70910–900, Brasilia-DF, Brazil
Yahaya Nazoumou	IRD, UMR HydroSciences Montpellier, 276 Av. Maradi, BP 11416, Niamey, Niger
Jarkko Okkonen	Geological Survey of Finland, Kokkola, Finland
Gualbert Oude Essink*	Deltares, Subsurface and Groundwater Systems, PO Box 85467, 3508 AL Utrecht, The Netherlands, email: gualbert.oudeessink@deltares.nl
Marta Paris	Grupo de Investigaciones Geohidrológicas. Facultad de Ingeniería y Ciencias Hídricas. Universidad Nacional del Litoral, Ciudad Universitaria (3000) Santa Fe, Argentina
Marcela Pérez	Grupo de Investigaciones Geohidrológicas. Facultad de Ingeniería y Ciencias Hídricas. Universidad Nacional del Litoral, Ciudad Universitaria (3000) Santa Fe, Argentina
Martine van der Ploeg	Wageningen University, Dep. Environmental Sciences, P.O. Box 47, 6700AA, Wageningen, The Netherlands.
Guillaume Ramillien	GET, GRGS, Observatoire Midi-Pyrénées, 14 Avenue Edouard Belin, 31400 Toulouse Cedex 01, France
Pekka Rossi	Water Resources and Environmental Engineering Laboratory, Department of Process and Environmental Engineering, University of Oulu, Finland

Eyal Shalev   Geological Survey of Israel, 30 Malkhe Israel, Jerusalem, 95501, Israel

Robert Sparkes   School of Environmental Sciences, University of East Anglia, Norwich, NR4 7TJ, UK

Sean Swenson   Climate and Global Dynamics Division, National Center for Atmospheric Research, Boulder, CO, USA

Tarik Tagma   Applied Geology and Geo-Environment Laboratory, Ibn Zohr University, BP. 8106 Cite Dakhla, 80000 Agadir, Morocco

Makoto Taniguchi*   Research Institute for Humanity and Nature (RIHN), Kyoto, Japan email: taniguchispot@gmail.com

Richard Taylor*   Department of Geography, University College London, London, UK, email: r.taylor@geog.ucl.ac.uk

Callist Tindimugaya Directorate of Water Resources Management, Ministry of Water and Environment, Entebbe, Uganda

Paul Tregoning   Research School of Earth Sciences, The Australian National University, Canberra, ACT 0200, Australia

Ofelia Tujchneider* Grupo de Investigaciones Geohidrológicas. Facultad de Ingeniería y Ciencias Hídricas. Universidad Nacional del Litoral, Ciudad Universitaria (3000) Santa Fe, Argentina/ Consejo Nacional de Investigaciones Científicas y Técnicas. Argentina, email: ofeliatujchneider@yahoo. com.ar

Sarah Tweed   School of Earth and Environmental Sciences, James Cook University, Cairns, QLD 4878, Australia

Wolf von Igel   Department of Geotecnical Engineering and Geoscience- Universitat Politècnica de Catalunya (UPC), C/Gran Capitán s.n., Barcelona, Spain/Amphos XXI Consulting S.L. Pg. de Rubí, 29-31, 08197, Valldoreix, Spain

Liya Wang   China University of Geosciences, Beijing, China & Beijing Geo-environmental Monitoring Station, China

Ian White*   Fenner School of Environment and Society, Australian National University, Canberra, Australia, email: ian. white@anu.edu.au

Chao Ye   Beijing Geo-environmental Monitoring Station, China

Yoseph Yechieli*   Geological Survey of Israel, 30 Malkhe Israel, Jerusalem, 95501, Israel, email: yechieli@gsi.gov.il

Yangxiao Zhou*   UNESCO-IHE Institute for Water Education, Delft, The Netherlands

Subject index

For Product Safety Concerns and Information please contact our EU
representative GPSR@taylorandfrancis.com
Taylor & Francis Verlag GmbH, Kaufingerstraße 24, 80331 München, Germany

www.ingramcontent.com/pod-product-compliance
Lightning Source LLC
Chambersburg PA
CBHW080006210526
45170CB00015B/1852